Encyclopaedia of Herbal Antioxidants

Encyclopaedia of Herbal Antioxidants

– Volume 3 –

Editor
T. Pullaiah
Department of Botany
Sri Krishnadevaraya University
Anantapur – 515 003
Andhra Pradesh

2012
REGENCY PUBLICATIONS
New Delhi - 110 002

ISBN ... et)
ISBN ... olume 1)
ISBN ... olume 2)
ISBN ... olume 3)

Published by : **REGENCY PUBLICATIONS**
(*Division of Daya Publishing House*)
4760-61/23, Ansari Road, Darya Ganj,
New Delhi - 110 002
Phone: 23245578, 23244987
Fax: (011) 23260116
e-mail : info@regency-books.com
website : www.dayabooks.com
website : www.regencybooks.com

Laser Typesetting : **Classic Computer Services**
Delhi - 110 035

Printed at : **Chawla Offset Printers**
Delhi - 110 052

PRINTED IN INDIA

Preface

The role of free radicals in many disease conditions has been well established. Several biochemical reactions in our body generate reactive oxygen species and these are capable of damaging crucial bio-molecules. If they are not effectively scavenged by cellular constituents, they lead to disease conditions, *e.g.*, cerebrovascular disease, cancer, Arteriosclerosis, Atherosclerosis, Heart disease, Senelity, Ageing, Behcet's disease, Crohn's disease, Cataracts, Sunburn, Ulcers, Osteoporosis, Rheumatoid Arthritis, Diabetes Mellitus, Emphysema, Stroke, Haemorrhagic Shock, Cardiovascular Disorders, Cystic Fibrosis, Neurodegenerative Diseases (*e.g.*, Parkinsonism, Alzheimer's disease), Gastrointestinal Ulcerogenesis, AIDS and even early Senescence. Free radicals are involved in numerous skin diseases, especially inflammatory reactions and photosenescence. Antioxidants reduce blood clotting and cholesterol level. Antioxidants have antistress, immunomodulatory, cognition-facilitating, anti-inflammatory and anti-aging properties.

The interest on the role of natural antioxidants as a tool to prevent aging and degenerative diseases process is growing due to the toxicity of synthetic antioxidants such as butylated hydroxyl anisole (BHA) and butylated hydroxyl toluene (BHT) at fairly high doses which limits their therapeutic usage. In the last few years there is a flurry of research on antioxidant plants.

The book, *Encyclopaedia of Herbal Antioxidants*, gives information on antioxidant activity of different plant species. All the species are listed in alphabetical sequence. Angiosperms are given first followed by other groups of plants. Plant part used, model system studied, methods of assay and chemical constituents are given for each and every species studied.

I tried to make this book as comprehensive as possible. However for a book of this magnitude there are bound to be omissions. Had I tried for perfection I could have never completed this book. Readers are requested to bring to the notice of the author for inclusion in the revised edition.

T. Pullaiah

Contents

Abbreviations

AAPH: 2, 2 – Azobios-2-amino-propane dihydro Chloride
ABTS: 2,2′-azino-bis-3-ethyl benzothiazoline-6-sulphonic acid
ALP: Alkaline Phosphatase
CAT: Catalase
DMSO: Dimethyl Sulphoxide
DPPH: 1,1,Diphenyl-2-Picryl-hydrazyl
FRAP: Ferric Reducing and Antioxidant Power
GAE: Gallic acid Equivalent
GPx: Glutathione Peroxidase
GSH: Glutathione
IC: Inhibitory Concetration
MDA: Malondialdehyde
ONOO: Peroxynitrite
ORAC: Oxygen Radical Absorbance Capacity
RNS: Reactive Nitrogen Species
ROS: Reactive Oxygen Species
SGOT: Glutamate Oxaloacetate transaminase
SGPG: Glutamate Pyruvate transaminase
SOD: Superoxide Dismutase
TEAC: Trolox Equivalent Antioxidant Capacity

***Ullucus tuberosus* Caldas Family: Basellaceae**

English Name: Ulluco

Penarrieeta *et al.* (2005) measured the Total Antioxidant Capacity (TAC) in some Andean foods. Eight Andean foods, including *Ullucus tuberosus* were analyzed by two methods ABTS and FRAP to assess TAC. TAC value of *Ullucus tuberosus* was good in both water-soluble fraction and water-insoluble fraction.

Campos *et al.* (2006) and Salluca *et al.* (2008) evaluated the phenolic compounds and antioxidant capacity of tuber of *Tropaeolum tuberosum*. Total antioxidant capacity was measured using ABTS and FRAP methods.The antioxidant activity was moderate. Data from FRAP displayed a linear correlation with TPH data. Ulluco contained betalins in the acid form of betaxanthins (22-96µg/g) and betacyanins (64µg/g) with no presence of carotenoids or anthocyanins.

***Ulmus davidiana* Nakai Family: Ulmaceae**

The antioxidant activities of water, ethanol, methanol, and chloroform extracts of *Ulmus davidiana* were evaluated by Guo and Wang (2007) using various antioxidant assays: DPPH free radical scavenging; hydroxyl radical scavenging; lipid peroxidation scavenging; and reducing power assays. All extracts, except the chloroform extract, demonstrated strong antioxidant activity in all assays. The chloroform extract had the highest hydroxyl radical scavenging activity, and its activity was equivalent to α-tocopherol at a concentration of 0.5 mg/ml.

Hypolipidemic effect and antioxidant activity of glycoprotein isolated from *Ulmus davidiana* in Triton WR-1339-treated mouse was investigated by Ko *et al.* (2007). *Ulmus davidiana* glycoprotein significantly decreased the level of TBARS and improved activities of catalase and glutathione peroxidase, without an apparent change of superoxide dismutase (SOD) activity. Also the glycoprotein significantly increased nitric oxide production in Triton WR-1339-induced mice.

A methanolic (MeOH) extract of *Ulmus davidiana* was analyzed by Jung *et al.* (2008) for antioxidant activity using model systems, including 1,1-diphenyl-2-picrylhydrazyl (DPPH) radical scavenging, hydroxyl radical (·OH) scavenging, reducing power, and total phenolic content. The MeOH extract exhibited strong antioxidant activity in the tested model systems. Among fractions using several solvents, the ethyl acetate (EtOAc)-soluble fraction, which exhibited strong antioxidant activity, was further purified by silica–gel and Sephadex LH-20 column chromatography. The (")-Catechin (1) and (")-catechin-7-*O*-β-d-apiofuranoside (2) were isolated as the active principles. Compounds 1 and 2 exhibited strong antioxidant activity on DPPH radicals, with IC_{50} values of 6.37 µM and 5.41 µM, respectively, and strong activity on ·OH radicals at 10 µg/ml, with 53.65 and 52.56 per cent inhibition.

Jung *et al.* (2010) investigated the antioxidative activities of (")-catechin (1), (")-catechin-7-*O*-β-d-apiofuranoside (2) and (")-catechin-7-*O*-β-d-xylopyranoside (3) in terms of their abilities to promote metal chelation, prevent lipid peroxidation, and inhibit DNA cleavage. In addition, high-performance liquid chromatography (HPLC) was used to quantify compounds (1–3) in each of various solvent extracts from *Ulmus davidiana*. Regarding metal-chelating activity, (")-catechin (1) exhibited moderate activity at a concentration of 180 µg/ml, with 57.5 per cent of metal ions chelated.

However, (")-catechin-7-*O*-β-d-apiofuranoside (2) and (")-catechin-7-*O*-β-d-xylopyranoside (3) exhibited weak chelating activity. The values obtained for compounds (2 and 3) were comparable to that of α-tocopherol at a concentration of 180 μg/ml (9.11 ± per cent). In addition, (")-catechin-7-*O*-β-d-xylopyranoside (3) showed inhibitory activity similar to α-tocopherol (inhibition of lipid peroxidation). Moreover, compounds (1–3) exhibited moderate protective activity against DNA cleavage.

Ulmus pumila L. Family: Ulmaceae

The antioxidant properties of ten Korean medicinal plants were analyzed by Kim *et al.* (2010) using various antioxidant assays, including their ability to scavenge the DPPH free radical, ABTS radical, superoxide anion, and nitrite, and their reducing powers. The contents of total phenolic compounds and flavonoids were also determined. Of the ten plants, *Ulmus pumila* (UP) possessed strong antioxidant activity for all models tested. The half-maximal inhibitory concentration (IC_{50}) for DPPH free radicals of UP was 4.16 μg/ml. The IC_{50} for superoxide anion of UP was 145.7μg/ml. In addition, the ABTS radical scavenging activity of these materials was higher than that of Trolox. The total phenol content was also significantly higher than in the other herbs. UP had the highest flavonoid content of the ten materials. A correlation was observed between the antioxidant activity and total phenol content.

Uncaria macrophylla Wall. Family: Rubiaceae

Cai *et al.* (2004) evaluated the antioxidant activity and phenolic compounds in traditional Chinese medicinal plants associated with anticancer. The improved ABTS.+ method was used to systematically assess the total antioxidant capacity (Trolox equivalent antioxidant capacity, TEAC) of the medicinal extracts. The TEAC values and total phenolic content for methanolic extract of root of *Uncaria macrophylla* were 500.6 μmol Trolox equivalent/100g dry weight (DW), and 2.39 g of gallic acid equivalent/100g DW respectively. A positive significant linear relationship between antioxidant activity and total phenolic content showed that phenolic compounds were the dominant antioxidant components in the tested herbs. Major types of phenolic compounds in *Uncaria macrophylla* include flavonoids (hyperin, catechins), tannins.

Uncaria rhynchophylla (Miq.) Jacks. Family: Rubiaceae

Liu and Mori (1992) investigated the antioxidant effects of *Uncaria rhynchophylla in vivo* and their free radical scavenging effects *in vitro*. Extracts significantly inhibited the increase in levels of lipid peroxide in the ipsilateral cortex. In *in vitro* experiments, the extract exhibited significant dose-dependent scavenging effects on free radicals.

The aqueous extract of *Uncaria rhynchophylla* inhibited lipopolysaccharide (LPS)-induced nitric oxide and interleukin-1β secretion as well as inducible NO synthase expression, without affecting cell viability (Kim *et al.*, 2010).

Uncaria sinensis (Oliv.) Havil Family: Rubiaceae

The antioxidant properties of Choto-san (containing hooks and stems of *Uncaria sinensis* and its related constituents such as Chotoko (Hooks) and Choto-san without Chotoko, and phenolic compounds contained in Chotoko such as epicatechin, caffeic acid and quercetin were evaluated by Mahakundakorn *et al.* (2004). In the DPPH radical-scavenging assay the scavenging activity of Chotoko (IC_{50} 14.3 μg/ml) was found to be higher than that of Choto-san (IC_{50} 206.2 µg/ml) and Choto-san without Chotoko (IC_{50} 244.3 µg/ml). Epicatechin, caffeic acid and quercetin also revealed scavenging activity against DPPH radicals. Choto-san (IC_{50} 92.42 µg/ml) but weaker activity than Chotoko (IC_{50} 18.3 µg/ml). IN NG 108-15 cells when added simultaneously Chotosan protected cells from oxidative damage. Results provided evidence that Choto-san acts as an antioxidant cytoprotective agent against oxidative damage which is due at least partly to the phenolic compounds contained in Chotoko (Mahakundakorn *et al.*, 2004).

Uncaria tomentosa (Willd.) DC. Family: Rubiaceae

Different extracts of *Uncaria tomentosa* were tested *in vitro* for their antioxidant activity utilizing tert-butyl-hydroperoxide-initiated chemiluminiscence in rat liver homogenates (Desmarchelier *et al.*, 1997). Methanol fractions of both stem-bark and roots were capable of exerting antioxidant activity by this technique. The presence of different concentrations of bark and root methanol extracts also prevented TBARS production and free radical-mediated DNA-sugar damage.

Bark decoction prepared according to the traditional Peruvian medicine presented a potent radical scavenger activity, as suggested by its high capacity to reduce the free radical DPPH, and by its reaction with superoxide anion, peroxyl and hydroxyl radicals as well as with the oxidant species, hydrogen peroxide and hypochlorous acid (Goncalves *et al.*, 2005). It also protected membrane lipids against peroxidation induced by the iron/ascorbate system, as evaluated by the formation of thiobarbituric acid reactive substances. HPLC/DAD and TLC revealing essentially the presence of anthocyanidins (oligomeric procyanidins) and phenolic acids, mainly caffeic acid. Results provide evidence for an antioxidant mechanism underlying the anti-inflammatory activity of cat's claw and support some of the biological effects of proanthocyanidins, more exactly its antioxidant and radical scavenging activities (Goncalves *et al.*, 2005).

Pilarski *et al.* (2006) investigated the antioxidant activity of ethanolic and aqueous extracts of *Uncaria tomentosa*. The analysis included TEAC, peroxyl radical trapping capacity (PRTC), superoxide radical scavenging activity (SID) and quantitation of total tannins (TT) and total phenolic compounds (TPC). The obtained results indicate high antioxidant capacity of the aqueous and ethanolic extracts bark in comparison to the other extracts of fruits, vegetables, cereals and medicinal plants. Higher antioxidant activity and total phenolic compounds of the alcoholic preparations TEAC=0.57 mmol of Trolox/g, PRTC=0.52 mmol of Trolox/g and SOD=0.39 U/mg than the aqueous preparation TEAC=0.34 mmol of Trolox/g, PRTC=0.19 mmol of Trolox/g and SOD=0.10U/mg were observed. These results might suggest higher medicinal suitability of alcoholic extracts. However highly elevated level of tannins in alcoholic extract may cause undesirable gastric effects.

Wojcikowski *et al.* (2007) evaluated the antioxidant activity of aerial parts of *Uncaria tomentosa*. Oxygen radical absorbance capacity of the ethyl extract, methanol extract and aqueous extracts of leaf were 23.24, 139.36 and 60.92µmol Trolox equivalent per g of dried starting material, respectively.

Uraria crinita Desv. Family: Fabaceae (Papilionaceae)

The antioxidant activity and the nitric oxide-scavenging effect of methanol extracts (MEUR) and their ethyl acetate fraction (EAF), from *Uraria crinita* root, were investigated by Yen *et al.* (2001). The antioxidant activity of the extracts was determined using the TEAC assay. The TEAC values of MEUR and EAF were 0.2 and 0.4 respectively, at a concentration of 200μg/ml. The antioxidant activity and nitric coxide-scavenging effect of MEUR and EAF showed concentration-dependence. In addition, the results also showed a decreasing effect on nitric oxide production of lipopolysaccharide-induced RAW 264.7, for both extracts. One of the major antioxidative components was isolated from EAF and was identified as genistein.

Urena lobata L. Family: Malvaceae

The methanolic extract of *Urena lobata* roots were found to inhibit lipid peroxidation, scavenge hydroxyl and superoxide radicals *in vitro* (Lissy *et al.*, 2006). IC_{50} of root extract of *Urena lobata* was 470.6 μg/ml, 1627.35μg/ml and 1109.24μg/ml for superoxide radical scavenging, hydroxyl radical scavenging and lipid peroxidation respectively.

The leaves extract of *Urena lobata* were examined by Adeloye *et al.* (2007) for their antioxidant activity. The ethyl acetate and n-butanol fractions had a fast antioxidant reaction with DPPH solution, while n-hexane and dichloroethane fractions gave no reaction. Three compounds were isolated from the ethyl acetate fraction include kaemmpferol, quercetin and 3-O-β-D-(6″-O-trans-*p*-coumaoyl)-α-L-glucopyranosyl-kaempferol (tiliroside). The effects of aqueous extract of *Urena lobata* root on the oxidative status of normal rabbits were monitored by Omonkhua and Onoagbe (2008). They seemed protective against lipid peroxidation.

Pieme *et al.* (2010) evaluated the antioxidant activity of *Urena lobata* extracts. The results of the antioxidant properties showed that these extracts induced significantly increase SOD, CAT and GsT activity after 48h.

Urginea indica Kunth Family: Liliaceae

English Name: Wild Onion.

Murthy *et al.* (2011) investigated the phytochemistry and antioxidant activities of white bulbs of *Urginea indica* by using DPPH free radical scavenging assay. The activity increased with increase in concentration of methanolic fraction.

Urginea maritima (L.) Baker Family: Liliaceae

Determination of antioxidant activities of different *Urginea maritima* plant extracts was carried out by Mommadov *et al.* (2010) using β-carotene-linoleic acid system and DPPH radical scavenging assay. The results demonstrated that the highest antioxidant activity in the *Urginea maritima* showed in the tubers with ethanol solvent (72.67 per cent) and the lowest in benzene extract 931.12 per cent) from the tubers. The free radical scavenging activity of the methanol extract from the tubers of *U.maritima* plant (66.89 per cent) was found lower than the BHT value (91.12 per cent) but still the highest. However, the amount of total phenolic compounds of acetone extract of *U. maritima* tubers was 6 times more than ethanol extract.

Urophyllum ferrugineum **King et Gamble Family: Rubiaceae**

Ahmad *et al.* (2010) evaluated the antioxidant potential of 22 species of medicinal plants from Malaysian Rubiaceae including *Urophyllum ferrugineum*. FTC, TBA, TPC and the DPPH assays were employed. The tested extracts showed strong antioxidant potential with percent inhibition of 97.3 per cent in the FTC, and 93.2 per cent in TBA assays. The TPC of the extract was 27.59mg GAE/g PE. A good correlation was observed between total phenolic content and radical-scavenging activities.

Urophyllum glabrum **Wall. Family: Rubiaceae**

Ahmad *et al.* (2010) evaluated the antioxidant potential of 22 species of medicinal plants from Malaysian Rubiaceae including *Urophyllum glabrum*. FTC, TBA, TPC and the DPPH assays were employed. The tested extracts showed strong antioxidant potential with percent inhibition of 98.0 per cent in the FTC, and 95.1 per cent in TBA assays. The TPC of the extract was 15.60mg GAE/g PE. A good correlation was observed between total phenolic content and radical-scavenging activities.

Urophyllum griffithianum **(Wight) Hook.f. Family: Rubiaceae**

Ahmad *et al.* (2010) evaluated the antioxidant potential of 22 species of medicinal plants from Malaysian Rubiaceae including *Urophyllum griffithianum*. FTC, TBA, TPC and the DPPH assays were employed. The tested extracts showed moderate antioxidant potential with per cent inhibition of 97.4 per cent in the FTC, and 97.3 per cent in TBA assays. The TPC of the extract was 27.59mg GAE/g PE and strong DPPH radical-scavenging activity with IC_{50} value of 43.24. A good correlation was observed between total phenolic content and radical-scavenging activities.

Urtica dioica **L. Family: Urticaceae**

English Name: Nettle.

A total of 27 extracts from non-cultivated and weedy vegetables traditionally consumed by ethnic Albanians in southern Italy were tested for their free radical scavenging activity (FRSA) in the DPPH screening assay, for their *in vitro* non-enzymatic inhibition of bovine brain lipid peroxidation and for their inhibition of xanthine oxidase. In the lipid peroxidation assay, extracts from leaves of *Urtica dioica* showed a remarkable inhibitory activity (>50 per cent).

The 50, 100 and 250 µg amounts of water extract of *Urtica dioica* showed 39, 66 and 98 percent inhibition on peroxidation of linoleic acid emulsion respectively, while 60 micro g/ml of α-tocopherol, exhibited only 30 per cent inhibition. Moreover the extract had effective reducing power, free radical scavenging, superoxide anion radical scavenging, hydrogen peroxide scavenging and metal chelating activities at the same concentrations. Those various antioxidant activities were compared to standard antioxidants such as BHA, BHT, quercetin and α-tocopherol (Gülçin *et al.*, 2004).

Kanter *et al.* (2003, 2005) evaluated the hepatoprotective effect of *Urtica dioica* on lipid peroxidation, antioxidant enzyme systems and liver enzymes in carbon tetrachloride-treated rats. The CCl_4 treatment for 60 d increased the lipid peroxidation and liver enzymes, and also decreased the antioxidant enzyme levels. *Urtica dioica* treatment for 60 d decreased the elevated lipid peroxidation and liver enzyme levels and also increased the reduced antioxidant enzyme levels.

Antioxidant and radical scavenging activities, reducing powers and the amount of total phenolic compounds of aqueous and methanolic extracts of *Urtica dioica* were studied by Mavi *et al.* (2004). The aqueous extract showed moderate antioxidant activity.

Avci *et al.* (2006) evaluated the hypercholesterolaemic and antioxidant activities of *Urtica dioica*. The ethanol extract increased the serum HDL concentration and decreased serum LDL-C concentration.

Pourmorad *et al.* (2006) carried out a systematic record of the relative antioxidant activity in selected Iranian medicinal plant species. The total phenol content of *Urtica dioica*, measured by Folin Ciocalteu method was 24.1 mg/g GAE. Flavonoid contents in terms of quercetin equivalent was 43.3 mg/g. DPPH radical scavenging effect of the extracts was detemined spectrophotometrically. The radical scavenging activity of *Urtica dioica* at 0.0254mg/ml concentration was 95.64 per cent.

In a screening of South African indigenous food plants for antioxidant activity by testing for inhibition of lipid peroxidation the pot herb *Urtica dioica* had very high activity (Lindsey *et al.*, 2006).

A study was designed by Celik and Tuluce (2007) to investigate the protective effect of *Urtica dioica*, beverage used against chemical carcinogen trichloroacetic acid (TCA)-exposure in rats. The prevention potential of the plant infusion was evaluated by measuring the level of antioxidant defense systems, reduced glutathione (GSH), glutathione reductase (GR), superoxide dismutase (SOD), glutathione-S-transferase (GST) and catalase (CAT); and lipid peroxidation (malondialdehyde, MDA) content in rats. The study revealed that the constituents present in *Urtica dioica* impart protection against carcinogenic chemical induced oxidative injury that may result in the development of cancer.

Wojcikowski *et al.* (2007) evaluated the antioxidant activity of aerial parts of *Urtica dioica.* Oxygen radical absorbance capacity of the methanol extract and aqueous extracts of leaf were 71.12 and 359.30µmol Trolox equivalent per g of dried starting material, respectively.

Kamkar *et al.* (2010) evaluated the antioxidant capacities of the liquid and organic extracts from *Urtica dioica* in pre-flowering an post-flowering stages. Antioxidant and radical-scavenging properties were tested by means of DPH assay and β-carotene/linoleic acid system. IC_{50} showed a remarkable decrease in response to both liquid an the organic extracts in pre-flowering (11.7 and 22.5µg/ml, respectively) and post-flowering (8.4 and 12.4µg/ml, respectively) stages. In this respect, β-carotene/linoleic acid assay showed 91.6, 68 per cent and 93.4, 69.3 per cent inhibition in pre-flowering and post flowering stages, respectively. Pre-flowering extracts showed higher antioxidative properties compared to post-flowering.

Urtica parviflora Roxb. Family: Urticaceae

Antioxidant activity of hydromethanolic extract of *Urtica parviflora* was investigated by Pandey *et al.* (2010) by different *in vitro* methods, namely, nitric oxide scavenging, DPPH scavenging, and reducing power assay. Plant extract exhibited dose dependent free radical scavenging and reducing activity. Phytochemical examination revealed the presence of alkaloids, polysaccharides, saponins, flavonoids, phenolic compounds, glycosides and tannins.

Urtica pilulifera L. Family: Urticaceae

Mahmoud *et al.* (2005) carried out study on some antioxidant parameters in mice livers affected with *Urtica pilulifera* extracts. The parameters studied were protein, glutathione reduced form, lipid peroxidation levels as well as the activities of glutathione transferase, glutathione reductase, glutathione peroxidase and superoxide dismutase. Methanolic extracts induced greater effect on the measured antioxidant parameters. Among all plant parts, the methanolic extract of the herb showed the best effect; where as the antioxidant activity enzymes were elevated and the lipid peroxidation was decreased.

Ozen *et al.* (2010) evaluated the antioxidative properties of hydroalcoholic (80 per cent) extracts from different parts of *Urtica pilulifera*, including leaf (UPL), flower (UPF), seed (UPS), and root (UPR). Antioxidative activity of the extracts was measured using the ferric thiocyanate method, thiobarbituric acid method, reductive potential, metal chelating, free radical, superoxide anion radical, and hydrogen peroxide scavenging activity. In addition, the results were compared with antioxidants such as tert-butylated hydroxyanisole (BHA), tert-butylated hydroxytoluene (BHT), and α-tocopherol. Total antioxidant activities of UPS, UPF, UPL, UPR, BHA, BHT, and α-tocopherol were 88.79 per cent, 85.13 per cent, 86.72 per cent, 78.46 per cent, 81.31 per cent, 76.12 per cent, and 46.28 per cent, respectively. Like the antioxidant activity, the reducing power and the superoxide anion radical and free radical scavenging activities of UPL, UPF, UPS, and UPR are concentration dependent. A correlation between higher antioxidant activity and the amount of total phenolics was found in the extracts.

Urtica urens L. Family: Urticaceae

Abdel Wahab *et al.* (2006) in their investigation reported that patuletin from *Urtica urens* possesses an antioxidant activity and free radical scavenging effects in AFB1 treated rats that can be observed at a dose as low as 7.5 mg/kg b.w.

Uvaria grandiflora Roxb. Family: Annonaceae

The hexane, chloroform, ethanol extracts of leaves and barks of *Uvaria grandiflora* were studied by Aminimoghadamfarouj *et al.* (2011) for their contents, antioxidant and antibacterial activities. Among all the 6 different extracts obtained, the highest amount of flavonoids was found in the bark ethanol extract which displayed profound antioxidant and antibacterial activities which warrant further phytochemical work. According to the results obtained from antibacterial assays flavonoids can be responsible for the strong antibacterial activity existed in the bark ethanol extract.

Uvaria semecarpinifolia Family: Annonaceae

Puvanendran *et al.* (2007) evaluated the antioxidant activity of Dichloromethane (DCM) and MeOH extract of stem bark and leaves of *Uvaria semecaripinifolia* using DPPH assay. The extract showed high antioxidant activities. Percentage of inhibition of stem bark was 33.23 (DCM extract) and 12.43 (MeOH extract); leaf extract was 57.33 (DCM extract) and 19.31 (MeOH extract).

Uvaria sphenocarpa Hook.f. et Thoms. Family: Annonaceae

Puvanendran *et al.* (2007) evaluated the antioxidant activity of Dichloromethane (DCM) and MeOH extract of stem bark, stem and leaves of *Uvaria sphenocarpa* using DPPH assay. The extract showed high antioxidant activities. Percentage of inhibition of stem bark was 23.66 (MeOH extract) and stem extract was 42.92 (MeOH extract).

Vaccinium angustifolium Aiton Family: Ericaceae

English Name: Low bush blue berries.

Different cultivars of *Vaccinium angustifolium* were analyzed by Prior *et al.* (1998) for total phenolics, total anthocyanins, and antioxidant capacity (ORAC). The total antioxidant capacity of different berries studied ranged from a low of 13.9 to 45.9µmol Trolox equivalents(TE)/g of fresh berry (63.2-282.3µmol TE/g of dry matter) in different cultivars of *Vaccinium angustifolium*. A linear relationship existed between ORAC and anthocyanin or total phenolic phytonutrients of the fresh fruits.

Moyer *et al.* (2002) analyzed the anthocyanins, phenolics and antioxidant capacity in diverse small fruits including *Vaccinium angustifolium*. Total anthocynins was 208mg/100g, total phenolics was 692mg/100g, Oxygen radical absorbance capacity (ORAC) was 87.8µmol TE/g, FRAP was 97.9 µmol/g.

Vaccinium arctostaphylos L. Family: Ericaceae

English Name: Lowbush blue berries.

Six lowbush blueberries were analyzed for total anthocyanins, total phenolics, and antioxidant activity as ferric reducing antioxidant power (FRAP) in this study. The respective ranges of total anthocyanin and total phenolic contents of the tested samples were: blue berries 0.18-2.94 and 0.77-5.42 mg/g. FRAP values varied from 7.41 to 57.92µmol/g for blueberries. A linear relationship was observed between FRAP values and total phenolics for blueberries (Koca and Karadeniz, 2009).

Vaccinium ashei Reade Family: Ericaceae

English Name: Rabbiteye blueberries.

Different cultivars of *Vaccinium ashei* were analyzed by Prior *et al.* (1998) for total phenolics, total anthocyanins, and antioxidant capacity (ORAC). The total antioxidant capacity of different berries studied ranged from a low of 13.9 to 45.9µmol Trolox equivalents(TE)/g of fresh berry (63.2-282..3µmol TE/g of dry matter) in different cultivars of *Vaccinium ashei*. Brightwell and Tiflue cultivars of rabbiteye blueberries were harvested at 2 times, 49 days apart. Increased maturity at harvest increased the ORAC the anthocyanin, and the total phenolic content. A linear relationship existed between ORAC and anthocyanin or total phenolic phytonutrients of the fresh fruits.

Moyer *et al.* (2002) analyzed the anthocyanins, phenolics and antioxidant capacity in diverse small fruits including *Vaccinium ashei*. Total anthocynins was 406mg/100g, total phenolics was 875mg/100g, Oxygen radical absorbance capacity (ORAC) was 123.4µmol TE/g, FRAP was 146.5 µmol/g.

A variety of Rabbiteye blueberry (*Vaccinium ashei*) cultivars and Southern highbush blueberry (*Vaccinium corymbosum*) cultivates were collected and analyzed for flavonoids, phenolic acids, total anthocyanins, total polyphenols, and TEAC (Sellappan *et al.*, 2002). Phenolic acids measured were gallic acid, *p*-hydroxybenzoic acid, caffeic acid, *p*-coumaric acid, ferulic acid and ellagic acid. Flavonoids measured were catechin, epicatechin, myricetin, quercetin an kaempferol. Rabbiteye blueberry, Tifblue, had the highest concentration of gallic acid (258.90 mg/100g) and ferulic acid (16.97 mg/100g). Some cultivars contained ellagic acid, which ranged from 0.75-6.65 mg/100g FW in the southern highbush and from 0.19-6.02mg/100g in the rabbiteye blueberries.

Vaccinium constablaei Gray X *V. ashei* Family: Ericaceae

Moyer *et al.* (2002) analyzed the anthocyanins, phenolics and antioxidant capacity in diverse small fruits including hybrid *Vaccinium constablaei* X *V.ashei*. Total anthocynins was 259mg/100g, total phenolics was 583mg/100g, Oxygen radical absorbance capacity (ORAC) was 24.6µmol TE/g, FRAP was 146.5 µmol/g.

Vaccinum corymbosum L. Family: Ericaceae

English Name: Blueberry, Highbush bleuberry.

Different cultivars of *Vaccinium corymbosum* were analyzed by Prior *et al.* (1998) for total phenolics, total anthocyanins, and antioxidant capacity (ORAC). The total antioxidant capacity of different berries studied ranged from a low of 13.9 to 45.9µmol Trolox equivalents(TE)/g of fresh berry (63.2-282.3µmol TE/g of dry matter) in different cultivars of *Vaccinium corymbosum*. The growing location (Oregon vs. Michigan vs New Jersey) did not affect ORAC, anthocyanin or total phenolic content of the cv. Jersey of highbush blueberries.A linear relationship existed between ORAC and anthocyanin or total phenolic phytonutrients of the fresh fruits.

A hyperoxia-induced redistribution of proteins and antioxidants, between blood stream, lung, and pleural effusion was partially blocked in lab animals fed a diet supplemented with blueberry extract for eight weeks (Cao *et al.*, 1999).

The fruit juice from Bluecrop and Elliot cultivars of blueberries and other berries, was evaluated for antioxidant activity against superoxide radicals, hydrogen peroxide, hydroxyl radicals and singlet oxygen. Juice from Elliot blueberries was among the juices with the highest antioxidant capacity against superoxide radicals, hydrogen peroxide, hydroxyl radicals and singlet oxygen (Wang and Jiao, 2000).

Antioxidant capacity as measured by Oxygen radical absorbance capacity (ORAC) and total phenolic and total anthocyanin contents were evaluated in fruit tissues of 87 highbush blueberry and species introgressed highbush bluberry cultivars. ORAC and phenolic levels were evaluated in leaf tissues of the same materials. Average values of ORAC, phenolics and anthocyanins in fruit were 15.9 ORAC units 1.79 mg/g (gallic acid equivalents) and 0.95 mg/g (cyanidin-3-glucoside equivalents) respectively. Cv. Rubel had the highest ORAC per gram of fresh weight values, at 31.1 units and cv. Elliot had the highest values on the basis of ORAC per square centimeter of surface area. In leaf tissue, values for both ORAC and phenolics were significantly higher than in fruit tissue, with mean values of 490 ORAC units and 44.80mg/g (gallic acid equivalent), respectively. Leaf ORAC has a low, but significant correlation with fruit phenolics and anthocyanins but not with ORAC (Ehlenfeldt and Prior, 2001).

Moyer *et al.* (2002) analyzed the anthocyanins, phenolics and antioxidant capacity in diverse small fruits including *Vaccinium corymbosum* and hybrids. Total anthocynins was 208mg/100g, total phenolics was 444mg/100g, Oxygen radical absorbance capacity (ORAC) was 52.3µmol TE/g, FRAP was 58.6 µmol/g.

A variety of Rabbiteye blueberry (*Vaccinium ashei*) cultivars and Southern highbush blueberry (*Vaccinium corymbosum*) cultivates were collected and analyzed for flavonoids, phenolic acids, total anthocyanins, total polyphenols, and TEAC (Sellappan *et al.*, 2002). Phenolic acids measured were gallic acid, *p*-hydroxybenzoic acid, caffeic acid, *p*-coumaric acid, ferulic acid and ellagic acid. Flavonoids measured were catechin, epicatechin, myricetin, quercetin and kaempferol. Rabbiteye blueberry, Tifblue, had the highest concentration of gallic acid (258.90 mg/100g) and ferulic acid (16.97 mg/100g). Some cultivars contained ellagic acid, which ranged from 0.75-6.65 mg/100g FW in the southern highbush and from 0.19-6.02mg/100g in the rabbiteye blueberries.

Howard *et al.* (2003) evaluated the antioxidant capacity and phenolic content in blueberries as affected by genotype and growing season. This study compared the total phenolics (TPH), total anthocyanins (ACY), total hydroxycinnamic acids (HCA), total flavonols (FLA), fruit weight and oxygen radical-absorbing capacity(ORAC) of 18 blueberry genotypes grown over two growing seasons at the same locations. ORAC, TPH, ACY, HCA, FLA and fruit weight levels were affected more by genotype than growing seasons. However, the ORAC and phenolic content of some genotypes varied over the two growing seasons, suggesting that environmental growing conditions can affect levels of phenolics and ORAC in blueberries. The authors suggest that blueberry genotypes should be screened over several growing seasons in order to identify germplasm rich in antioxidants and phenolics.

Antioxidant activity of phenolic compounds in blueberries and other berries and the activity structure relationships of flavonoids and phenolic acids using the ORAC were investigated by Zheng and Wang (2003). ORAC, anthocyanin and total phenolic content of blueberries (cv. Serr) was 28.9 µmol of TE/g, 1.20mg cyanidin 3-glucoside equivalents/g and 4.12 mg GAE equivalents/g respectively. Blueberries contained the phenolic compounds chlorogenic acid, myricetin, quercetin and kaempfeol. Chlorogenic acid (20.9 per cent of the ORAC value) was a major contributor to antioxidant activity due to its high concentration in blueberries. The combination of the 11 anthocyanins found in blueberries accounted for 56.3 per cent of the total ORAC value.

The phenolic profile and quantitative composition of blueberries as well as the corresponding antioxidant activity of blueberries during fruit maturation was investigated by Castrejon *et al.* (2008). Anthocyanins of all varieties increased during successive harvest stages; meanwhile flavonoils and hydroxycinnamic acids decreased from unripe green to ripe blue stage of berry ripening. Blueberry antioxidant activity, as well as total phenolic content tended to decrease during ripening.

Varieties of cultivated blueberries (*Vaccinium corymbosum*) originating from the Modena region in Italy and protected by the mark of origin, were examined in order to determine their antioxidant activity as related to their phenolic composition. Results showed that total phenolics and total anthocyanin concentrations were, respectively, two fold and three fold higher in the wild fruits, which also had a higher anthocyanin-to-total phenolic ratio. The antioxidant activity was much higher in wild blueberries than in the cultivated ones, and it was more related to the total phenolic rather than to the anthocyanin concentration (Giovanelli and Buratti, 2008).

Four highbush blueberries were analyzed for total anthocyanins, total phenolics, and antioxidant activity as ferric reducing antioxidant power (FRAP) in this study. The respective ranges of total anthocyanin and total phenolic contents of the tested samples were: blue berries 0.18-2.94 and 0.77-5.42 mg/g. FRAP values varied from 7.41 to 57.92µmol/g for blueberries. A linear relationship was observed between FRAP values and total phenolics for blueberries.

The antioxidant capacity (AOC) of blueberries was determined by Borges *et al.* (2010) using FRAP assay. In addition, the vitamin C content of the berries was determined and phenolic and polyphenolic compounds in the extracts were analyzed by reversed-Phase HPLC-PDA-MS and by reversed-phase HPLC-PDA with online antioxidant detection system. A complex spectrum of anthocyanins was a major contributor to the AOC of blueberries. Vitamin C did not contribute to the AOC of the blueberry extract that was examined.

Vaccinium macrocarpon Ait. Family: Ericaceae

English Name: Cranberry.

Drinking cranberry juice increased plasma antioxidant status, protected RBC against hemolysis, but had no positive effect on bone quality or bone mineral status (Villareal *et al.*, 2007). Akroum *et al.* (2009) reported that methanolic extract of *Vaccinium macrocarpon* showed high free radical scavenging activity.

Vaccinium membranaceum Douglas ex Torr. Family: Ericaceae

Moyer *et al.* (2002) analyzed the anthocyanins, phenolics and antioxidant capacity in diverse small fruits including *Vaccinium membranaceum*. Total anthocynins was 124mg/100g, total phenolics was 6335g/100g, Oxygen radical absorbance capacity (ORAC) was 38.7µmol TE/g, FRAP was 43.7 µmol/g.

Vaccinium myrtilloides Michx. Family: Ericaceae

Moyer *et al.* (2002) analyzed the anthocyanins, phenolics and antioxidant capacity in diverse small fruits including *Vaccinium myrtilloides*. Total anthocynins was 298mg/100g, total phenolics was 656mg/100g, Oxygen radical absorbance capacity (ORAC) was 73.0µmol TE/g, FRAP was 94.8 µmol/g.

Vaccinium myrtillus L. Family: Ericaceae

English Name: Bilberry.

Different cultivars of *Vaccinium myrtillus* were analyzed by Prior *et al.* (1998) for total phenolics, total anthocyanins, and antioxidant capacity (ORAC). The total antioxidant capacity of different berries studied ranged from a low of 13.9 to 45.9µmol Trolox equivalents (TE)/g of fresh berry (63.2-282.3µmol TE/g of dry matter) in different cultivars of *Vaccinium myrtillus*. A linear relationship existed between ORAC and anthocyanin or total phenolic phytonutrients of the fresh fruits.

The antioxidative activity of phenolic extracts from fruits of *Vaccinium myrtillus* was examined by autoxidation of methyl linoleate by Kahkonen *et al.* (1999). Extracts showed 62 per cent antioxidant activity and high total phenolic content were found in bilberry. In addition, berry wines made of berries showed stronger antioxidative activity in the oxidation of MeLo. According to Hakkinen *et al.* (1998) bilberry is rich in flavonols, mainly quercetin. In addition, bilberries are also especially rich in hydroxycinnamic acid derivatives, which have been shown to exert antioxidant activity (Chen and Ho, 1998; Meyer *et al.*, 1998).

Bilberry anthocyanosides and some related aglycones are reported to be potent scavengers of free radicals when tested *in vitro* in a superoxide anion generating system such as hypoxanthine/xanthine oxidase (Salvayre *et al.*, 1982). Meunier *et al.* (1989) confirmed that anthocyanosidic extract of *V.myrtillus* fruits behaves both as scavenger against superoxide anion, generated by reaction of phenazine methasulfate in presence of NADH, H^+ and molecular oxygen, and as inhibitor of lipid peroxidation, induced by ADP/Fe^{2+} and ascorbate in rat liver microsomes.

Morazzoni and Malandrino (1988) studied the radical scavenging and chain breaking antioxidative properties of two anthocyanosides present in Myrtocyan and their aglycons, in comparison with quercetin. Anthocyanosides appeared slightly less active than their aglycons. Cyanidin and delphinidin chlorides proved to be potent scavengers, interacting with DPPH free radical: the IC_{50} values were 2.5 and 4.0 µM, respectively, comparable to that of quercetin. Cyanidin was the most active compound in CCl_4 induced lipoperoxidation. In addition, anthocyanosidic extract was reported to inhibit the K^+ loss induced by free radicals in human erythrocytes as well as the cellular reactions induced by oxidant compounds as daunomycin and paraquat (Maridonneau *et al.*, 1982; Navelli *et al.*, 1983).

Varieties of wild crop (*Vaccinium myrtillus*) originating from the Modena region in Italy and protected by the mark of origin, were examined in order to determine their antioxidant activity as related to their phenolic composition. Results showed that total phenolics and total anthocyanin concentrations were, respectively, two fold and three fold higher in the wild fruits, which also had a higher anthocyanin-to-total phenolic ratio. The antioxidant activity was much higher in wild blueberries than in the cultivated ones, and it was more related to the total phenolic rather than to the anthocyanin concentration (Giovanelli and Buratti, 2008).

The effect of bilberry on early atherosclerosis in hamsters was investigated by Rouanet *et al.* (2010). They received an atherogenic diet at the same time a juice at a daily dose corresponding to the consumption of 275 ml by a 70kg human. After 12 weeks berry juices inhibited aortic lipid deposition by 79-96 per cent and triggered reduced activity of hepatic antioxidant enzymes, not accompanied by a lowered plasma cholesterol. These findings suggest that moderate consumption of berry juices and teas can help prevent the development of early atherosclerosis.

The antioxidative activity of the 80 per cent ethanol extract obtained from eleven commonly consumed wild edible plants was determined according to the phosphomolybdenum method, reducing power, metal chelating, superoxide anion and free radical scavenging activity (Ozen, 2010). *Vaccinium myrtillus* had the highest antioxidant capacities. The results indicated that the antioxidant activity of vegetables seems to be due to the presence of polyphenols, flavonoid and anthocyanoside. Total phenolics in *Vaccinium myrtillus* were 110.8mg pyrocatechol equivalent/g dw, Flavonoids were 61.87 mg quercetin equivalent/g dw and anthocyanins 24.7 mg cyanidin glycoside equivalent/dw.

Vaccinium ovalifolium Smith Family: Ericaceae

Moyer *et al.* (2002) analyzed the anthocyanins, phenolics and antioxidant capacity in diverse small fruits including *Vaccinium ovalifolium*. Total anthocynins was 266mg/100g, total phenolics was 678mg/100g, Oxygen radical absorbance capacity (ORAC) was 48.0μmol TE/g, FRAP was 80.4 μmol/g.

Vaccinium ovatum Pursh Family: Ericaceae

Moyer *et al.* (2002) analyzed the anthocyanins, phenolics and antioxidant capacity in diverse small fruits including *Vaccinium ovatum*. Total anthocynins was 357mg/100g, total phenolics was 842mg/100g, Oxygen radical absorbance capacity (ORAC) was 69.8μmol TE/g, FRAP was 115.1 μmol/g.

Vaccinium oxycoccus L. Family: Ericaceae

English Name: Cranberry.

The antioxidative activity of phenolic extracts from fruits of cranberry was examined by autoxidation of methyl linoleate by Kahkonen *et al.* (1999). Extracts showed 96 per cent antioxidant activity and high total phenolic content were found in cranberry. In addition, berry wines made of berries showed stronger antioxidative activity in the oxidation of MeLo. According to Hakkinen *et al.* (1998) cranberry is rich in flavonols, mainly quercetin.

The antioxidant capacity (AOC) of cranberry was determined by Borges *et al.* (2010) using FRAP assay. In addition, the vitamin C content of the berries was determined and phenolic and polyphenolic compounds in the extracts were analyzed by reversed-Phase HPLC-PDA-MS and by reversed-phase HPLC-PDA with online antioxidant detection system. The lower AOC of cranberries was due to mainly to a reduced anthocyanin content. Vitamin C was responsible for 18-23 per cent of the HPLC-AOC of cranberries.

Vaccinium parvifolium Smith Family: Ericaceae

English Name: Red huckleberry

Moyer *et al.* (2002) analyzed the anthocyanins, phenolics and antioxidant capacity in diverse small fruits including *Vaccinium parviflorum*. Total anthocynins was 34mg/100g, total phenolics was 228mg/100g, Oxygen radical absorbance capacity (ORAC) was 78.0μmol TE/g, FRAP was 64.6 μmol/g.

Vaccinium uliginosum L. Family: Ericaceae

English Name: Whortleberry.

The antioxidative activity of phenolic extracts from fruits of *Vaccinium uliginosum* was examined by autoxidation of methyl linoleate by Kahkonen *et al.* (1999). Remarkable high antioxidant activity and high total phenolic content were found in whortleberry. In addition, berry wines made of berries showed stronger antioxidative activity in the oxidation of MeLo.

Two new *ortho*-benzoyloxyphenyl acetic acid derivatives, vacciuligins A (1) and B (2), were isolated, along with five known compounds, quercetin 3-*O*-α-arabinopyranoside (3), syringetin 3-*O*-β-glucopyranoside (4), syringetin 3-*O*-β-galactopyranoside (5), protocatechuic acid (6), and syringic acid (7) from the methanol extract of the fruit of *Vaccinium uliginosum* by Masuoka *et al.* (2007). Compounds 1-3 displayed almost identical antioxidative activity as that of α-tocopherol, and 4-7 were identified as stronger antioxidants than α-tocopherol by the ferric thiocyanate method. Moreover, 1, 3, 6, and 7 exhibited stronger radical-scavenging effect than that of α-tocopherol. In addition, twelve anthocyanins in the methanol extract were identified by high performance liquid chromatographic analysis.

Vaccinium vitis-idaea L. Family: Ericaceae

English Name: Cowberry.

The antioxidative activity of phenolic extracts from fruits of cowberry was examined by autoxidation of methyl linoleate by Kahkonen *et al.* (1999). Extracts showed 91 per cent antioxidant activity and high total phenolic content were found in cowberry. In addition, berry wines made of berries showed stronger antioxidative activity in the oxidation of MeLo. According to Hakkinen *et al.* (1998) cowberry is rich in flavonols, mainly quercetin.

Cai *et al.* (2006) reported structure – radical scavenging activity relationships of phenolic compounds from traditional Chinese medicinal plants including *Vaccinium vitis-idaea.*

Valeriana officinalis L. Family: Valerianaceae

English Name: Valerian.

Sudati *et al.* (2009) examined the protective effect of *Valeriana officinalis* on lipid peroxidation (LPO) induced by different pro-oxidant agents with neuropathological importance. Ethanolic extract of valerian (0-60 microg/ml) was tested against quinolinic acid (QA); 3-nitropropionic acid; sodium nitroprusside; iron sulfate (FeSO4) and Fe^{2+}/EDTA induced LPO in rat brain homogenates. The effect of *V. officinalis* in deoxyribose degradation and reactive oxygen species (ROS) production was also investigated. In brain homogenates, *V. officinalis* inhibited thiobarbituric acid reactive substances induced by all pro-oxidants tested in a concentration dependent manner. Similarly, *V. officinalis* caused a significant decrease on the LPO in cerebral cortex and in deoxyribose degradation. QA-induced ROS production in cortical slices was also significantly reduced by *V. officinalis*. Their results suggest that *V. officinalis* extract was effective in modulating LPO induced by different pro-oxidant agents.

Valeriana prinophylla Family: Valerianaceae

8-hydroxypinoresinol and prinsepiol isolated from the roots of *Valeriana prinophylla* displayed powerful antioxidant activity (Piccinelli *et al.*, 2004). The antioxidant activity was evaluated in Trolox equivalent antioxidant activity (TEAC) and Chemiluminiscence (CL) assays.

Valeriana wallichii DC. Family: Valerianaceae

Kalim *et al.* (2010) evaluated the antioxidant activity of Methanol (50 per cent) extract of *Valeriana wallichii*. IC_{50} values for DPPH, ABTS, OH, O_2 and ONOO were 86.61, 21.26, 37.92, 78.35 and 943.12µg/ml. The extract contained significant amount of polyphenols.

Vanda caerulea Griff. ex. Lindl Family: Orchidaceae

A preliminary biological screening revealed that *Vanda coerulea* crude hydro-alcoholic stem extract displayed the best DPPH/(•)OH radical scavenging activity (Simmler *et al.*, 2010). Bio-guided fractionation and phytochemical analysis led to the isolation of five stilbenoids: imbricatin (1), methoxycoelonin (2), gigantol (3), flavidin (4), and coelonin (5). Stilbenoids (1-3) were the most concentrated in crude hydro-alcoholic stem extract and were considered as *Vanda coerulea* stem biomarkers. Dihydro-phenanthropyran (1) and dihydro-phenanthrene (2) displayed the best DPPH/(•)OH radical scavenging activities.

Vangueria infausta Burch. Family: Rubiaceae

Abeer (2011) isolated five flavonoids from the methanolic extract of leaves of *Vangueria infausta* – luteolin-7-O-rutinoside, apigenin-7-O-rutinoside, luteolin-4'-O-glucoside, quercetin-7-O-glucoside and quercetin. The methanolic extract revealed both free radical (DPPH), H_2O_2 scavenging activities and reducing power potential as compared to rutin.

Vanilla planifolia Andr. Family: Orchidaceae

Vanilla extract was prepared by Shyamala *et al.* (2007) by extraction of cured vanilla beans with aqueous ethyl alcohol (60 per cent). The extract was profiled by HPLC, wherein major compounds, *viz.*, vanillic acid, 4-hydroxybenzyl alcohol, 4-hydroxy-3-methoxybenzyl alcohol, 4-hydroxybenzaldehyde and vanillin, could be identified and separated. Extract and pure standard compounds were screened for antioxidant activity using beta-carotene-linoleate and DPPH *in vitro* model systems. At a concentration of 200 ppm, the extract showed 26 per cent and 43 per cent of antioxidant activity by beta-carotene-linoleate and DPPH methods, respectively, in comparison to corresponding values of 93 per cent and 92 per cent for BHA. Interestingly, 4-hydroxy-3-methoxybenzyl alcohol and 4-hydroxybenzyl alcohol exhibited antioxidant activity of 65 per cent and 45 per cent by β-carotene-linoleate method and 90 per cent and 50 per cent by DPPH methods, respectively. In contrast, pure vanillin exhibited much lower antioxidant activity (Shyamala *et al.*, 2007).

Varthemia iphionoides Boiss Family: Asteraceae (Compositae)

Various extracts of aerial parts of *Varthemia iphionoides* were investigated by Al-Dabbas *et al.* (2006a) for radical-scavenging activity and antioxidative activity. The ethanol and water extracts showed a pronounced DPPH radical-scavenging activity, with inhibition of about 90 per cent at a concentration of 100μg/ml. 5,7,4′-trihydroxy-3,6-dimethoxyflavone, 5,7,4′-trihydroxy-3,3′-dimethoxyflavone and 5,4′-dihydroxy-3,7,3′trimethoxyflavone (3,7,3′-tri-O-methyl-quercetin, isolated from the ethanol extract exhibited pronounced radical-scavenging activity. The hexane, ethyl acetate, chloroform, ethanol and water extracts of aerial parts of Varthemia, *Varthemia iphionoides*, were investigated for cytotoxic activity against human myelocytic leukemia (HL-60) cells; DPPH radical-scavenging activity; antioxidative activity in the linoleic acid system; reducing power; the contents of phenolic compounds. High DPPH radical-scavenging activity, antioxidative activity in the linoleic acid system and reducing power were found in the water and ethanol extracts, and were correlated to the contents of phenolic compounds (Al-Dabbas *et al.*, 2006b).

Al-Mustafa and Al-Thunibat (2008) evaluated the antioxidant activity of Jordanian medicinal plant *Varthemia iphionoides* shoot. The level of antioxidant activity was determined by DPPH and ABTS assay in relation to total phenolic contents of the medicinally used part. They concluded that *Varthemia iphionoides* shoot possesses moderate antioxidant capacity (DPPH-TEAC range 20-80mg/g).

Ventilago maderaspatana Gaestn Family: Rhamnaceae

Basu and Hazra (2006) evaluated the nitric oxide scavenging activity of extracts of *Ventilago maderaspatana* with different solvents. Notable Nitric oxide scavenging activity was exhibited *in vitro* by some extracts (IC_{50}< 0.2mg/ml). The extracts showed marked inhibition (60-80 per cent) *ex vivo* at a dose of 80μg/ml without appreciable cytotoxic effect on the cultured macrophages.

Vepris heterophylla R.Letouzey Family: Rutaceae

Antioxidant activity test using two different methods namely 2,2-diphenyl-1-picrylhydrazyl and 2,2'-azinobis(3-ethylbenzothialozinesulfonate) diammonium salt free radical scavenging test has been carried out by Momeni *et al.* (2010) on three Cameroonian plant extracts. Results of this study in the 2,2-diphenyl-1-picrylhydrazyl scavenging test showed that the methanol extract of *V. heterophylla* exhibit high free radical scavenging activities with IC_{50} value of 204.69 μg/ml as compared to Trolox (939.19 μg/ml) used as standard. In the same manner, 2,2'-azinobis(3-ethylbenzothialozinesulfonate) diammonium salt radical scavenging test of these extracts was in accordance of the result of 2,2-diphenyl-1-picrylhydrazyl test.

Verbascum alceoides Boiss. et Hausskn. Family: Scrophulariaceae

Souri *et al.* (2004) evaluated the antioxidative activity of *Verbascum alceoides* by linoleic acid peroxidation test using 1,3-diethyl-2-thiobarbituric acid as the reagent. Methanolic extract showed 75.99 per cent inhibition at 40μg concentration with IC_{50} value of 0.87μg which is approximately in the range of α-tocopherol (IC_{50}=0.60μg).

Verbascum bottae (Defless) Hub.-Mor. Family: Scrophulariaceae

Methanolic extracts of *Verbascum bottae* showed high free radical scavenging activity (Mothana *et al.*, 2010).

Verbascum densiflorum L. Family: Scrophulariaceae

Mariassyova (2006) evaluated the antioxidant activity of *Verbascum densiflorum* in rapeseed and sunflower oils. The per cent inhibition of DPPH was 49.92. Flavonoids were 568mg/l and carotenoids 62mg/l.

Verbascum macrurum Ten. Family: Scrophulariaceae

The antioxidant properties of various fractions of a methanolic extract obtained from the aerial parts of *Verbascum macrurum* have been determined by Aligiannis *et al.* (2003) by monitoring their capacity to scavenge the stable free-radical DPPH. They were also evaluated as natural preservatives against oxidative rancidity using the accelerated Rancimat method. Their activities expressed as protection factor indicated that the fractions rich with phenylpropanoid glycosides were more potent compared to α-tocopherol and of the same magnitude as BHT, which were used as reference standards. Ten natural compounds were identified as components of this methanolic extract. Antioxidant activities established that acetoside, a polyhydroxylated phenylpropanoid glycoside derivative, was the most potent free radical scavenger and showed the highest protection factor (PFr) against sunflower-oil-induced oxidative rancidity.

Verbascum phlomides L. Family: Scrophulariaceae

Spiridon *et al.* (2011) evaluated the total phenolic content and antioxidant activity of plants used in Romanian herbal medicine. *Verbascum phlomides* extracts showed good antioxidant activity.

Verbascum thapsus L. Family: Scrophulariaceae

Wojcikowski *et al.* (2007) evaluated the antioxidant activity of aerial parts of *Verbascum thapsus.* Oxygen radical absorbance capacity of the ethyl acetate extract, methanol extract and aqueous extracts of aerial parts were 117.55, 300.65 and 189.20μmol Trolox equivalent per g of dried starting material, respectively.

Narayanaswamy and Balakrishnan (2011) evaluated the antioxidant activity of 13 medicinal plants including leaf of *Verbascum thapsus*. The aqueous extract showed 40 per cent inhibition of DPPH radical while ethanolic extract showed 90 per cent inhibition of DPPH radical.

Verbascum wiedemannianum Fisch. et Mey. Family: Scrophulariaceae

Tepe *et al.* (2006) screened the antioxidative properties of the methanolic extracts of *Verbascum wiedemannianum* by two complementary test systems, namely DPPH free radical-scavenging and β-carotene/linoleic acid. The extract showed good antioxidant activity. In the β-carotene/linoleic acid test system, *V. wiedemannianum* showed 52.5 per cent inhibition rate. Antioxidant activities of curcumin and ascorbic acid were also determined as positive controls in parallel experiments.

Verbena littoralis H.B.K. Family: Verbenaceae

Mensor *et al.* (2001) evaluated the antioxidant activity of ethanol extracts of *Verbena littoralis* and partitions by DPPH radical scavenging assay. The extract exhibited significant free radical scavenging properties. Among the partitions the more polar ones (ethyl acetate and n-butanol) are those that generally have higher antioxidant activity.

Two iridoids, 6S-hydroxy-8S-methyl-4-methylene-hexahydro-cyclopental{c}pyran-3-one and 6S, 9S-dihydroxy-8S-methyl-4-methylene- hexahydro-cyclopenta{c}pyran-3-O ne, were isolated from the aerial parts of *Verbena littoralis* by Castro-Gamboa and Castro (2004). The iridoids showed moderate free radical scavenging activity against DPPH as well as antioxidant activity, the latter being evidenced by redox properties measured using EICD-HPLC.

Verbena officinalis L. Family: Verbenaceae

The scavenging activity against DPPH radical and the antifungal effect against chloroform, ethyl acetate and 50 per cent methanolic extracts of *Verbena officinalis* leaves were investigated by Casanova *et al.* (2008). The activity of different fractions of 50 per cent methanolic extract and some isolated compounds were also investigated. The results suggest that 50 per cent methanolic extract and caffeoyl derivatives could potentially be considered as excellent and readily available sources of natural antifungal and antioxidant compounds.

Aqueous and hydroalcoholic extracts from *Verbena officinalis* were obtained and characterized by Rehecho *et al.* (2011). The analysis allowed the detection and identification of three iridoids, fifteen flavonoids and four phenolic acid derivatives. Four flavonoids, scutellarein 7-diglucuronide (9), scutellarein 7-glucuronide (13), pedalitin 6-galactoside (15) and scutellarein 7-glucoside (19) were reported for the first time from this plant. In addition, three new flavonoids have been isolated: scutellarein 7-*O*-(2-*O*-feruloyl)-diglucuronide (5), pedalitin 6-*O*-diglucuronide (6) and pedalitin 6-*O*-(2-*O*-feruloyl)-diglucuronide (13). Both extracts showed significant antioxidant activity using three *in vitro* model systems and the results have been correlated with total phenolic and total flavonoid contents. The results have allowed establishing an important relation structure–activity and significant correlations have also been found between the mineral content and the flavonoids present in both extracts.

Verbena rigida L. Family: Verbenaceae

El-Hela and Abdullah (2010) carried out the *in vitro* evaluation of the antioxidant activity in relation to polyphenolic content of methanolic extract of *Verbena rigida*. The tested extract showed significant antioxidant effect which upon investigation proven that there is a significant relation between the free radical scavenging potency and the total phenolic and phenylpropanoid glycoside content of the extract but no relation were noticed between the lipid peroxidation potency and the individual extract composition. The antioxidant activity, assessed by DPPH method was IC_{50} = 0.02 mg/ml. At 0.1mg/ml concentration the antioxidant activity was 98 per cent. The phenolic content was 43μg/g, Flavonoid content was 17.1μg/g, and phenylpropanoid content was 167μg/g. The potency of radical scavenging observed in *Verbena rigida* extract has greater than synthetic antioxidant BHT. Therefore the antioxidant effect of *V. rigida* was 4 times than that of the synthetic antioxidant BHT.

Verbena tenera Spreng. Family: Verbenaceae

El-Hela and Abdullah (2010) carried out the *in vitro* evaluation of the antioxidant activity in relation to polyphenolic content of methanolic extract of *Verbena tenera*. The tested extract showed significant antioxidant effect which upon investigation proven that there is a significant relation between the free radical scavenging potency and the total phenolic and phenylpropanoid glycoside content of the extract but no relation were noticed between the lipid peroxidation potency and the individual extract composition. The antioxidant activity, assessed by DPPH method was IC_{50} = 0.18 mg/ml. At 0.2mg/ml concentration the antioxidant activity was 89 per cent. The phenolic content was 35µg/g, Flavonoid content was 9.55µg/g, and phenylpropanoid content was 185µg/g.

Verbena venosa Gill. et Hook. Family: Verbenaceae

El-Hela and Abdullah (2010) carried out the *in vitro* evaluation of the antioxidant activity in relation to polyphenolic content of methanolic extract of *Verbena venosa*. The tested extract showed significant antioxidant effect which upon investigation proven that there is a significant relation between the free radical scavenging potency and the total phenolic and phenylpropanoid glycoside content of the extract but no relation were noticed between the lipid peroxidation potency and the individual extract composition. The antioxidant activity, assessed by DPPH method was IC_{50} = 0.02 mg/ml. At 0.2 mg/ml concentration the antioxidant activity was 82 per cent. The phenolic content was 49.4µg/g, Flavonoid content was 12.03µg/g, and phenylpropanoid content was 211µg/g.

Verbesina encelioides (Cav.) Benth. et Hook. f. ex Gray. Family: Asteraceae (Compositae)

Jain *et al.* (2008) evaluated the antioxidant activity of *Vebesina enceloides*. Hot water extract demonstrated 20.04 per cent inhibition of DPPH at 80 µg concentration.

Singh *et al.* (2009) evaluated the antioxidant activity of dichloromethane and methanolic extracts of 12 arid zone medicinal plants. *Verbesina enceloides* showed appreciable antioxidant activity and per cent of inhibition of DPPH where RC_{50} (µg/ml) was 60 (Dichloromethane extract) and 7.5 (MeOH extract). per cent of inhibition of DPPH was 94.32 per cent in methanol extract of *Verbesina enceloides*. Key phytochemicals include galegine, n-triacontane, taraxasterol and phytol.

Vernonia ambigua Kotschy et Peys. Family: Asteraceae (Compositae)

Builders *et al.* (2011) evaluated the antimalarial and antioxidant activities of *Vernonia ambigua*. The freeze-dried decoctions of the whole plant of *V. ambigua* were used for the study. The phytochemical components and antioxidant activity using 2, 2-Diphenyl-1-picryl-hydrazyl radical (DPPH) were determined. The results showed good antioxidant and antimalarial activity.

Vernonia amygdalina Del. Family: Asteraceae (Compositae)

Three flavonols – luteolin, luteolin-7-O-β-glucuronoside and luteolin-7-O-β-glucoside were isolated from *Vernonia amygdalina* by Igile *et al.* (1994). The antioxidant activity of three flavones was determined by measuring coupled oxidation of β-carotene and linoleic acid. It was observed that luteolin was a significantly more potent antioxidant than the synthetic antioxidant BHT at the same concentration (15mg/l). The two glycosides showed similar activities but significantly lower activities than luteolin or BHT.

Oboh and Akindahunsi (2004) investigated the effect of preservation of green leafy vegetables of Nigeria on the antioxidant phytoconstituents (Vitamin C and Total phenol) and activity (reducing property and free radical scavenging ability) including *Vernonia amygdalina*. The result of the study revealed that sun-drying of green leafy vegetables cause a significant decrease in the vitamin C content (16.67-64.68 per cent loss). Conversely it leads to a significant increase in the total phenol content (6.45 -223.08 per cent grain) reducing property (16.00-362.50 per cent grain) and free radical scavenging ability (126.00-5757.00 per cent gain) of green leafy vegetables.

Iwaleva *et al.* (2005) reported antioxidant and cytoprotective activities of boiled, cold and methanolic extracts of nine edible vegetables in South West Nigeria which were evaluated in the DPPH assay and hemeagglutination assay in bovine erythrocytes, respectively. *Vernonia amygdalina* exhibited significant antioxidant activity. Iwalokum *et al.* (2006) reported the antioxidant effects of aqueous extract of *Vernonia amygdalina* leaves against acetaminophen-induced hepatotoxicity and oxidative stress in mice.

Odukoya *et al.* (2007) evaluated the antioxidant activity and total phenolic contents of Nigerian green leafy vegetables including *Vernonia amygdalina*. Ascorbic acid content was 13.41 mg/100g dry weight in *V. amygdalina*.

Ayoola *et al.* (2008) carried phytochemical screening and antioxidant activities of the ethanolic extract of leaves of *Vernonia amygdalina*. The crude extract was least potent (42 times less potent than vitamin C) with a maximum inhibition of 77.7 per cent at 5mg/ml. They are of the opinion that the antioxidant activity of this plant may contribute to their claimed antimalarial activity.

Owolabi *et al.* (2008) evaluated the *in vitro* antioxidant activity of water (VAWE) and ethanol (VAEE) extracts of the leaf of *Vernonia amygdalina*. VAEE exhibited antioxidant scavenging potential comparable to BHA, but higher than BHT or VAWE. The total antioxidant scavenging activity of the extracts measured as the millimolar (mM) equivalent of vitamin C gave VAEE a 10.09, which was comparable to BHA with a 9.31, but higher than VAWE with an 8.75 or BHT with a 6.14. Extracts inhibited bleaching of β-carotene, oxidation of linoleic acid and lipid peroxidation induced by Fe^{2+}/ascorbate in a rat liver microsomal preparation. The total flavonoid and phenolic contents correlated positively with total antioxidant activity. The total flavonoid content also correlated well with total phenolic.

Salawu *et al.* (2011) evaluated the cellular antioxidant activities of green leafy vegetables including *Vernonia amygdalina* against peroxyl radical-induced oxidation in HepG2 cells. HPLC/DAD/MS investigation of the ethanolic extract of the vegetal matter revealed the presence of 7 phenolic compounds. At a concentration of 1 mg/ml the quenching of peroxyl radical in HepG2 cells revealed that *Vernonia amygdalina* possessed antioxidant activity. Prooxidant activities were observed at lower concentration (0.1mg/ml).

Vernonia anthelmintica (L.) Willd. Family: Asteraceae (Compositae)

Sabu and Kuttan (2003) investigated the antioxidant activity of Indian herbal drugs, including *Vernonia anthelmintica*, in rats with alloxan-induced diabetes. The extract of *Vernonia anthelmintica* showed antidiabetic and antioxidant activity.

Vernonia canescense Kunth Family: Asteraceae(Compositae)

Mosquera *et al.* (2007) studied the antioxidant activity of the extracts using DPPH free-radical scavenging method. The percentage of the antioxidant activity of the n-Hexane fraction was 21.77, Dichloromethane fraction was 46.9 and methanol fraction was 43.193.

Vernonia cinerea L. Family: Asteraceae (Compositae)

Kumar *et al.* (2009) evaluated the anti-oxidant and anti-inflammatory activities of *Vernonia cinerea* using *in vitro* as well as *in vivo* models. Methanolic extract of *Vernonia cinerea* was found to scavenge the hydroxyl radical generated by Fenton reaction (IC_{50}130 µg/ml), Superoxide generated by photo reduction of riboflavin (IC_{50}190 µg/ml) and inhibited lipid peroxidation significantly (IC_{50}130.5 µg/ml). The drug also scavenged nitric oxide (IC_{50}210 µg/ml). Intraperitoneal administration of *Vernonia cinerea* was found to inhibit the PMA induced Superoxide generation in mice peritoneal macrophages. The administration of *Vernonia cinerea* to mice significantly increased the levels of catalase, superoxide dismutase, glutathione, glutathione peroxidase and glutathione-S transferase in blood and liver, whereas lipid peroxidation activity was significantly decreased. It was also found that *Vernonia cinerea* extract significantly inhibited carrageenan induced inflammation, compared with control models. Down regulation of pro-inflammatory cytokine level and gene expression also support the above result.

Vernonia cognata Less. Family: Asteraceae (Compositae)

Mota *et al.* (2011) evaluated the effect of *Vernonia cognata* extract on oxidative damage induced by ethanol in rats. Rats received *Vernonia cognata* extract (0, 1 and 2 g/kg bw, by gavage) 1 hour after EtOH had been administered (0 or 70 per cent, 0.5 ml/100 g bw, by gavage) and were killed 1 hour after *Vernonia cognata* extract administration. The stomach was removed for macroscopic and histopathological evaluation, as well as, oxidative stress markers such as lipoperoxidation (LPO) and non-protein thiol groups (NPSH) levels and catalase (CAT) activity. EtOH acute exposure increased LPO and decreased NPSH levels and CAT activity along with macroscopic and microscopic lesions in gastric tissue, confirming the involvement of oxidative stress in EtOH toxicity. *Vernonia cognata* extract attenuated oxidative and histopathological features induced by EtOH at all evaluated doses. Moreover, both studied doses of *Vernonia cognata* extract caused an increase in NPSH levels per se. However, only the dose of 2 g/kg reverted ali macroscopic changes caused by EtOH toxicity. The protective effect of the extract could be attributed to antioxidant molecules present in the extract, such as flavonoids and phenolic acids.

Veronica anagallis-aquatica L. Family: Plantaginaceae

Harput *et al.* (2011) evaluated the antioxidant activity of 14 species of *Veronica*. All the tested extracts were found to show dose dependent DPPH radical scavenging ability from the concentration of 50µg/ml. For *Veronica anagallis-aquatica* IC_{50} values of DPPH radical scavenging was 267.83µg/ml, NO scavenging was <900µg/ml, Superoxide radical was 393.09µg/ml. Phenolic content was 50.87 mg GAE/g dry extract.

Veronica baranetzkii Bordz Family: Plantaginaceae

Harput *et al.* (2011) evaluated the antioxidant activity of 14 species of *Veronica*. All the tested extracts were found to show dose dependent DPPH radical scavenging ability from the concentration of 50µg/ml. For *Veronica baranetzkii* IC_{50} values of DPPH radical scavenging was 99.04µg/ml, NO scavenging was <900µg/ml, Superoxide radical was 330.40µg/ml. Phenolic content was 83.15 mg GAE/g dry extract.

Veronica chamaedrys L. Family: Plantaginaceae

Harput *et al.* (2011) evaluated the antioxidant activity of 14 species of *Veronica*. All the tested extracts were found to show dose dependent DPPH radical scavenging ability from the concentration of 50µg/ml. For *Veronica chamaedrys* IC_{50} values of DPPH radical scavenging was 57.31µg/ml, NO scavenging was 651.98µg/ml, Superoxide radical was 113.39µg/ml. Phenolic content was 79.50 mg GAE/g dry extract.

Veronica cuneifolia D.Don Family: Plantaginaceae

Harput *et al.* (2011) evaluated the antioxidant activity of 14 species of *Veronica*. All the tested extracts were found to show dose dependent DPPH radical scavenging ability from the concentration of 50µg/ml. For *Veronica cuneifolia* IC_{50} values of DPPH radical scavenging was 76.09µg/ml, Superoxide radical was 214.33µg/ml. Phenolic content was 91.02 mg GAE/g dry extract.

Veronica cymbalaria Bordz Family: Plantaginaceae

Harput *et al.* (2011) evaluated the antioxidant activity of 14 species of *Veronica*. All the tested extracts were found to show dose dependent DPPH radical scavenging ability from the concentration of 50µg/ml. For *Veronica cymbalaria* IC_{50} values of DPPH radical scavenging was 390.09µg/ml, NO scavenging was <900µg/ml, Superoxide radical was 588.89µg/ml.

Veronica fuhsii Freyen et Sint Family: Plantaginaceae

Harput *et al.* (2011) evaluated the antioxidant activity of 14 species of *Veronica*. All the tested extracts were found to show dose dependent DPPH radical scavenging ability from the concentration of 50µg/ml. For *Veronica fuhsii* IC_{50} values of DPPH radical scavenging was 117.99µg/ml, NO scavenging was <900µg/ml, Superoxide radical was 697.05µg/ml. Phenolic content was 46.52 mg GAE/g dry extract.

Veronica hederifolia L. Family: Plantaginaceae

Harput *et al.* (2011) evaluated the antioxidant activity of 14 species of *Veronica*. All the tested extracts were found to show dose dependent DPPH radical scavenging ability from the concentration of 50µg/ml. For *Veronica hederifolia* IC_{50} values of DPPH radical scavenging was 274.1µg/ml, NO scavenging was <900µg/ml, Superoxide radical was 540.54µg/ml. Phenolic content was 40.90 mg GAE/g dry extract.

Veronica longifolia L. Family: Plantaginaceae

From *Veronica longifolia* Jensen *et al.* (2010) isolated three chlorinated iridoid glucosides, namely, asystasioside E (6) and its 6-O-esters 6a and 6b, named longifoliosides A and B, respectively. The structures of 6a and 6b were proved by analysis of their spectroscopic data and by conversion to the catalpol ester verproside (5a) or to catalpol (5), respectively. The configuration of the previously known vanilloyl analogue, urphoside B, was shown to be the 6β-epimer (6c) of the structure originally reported. Longifoliosides A (6a) and B (6b) were found to exhibit radical-scavenging activity against nitric oxide, superoxide, and 2,2-diphenyl-1-picrylhydrazyl radicals.

Veronica officinalis L. Family: Plantaginaceae

Harput *et al.* (2011) evaluated the antioxidant activity of 14 species of *Veronica*. All the tested extracts were found to show dose dependent DPPH radical scavenging ability from the concentration of 50µg/ml. For *Veronica officinalis* IC_{50} values of DPPH radical scavenging was 40.93µg/ml, NO scavenging was 570.33µg/ml, Superoxide radical was 223.44µg/ml. Phenolic content was 200.20 mg GAE/g dry extract.

Veronica orientalis Miller Family: Plantaginaceae

Harput *et al.* (2011) evaluated the antioxidant activity of 14 species of *Veronica*. All the tested extracts were found to show dose dependent DPPH radical scavenging ability from the concentration of 50µg/ml. For *Veronica orientalis* IC_{50} values of DPPH radical scavenging was 103.63µg/ml, NO scavenging was 824.24µg/ml, Superoxide radical was 198.91µg/ml. Phenolic content was 127 mg GAE/g dry extract.

Veronica peduncularis Bieb Family: Plantaginaceae

Harput *et al.* (2011) evaluated the antioxidant activity of 14 species of *Veronica*. All the tested extracts were found to show dose dependent DPPH radical scavenging ability from the concentration of 50µg/ml. For *Veronica peduncularis* IC_{50} values of DPPH radical scavenging was 54.19µg/ml, NO scavenging was 770.01µg/ml, Superoxide radical was 186.31µg/ml. Phenolic content was 139.92 mg GAE/g dry extract.

Veronica peregrina L. Family: Plantaginaceae

Eight iridoid glycosides and four phenolic compounds were isolated from the EtOAc soluble fraction of *Veronica peregrina* MeOH extract as the radical scavengers for antioxidant activity (Kwak *et al.*, 2009). The compounds were identified as protocatechuic acid (1), luteolin (2), veronicoside (3), minecoside (4), specioside (5), amphicoside (6), catalposide (7), 6-O-cis-p-coumaroyl catalpol (8), p-hydroxy benzoic acid methyl ester (9), verproside (10), verminoside (11), and chrysoeriol 7-glucuronide (12) by spectroscopic analysis. The antioxidant activity was evaluated by the ORAC(Oxygen Radical Absorbance Capacity) assay, which measures scavenging activity against peroxy radicals induced by 2,2'-azobis (2-methoxypropion-amidine) dihydrochloride, and the ORAC value is expressed as relative trolox equivalent. Compounds 2, 4, 5, 6, 8, and 12 exhibited potent antioxidant activity, and compounds 1, 11 had similar activity with trolox, whereas the other compounds showed weaker activity than trolox.

Veronica persica Bordz Family: Plantaginaceae

Harput *et al.* (2011) evaluated the antioxidant activity of 14 species of *Veronica*. All the tested extracts were found to show dose dependent DPPH radical scavenging ability from the concentration of 50µg/ml. For *Veronica persica* IC_{50} values of DPPH radical scavenging was 154.49µg/ml, NO scavenging was 811.43µg/ml, Superoxide radical was 142.81µg/ml. Phenolic content was 82.04 mg GAE/g dry extract.

Veronica polita Fries Family: Plantaginaceae

Harput *et al.* (2011) evaluated the antioxidant activity of 14 species of *Veronica*. All the tested extracts were found to show dose dependent DPPH radical scavenging ability from the concentration of 50µg/ml. For *Veronica polita* IC_{50} values of DPPH radical scavenging was 161.61µg/ml, NO scavenging was 700.10µg/ml, Superoxide radical was 143.91µg/ml. Phenolic content was 63.75 mg GAE/g dry extract.

Veronica serpyllifolia L. Family: Plantaginaceae

Harput *et al.* (2011) evaluated the antioxidant activity of 14 species of *Veronica*. All the tested extracts were found to show dose dependent DPPH radical scavenging ability from the concentration of 50µg/ml. For *Veronica serpylllifolia* IC_{50} values of DPPH radical scavenging was 53.96µg/ml, NO scavenging was 697.09µg/ml, Superoxide radical was 475.23µg/ml. Phenolic content was 78.11 mg GAE/g dry extract.

Vetiveria zizanioides (L.) Nash Family: Poaceae (Graminae)

English Name: Vetiver.

Antioxidant capacities of vetiver (*Vetiveria zizanioides*) oil were evaluated by Kim *et al.* (2005) by two different *in vitro* assays: the DPPH free radical scavenging assay and the Fe^{2+}-metal chelating assay. Results showed that the vetiver oil (VO) possessed a strong free radical scavenging activity when compared to standard antioxidants such as BHT and α-tocopherol. However, its metal chelating capacity was relatively weak. VO (10 microl/ml) dissolved in methanol exhibited approximately 93 per cent free radical scavenging activity in the DPPH assay and approximately 34 per cent Fe^{2+} chelating activity in the metal chelating assay. By contrast, 10 mM BHT and 0.1 mM α-tocopherol exhibited 93 and 89 per cent free radical scavenging activities in the DPPH assay, respectively, and 1 mM EDTA exhibited approximately 97 per cent activity in the metal chelating assay. Among the complex constituents in the crude VO, beta-vetivenene, beta-vetivone, and alpha-vetivone, which had shown strong antioxidant activities, were isolated.

The essential oil of vetiver root has been shown to possess antioxidant activity. Ferric reducing, free radical scavenging and antioxidant activity of two genotypes namely KS1 and gulabi of *V. zizanioides* root were investigated by Luqman *et al.* (2009) using *in vitro* assays - FRAP, DPPH, total phenolic content (TPC), total antioxidant capacity (TAC) and reducing power (RP). KS1 genotype showed higher FRAP values, DPPH inhibition, TPC and RP potential compared to gulabi and the antioxidant activity increased with the concentration of the extract (10-1000 microg/ml). A significant protective effect of cv KS1 (100 µg/ml) extract was also observed in reduced glutathione (GSH) and malondialdehyde (MDA) concentrations of erythrocytes subjected to oxidative stress by tert-butyl hydroperoxide (t-BHP) and hydrogen peroxide (H_2O_2). The cv KS1 showed better antioxidant activity, compared to cv gulabi indicating the possibility of exploring the presence of different phytoconstituents in the two varieties.

Viburnum dilatatum Thunb. Family: Caprifoliaceae

English Name: Gamazumi.

The antioxidant activity and antiulcer effects of Gamazumi (*Viburnum dilatatum*) fruit squeezing solution (GSS) were investigated. GSS including no antioxidant additive showed strong antioxidant activity by the XYZ-dish method and the electron spin resonance (ESR) method (Iwai *et al.*, 2001). Iwai *et al.* (2003) reported that ingestion of gamagumi crude extract (GCE) does not induce antioxidant enzymes and that the absorbed antioxidant components of GCE have a direct effect on the oxidative injury in the body. Two major anthocyanins were isolated, as cyanidin 3-*O*-(2-*O*-β-d-xylopyranosyl)-β-d-glucopyranoside, namely, cyanidin 3-sambubioside (*I*) and that of cyanidin 3-*O*-β-d-glucopyranoside, namely, kuromanin (*II*). By the electron spin resonance method, the superoxide anion radical scavenging activities (IC_{50}) of *I* and *II* were 17.3 and 69.6 μM, and their activities on hydroxyl radicals were 4.3 and 53.2 mM. As a positive control, the activities of ascorbic acid were 74.2 μM on superoxide anion radicals and 3.0 mM on hydroxyl radicals. Their results suggest that these anthocyanins having radical scavenging properties may be key compounds contributing to the antioxidant activity and physiological effects of *V. dilatatum* fruits.

Viburnum lantana L. Family: Caprifoliaceae

Serteser *et al.* (2009) studied the antioxidant activity of 50 per cent aqueous methanol extracts of fruits by various antioxidant assays, including free radical scavenging, hydrogen peroxide scavenging and metal (Fe^{2+}) chelating activities. The DPPH radical scavenging effect of fruit extract was 0.657. Fe^{2+} chelating activity (per cent) of fruit extract is 39.43 and Hydrogen peroxide inhibition activity of fruit extract was 43.37 per cent.

Altun *et al.* (2008) investigated the antioxidant properties of different parts of *Viburnum lantana* using DPPH scavenging and superoxide anion scavenging methods. The extracts were prepared from the fruits, branches and leaves. The branch extracts inhibited superoxide anion in a concentration dependent manner. Compared with α-tocopherol, the fruit extract of *Viburnum lantana* did not show any scavenging on superoxide anion formation. *V. lantana* leaf extracts, however, showed moderate scavenging effect on superoxide anion formation, whereas *V. lantana* branch extracts showed a strong scavenging effect (IC_{50} = 3.1 mg/ml) on superoxide anion in higher concentration. On the other hand, all extracts exhibited a scavenging effect on the DPPH radical with various potencies. When compared with BHT, *V. lantana* leaf, branch and fruit extracts, showed strong DPPH radical scavenging activity with IC_{50} values of 0.035, 0.052 and 0.085 mg/ml, respectively.

***Viburnum opulus* L. Family: Caprifoliaceae**

English Name: Cranberry

Andreeva *et al.* (2004) reported antioxidant activity of cranberry tree bark. Rop *et al.* (2010) determined the contents of polyphenols, antioxidant activity, flavonoids and vitamin C in the fruit of *Viburnum opulus* var. *edule*. In the case of polyphenolics high contents (up to 8.29 g of GAE/kg of fresh mass) were observed. The DPPH and ABTS tests were applied to determine the antioxidant activity, which was also high in comparison with other fruit species. The fruit extracts showed moderate inhibitory ability on nitric oxide (21.89-25.44 per cent), superoxide anion (25.13-28.50 per cent), hydroxyl radical (19.40-23.94 per cent) and lipid peroxidation (11.20 – 13.90 per cent).

Antioxidant activities of dried fruit extract of Cranberry (*Viburnum opulus*) were determined by Sagdic *et al.* (2006). The total phenolic content was found to be 131.99 mg GAE/g in cranberry fruit extract. The antioxidant activity of the extract was found to be 315.50 mg/g in dried methanol extract.

Altun *et al.* (2008) investigated the antioxidant properties of different parts of *Viburnum opulus* using DPPH scavenging and superoxide anion scavenging methods. The extracts were prepared from the fruits, branches and leaves. The branch extracts inhibited superoxide anion in a concentration dependent manner. When compared with BHT, *V. opulus* branch and fruit extracts, showed strong DPPH radical scavenging activity with IC_{50} values of 0.014, and 0.057 mg/ml, respectively.

***Viburnum propinquum* Hemsl. Family: Caprifoliaceae**

A new chalcone glycoside and a new tetrahydrofuranoid lignan along with 12 known compounds, were isolated from the stems and leaves of *Viburnum propinquum*. Antioxidnat activities of most phenolic compounds were evaluated by Wang *et al.* (2009). Compounds 1(3,4,2′,4′-tetrahydroxy-trans-chalcone), 2(3,r,2′,4′-tetrahydroxy-trans-chalcone-2′-O-β-D-glucoside, 12 (quercetin), 13(+)dihydroquercetin), and 14(eriodictyol) showed antioxidative capacities in the DPPH and hydroxyl free-radical assays, with IC_{50} values of 3.80-6.12 µg/ml, and 9.24-11.87 µg/ml, respectively. Compounds 1,2,12,13 and 14 also exhibited inhibitory activities against lipid peroxidation in rat liver homogenate, with an inhibitory rate of 10.8-39.9 per cent at 20µg/ml, 38.8-57.2 per cent at 100µg/ml, and 44.2-72.4 per cent at 200µg/ml respectively.

***Vicia canescens* L. Family: Fabaceae (Papilionaceae)**

Antioxidant and antihemolytic activities of aerial parts of *Vicia* were investigated by Nabavi (2009). Antioxidant activity was investigated in three different models. In addition, antihemolytic effect of extract was determind in H_2O_2-induced hemolysis. The extract showed different level of antioxidant activity in studied models. Extract showed moderate Fe^{2+} chelating ability. IC_{50} was 846 µg/ml. Extract also exhibited low antioxidant activity in hemogolobine-induced linoleic acid system but were capable of scavenging hydrogen peroxide in a concentration dependent manner. Extract showed moderate antihemolytic activity againts H_2O_2-induced hemolysis. The total amount of phenolic compounds in extract was determined as gallic acid equivalents (55.5) and total flavonoid contents were calculated as quercetin equivalents (16.12) from a calibration curve.

Vicia faba L. Family: Fabaceae (Papilionaceae)

Akroum *et al.* (2009) reported that methanolic extract of *Vicia faba* showed high free radical scavenging activity.

The antioxidant capacity of thirteen faba bean genotypes have been evaluated by Chaieb *et al.* (2011) by means of the total phenolic and total flavonoid contents (TPC and TFC respectively), FRAP and DPPH radical-scavenging assays. The results showed different antioxidant properties for the same genotype when considering the different parts or stages of the plant. High content of both total phenolics (> 92.85 mg GAE)/g of dried plant) and flavonoids contents (> 11.87 mg as rutin equivalent (RE)/ g of dried plant) were found for all genotypes during the vegetative and the reproductive stages, which also showed the highest antioxidant activity (FRAP value ≥ 1.157 mmol/g and DPPH radical-scavenging capacity > 43.49 per cent), indicating their potential as natural sources of antioxidant foods. For the maturity stage, the highest values for the TPC, TFC, FRAP and DPPH scavenging capacity were observed for the pods and the lowest for the cotyledons.

Vicia sativa L. Family: Fabaceae (Papilionaceae)

English Name: Vetch.

Gamal-Eldeen *et al.* (2004) studied antioxidant activity of aerial parts of *Vicia sativa* and its flavonoids. The findings indicated a potent antioxidant activity against carbon tetrachloride-induced lipid peroxidation.

Amarowicz *et al.* (2008) evaluated the antioxidant and antiradical activity of Vetch extract and its fractions. The acetone extract, of seeds and fractions, comprising low molecular-weight phenolics and tannins, displayed good antioxidant activity.

The metal chelating activity, antioxidant properties, and the effect on cell growth of a polyphenol extract from *Vicia sativa* have been investigated by Megias *et al.* (2009), and compared to those of soybean. The extracts from *V. sativa* seeds contained three and five times more polyphenols and flavonoids than soybean, respectively. The soybean polyphenol extracts showed higher copper and iron chelating activity than those from *V. sativa*, although polyphenols from *V. sativa* were more effective in preventing carotene oxidation and showed higher reducing power and scavenging activity than soybean polyphenols.

Vigna mungo (L.) Hepper Family: Fabaceae (Papilionaceae)

English Name: Black gram.

Sreeramulu *et al.* (2009) evaluated the antioxidant activity of black gram dhal by DPPH scavenging assay, FRAP assay and reducing power. Phenolic content was 418.34 mg/100g. DPPH radical scavenging activity expressed as trolox equivalent, FRAP (μmol/g) and reducing power (mg/g) were 0.27, 61 and 0.77, respectively in blackgram dhal.

Vigna radiata (L.) Wilczek Family: Fabaceae (Papilionaceae)

English Name: Green gram.

Sreeramulu *et al.* (2009) evaluated the antioxidant activity of green gram, whole and dhal by DPPH scavenging assay, FRAP assay and reducing power. Phenolic content of the whole seed was 327.64 mg/100g. DPPH radical scavenging activity expressed as trolox equivalent, FRAP (µmol/g) and reducing power (mg/g) were 0.86, 152 and 1.60, respectively in green gram whole. Greengram dhal phenolic content was 62.35 mg/100g. For green gram dhal the DPPH radical scavenging activity expressed as trolox equivalent, FRAP (µmol/g) and reducing power (mg/g) were 0.30, 53 and 1.04 respectively.

Vigna subterraneana (L.) Verdc. Family: Fabaceae (Papilionaceae)

English Name: Bambara groundnut.

The effect of fermentation on the polyphenol distribution and antioxidant activity of Bambara groundnut was investigated by Oboh *et al.* (2009). The results of the study revealed that fermentation caused a significant decrease in the bound phenol content of the legume. Free soluble phenol from both the fermented and unfermented legumes had a significantly higher reducing power.

Vigna umbellata Ohwi et Ohashi Family: Fabaceae (Papilionaceae)

Gan *et al.* (2010) evaluated the antioxidant activity of *Vigna umbellata* using the FRAP and TEAC assays, and its total phenolic content was measured by the Folin-Ciocalteu method. For FRAP assay the value was 24.74 µmol Fe(II)/g dry weight. For TEAC assay, the value was 32.82µmol Trolox/g dry weight. Total phenolic content was 3.01 mg gallic acid equivalent/g dry weight of plant material.

Vigna unguiculata (L.) Walp. Family: Fabaceae (Papilionaceae)

English Name: Cow pea

The antioxidant properties of some commonly consumed cow pea (*Vigna unguiculata*) were assessed with regard to its vitamin C, total phenol and phytate content, as well as antioxidant activity as typified by its reducing power and free radical scavenging ability (Oboh, 2006a). Vitamin C content of the cow peas ranged from 0.5 to 0.9 mg/100g. The phenol content of the cowpeas were 0.3-1.0 mg/g. The phytate content of cow pea was 2.0-2.9 per cent analyzed were generally high while the reducing power and free radical scavenging ability was 29.9 per cent (cowpea brown).

Viola mandshurica W. Becker Family: Violaceae

Lee *et al.* (2008) investigated the physiological activities of *Viola mandshurica*. Antioxidant activity was evaluated by measuring total phenolic contents, reducing power, DPPH radical scavenging activity, ABTS radical scavenging. *V. mandshurica* extracts were prepared by extracting with four different solvents (methanol, ethanol, acetone, and water). The methanol extract showed the highest total phenol content (34.49 mg/g gallic acid equivalents) among the extracts. The water extract showed the highest reducing power (0.454) at the concentration of 1,000 3/ml. The acetone extract showed the most potent radical scavenging activity. DPPH and ABTS radical scavenging activity of the acetone extract at the concentration of 1,000µg/ml were 21.13 per cent and 43.53 per cent, respectively.

Viola odorata L. Family: Violaceae

Ebrahimzadeh *et al.* (2010) evaluated the antioxidant activity of *Viola odorata*. The IC_{50} value of DPPH radical scavenging was 245.1µg/ml, Fe^{2+} chelating ability was 188µg/ml. The extract showed weak nitric oxide-scavenging activity. IC_{50} for scavenging of H_2O_2 was 640µg/ml.

Viola serpens Wall. Family: Violaceae

Kumar *et al.* (2011) investigated the phytochemistry and antioxidant activity of ethanolic extract of *Viola serpens*. The antioxidant screening of ethanolic extract showed the presence of enzymatic antioxidants such as catalase, peroxidase and ascorbate oxidase and non-enzymatic antioxidants such as ascorbic acid. Ethanolic extract of *Viola serpens* showed the absence of alkaloids, saponins, tannin, terpenoids and glycosides, flavonoids, aminoacids and reducing sugars.

Viola tianshanica Maxim. Family: Violaceae

The composition and *in vitro* antioxidant activities of the essential oil and methanol extract of the aerial parts of *Viola tianshanica* were evaluated by Yang *et al.* (2011). GC–MS analysis of the essential oil resulted in the identification of 15 constituents, representing 89.67 per cent of the oil. The major compounds detected in the essential oil were dibutyl phthalate (15.19 per cent), hexadecanoate methyl (8.65 per cent), *n*-hexadecanoic acid (3.07 per cent) and 2,3-pentanedione (2.62 per cent). Essential oil and methanol extract were tested for their antioxidant activities using 1,1-diphenyl-2-picryl-hydrazyl free radical scavenging and β-carotene linoleic acid assay.

Viola yedoensis Makino Family: Violaceae

Cai *et al.* (2004) evaluated the antioxidant activity and phenolic compounds in traditional Chinese medicinal plants associated with anticancer. The improved ABTS.+ method was used to systematically assess the total antioxidant capacity (Trolox equivalent antioxidant capacity, TEAC) of the medicinal extracts. The TEAC values and total phenolic content for methanolic extract of whole plant of *Viola yedoensis* were 537.5 µmol Trolox equivalent/100g dry weight (DW), and 2.40 g of gallic acid equivalent/100g DW respectively. A positive significant linear relationship between antioxidant activity and total phenolic content showed that phenolic compounds were the dominant antioxidant components in the tested herbs. Major types of phenolic compounds in *Viola yedoensis* include phenolic acids (*p*-hydroxybenzoic acid, p-coumaric acid), flavonols (kaempferol glycosides).

Li *et al.* (2008) evaluated the antioxidant properties of 45 medicinal plants using FRAP and TEAC assays and the total phenolic contents of these plants were measured. *Viola yedoensis* showed high antioxidant activity with 199.10 µmol Fe(II)/g FRAP value, 91.13µmol Trolox/g of TEAC value and 15.48 mg GAE/g of Phenolic content.

Virola sebifera Aubl. Family: Myristichaceae

Aryltetralone lignans bearing methylenedioxy groups (1a-b; 2a-b) were isolated from seeds of *Virola sebifera*. Their antioxidant activities were evaluated by inhibition of lipid peroxidation as indicated by TBARS and chemiluminescence emission (CL) assays. The lignan 1c, 'having a 2'-hydroxy-4',5'-methylenedioxyphenyl group, was the most active compound with TBARS/CL Q 1/2 values of 0.89 and 0.10 μg/ml, respectively. The catechol derivatives 3 and 4, obtained by demethylenation of lignans 1a and 2a, were of similar activity to 1c, and all were much more effective as antioxidants than α-tocopherol (Rezende *et al.*, 2005).

Viscum album L. Family: Loranthaceae

The antioxidant activity of water and ethanolic extracts of three *Viscum album* subspecies, ssp. *album*, ssp. *austriacum* and ssp. *abietis* was investigated by Orhan *et al.* (2005) in liver, kidney and heart tissues of streptozotocin-induced diabetic rats. The findings obtained in the experiments demonstrated that *Viscum album* subspecies possess potent antihyperglycemic and antioxidant activity depending on host plant.

Avci *et al.* (2006) evaluated the hypercholesterolaemic and antioxidant activities of *Viscum album* var. *album*. The ethanol extract increased the serum HDL concentration, whereas the ethanolic extracts decreased serum LDL-C concentration.

Wojcikowski *et al.* (2007) evaluated the antioxidant activity of aerial parts of *Viscum album* Oxygen radical absorbance capacity of the ethyl acetate extract, methanol extract and aqueous extracts of aerial parts were 35.43, 219.88 and 50.06μmol Trolox equivalent per g of dried starting material, respectively.

Janakat and Al-Thnaibat (2008) evaluated the antilipoperoxidative activity of the aqueous, boiled and non-boiled and methanolic extracts of *Viscum album* (leaves, stems and fruits) against lipid peroxidation (LPO) induced by a rat liver homogenate model. Non-boiled aqueous extracts exhibited the highest LPO inhibition percent followed by methanolic and boiled aqueous extracts. Moreover, LPO inhibition per cent was dose dependent, where 1.65 x 10-2 mg/ml caused the highest inhibition per cent in all aqueous extracts. Antioxidative capacity was the highest in stems and leaves in all methods of extraction, and was comparable to the *Rosmarinus officinalis* results while fruits possessed approximately 50 per cent of antioxidant capacity using all methods of extractions.

Methanolic extracts of *Viscum album* leaves from two hosts (Cocoa and cashew trees) were subjected to antioxidant assay. *V. album* from cocoa tree had higher total phenol content (182 mg/100g) as against that of cashew tree (160 mg/100g). The result of the study also revealed that the ferric reducing ability, free radical scavenging ability as well as the Fe^{2+} chelation of the extracts were all dose-dependent (0- 1.0 mg/ml). *Viscum album* from cocoa tree had significantly higher ferric reducing and free radical scavenging ability than that from cashew trees, while *V. album* from cashew tree has significant higher Fe^{2+} chelating ability than *V. album* from cocoa tree (Oluwaseum and Oboh, 2008).

Crude extract from *Viscum album* was screened by Sengul *et al.* (2009) for its *in vitro* antioxidant and antimicrobial properties. Total phenolic content of extract of this plant was also determined. β-carotene bleaching assay and Folin-Ciocalteu reagent were used to determine total antioxidant activity and total phenols of plant extract. Antioxidant activity was 82.23 per cent and the total phenolic content was 42.29 mgGAE/g DW).

Alpsoy *et al.* (2010) evaluated the antioxidant and animutagenic potencies of polyphenolic compounds of *Viscum album* against trichloroethylene (TCE)-induced oxidative and genotoxic damage. *V. album* extract (VAE 0.5 µg/ml) protected human lymphocytes against TCE. VAE(0.5 µg/ml) reduced the levels of malondialdehyde (MDA) significantly whereas VAE (1.0 and 2.0 µg/ml) samples increased MDA concentrations significantly.

***Viscum album* L. ssp. *album* Family: Loranthaceae**

English Name: Mistletoe.

Methanolic extracts of mistletoe grown on different host trees were investigated by Onay-Ucar *et al.* (2006) for their potential antioxidant activity. Scavenging activity was tested by DPPH method and the inhibitory effect on lipid peroxidation was examined by ferric thiocyanate and thiobaribituric acid methods. The extract from mistletoe grown on lime tree in summer showed the highest activity. It was found that antioxidant capacity of the plant differed according to the harvesting time as well as the host tree.

***Viscum coloratum* (Kom.) Nakai Family: Loranthaceae**

Cai *et al.* (2004) evaluated the antioxidant activity and phenolic compounds in traditional Chinese medicinal plants associated with anticancer. The improved ABTS.+ method was used to systematically assess the total antioxidant capacity (Trolox equivalent antioxidant capacity, TEAC) of the medicinal extracts. The TEAC values and total phenolic content for methanolic extract of stem and branch of *Viscum coloratum* were 559.8 µmol Trolox equivalent/100g dry weight (DW), and 2.68 g of gallic acid equivalent/100g DW respectively. A positive significant linear relationship between antioxidant activity and total phenolic content showed that phenolic compounds were the dominant antioxidant components in the tested herbs. Major types of phenolic compounds in *Viscum coloratum* include flavonols (avicalarin, quercetin), flavanols (+)-catechin).

***Vitellaria paradoxa* C.F.Gaertn. Family: Sapotaceae**

Polyphenol content and antioxidant activity of fourteen wild edible fruits from Burkina Faso were investigated by Lamien-Meda *et al.* (2008). The data obtained show that the total phenolic and total flavonoid levels were significantly higher in the acetone than in the methanol extracts. *Vitellaria paradoxa* fruit had high flavonoid content and high antioxidant activity.

***Vitex agnus-castus* L. Family: Verbenaceae**

Sarikurkcu *et al.* (2009) examined the chemical composition and antioxidant activity of the essential oil and different solvent extracts of *Vitex agnus-castus*. Antioxidant activities of the samples were determined by three different test systems, DPPH, β-carotene/linoleic acid and reducing power assays. In all systems, water extract exhibited excellent activity potential than those of other extracts and the oil. As expected, amount of total phenolics was very high in this extract.

Vitex altissima L.f. Family: Verbenaceae

Six new iridioid glucosides 6'-*O*-trans-ferulyl negundoside (1), 6'-*O*-trans-caffeoylnegundoside (2), 2-*O*-*p*-hydroxybenzoyl-6'-*O*-trans-caffeoylgardoside (3), 2-*O*-*p*-hydroxybenzoyl-6'-*O*-trans-caffeoyl-8-epiloganic acid (4), 2-*O*-*p*-hydroxybenzoyl gardoside (5) and 2-*O*-*p*-hydroxybenzoyl-8-epiloganic acid (6) along with two known iridoids, agnuside and negundoside, have been isolated from the ethyl acetate extractive of the leaves of *Vitex altissima*. These isolates did not exhibit significant 5-lipoxygenase enzyme inhibitory activity by both the superoxide (NBT riboflavin reduction) free-radical-scavenging and DPPH-radical-scavenging methods (Sridhar *et al.*, 2004).

Vitex doniana Sweet Family: Verbenaceae

Muanda *et al.* (2011) evaluated the antioxidant activity of three Malian medicinal plants including *Vitex doniana* employing phosphomolybdenum assay (PPM), DPPH and ABTS tests. The content of TPC and TFC were higher in the stem barks. The PPM assay showed that the highest value was 194.0mg/g dw for the root bark, followed by leaves (180 mg/g dw) and stem bark (74mg/g dw). The values of ABTS were 129.6, 126.2 and 127.1 mg/g d w for stem bark, root bark and leaves, respectively. The values for DPPH were 205.5, 200.1 and 195.0 mg/g dw, for stem bark, root bark and leaves respectively.

Vitex mollis H.B.-K. Family: Verbenaceae

Ruiz-Terán *et al.* (2008) evaluated the antioxidant activity of aerial parts of *Vitex mollis* using DPPH and β-carotene-linoleic acid bleaching assay. The extract displayed antioxidant activity comparable to the commercial antioxidant BHA. For the methanolic extract analysed, a clear relation between the total phenolic content of the extracts and their antioxidant activity was found.

Vitex negundo L. Family: Verbenaceae

A new phenyldihydronaphthalene-type lignan, vitedoin A (1), a new phenylnaphthalene-type lignan alkaloid, vitedoamine A (2), and a new trinorlabdane-type diterpene, vitedoin B (3), were isolated from the seeds of *Vitex negundo* along with five known lignan derivatives (4-8) (One *et al.*, 2004). Compounds 1,2 and 4-7 showed stronger antioxidant activity than α-tocopherol using the ferric thiocyanate method. Compounds 1,2, and 4-7 showed higher radical scavenging activity on the stable free radical DPPH than L-cysteine (Ono *et al.*, 2004).

Total methanl extract of *Vitex negundo* exhibited a strong free radical scavenging activity by DPPH method and caused a significant reduction in the formation of TBARS when evaluated for its lipid peroxidation inhibitory activity (Kulkarni *et al.*, 2008).

The antioxidant potential of the ethanol extract of *Vitex negundo* was evaluated by using DPPH, hydrogen peroxide, total antioxidant capacity and erythrocyte membrane stabilization assays. The extract showed significant activities in all antioxidant assays compared to the reference antioxidant ascorbic acid in a dose dependent manner. In DPPH and hydrogen peroxide scavenging assays the IC_{50}

values of the extract were found to be 178.43 µg/ml and 158.93µg/ml respectively. Total antioxidant activity was also increased in a dose dependent manner (Alam *et al.*, 2009).

Guha *et al.* (2010a) investigated the phenolic constituents of methanolic and aqueous extracts of *Vitex negundo* in context of antioxidant potential and inhibition of oxidative stress-induced cytotoxicity. Antioxidant efficacies of both extracts were estimated by their abilities to scavenge DPPH and to reduce Fe(+3) to Fe(+2). Their protective potential against H_2O_2-driven oxidative damage to cells were evaluated by XTT assay in Hep3B (human hepatoma) cells. Polyphenolic levels in both extracts showed significant positive correlations with DPPH scavenging, Fe(+3) – Fe (+2) reduction and percentage decrease in H_2O_2-induced cytotoxicity.

Zargar *et al.* (2011) assessed the antioxidant activity of methanol and hexane extract and essential oil from *Vitex negundo* leaf using different *in vitro* antioxidant assays. Antioxidant property was tested using DPPH free radical scavenging capacity, FRAP and β-carotene-linoleic acid assays. Total phenolic contents (TPC) were measured using Folin-Ciocalteu method. Flavonoids, tocopherols, β-carotene and lycopene were analyzed. Results of this study showed that methanol extract of *V. negundo* leaf exhibited significantly higher antioxidant activity in terms of measurements of DPPH free radical (IC_{50}), FRAP and β-carotene-linoleic assays than those of hexane extract and essential oil. Methanol extract of *V. negundo* leaf also contained high amounts of bioactive compounds including total phenolic compounds (363 mg GAE/g), epicatechin (16.98 mg/g), quercetin (13.45 mg/g), catechin (8.95 mg/g) and myricetin (3.32 mg/g) while the concentrations of tocopherol, β-carotene and lycopene were found to be lower.

Vitex obovata ssp. *obovata* and *Vitex obovata* ssp. *wilmsii* Family: Verbenaceae

In vitro phytochemical and antioxidant activity of five South African *Vitex* species were investigated by Nyligira *et al.* (2008) using DPPH assay. Extracts of *Vitex obovata* ssp. *obovata* and *Vitex obovata* ssp. *wilmsii* exhibited scavenging activity (IC_{50} = between 22.14 to 33.06) in the antioxidant assay.

Vitex peduncularis Wall. ex Schayer Family: Verbenaceae

The chloroform extract of dried stem bark of *Vitex peduncularis*, on column chromatography afforded β-sitosterol, ursolic acid, 5,7-dihydroxy flavone (chrysin), 5,7- dihydroxy-6-methoxy flavone (oroxylin) and 5,7,4'- trihydroxy-6-methoxy flavone (hispidulin). Antioxidant activities of the isolated compounds were studied by inhibition of lipid peroxidation in liver homogenate of adult male Wistar rats induced by Fe^{2+}/ascorbate system and quantified spectrometrically. The results showed that all the tested flavonoids exhibited strong antioxidant activity (Mathew *et al.*, 2004).

Vitex polygama Cham. Family: Verbenaceae

Mensor *et al.* (2001) evaluated the antioxidant activity of ethanol extracts of *Vitex polygama* and partitions by DPPH radical scavenging assay. The extract exhibited significant free radical scavenging properties. Among the partitions the more polar ones (ethyl acetate and n-butanol) are those that generally have higher antioxidant activity.

Vitex pooara Corbishley Family: Verbenaceae

In vitro phytochemical and antioxidant activity of five South African *Vitex* species were investigated by Nyligira *et al.* (2008) using DPPH assay. Extracts of *Vitex pooara* exhibited scavenging activity (IC_{50} = between 22.14 to 33.06) in the antioxidant assay.

Vitex rehmannii Gurke Family: Verbenaceae

In vitro phytochemical and antioxidant activity of five South African *Vitex* species were investigated by Nyligira *et al.* (2008) using DPPH assay. Extracts of *Vitex rehmannii* exhibited scavenging activity (IC_{50} = between 22.14 to 33.06) in the antioxidant assay.

Vitex rotundifolia L. Family: Verbenaceae

Antioxidant capacities of 56 selected Chinese medicinal plants, including *Vitex rotundifolia* were evaluated by Song *et al.* (2010) using the TEAC and FRAP assays, and their total phenolic content was measured by the Folin-Ciocalteu method. *Vitex rotundifolia* showed moderate antioxidant activity with TEAC value of 37.18µmol Trolox/g, FRAP value of 9.35µmol Fe^{2+}/g and phenolic content was 5.66mg GAE/g.

Vitex trifolia L. Family: Verbenaceae

The chloroform and methanolic extracts of 124 Egyptian plant species belonging to 56 families were investigated by Moussa *et al.* (2011) and compared their antioxidant activity by DPPH scavenging assay. Among the 124 plant species tested 18 exhibited extremely high antiradical activity (>80 per cent inhibition). The IC_{50} value of DPPH radical scavenging of *Vitex trifolia* was 22.94µg/ml, while total phenolic and flavonoid contents were 135.66 mg Tannic acid equivalent/g extract and 162.06 mg Rutin equivalent/g extract, respectively. Correlation coefficient between DPPH radical scavenging activity and the total phenolic and flavonoid contents suggest that phenolics and flavonoids in the extracts were partly responsible for the antiradical activities.

Vitex trifolia L. var. _simplicifolia_ Cham. Family: Verbenaceae

Cai *et al.* (2004) evaluated the antioxidant activity and phenolic compounds in traditional Chinese medicinal plants associated with anticancer. The improved ABTS.+ method was used to systematically assess the total antioxidant capacity (TEAC) of the medicinal extracts. The TEAC values and total phenolic content for methanolic extract of fruit of *Vitex trifolia* var. *simplicifolia* were 93.6 µmol Trolox equivalent/100g dry weight (DW), and 0.57 g of gallic acid equivalent/100g DW respectively.

Vitis thunbergii Siebold. et Zucc. Family: Vitaceae

Investigating the constituents of the roots of *Vitis thunbergii* led to the isolation of four new resveratrol derivatives, vitisinols A-D (1-4), together with (+)-epsilon-viniferin, (-)-viniferal, ampelopsin, C, miyabenol A, (+)-vitisin A, and (+)vitisin C. All of the isolated compounds except 1, were tested for their antiplatelet and antioxidative activities (Huang *et al.*, 2005).

Shyur *et al.* (2005) examined antioxidant activities of 26 medicinal herbal extracts that have been popularly used as folk medicines in Taiwan. The results of scavenging DPPH radical activity show that, among the 26 tested medicinal plants, *Vitis thunbergii* exhibited strong activities and its IC_{50} value for DPPH radical was 24µg/ml. Superoxide anion scavenging activity was IC_{50} = 58µg/ml.

Vitis vinifera L. Family: Vitaceae

English Name: Grapes.

The total antioxidant activity of grape was measured by Wang *et al.* (1996) using automated oxygen radical absorbance capacity (ORAC) assay. On the basis of the wet weight of the fruits (edible portion), grape (red) had the good ORAC activity (7.39 micromoles of Trolox equivalents per gram). On the basis of dry weight of the fruits grape had the highest ORAC activity (36.0µmoles/g Trolox equivalent). The contribution of the fruit pulp fraction (extracted with acetone) to the total ORAC activity of a fruit was usually less than 10 per cent.

Administration (oral) of grape seed extract (100mg/kg/mol wt) to aged rats for 30 days decreased the oxidant generation, lipid peroxidation, protein oxidative stress mediated macromolecular damage in brain regions of aged rats (Balu and Pannerselvan, 2004).

Lee and Talcott (2004) reported that fruit maturity and juice extraction influences ellagic acid derivatives and other antioxidant polyphenolics in muscadine grapes. All polyphenolics generally increased as fruit ripened and the highest concentrations were located in the skins. Free ellagic acid, ellagic acid glycosides, and total ellagic acid ranged from 8 to 162, 7 to 115 and 587 to 1900 mg/kg, respectively in skin of ripe grapes. Hot-pressed juices contained considerable lower polyphenolic concentrations that were present in whole grapes. Antioxidant capacity was appreciably influenced by cultivar, maturity, and location in the fruit with good correlations to soluble phenolics found in both methanolic and ethyl acetate extracts.

The contribution of the major monomeric flavonols and phenolic acid to the total antioxidant capacity of grape seeds and skins was determined by Yilmaz and Toledo (2004). Peroxyl radical scavenging activities of phenolics present in grape seeds or skins in decreasing order were resveratrol>catechin>epicatechin = gallocatechin > gallic acid = ellagic acid. The results indicated that dimeric, trimeric, oligomeric or polymeric procyanidns account for most of the superior antioxidant capacity of grape seeds (Yilmaz and Toledo, 2004).

Sendogdu *et al.* (2006) investigated antidiabetic and antioxidant effects of *Vitis vinifera* leaves in streptozotocin-diabetic rats. Ethanolic extract of leaves at 250 mg/kg dose was found to possess a high antidiabetic and antioxidant activity. Mainly condensed tannins and flavonoids were suggested to contribute in the activities.

The OPC (Oligomeric proanthocynidine) is a powerful component present in grape seed. It acts as superoxidant, protecting cells against free radical formation. It reduces LDL cholesterol and the risk of cardiovascular tissue. It enhances the ability of collagen to repair itself preventing degeneration. It also increases the strength and elasticity of blood vessels, protecting against rupture, leakage and degeneration.

OPC, reduces High Blood Pressure by inhibiting the activity of Angiotensin converting enzyme (ACE). It improves the strength of capillaries and blood vessels. In comparison to Vitamin C, OPC is more powerful antioxidant. It has sparing effect on vitamin C by blocking the action of ascorbate oxydase enzyme that destroys vitamin C. Alanine, α-tocopherol, ascorbic acid, β-carotene, β-sitosterol, histidine, methionine, OPC, palmitic acid and selenium are the main antioxidative constituents found in grape (Mishra *et al.*, 2007).

Kedage *et al.* (2007) evaluated the antioxidant potential of 11 grape varieties from India and nearby Asian countries employing FRAP, DPPH, ferromyoglobin/2,2'-azobios-3-ethylbenzthiazoline-6-sulfoniic acid, ORAC and inhibition of lipid peroxidation. The total phenolic and flavonoids contents were also estimated. Their study indicated that cv. Mango is the most potent followed by Sharad seedless. Ethanolic extracts were found to be more effective than aqueous extracts. Cv. Sharad seedless, Mango and Manikchaman also had high ORAC values. Their HPLC analysis showed the presence of various antioxidant polyphenols.

The hepatoprotective effect of ethanolic extract and its four different fractions of *Vitis vinifera* leaves was investigated against CCl_4-induced acute hepatotoxicity in rats by Orhan *et al.* (2007). n-BuOH fraction in 83 mg/kg dose possessed remarkable antioxidant and hepatoprotective activities.

Parker *et al.* (2007) investigated the antioxidant capacity of phenolic content of grapes, sun-dried raisins, and golden raisins and their effect on *ex vivo* serum antioxidant activity.

Kelen and Tepe (2007) examined the *in vitro* antioxidant activities and total phenolic contents of various extracts prepared by using solvents of varying polarity from three different seed of grape varieties. In DPPH system, grape seeds, exerted two-fold greater antioxidant activity than that of synthetic antioxidant BHT. The amount of total phenolics was highest in polar and non-polar extracts of grape seeds. Especially, a positive correlation was observed between total phenolic content and antioxidant activity of polar extracts.

Whole body exposure to ionizing radiation induces the formation of reactive oxygen species (ROS) in different tissues provoking oxidative damage, organ dysfunction and metabolic disturbances. Saada *et al.* (2008) reported that grape (*Vitis vinifera*) seed extract protects radiation-induced oxidative damage and metabolic disorders in rats. Grape seed extract treatment also attenuated the oxidative stress in pancreas tissues which was associated with a significant improvement in radiation induced hyperglycemia and hyperinsulinemia. In conclusion they opined that grape seed extract would protect the heart and pancreas tissue from oxidative damage induced by ionizing radiation.

Katalinic *et al.* (2010) determined the polyphenolic composition, related antioxidant and antimicrobial properties of grape skin extracts from 14 grape varieties (seven white and seven red grape) grown in Dalmatia (Croatia). The content of total phenols, flavonoids, catechins, flavanols and individual polyphenols ((+)-catechin, (-)-epicatechin, epicatechin gallate, procyanidin B1 and procyanidin B2, quercetin glucoside, resveratrol monomers, piceid and astringin was variety dependent. The high antioxidant capacity of all extracts both red and white, has been observed and related to the relative amounts of polyphenolic compounds with good antioxidant properties.

Breska *et al.* (2010) analyzed six raisin grape cultivars and 10 new raisin grape selections for antioxidant activity (ABTS assay) and for total and individual phenolic compounds. Total phenolic content, determined in gallic acid equivalents using Folin-Ciocalteu assay, ranged from 316.3 to 1141.3 mg gallic acid/100g DW and was strongly correlated with antioxidant results.

Mitic *et al.* (2010) evaluated the antioxidant activities using DPPH radical scavenging assay. The radical scavenging activity and total antioxidant activities were 15.30 per cent and 1.055mM Trolox equivalent, respectively. Total phenolic and total flavonoid contents ranged from 238.3 to 420.6 mg GAE per L wines and 42.64 to 81.32 mg catechin equivalent respectively. A high significant correlation between antioxidant activity and total phenolic content was found.

Voacanga africana **Stapf Family: Apocynaceae**

Oke and Hamburger (2002) studied the antioxidant activity of ethanol extract of leaf using DPPH assay. The extract showed high antioxidant activity.

Ayoola *et al.* (2008) carried out phytochemical and antioxidant screening of *Voacanga africana*. The extract tested positive for tannins, flavonoids, glycosides and anthraquinones. DPPH free radical scavenging assay was followed and the extracts showed good antioxidant activity with IC_{50} value of 0.048 and 0.054mg/ml for *V.africana* and Vitamin C, respectively.

Vochysia discolor **Mart. Family: Vochysiaceae**

Khalil *et al.* (2006) evaluated the antioxidant activity of ethanol extract of flower of *Vochysia discolor* by ABTS and TNB methods. Ethanol extract showed good antioxidant activity.

Vochysia tuconorum **Mart. Family: Vochysiaceae**

Gomes *et al.* (2009) evaluated the antioxidative action of methanolic extract and butanolic fraction of *Vochysia tuconorum*. Both MeOH extract and Butanolic fraction present elevated gastroprotective action in all *in vivo* experimental models and this was related to antioxidant activity.

Waldsteinia geoides **Willd. Family: Rosaceae**

Oszmianski *et al.* (2007) investigated the antioxidant tannins from *Waldsteinia geoides* roots. The predominant constitutive units of the procyanidins of roots were (+)catechin. The antioxidant activity, measured by the DPPH method was 4.40 mM trolox equivalent/kg dried roots, measured by the ABTS method 6.60 tolox equivalent per kg of dried roots.

Walsura robusta **Roxb. Family: Meliaceae**

3,4,5-trimethoxyphenyl beta-D-glucopyranoside (1), turpinionoside A (2), (+)-lyoniresinol 3alpha-O-beta-D-glucopyranoside (3) and (-)-lyoniresinol 3alpha-O-beta-

D-glucopyranoside (4), were isolated from the leaves and twigs of *Walsura robusta* by Voravuthikunchai *et al.* (2010). They investigated antioxidant effects of the crude extracts as well as the isolated compounds using DPPH, hydroxyl radical (OH), and superoxide anion (O(2)) scavenging assays. Three phenolic glucosides (1, 3 and 4) were found to possess strong antioxidant activity. They scavenged DPPH(*) with IC(50) values in the range of 51.5-86.6 microM. They also detected the superoxide dismutase-like activities in compounds 3 and 4 which are lignan glucosides, demonstrating potent superoxide scavenging activity with IC(50) values in the range of 0.8 and 0.7 microM, respectively.

Walsura yunnanenis C. Y. Wu. Family: Meliaceae

Four new polyphenolic glycosides (1-4) were isolated from the BuOH extract of the bark of *Walsura yunnanenis*. These compounds comprise two lignans, *i.e.*, the 9-O-alpha-L-arabinopyranosides 1 and 2 of (-)-isolariciresinol and of (+)-5-methoxyisolariciresinol, respectively, and the two phenolic glycosides 3,4,5-trimethoxyphenyl 2-O-(alpha-L-fucopyranosyl)-beta-D-glucopyranoside (3) and 3,5-dihydroxyphenyl 6-O-(4-hydroxy-3,5-dimethoxybenzoyl)-beta-D-glucopyranoside (walsuraside; 4). In addition, three known compounds, 3,4,5-trimethoxyphenyl beta-D-glucopyranoside, asperglaucide, and butyl alpha-D-fructofuranoside were identified. The new compounds 1, 2, and 4 displayed significant antioxidant activities, with IC(50) values in the range of ca. 42 to 49 microg/ml (Luo *et al.*, 2006).

Warburgia ugandensis Sprague Family: Canellaceae

Kuglerova *et al.* (2008) investigated the antioxidant activity of seven Ugandan tree species. The crude ethanolic extract of bark of *Warburgia ugandensis* showed the most promising result, which exhibited DPPH scavenging activity with IC_{50} value (7µg/ml) very close to the effect achieved by reference compounds (Trolox IC_{50} = 4µg/ml) and ascorbate (IC_{50} = 2µg/ml).

Washingtonia filifera (Linden ex Andre) Wendl. Family: Arecaceae (Palmae)

Two new flavonoids, 8-hydroxyisoscoparin (1) and luteolin 7-O-glucoside 4''-sulfate (2), along with eight known flavonoids, including luteolin 7-O-glucoside 2''-sulfate, were isolated and identified from *Washingtonia filifera*. All compounds were characterized by (1)H-NMR, (13)C-NMR, CI-MS, FABMS and UV. The antioxidant activities of various *W. filifera* extracts were determined (El-Sayed *et al.*, 2006).

Chen *et al.* (2009) studied the antioxidant activity of leaves extract towards xanthine oxidase, lipoxygenase and tyrosinase activities. The inhibitory effect of acetone extract on xanthine oxidase activity was less than 10 per cent, tyrosine activity was less than 20 per cent and lipoxygenase activity was about 40 per cent.

Wattakaka volubilis (L.f.) Stapf Family: Asclepiadaceae

The ethanol extract of leaves showed potent *in vitro* inhibition of $FeCl_2$-ascorbic acid-stimulated mice liver peroxidation (Divya *et al.*, 2009). Phytochemical reports have shown the presence of glucosides and alkaloids in roots which are considered to produce significant antioxidant, anti-inflammatory and analgesic effects.

Wedelia chinensis (Osbeck.) Merr. (Syn.: Wedelia calendulacea Less. non-Rich.) Family: Asteraceae (Compositae)

Using DPPH, nitric oxide, superoxide radical scavenging activity and Ferric reducing antioxidant power assay to investigate the antioxidant potential of crude ethanolic extracts from the leaves of *Wedelia chinensis*, Gurusamy and Saranya (2010) found that the ethanolic extracts exhibited a stronger antioxidant activity. The extracts showed the inhibition in a dose dependant manner. The results were expressed as IC_{50}. *Wedelia chinensis* showed higher reducing activity in DPPH assay and FRAP assay. *Wedelia chinensis* showed potent radical scavenging activity for Nitric oxide and Superoxide radicals.

Kataki *et al.* (2012) evaluated the *in vitro* antioxidant profile of methanolic extract of the leaves of *Wedelia calendulacea* using antioxidant activity, reducing power assay, DPPH radical scavenging activity, superoxide anion radical scavenging activity, nitric oxide scavenging activity and hydrogen peroxide scavenging activity. The total phenolic content was also estimated. Hydrogen peroxide scavenging activity of the extract was found to be the most potent when compared to the standard antioxidant. The IC_{50} values of the extract were found to be 110.881, 175.292, 207.615 and 139.824 µg/ml for DPPH radical scavenging activity, superoxide anion radical scavenging, nitric oxide scavenging activity and hydrogen peroxide scavenging activity, respectively.

Wedelia trilobata (L.) Hitch. Family: Asteraceae (Compositae)

Ethanol extract of leaf, stem and flower of *Wedelia trilobata* was assessed by Govindappa *et al.* (2011) for its antioxidant activity by measuring the scavenging activity of DPPH radical and FRAP assay. The leaf and stem ethanol extract possessed strong scavenging activity in both DPPH and FRAP methods. In DPPH and FRAP method, the leaf and stem showed free radical inhibition of 86, 82 and 630.72, 506.81, respectively.

Wendlandia formosana Cowan Family: Rubiaceae

Raju *et al.* (2004) evaluated the antioxidant activity of the iridoid glucosides isolated from the leaves of *Wendlandia formosana* against DPPH, hydroxyl radical and peroxynitrite.

Wendlandia wallichii Wight et Arn. Family: Rubiaceae

Kshirsagar and Upadhyay (2009) reported antioxidant activity of the methanol extract of stem and twig. Their ability of scavenging radicals was measured by DPPH reduction spectrophotometric assay.

Wickstroemia indica (L.) C.A.Mey. Family: Thymeleaceae

The total flavonoid extract from *Wickstroemia indica* in concentration ranges from 13.16 to 78.96µg/ml and from 0.43 to 52.64µg/ml could scavenge both DPPH and hydroxyl free radicals in a significant concentration dependent way, with scavenging ratios ranging from 20.87 per cent to 71.11 per cent and from 9.04 per cent to 52.21 per cent, respectively.

Wilbrandia ebracteata Cogn. Family: Cucurbitaceae

Menezes *et al.* (2004) investigated the antioxidant activity of hydroalcoholic extracts of *Wilbrandia ebracteatai* by DPPH assay and phosphomolybdenum method. The relative activity in phosphomolybdenum complex was 0.28 while scavenging activity against DPPH in IC_{50} (µg/ml) was 170.41.

Withania qaraitica Miller et Biagi Family: Solanaceae

Marwah *et al.* (2007) investigated the antioxidant activity of wound-healing herbs using *in vitro* DPPH and phosphomolybdenum assay methods. Out of the 19 plants screened, the aqueous ethanol extract of *Withania qaraitica* showed the best DPPH scavenging activities with per cent IP value 60.1 per cent, after 15 min of incubation at an effective test concentration of 50µg/ml. The Total Antioxidant Capacity as gallic acid equivalent of 250mg/g of ethanol extract was obtained in the phosphomolybdenum assay. Total phenolic content was 67.8 mg GAE/g of ethanol extract.

Withania somnifera (L.) Dunal. Family: Solanaceae

English Name: Ashwagandha.

Antioxidant activity of active principles of *Withania somnifera*, consisting of equimolar concentrations of sitoindosides VII-X and withaferin A, was investigated for their effects on rat brain frontal cortical and striatal concentrations of superoxide dismutase, catalase and glutathione peroxidase (Dhuley, 1998). Active glycowithanolides, administered once daily for 21 days, induced a dose-related increase in SOD, CAT and GPX activity in frontal cortex and striatum, which was statistically significant on days 14 and 21. At the lower dose of WSG on GPX, the effect was evident only on day 21 (Bhattacharya *et al.*, 1997). Antioxidant effect of active principles of *Withania somnifera* may explain, at leaset in part, the reported antistress, immunomodulatory, cognition produced by them in experimental animals, and in clinical situations (Surh, 1999).

Scartezzini and Speroni (2000) are of the opinion that this herb contains antioxidant principles, that can explain and justify their use in traditional medicine in the past as well as the present.

The antioxidant activity of *Withania somnifera* (WS) glycowithanolides was assessed in chronic foot shock stress induced changes in rat brain frontal cortex and striatum (Bhattacharya *et al.*, 2001). The stress procedure given once daily for 21 days, induced an increase in SOD and LPO activity, with concomitant decrease in CAT and GPx activities in both the brain regions. WS glycowithanolides administered orally 1 h prior to the stress procedure for 21 days, in the doses of 10, 20 and 50 mg/kg, induced a dose-related reversal of the stress effects. Thus, WSG tended to normalize the augmented SOD and LPO activities and enhanced the activities of CAT and GRx.

Leaves, roots and berries of the plant possess antioxidative properties. It prevents premature ageing and arthritis, enhances psychomotor co-ordiantion and immune system. Berries are blood tonifier that improves circulation and absorption of nutrients by cells. Leaves are used for treatment of tumors. Withaferin A, Viscosalactone B (12) inhibit the tumor formation (Mishra *et al.*, 2007).

Lipid peroxidation, enhanced by stress exposure was decreased following treatment with *Withania somnifera* root powder (250mg/kg) (Zafir and Banu, 2007).

Wojcikowski *et al.* (2007) evaluated the antioxidant activity of aerial parts of *Withania somnifera*. Oxygen radical absorbance capacity of the ethyl acetate extract, methanol extract and aqueous extracts of aerial parts were 1.48, 47.58 and 35.80μmol Trolox equivalent per g of dried starting material, respectively.

An investigation was carried out by Jatwa and Kar (2009) to reveal the possible ameliorative role of *Withania somnifera* plant extract on an antidiabetic drug-induced hypothyroidism in Type 2 diabetic animals. The results revealed that the plant extract showed antioxidant activity.

Jain and Agrawal (2009) reported antioxidant activity of methanolic extract of *Withania somnifera*. Mahour and Saxena (2010) studied free radical scavenging activity of Indian ginseng (*Withania somnifera*) under *in vitro* models *viz.*, DPPH, nitric oxide and Lipid peroxidation method. IC_{50} values were 523.89, 7.629 and 270.43 μg/ml with DPPH, nitric oxide and lipid peroxidation groups, respectively.

Woodfordia fruticosa Kurz. Family: Lythraceae

Antioxidant activity of *Woodfordia fruticosa* was evaluated by Kumaraswamy and Satish (2008) using DPPH, ABTS and lipoxygenase inhibition assay. In both DPPH and ABTS scavenging activities methanol and water extract showed significant scavenging activities (4.96, 3.08μg/ml IC_{50} for DPPH and 6.4, 7.15μg/ml IC_{50} for ABTS assay and good activity against lipoxygenase inhibition activity (45.22 and 74.24 per cent inhibition at 1.0 mg/ml).

The dried flower of *Woodfordia fruticosa* was successively extracted with petroleum ether, chloroform, methanol and water depending on the increasing order of polarity. The successive methanolic extract which recorded promising antioxidant activity was subjected to activity guided fractionation. DPPH and ABTS free radical scavenging assay revealed that isolated gallic acid has strong antioxidant activity compared to that of ascorbic acid and gallic acid, respectively (Kumaraswamy *et al.*, 2010).

Xanthium indicum Koenig Family: Asteraceae (Compositae)

Akter *et al.* (2009) investigated the antioxidant activity of *Xanthium indicum* leaves using DPPH radical scavenging and nitric oxide scavenging assays. The extract displayed a dose dependent scavenging of DPPH radical and NO.

Hydromethanol extracts of 15 Bangladeshi medicinal plants, including *Xanthium indicum*, were evaluated by Hasan *et al.* (2009) for antioxidant potential using DPPH radical scavenging assay. *Xanthium indicum* leaves exhibited radical scavenging activity with an IC_{50} value of 23.44 μg/ml compared to the IC_{50} value of 5.15μg/ml as shown by the reference antioxidant ascorbic acid, in a dose dependent fashion.

Xanthium sibiricum Patr. Family: Asteraceae (Compositae)

Cai *et al.* (2004) evaluated the antioxidant activity and phenolic compounds in traditional Chinese medicinal plants associated with anticancer. The improved ABTS.+ method was used to systematically assess the total antioxidant capacity

(Trolox equivalent antioxidant capacity, TEAC) of the medicinal extracts. The TEAC values and total phenolic content for methanolic extract of fruit of *Xanthium sibiricum* were 138.6 µmol Trolox equivalent/100g dry weight (DW), and 1.02 g of gallic acid equivalent/100g DW respectively. A positive significant linear relationship between antioxidant activity and total phenolic content showed that phenolic compounds were the dominant antioxidant components in the tested herbs. Major types of phenolic compounds in *Xanthium sibiricum* include phenolic compounds (cinnamic acid), phenolic volatile oils.

Antioxidant capacities of 56 selected Chinese medicinal plants, including *Xanthium sibiricum* were evaluated by Song *et al.* (2010) using the TEAC and FRAP assays, and their total phenolic content was measured by the Folin-Ciocalteu method. *Xanthium sibiricum* showed moderate antioxidant activity with TEAC value of 31.42µmol Trolox/g, FRAP value of 23.63µmol Fe^{2+}/g and phenolic content was 6.6mg GAE/g.

Xanthosoma sagittifolium (L.) Schott. Family: Araceae

Arruda *et al.* (2004) reported that ingestion of *Xanthosoma sagittifolium* leaves have a protective effect against oxidative stress caused by vitamin A-deficiency. Schmourlo *et al.* (2007) evaluated the chemical composition, antioxidant and antimicrobial activity of aqueous extract of *Xanthosma sagittifolium* fruits. The antioxidant activity was evaluated by DPPH method. The flavonoid content influenced the antioxidant activity

Antioxidant activities of eight leafy vegetables of Ghana including *Xanthosma sagittifolium* were assessed by Morrison and Twumasi (2010). The total antioxidant capacity (TAC) and Total Phenol content (TPC) in the methanol extracts (METE) and hydro-ethanol extracts (HETE) were measured. The TPC of the methanol extract was 0.485 mg/ml TAE and TAE was 0.488 mg/ml Ascorbic acid equivalent (AAE) while TPC of hydro-ethanol extract was 0.706 mg/ml TAE and TAE was 0.681 mg/ml Ascorbic acid equivalent (AAE). Radical scavenging activity of methanol extract was EC_{50} = 0.1294mg/ml and HETE was 0.1186 mg/ml while Fe^{3+} reducing potential was EC_{50} = 0.1186 (METE) and 0.1186 (HETE).

Xanthosoma violaceum Schott. Family: Araceae

An extract of *Xanthosoma violaceum* leaves was subjected by Picerno *et al.* (2003) to a polyphenol profile determination including total polyphenols, and antioxidant activity evaluation. The antioxidant activity of the extract was assessed by means of two different *in vivo* tests: bleaching of the stable DPPH radical and peroxidation induced by the water soluble radical initiation 2,2′-azobis(2-aminopropane)hydrochloride (LP-LUV test). In both tests used, the extract and fraction II showed a significant antioxidant/free radical scavenging effect (fraction II, EC(50) = 11.6 µg/ml) in comparison to α-tocopherol (EC_{50} = 10.1)

Xeropta viscosa Baker Family: Velloziaceae

Mowla *et al.* (2002) reported the identification of cDNA corresponding to 1-Cys peroxiredoxin, an evolutionarily conserved thiol specific antioxidant enzyme from *Xeropta viscosa*, a South African medicinal plant.

Ximenia americana L. Family: Olacaceae

Polyphenol content and antioxidant activity of fourteen wild edible fruits from Burkina Faso were investigated by Lamien-Meda *et al.* (2008). The data obtained show that the total phenolic and total flavonoid levels were significantly higher in the acetone than in the methanol extracts. *Ximenia americana* fruit had higher flavonoid content and higher antioxidant activity.

Xylocarpus granatum Koenig Family: Meliaceae

Vadlapudi and Naidu (2009) evaluated the antioxidant potential of *Xylocarpus granatum*, a mangrove plant. The SOD was 0.44U/mg, Catalase was 0.24U/mg, Ascorbic acid was 390mg/100g, DPPH radical inhibition was 55.5 per cent and FRAP units were 350.

Xylocarpus moluccensis (Lamk.) Roem. Family: Meliaceae

Antioxidant activity of *Xylocarpus moluccensis* and four other species have been investigated by Adib *et al.* (2010) by DPPH assay. *X. moluccensis* was found to have the highest antioxidant activity with an IC_{50} value of 22μg/ml.

Xylopia aethiopica (Dun.) A.Rich. Family: Annonaceae

Odukoya *et al.* (2005) evaluated the antioxidant activity of Nigerian dietary spices including *Xylopia aethiopica*. The antioxidant activity (expressed as per cent inhibition of oxidation) was 72.47 per cent, reducing power was 13.89 per cent and Total phenolic content was 152.24 mg/100g Tannic acid Equivalent. Antioxidant activity correlated significantly and positively with total phenolics while there was no linear correlation between total antioxidant activity and reducing power neither between reducing power and total phenolic content.

Xylopia championii Hook.f. et Thoms. Family: Annonaceae

Puvanendran *et al.* (2007) evaluated the antioxidant activity of Dichloromethane (DCM) and MeOH extract of stem bark, leaves and seeds of *Xylopia championii* using DPPH assay. The extract showed high antioxidant activities. Percentage of inhibition of stem bark was 79.03 (MeOH extract), 67.05 (DCM extract); leaf extract was 27.82 (DCM extract) and 48.45 (MeOH extract).

The alkaloids (+)-laudanidine and (-)-discretine isolated from *Xylopia championii* stem bark and stem at a concentration of 0.5 mg/ml, exhibited exceptionally high antioxidant activity whereas norcicentrine and dehydrocorytechine showed moderate activity as compared with standard antioxidant DL-α-tocopherol in the DPPH assay (Puvanendran *et al.*, 2008).

Xylopia nigricans Hook.f. et Thoms. Family: Annonaceae

Puvanendran *et al.* (2007) evaluated the antioxidant activity of Dichloromethane (DCM) and MeOH extract of stem bark, leaves and seeds of *Xylopia nigricans* using DPPH assay. The extract showed high antioxidant activities. Percentage of inhibition of stem bark was 25.58 (MeOH extract); leaf extract was 21.03 (DCM extract) and 33.92 (MeOH extract) while that of seed was 62.06 (MeOH extract) and 14.33 (DCM extract).

Yucca schidigera **Roeze ex Ortgies Family: Agavaceae**

Two new phenolic constituents with unusual spirostructures, named yuccaols D and E, were isolated from the MeOH extract of *Yucca schidigera* bark. Their structures were established by spectroscopic (ESIMS and NMR) analysis. The new yuccaols D and E, along with resveratrol, trans-3′,5,5′-tetrahydroxy-4-methoxystilbene, yuccaols A-C, yuccaone A, larixinol. The MeoH extract of *Yucca schidigera* bark, and the phenolic portion of this extract were assayed for antioxidant activity by measuring the free radical scavenging effects using two different assays (Piacente *et al.*, 2004).

Zanthoxylum alatum **Family: Rutaceae**

The methanolic extract of leaves of *Zanthoxylum alatum* was found to show high antioxidant potential (Dhobal *et al.*, 2008).

Zanthoxylum americanum **P. Mill. Family: Rutaceae**

Borchardt *et al.* (2008) evaluated the antioxidant activity of methanol extracts of seeds of *Zanthoxylum americanum* using DPPH antioxidant assay.The antioxidant value reported in µM Trolox/100g (TE) from DPPH radical scavenging activity of crude seeds extract was 47,367.

Zanthoxylum armatum **DC. Family: Rutaceae**

Ranawat *et al.* (2010) investigated the hepatoprotective activity of the ethanolic extracts of bark of *Zanthoxylum armatum* in CCl_4 induced hepatic damage in rats. Bark extracts significantly increased the levels of antioxidant enzymes: SOD, catalase and glutathione. Phytochemical analysis revealed presence of isoquinoline alkaloid berberine, as well as flavonoids and phenolic compounds.

Zanthoxylum chalybeum **Engl. Family: Rutaceae**

Kuglerova *et al.* (2008) investigated the antioxidant activity of seven Ugandan tree species. The crude ethanolic extract of bark of *Zanthoxylum chalybeum* exhibited DPPH scavenging activity with IC_{50} value 23µg/ml.

Zanthoxylum nitidum **(Roxb.) DC. Family: Rutaceae**

Cai *et al.* (2004) evaluated the antioxidant activity and phenolic compounds in traditional Chinese medicinal plants associated with anticancer. The improved ABTS.+ method was used to systematically assess the total antioxidant capacity (Trolox equivalent antioxidant capacity, TEAC) of the medicinal extracts. The TEAC values and total phenolic content for methanolic extract of root and stem of *Zanthoxylum nitidum* were 191.2 µmol Trolox equivalent/100g dry weight (DW), and 1.27g of gallic acid equivalent/100g DW respectively. A positive significant linear relationship between antioxidant activity and total phenolic content showed that phenolic compounds were the dominant antioxidant components in the tested herbs. Major types of phenolic compounds in *Zanthoxylum nitidum* include phenolic acids (nitidin, dihydronitidine), lignans.

Shyur *et al.* (2005) examined antioxidant activities of 26 medicinal herbal extracts that have been popularly used as folk medicines in Taiwan. The results of scavenging DPPH radical activity show that, among the 26 tested medicinal plants, *Zanthoxylum nitidum* exhibited strong activities and its IC_{50} value for DPPH radical was 50µg/ml.

The methanol extract of root was evaluated by Bhattacharya *et al.* (2010) for its antioxidant properties by DPPH radical scavenging assay and *in vitro* lipid peroxidation induced by the Fe^{2+} ascorbate system in rat liver homogenate. In DPPH radical scavenging assay the extract demonstrated marked and dose dependent free radical scavenging effect and the mean inhibitory concentration of the extract was found to be 75.2µg/ml, while the ascorbic acid (reference) exhibited 43.7 µg/ml.The extract effectively inhibited the lipid peroxidation in a dose related manner showing the IC_{50} value of 279.1µg/ml, whereas the quercetin (reference) showed 46.6 µg/ml. These findings revealed that the *Z. nitidum* root had *in vitro* antioxidant action in the tested models.

Gan *et al.* (2010) evaluated the antioxidant activity of *Zanthxylum nitidum* using the FRAP and TEAC assays, and its total phenolic content was measured by the Folin-Ciocalteu method. For FRAP assay the value was 198.25 µmol Fe(II)/g dry weight. For TEAC assay, the value was 37.45µmol Trolox/g dry weight. Total phenolic content was 5.44 mg gallic acid equivalent/g dry weight of plant material.

Zanthoxylum piperitum DC. Family: Rutaceae

Methanol aqueous extract of *Zanthoxylum piperitum* was screened by Kim *et al.* (1997) for anti-oxidative activity using Fenton's reagent/ethyl lineolate system. The extract showed antioxidant activity.

Chon *et al.* (2009) evaluated the antioxidant activities and total phenolics level and cytotoxicity of young sprouts of some traditional Korean salad plants. Methanol extract of *Zanthoxylum piperitum* at 63 mg/kg exhibited the highest dose-dependent DPPH radical scavenging activity by 83.9 per cent and highest amount of phenolics (194.1 mg/kg).

Zataria multifora Boiss. Family: Lamiaceae (Labiatae)

English Name: Saatar.

Souri *et al.* (2008) investigated the antioxidant activity and radical scavenging activity of methanolic extract of leaves and twig. The antioxidant activity of this plant extract against linoleic acid peroxidation expressed as IC_{50} was 3.23ng/ml and radical scavenging activity against DPPH expressed as IC_{50} was 1.80 µg/ml.

Chemical composition of essential oils in *Zataria multiflora* from different parts of Iran and their radical scavenging and antimicrobial activity were evaluated by Saei-Dehkordi *et al.* (2010). In the DPPH antioxidant assay, all samples exhibited a remarkable activity with a higher degree of IC_{50} = 19.7 µg/ml, almost similar to BHT (18.1µg/ml).

Zea mays L. Family: Poaceae

English Name: Maize.

Wojcikowski *et al.* (2007) evaluated the antioxidant activity of aerial parts of *Zea mays*. Oxygen radical absorbance capacity of the methanol extract and aqueous extracts of aerial parts were 17.84 and 41.90µmol Trolox equivalent per g of dried starting material, respectively.

Sreeramulu *et al.* (2009) evaluated the antioxidant activity of maize by DPPH, scavenging assay, FRAP assay and reducing power. Phenolic content was 112.68 mg/100g. DPPH radical scavenging activity expressed as trolox equivalent, FRAP (μmol/g) and reducing power (mg/g) were 1.39, 86 and 1.20, respectively in maize. α-tocopherol present in corn oil could serve as antioxidant (Nariman and Omaye, 2007).

Chemical composition of the volatile extract and antioxidant activities of the volatile and nonvolatile extracts of Egyptian corn silk was investigated by El-Ghorab *et al.* (2007). All extracts from solvent extraction and the volatile extract exhibited clear antioxidant activities at levels of 50-400μg/ml in the DPPH/linoleic acid assay. The ethanol extract inhibited DPPH activity by 84 per cent at a level of 400μg/ml. All samples tested via the β-carotene bleaching assay also exhibited satisfactory antioxidant activity with clear dose responses.

Ground corn tassels, a by-product of corn, were used as a source of phenolic compounds. Water, ethanol, methanol, acetone, hexane, chloroform, butanol, petroleum ether and methylene chloride were evaluated by Maohsen and Ammar as different polarity solvents to extract these phenolic compounds. Ethanol exhibited the highest extraction ability for such phenolic compounds, followed by methanol and water, where the total phenols were 0.1575 per cent, 0.1125 per cent and 0.0737 per cent, respectively. Antioxidant activity of corn tassels ranged from 83.0 per cent to 85.2 per cent, 69.9 per cent to 83.7 per cent, 69.8 per cent to 80.4 per cent, 22.2 per cent to 49.1 per cent and 14.8 per cent to 19.3 per cent radical scavenging activity (per cent RSA) for ethanol, methanol, acetone, butanol and water extracts, respectively (Mohsen and Ammar, 2009).

Zelkova serrata **Makino Family: Ulmaceae**

Effects of 7-hydroxy-3-methoxy-cadalene (cadalene) extracted from *Zelkova serrata* on 4-(-methylinitros amino)-1-(3-pyridyl)-1-butanone(NNK)-induced oxidative stress were examined by Kim *et al.* (2004) using A/J mice. NNK (65 μg/ml water) was orally administered to mice followed by free feeding on a commercial diet, not containing cadalene, for 2 weeks.The glutathione concentration of cadalene-treated (65 μg/ml water) group was significantly higher than that of the group treated only with NNK. The results of the study strongly indicate that cadalene exerts antioxidative effect on NNK-induced lung tumorigenesis in A/J mice (Kim *et al.*, 2004).

Zingiber cassumnar **Roxb. Family: Zingiberaceae**

Chanwitheesuk *et al.* (2005) investigated the antioxidant activity of rhizome of *Zingiber cassumnar* using a β-carotene bleaching method. The contents of plant chemicals such as vitamin C, Vitamin E, carotenoids, tannin and total phenolics, were also determined. Methanolic extract showed antioxidant activity. *Zingiber cassumnar* has antioxidant index of 10.9, Vitamin C 13.4 mg per cent, Vitamin E 0.0065mg per cent, Total carotenes 1.27 mg per cent, Total xanthophylls 0.54mg per cent, Tannins 1.18mg per cent and Total phenolics of 83.9mg per cent.

Leelapornpisid *et al.* (2007) reported antioxidant activity of the volatile oil of *Zingiber cassumnar* (IC_{50} = 1.0599 mg/ml).

Zingiber officinale Roscoe Family: Zingiberaceae

English Name: Ginger.

Ginger has a high content of antioxidants. This makes it a free radical scavenger. This means it has antimutagenic and anti-inflammatory properties (Lee and Ahn, 1985; Kikuzaki and Nakatani, 1993; Kikuzaki *et al.*, 1994).

In human aortic endothelial cells, zingerone demonstrated significant antioxidant effects on low density lipoproteins (Pearson *et al.*, 1997; Zhou and Xu, 1992). In human erythrocyte membranes, ginger extracts inhibited lipid peroxidation by 72 per cent (Sujatha and Srinivas, 1995). In human chondrocytes, ginger's volatile oil effectively prevented the production of hydrogen peroxide usually induced by fulvic acid (Guo *et al.*, 1997).

In rats fed with high fat diet, supplementation with ginger provided significant antioxidant effects, raising tissue concentrations of superoxide dismutase and catalase and reducing glutathione (Jeyakumar *et al.*, 1999).

Patricia *et al.* (2003) reported the antioxidant, anticancer and antimicrobial activities of extracts from ginger. The extracts were obtained using super critical CO_2 with and without ethanol or isopropyl alcohol as co solvent.

Hydrodistillated extracts from ginger were assessed for their total phenol content, and antioxidant (iron(III) reduction, inhibition of linoleic acid peroxidation, iron (II) chelation, DPPH radical-scavenging and inhibition of hydroxyl radical-mediated 2-deoxy-D-ribose degradation (site and nonsite-specific) activities by Hinneburg *et al.* (2005). Ginger showed lower total antioxidant activity than did basil and parsley in experiments by Wu *et al.* (2004). Rabbah *et al.* (2004) found a 60 per cent ethanol extract from ginger to be less effective in the inhibition of lipid peroxidation than equivalent amount of vitamin E. The IC_{50} value inlipid peroxidation was about 300 times higher for a 50 per cent ethanol extract from ginger than for BHT (Shobhana and Naidu, 2000). In investigations by Murcia *et al.* (2004) a 5 per cent water extract from ginger yielded nearly the same antioxidant activity toward lipid peroxidation as did BHT, both applied in a dose of 0.1 per cent. A similar result was found for a dichloromethane extract from ginger, it seems advisable to use extraction media which are able to extract the lipophilic antioxidant compounds. The results from Murcia *et al.* (2004) seems to contradict Hinneburg *et al.* (2005) findings, but Murcia *et al.* (2004) used an extraction method which might have preserved essential oils in the extracts. The antioxidant activity of essential oils from ginger has been shown by Dang *et al.* (2001).

Cai *et al.* (2004) evaluated the antioxidant activity and phenolic compounds in traditional Chinese medicinal plants associated with anticancer. The improved ABTS.+ method was used to systematically assess the total antioxidant capacity (TEAC) of the medicinal extracts. The TEAC values and total phenolic content for methanolic extract of rhizome of *Zingiber officinale* were 884.3 μmol Trolox equivalent/ 100g dry weight (DW), and 4.28 g of gallic acid equivalent/100g DW respectively. Major types of phenolic compounds in *Zingiber officinale* include pheolic volatile oils (gingerol analogues, gingerols, shogaol).

Odukoya *et al.* (2005) evaluated the antioxidant activity of Nigerian dietary spices including *Zingiber officinale*. The antioxidant activity (expressed as per cent inhibition of oxidation) was 64.33 per cent, reducing power was 13.92 per cent and Total phenolic content was 229.83 mg/100g Tannic acid Equivalent. Antioxidant activity correlated significantly and positively with total phenolics while there was no linear correlation between total antioxidant activity and reducing power neither between reducing power and total phenolic content.

The lipid lowering and antioxidant potential of ethanolic extract of *Zingiber officinale* was evaluated by Bhandari *et al.* (2005) in streptozotocin (STZ)-induced diabetic rats. The extract treatment lowered serum total cholesterol, triglycerides and increased HDL-cholesterol levels when compared with pathogenic diabetic rats. The extract treatment lowered the liver and pancreas TBRAS values as compared to pathogenic rats.

Chanwitheesuk *et al.* (2005) investigated the antioxidant activity of rhizome of *Zingiber officinale* using a β-carotene bleaching method. The contents of plant chemicals such as vitamin C, Vitamin E, carotenoids, tannin and total phenolics, were also determined. Methanolic extract showed antioxidant activity. *Zingiber officinale* has antioxidant index of 2.00, Vitamin C 9.38 mg per cent, Vitamin E 0.0039mg per cent, Total carotenes 0.64 mg per cent, Total xanthophylls 1.06mg per cent, Tannins 11.1mg per cent and Total phenolics of 60.1mg per cent.

Kruawan and Kangsadalampai (2006) investigated the antioxidant activity and phenolic compound contents of water extract of *Zingiber officinale*. The extract exhibited strong scavenging activity against DPPH radicals (84.88 per cent) and their FRAP value was 1030.5 µmol/g. The content of phenolic compound in the extracts was determined using the Folin-Ciocalteu reagent and calculated as gallic acid equivalent (GAE). The herbal extract gave a high phenolic compound content (GAE 59.86 mg/g).

Phytochemical flavonols, carotenoids and the antioxidant capacity of rhizome of *Zingiber officinale* was evaluated by Lako *et al.* (2006). *Zingiber officinale* rhizome had the highest Total antioxidant capacity (320 mg/100g) and were rich in Total polyphenol content (TPP) (200mg/100g).

The active compound of ginger is (6)-gingerol which is responsible for ginger flavour and has been found to inhibit human cancer growth. 6-Gingerol, alanine, ascorbic acid, histidine, aluric acid, methionine, myristic acid, palmitic acid and tryptophan are the main antioxidative constituents in the ginger plant (Mishra *et al.*, 2007).

Recently, it has been shown that [6]-gingerol is endowed with strong antioxidant action both *in vivo* and *in vitro*, in addition to strong anti-inflammatory and anti-apoptotic actions (Kim *et al.*,2007). Leelapornpisid *et al.* (2007) reported antioxidant activity of the volatile oil of *Zingiberofficinale* (IC_{50} = 4.385 mg/ml).

The protective effect of dietary feeding of ginger against lindane-induced oxidative stress was investigated by Ahmed *et al.* (2008) in male albino rats. The findings suggest that a diet containing naturally occurring compounds is effective in exerting protective effects by modulating oxidative stress.

Mariutti *et al.* (2008) evaluated the antioxidant activity of ginger employing ABTS and DPPH radical assay. DPPH EC_{50} value was 19.1g/kg while DPPH-TEAC value was 7.8mM/g. ABTS-TEAC value was 2305mM/g.

Khalaf *et al.* (2008) reported antioxidant activity in the methanolic crude extracts of powdered rhizomes of ginger using DPPH assay.

Ethanolic extract of *Zingiber officinale* was prepared and tested for antioxidant and cardioprotective effect against isoproterenol-induced myocardinal necrosis in rats. The recent study indicates that the ethanolic extract (200mg/kg body wt.p.o – for 20 days) post treatment showed a significant antioxidant and cardioprotective activity.

Jain and Agrawal (2009) reported antioxidant activity of methanolic extract of *Zingiber officinale*. El-Ghorab *et al.* (2010) evaluated the chemical composition and antioxidant activity of ginger. Fresh ginger showed the lowest yield for volatile oil (0.31 per cent). Maximum total phenolic contents were observed in the methanol extract of fresh ginger (95.2 mg/g dry extract) followed by the hexane extract of fresh ginger (87.5 mg/g dry extract). The DPPH method showed the highest antioxidant activity for dried ginger essential oil (83.87 per cent) followed by fresh ginger essential oil (83.03 per cent). The FRAP of essential oils showed almost comparable results with DPPH. Dried ginger essential oil was found better in reducing Fe^{3+} ions, followed by fresh ginger.

Suresh *et al.* (2010) investigated anti-lipidperoxidative and antioxidant effects of *Zingiber officinale* root extract in 7,12-dimethyl benz(A) anthracene-induced oral carcinogensis. The results showed that oral administration of *Zingiber officinale* aqueous root extract significantly prevented the tumor formation as well as decreased the levels of lipid peroxidation by products and enhanced the antioxidants defense mechanism in DMBA treated hamsters.

Zingiber ottensii Valeton Family: Zingiberaceae

Leelapornpisid *et al.* (2007) reported antioxidant activity of the volatile oil of *Zingiber ottensii*.

Zingiber zerumbet (L.) Smith Family: Zingiberaceae

Chanwitheesuk *et al.* (2005) investigated the antioxidant activity of rhizome of *Zingiber zerumbet* using a β-carotene bleaching method. The contents of plant chemicals such as vitamin C, Vitamin E, carotenoids, tannin and total phenolics, were also determined. Methanolic extract showed antioxidant activity. *Zingiber zerumbet* has antioxidant index of 3.00, Vitamin C 5.87 mg per cent, Vitamin E 0.0209mg per cent, Total carotenes 1.29 mg per cent, Total xanthophylls 4.23mg per cent, Tannins 17.7mg per cent and Total phenolics of 79.3mg per cent.

Phytochemical flavonols, carotenoids and the antioxidant capacity of rhizome of *Zingiber zerumbet* was evaluated by Lako *et al.* (2006). *Zingiber zerumbet* rhizome have low Total antioxidant capacity (18 mg/100g) and are rich in Total polyphenol content (TPP) (130mg/100g). *Zingiber zerumbet* (*Layalaya*), a widely used herb taken before meals has high levels of kaempferol (240mg/100g).

Zizania aquatica L. Family: Poaceae (Graminae)

English Name: Wild rice.

Qiu *et al.* (2010) reported that DPPH radical scavenging activities of soluble and insoluble fractions of wild rice suggest that the antioxidant activity of wild rice is partially attributed to its phenolic profile. Wild rice contained 2.2 per cent (w/w) phytic acid.

Zizania latifolia (Griseb.) Stapf Family: Poaceae (Graminae)

English Name: Manchurian wild rice; Japanese: Makomo.

Maeda *et al.* (2006) examined the phenolic contents of young stalk of *Zizania aquatica* and its radical scavenging activities for DPPH radical. The polyphenol content was 7.8 mg-gallic acid equivalent/100 g) and DPPH scavenging activity was 96.2 µmol-Trolox eq./100 g).

Ziziphora clinopodioides Lam. subsp. _rigida_ (Boiss.) Rech.f. Family: Lamiaceae (Labiatae)

Souri *et al.* (2004) evaluated the antioxidative activity of *Ziziphora clinopodioides* subsp. *rigida* by linoleic acid peroxidation test using 1,3-diethyl-2-thiobarbituric acid as the reagent. Methanolic extract showed 91.73 per cent inhibition at 40µg concentration with IC_{50} value of 1.45µg.

The free radical scavenging activity of MeOH extract of *Ziziphora clinopodioides* was superior to all other extracts (IC_{50} =30.7 µg/ml), while the essential oil was less effective (Salehi *et al.*, 2005). DPPH assay was used for antioxidant activity screening.

The aerial parts of *Ziziphora clinopodioides* were screened by Gursoy *et al.* (2009) for their possible antioxidant and antimicrobial activities in addition to their phenolic contents. Antioxidant activity was employed by two complementary test systems: DPPH free radical scavenging and β-carotene/linoleic acid. In the first case the inhibitory concentration IC_{50} value was 37.73µg/mg. Similar results were obtained from the β-carotene/linoleic acid system. Inhibition capacity of the linoleic acid was 83.56 per cent. The amount of total phenolics was 129.55 µg/mg.

Ziziphus amolle (Sesse et Moc.) M.C. Johnst. Family: Rhamnaceae

Ruiz-Terán *et al.* (2008) evaluated the antioxidant activity of aerial parts of *Ziziphus amolle* using DPPH and β-carotene-linoleic acid bleaching assay. The extract displayed antioxidant activity comparable to the commercial antioxidant BHA. For the methanolic extract analysed, a clear relation between the total phenolic content of the extracts and their antioxidant activity was found.

Ziziphus hajarensis L. Family: Rhamnaceae

Marwah *et al.* (2007) investigated the antioxidant activity of wound-healing herbs using *in vitro* DPPH and phosphomolybdenum assay methods. Out of the 19 plants screened, the aqueous ethanol extract of *Ziziphus hajarensis* showed moderate DPPH scavenging activities with per cent IP value 44.9 per cent, after 15 min of incubation at an effective test concentration of 50µg/ml. The Total Antioxidant Capacity as gallic acid equivalent of 309mg/g of ethanol extract was obtained in the phosphomolybdenum assay. Total phenolic content was 35.4 mg GAE/g of ethanol extract.

***Ziziphus joazeiro* Mart. Family: Rhamnaceae**

Alviano *et al.* (2008) evaluated the antioxidant activity of the aqueous extract of the inner bark of *Ziziphus joazeiro* by DPPH photometric assay. The extract showed good antioxidant potential. The antioxidant EC_{50} value is 821.4µg/ml (Alviano and Alviano, 2009).

***Ziziphus jujube* Mill. Family: Rhamnaceae**

English Name: Jujube.

Kamiloglu *et al.* (2009) determined the total phenolic content and antioxidant activity of methanol extract from fifteen selected jujube genotypes indigenous to the Mediterranean region of Turkey. Total phenolic content of the fruits was analyzed by Folin-Ciocalteu method while the total antioxidant activity was analyzed using the β-carotene bleaching, ferric ion reducing antioxidant power and DPPH radical scavenging assays. The highest total phenolic content was observed in MHS 6 and MHS 7 genotypes (42 and 40 mg GAE/g dw, while the lowest was found in MHS 5 and MHS 14 (28 and 25 mg GAE/g dw). MHS 13 was among the genotypes with the highest antioxidant capacity in all three methods tested (1237 µmol/g FRAP, 83 per cent in β-carotene bleaching method and 99 per cent in DPPH).

***Ziziphus mauritiana* Lam. Family: Rhamnaceae**

Bhuiyan *et al.* (2009) investigated the free radical scavenging activities of *Ziziphus mauritiana*. Ethanolic extracts of fruits powder were screened by DPPH method. It was found that all of the fruit extracts have remarkable antioxidant activities. The IC_{50} of the ethanolic extracts of *Ziziphus mauritiana* (Local) and *Z. mauritiana* (Naiikei kul) were 72 and 250µg/ml respectively. Between these two varieties local variety showed higher antioxidant activities.

Polyphenol content and antioxidant activity of fourteen wild edible fruits from Burkina Faso were investigated by Lamien-Meda *et al.* (2008). The data obtained show that the total phenolic and total flavonoid levels were significantly higher in the acetone than in the methanol extracts. *Ziziphus mauritiana* fruit had higher flavonoid content and higher antioxidant activity.

***Ziziphus rugosa* Lam. Family: Rhamnaceae**

Karuppusamy *et al.* (2011) evaluated the anthocyanins, ascorbic acid, total phenolics and flavonoids and antioxidant activity of methanol extracts of fruits of *Ziziphus rugosa*. The anthocyanins were 0.34 CGE/100g, Ascorbic acid 35.0 AAE/100g, Total phenolics 41.8GAE/100g, Total flavonoids 41.8QE/100g and antioxidant activity (DPPH) 116.9µg/ml.

***Zornia gibbosa* Span. Family: Fabaceae (Papilionaceae)**

Laxane *et al.* (2008) evaluated the antioxidant activity of *Zornia gibbosa*.

***Zygophyllum potalini* Maxim. Family: Zygophyllaceae**

Barma and Ambaga (2010) evaluated the antioxidant action of *Zygophyllum potalini* on liver hepatitis induced by carbon tetrachloride and the content of Selenium in this herb. The results indicated that Selenium present in *Zygophyllum potalini* can diminish liver injury fibrosis and this was attributed to the antioxidant function.

Herbal Formulations

Chinese Medicine Prescriptions

Lin *et al.* (2000) investigated the antioxidant and free radical scavenger effects of five traditional Chinese Medicinal prescriptions: Shiee FUU Jwu lu Tang (TCMP 1), Oh Yaw Shuen Chin Saan (TCMP 2), Buu Yang Hwan Wuu Tang (TCMP3), Sheau Shiuh Ming Tang (TCMP 4) and Chir Hwu Jia Long Guu Muu Lin Tang (TCMP 5). The results showed that TCMP 5 had greater anti-lipid peroxidation and anti-superoxide formation activity than the other prescriptions. TCMP 4 had the greatest free radical scavenging effect. TCMP5 showed the greatest superoxide radical scavenger activity and TCMP3 showed the greatest hydroxyl radical scavenger ability.

Cosmetic Cream and Lotion

The antioxidant activity of prepared skin care herbal cosmetic cream and lotions comprising ethanolic extracts of *Curcuma longa, Cassia tora, Glycyrrhiza glabra, Aloe vera* and *Acacia catechu* etc. in various concentration were determined by mean of reducing power (Ashawat and Saraf, 2007). The relative antioxidant activity was compared with standard antioxidant activity of L-ascorbic acid. Results indicated that the cream and lotion formulations comprising extracts possess antioxidant activity as compared to standard L-ascorbic acid. The lotion formulation showed 4.7 per cent more activity as compared to cream formulation in equal extract concentration (Ashawat and Saraf, 2007).

DHC-1

DHC-1, an herbal formulation derived from the popular plants *Bacopa monniera, Emblica officinalis, Glycyrrhiza glabra, Mangifera indica* and *Syzygium aromaticum* was studied by Bafna and Balaram (2005b) for its antioxidant activity. The cardial infarction and cisplatin-induced renal damage were studied. A significant reduction in the serum markers of heart and kidney damage and the extent of lipid peroxidation with comcomitant increase in the enzymatic (SOD and CAT) and non-enzymatic antioxidants (reduced glutathione) were observed in DHC-1 pretreated antimals compared with the isoprotenol or cisplatin alone treated animals. Bafna and Balaram (2005b) concluded that DHC-1 possesses a protective effect against both damaged heart and kidneys in rats. This beneficial effect may be attributed, at least in part, to its antioxidant activity.

ESP-102

ESP-102, a combined extract of *Angelica gigas, Saururus chinensis* and *Schisandra chinensis* protects against glutamate-induced toxicity in primary cultures of rat cortical cells (Ma *et al.*, 2009). ESP-102 decreased the cellular calcium concentration increased by glutamate and inhibited the subsequent overproduction of cellular nitric oxide and reactive oxygen species to the level of control cells. It also preserved cellular activities of enzymes such as superoxide dismutase, glutathione peroxidase and glutathione reductase reduced in the glutamate-injured neuronal cells.

Eviprostat

Eviprostat consists of plant extracts from the umbellate wintergreen (*Chimaphila umbellata*), the aspen (*Populus tremula*), the small pasque (*Pulsatilla pratensis*) and the horse-tail (*Equisetum arvense*) mixed with wheat-germ oil. Its effects were examined on the levels of the urinary oxidative stress marker 8-hydroxy-2′-deoxyguanosine in a rabbit model of surgical partial bladder outlet obstruction and in patients with lower urinary tract symptoms associated with Benign prostrate hyperplasia (Matsumoto *et al.*, 2010). The findings provide evidence that the antioxidant activity of Eviprostat is responsible for its beneficial effects in the treatment of BPH.

Hab-E-Jund (Unani Medicine)

In vitro antioxidant activity of Hab-E-Jund (Unani medicine) prescribed for febrile convulsions was evaluated by Anupama *et al.* (2010). It showed considerable *in vitro* antioxidant activity in a dose dependent manner.

Jerusalem Balsam

The Jerusalem Balsam, a remedy based on an ethnobotanic extract of a herbal mixture, was formulated in 1719 in the pharmacy of the Saint Savior Monastery in the old city of Jerusalem. Having gained fame, the Jerusalem Balsam was replicated and prepared in Europe. One of the formulas, found in a manuscript from the archive of the monastery, contains four plants: Olibanum *Boswellia* sp., myrrh *Commiphora* sp., aloe *Aloe* sp. and mastic *Pistasia lentiscus*. Pharmacological assays on this four-plant formula by Moussaieff *et al.* (2005) showed anti-inflammatory, as well as antioxidative and antiseptic properties.

Jubliang

Jubliang is a mixture of eight herbs, namely *Bombax ceiba* L., *Chrysanthemum morifolium*, *Imperata cylindrica* (L.) Beauv., *Lophatherum gracile* Brogn., *Nelumbo nucifera* Gaertn., *Oroxylum indicum* L., *Phragmites communis* Trin and *Prunella vulgaris*. Kruawan and Kangsadalampai (2006) investigated the antioxidant activity and phenolic compound contents of water extract of Jubliang. The extract exhibited strong scavenging activity against DPPH radicals (95.59 per cent) and their FRAP value was 2295.5 µmol/g. The content of phenolic compound in the extracts was determined using the Folin-Ciocalteu reagent and calculated as gallic acid equivalent (GAE). The herbal extract gave a high phenolic compound content (GAE 283 mg/g).

Jyuzentaihoto

The herb medicine prescription, jyuzentaihoto was studied by Abe *et al.* (1997) for its scavenging effects on the reactive oxygen species and on the lipid peroxidation of mitochondrial membrane, by the ESR spin trapping methods using 5,5,-dimethyl-1-pyrroline-1-oxide as a spin trap. Jyuzentaihoto scavenged both superoxide anion and hydroxyl radicals almost completely at 1 mg/ml, Jyuzentaihoto completely inhibited lipid the ADP/Fe^{2+} initiated peroxidation of rat liver mitochondria at 0.5 mg/ml, when determined by measuring the formation of thiobarbituric acid reactive substances.

Pepticare

Pepticare, a herbomineral formulation of the Ayurveda medicine consisting of the herbal drugs *Glycyrrhiza glabra, Emblica officinalis* and *Tinospora cordifolia* was tested by Bafna and Balaram (2005) for its anti-ulcer and anti-oxdant activity in rats. Effects of various doses (125, 250, 500 and 1000 mg/kg, p.o.) of pepticare were studied on gastric secretion and gastric ulcers in pylorus-ligation and on ethanol-induced gastric mucosal injury in rats. The reduction in ulcer index in both the models along with the reduction in volume and total acidity and an increase in the pH of gastric fluid in pylorus-ligated rats proved the anti-ulcer activity of pepticare. It was also found that pepticare was more potent than *G. glabra* along in protecting against pylorus-ligation and ethanol-induced ulcers. The increase in the levels of superoxide dismutase, catalase, reduced glutathione and membrane bound enzymes like Za^{2+}ATPase, Mg^{2+}ATPase and $Na^{+}K^{+}$ATPase and decrease in lipid peroxidation in both the models proved the anti-oxidant activity of the formulation. Bafna and Balaram (2005) concluded that pepticare possesses anti-ulcer activity, which can be attributed to its antioxidant mechanism of action.

PSE 222

A standard mixture of Chinese herbs has recently been demonstrated to be an effective treatment for chronic atopic eczema in placebo controlled trials in the UK. Aqueous decoctions of the formulated mixture (PSE 222), the placebo mixture, and their component herbs were examined for antioxidant activity to determine whether antioxidant activity could account for the anti-eczema activity. Two measures of antioxidant activity were employed: the DPPH assay for non-specific hydrogen atom (or electron) donating activity and a superoxide scavenging assasy. Antioxidant activity was detected in some components of both the active and placebo mixtures, but the formulated active mixture (PSE 222) was significantly more effective than the formulated placebo mixture. The Chinese herbs in the mixture included *Rehmannia glutinosa, Paeonia lactiflora, Dictamnus dasycarpus, Ledebouriella seseloides, Lophatherum gracile, Glycyrrhiza uralensis, Potentilla chinensis, Schizonepeta tenuifolia, Tribulus terrestris, Clematis armandii, Humulus lupulus, Thymus vulgaris* and *Rosmarinus officinalis.*

Qizhu tang (QZT)

QZT consists of 4 herbal constituents (Rhizome *Atractolydes macrocephala, Poria cocos,* Radix Notoginseng, and Radix Astragali). Each of the components and their combinations were examined by Wang and Feng (2000) *in vitro* for DPPH and hydroxyl radical scavenging activities and for the inhibition of TBARS formation in rat liver homogenate. Only the preparations having a higher antioxidant activity comparable to QZT in all three *in vitro* assays were relatively active *in vivo* both for TBARS inhibition and glutathione peroxidase preservation, although the activities were much lower than that of QZT as a whole.

Reatival

Reatival, a herbal formulation of seven well known medicinal plants – *Mentha piperita* L. (Lamiaceae 17.5 per cent, whole plant), *Artemisia absinthium* L. (Asteraceae; 5.5 per cent, leaves and flowers), *Salvia officinalis* L. (Lamiaceae; 7 per cent, leaves), *Prunus spinosa* L (Rosaceae; 17.5 per cent, flower), *Centaurium minus* Moench. (Gentianaceae; 17.5 per cent, whole plant), *Sambucus nigra* L. (Sambucaceae; 17.5 per cent, flower), *Polygonum aviculare* L. (Polygonaceae; 17.5 per cent, whole plant) was evaluated for oxygen radical scavenging activity by Stajner *et al.* (1997). They concluded that Reatival exerts a clear protective action against damage induced by toxic oxygen speies in human heamolysate, by scavenging O_2^- and OH., reducing the lipid peroxidation, and increasing SOD and C-ase activity.

Sheng Mai San (SMS)

Sheng Mai San (SMS), a compound formulation in traditional Chinese medicine comprising *Panax ginseng*, *Schisandra chinensis* and *Ophiopogon japonicus*, is used for the treatment of ischaemic heart disease. Ko *et al.* (2006) have demonstrated the *in vitro* and *in vivo* antioxidant activities in SMS, as indicated by its ability to protect against CCl_4 induced hepatotoxicity and to scavenge superoxide radicals generated in an *in vitro* system. Their results indicate that the antioxidant activity of SMS was mainly from *Schisandra chinensis*.

SR 10 (Radix Astragali, Radix Codonopsis and Cortex Lycii)

A formulation SR10, comprising Radix Astragali (the root of *Astragalus* sp), Radix Codonopsis (the root of *Codonopsis* sp.) and Cortex Lycii (bark of *Lycium* sp.) was examined by Chan *et al.* (2009) for its antidiabetic and antioxidative effects using the diabetic mouse model. The activities and expression of antioxidant enzymes, catalase and superoxide dismutase, were up-regulated when treated with SR10. More over, SR10 treatment did not exhibit any toxic effect to the host.

Toki-Shakuyaku-San (TJ-23)

The free radical scavenging activity of the Japanese herbal medicine, Toki-Shakuyaku-San (TJ-23; Tsumura and Co, Tokyo, Japan), was examined using ESR spectrometry, TJ-23 scavenged DPPH, superoxide and hydroxyl radicals dose-dependently. It also diminished carbon centred radicals generated by oxidative stress and inhibited TBARS formation in mouse cortex homogenate. In addition the effect of TJ-23 on the concentration of neurotransmitters and TBARS formation, and SOD activity in the cortex, hippocampus and striatum of the aged rat brain was studied. The concentration of the metabolites of monoamines, glutamate and glutamine were decreased by 4 weeks of oral administration of TJ-23. The SOD activity of mitochondrial fraction was increased and TBARS formation was significantly suppressed. These results suggest that TJ-23 has an antioxidant action and would have a prophylactic effect against free radical-mediated neurological diseases associated with aging.

Triphala

Shivaprasad *et al.* (2008) investigated the antioxidant and adaptogenic effect of Triphala, compound of *Terminalia chebula*, *T. belerica* and *Phyllanthus emblica* (=*Emblica officinalis*), an Indian ayurvedic medicinal preparation. Triphala was found to scavenge hydroxyl and nitric oxide radicals *in vitro*. The IC_{50} value for hydroxyl radical scavenging was 40.5µg/ml and that for nitric oxide radical scavenging was found to be 40 µg/ml, respectively.

VI-28

Mak *et al.* (2004) investigated the antioxidant activity of a Yang-promoting Chinese herbal suppository preparation VI-28 on the red cells in healthy male subjects. VI-28 treatment for 12 weeks produced in a time/dose-dependent alteration in red cell antioxidant status. The VI-28-induced change is characterized by a slight depletion in cellular reduced glutathione (GSH) level and a decrease in susceptibility to peroxide-induced lipid peroxidation as well as increase in catalase (CAT) and Cu-Zn-superoxide dismutase (SOD) activities. While a reversal trend of change was observed in cellular GSH level, the susceptibility to lipd peroxidation as well as the CAT activity after the cessation of treatment for 4 weeks, the SOD activity exhibited a protracted increase. The result indicated that VI-28 treatment enhances red cell antioxidant status in male subjects (Mak *et al.*, 2004).

Antioxidant Activity of Minimally Processed (In modified atmospheres), Dehydrated and Ready-to-Eat Vegetables

In a study carried out by Murcia *et al.* (2009), the antioxidant activity of vegetables subjected to minimal processing (in MAP), and of dehydrated and ready-to-eat vegetables such as soups and purees, using different antioxidant methods (the scavenging of lipoperoxyl and hydroxyl radicals and TEAC) were examined. The specific aim was to analyse MAP vegetables (intended for cooking or for salads) during commercial chilled storage and when submitted to different types of cooking (boiling, microwaving, pressure cooking, griddling, frying and baking) after storage for 0-8 days in a domestic refrigerator. The antioxidant activity of different vegetable products (dehydrated, condiments and ready-to-eat soup) was also evaluated. MAP vegetables had a good or very good antioxidant capacity, and showed no significant loss of antioxidant activity or scavenging capacity compared with fresh vegetables. The cooking treatments that keep the antioxidant activity of MAP vegetables are microwaving, sautéing and baking. The most aggressive method of cooking were steaming, boiling and frying. The dehydrated condiments (tablets) showed higher antioxidant activity than the ready-to-eat soup. The enrichment of stews and casseroles, with dehydrated vegetable tablets, and the consumption of soup or vegetable purees represent an increased antioxidant intake in our diet. Also "ready-to-eat" vegetable soups show antioxidant activity after they have been submitted to heat treatment to increase their shelf-life. They can be recommended as alternatives in our non-stop "life style".

Vegetables Antioxidant Losses during Industrial Processing and Refrigerated Storage

Vegetables are a major source of antioxidants although it is not clear whether analytical estimators of total antioxidant activity predict beneficial effects after consumption. It is therefore desirable to assess the antioxidant activity of different foods and to compare different storage or processing methods. Thus, the aim of the study carried out by Murcia *et al.* (2009) was to show how the health functionality of the diet (rich in antioxidants may be influenced by the selection of processed vegetables. Twenty-five vegetables (artichoke, asparagus, beetroot, broad bean, broccoli, Brussels sprout, carrot, cauliflower, celery, chicory, cucumber, eggplant, endive, garlic, green bean, leek, lettuce, maize, onion, pea, pepper, radish, spinach, Swiss chard and zucchini) were used to evaluate their antioxidant activity. All fresh vegetables studied were able to scavenge lipoperoxyl and hydroxyl radicals. All the vegetables also presented good total capacity antioxidant by TEAC assay except cucumber, endive, carrot and zucchini. Vegetables stored (7 days) in a home refrigerator recorded the same antioxidant activity as fresh samples, except cucumber and zucchini (lipid peroxidation) and broccoli, Brussels sprout and leek (TEAC). Canned vegetables showed a more pronounced loss of antioxidant activity than frozen vegetables, compared with fresh vegetables. During the shelf life of the processed vegetables (8 months for frozen and 18 months for canned vegetables), some products showed losses (19-48 per cent) of their lipoperoxyl radical scavenging capacity and total antioxidant activity. The results suggest that the consumption of raw vegetables will add a higher antioxidant capacity to diet than the consumption of identical amount of frozen or canned vegetables. The refrigeration storage showed no losses on vegetables antioxidant activity. However, the industrial processing conditions of some frozen vegetables, as broccoli, asparagus, garlic, spinach, green beans, Brussels sprouts, peas, should be improved. On the other hand, in the case of garlic the antioxidant activity is progressively lost during storage as frozen. Respect to the canned vegetables, industrial processing conditions would be revised for prevention of antioxidant losses.

High pressure processing (HPP) is a relatively new food preservation processing technology that enhances food safety and shelf-life without compromising organoleptic qualities. There has been little research on the impact of HPP on the nutritional and health-promoting properties of foods to date and most of it has focused on juices and purees of fruit such as oranges and tomatoes. The objective of the study conducted by McInerney *et al.* (2007) was to determine the effects of HPP treatment at two pressure levels (400 and 600 MPa) on antioxidant activity, total carotenoid content and carotenoid availability *in vitro*, of three commonly consumed vegetables, carrots, broccoli and green beans. Antioxidant capacity and total carotenoid content differed between vegetales but were unaffected by HPP treatment. *In vitro* availability of specific carotenoids also varied greatly between vegetables (3-35 per cent). HPP altered availability of carotenoids according to the type of vegetable treated and processing pressure applied, however the magnitude of the response was minor. The study provides further scientific evidence of the benefits of HPP in retaining the nutritional attributes of fresh foods. Antioxidant activity and levels of carotenoids before and after exposure to high pressures (up to 600MPa for 2 min) were essentially no different. Also, the data suggest that micronutrients and phytochemicals in certain vegetables may be made more bioavailable by high pressure treatment.

Influence of Natural Antioxidants and their Carry-through Property in Biscuit Processing

Bheema Rao *et al.* (2009) have explored the use of natural antioxidants from plant sources such as hexane (G-1) and chloroform (G-2) extracts of Garcinia, *Peltophorum ferrugineum* flower extract (FE), turmeric powder (C-1) and curcumin (C-2), in biscuit preparation in the place of synthetic ones and their carry-through properties were evaluated during processing and storage. Microstructure and electrophoretic characterization of biscuit dough were carried out to understand the influence of these natural antioxidants in the processing of biscuits. Biscuits prepared with C-1, C-2, g-1 and G-2 was well acceptable and FE biscuits were least acceptable. Among the antioxidants used, C-2 was found to retain only 11.23 per cent activity whereas G-2 had retained 51 per cent activity after baking, which was comparable to synthetic antioxidant butylated hydroxyanisole (BHA) and better than tert-butylhydroquinone (TBHQ). Microstructure studies of biscuit dough indicated that in the case of G-1, G-2, C-1 and BHA the gluten network formation was intact when compared to FE and TBHQ. Supportive electrophoresis studies indicated that there was no change in the wheat protein subunits of different samples except for FE biscuit dough. Thus based on the result of retention of antioxidant activity during baking and storage, hexane and chloroform extracts of Garcinia (G-1, G-2) and turmeric powder (C-1) were found to be suitable for use in biscuits as natural antioxidants and results were comparable with BHA. The study also indicated that protein subunits of biscuit dough were not affected by antioxidants. However, antioxidants influenced the protein-starch-lipid network during biscuit dough formation.

Total Antioxidant Content of Alternatives to Refined Sugar

Oxidative damage is implicated in the etiology of cancer, cardiovascular disease and other degenerative disorders. Recent nutritional research has focused on the antioxidant potential of foods, while current dietary recommendations are to increase the intake of antioxidant-rich foods rather than supplement specific nutrients. Many alternatives to refined sugar are available, including raw cane sugar, plant saps/syrups (*e.g.*, maple syrup, agave nectar), molasses, honey and fruit sugars (*e.g.*, date sugar). Unrefined sweeteners were hypothesized to contain higher levels of antioxidants, similar to the contrast between whole and refined grain products. Phillips *et al.* (2009) compared the total antioxidant content of natural sweeteners as alternatives to refined sugar. The ferric-reducing ability of plasma (FRAP) assay was used to estimate total antioxidant capacity. Major brands of 12 types of sweeteners as well as refined white sugar and corn syrup were sampled from retail outlets in the United States. Substantial differences in total antioxidant content of different sweeteners were found. Refined sugar, corn syrup, and agave nectar contained minimal antioxidant activity (<0.01mmol FRAP/100g); raw cane sugar has a higher FRAP (0.1mmol/100g). Dark and blackstrap molasses had the highest FRAP (4.6 to 4.9 mmol/100g), while maple syrup, brown sugar and honey showed intermediate antioxidant capacity (0.2 to 0.7 mmol FRAP/100g). Based on an average intake of 130g/day refined sugars and the antioxidant activity measured in typical diets, substituting alternative sweeteners could increase antioxidant intake an average of

2.6mmol/day, similar to the amount found in a serving of berries or nuts. Many readily available alternatives to refined sugar offer the potential benefit of antioxidant activity.

Algae

Aphanizomenon flos-aquae L. Family: Nostocaceae (Blue-Green algae)

Kuriakose and Kurup (2010) reported antioxidant and hepatoprotective activity of *Aphanizomenon flos-aquae* against paracetamol intoxication in rats. Antioxidant activity was measured by DPPH assay.

Aulosira fertilissima R. Ross. Family: Cyanophyceae

Gini *et al.* (2008) reported antioxidant property of *Aulosira fertilissima* on CCl_4 induced hepatic damage in rats.

Bifurcaria bifurcata Family: Fucaceae

Zubia *et al.* (2009) assessed the antioxidant activity of the crude extracts of *Bifurcaria bifurcata* by three methods (DPPH, reducing activity and β-carotene-linoleic acid system) and their phenolic contents. The extract exhibited antioxidant activity. Further to the fractionation of these crude extracts, a significant correlation was found in most fractions between the high antioxidant activity and the high phenolic content.

Callophyllis japonica Okamura Family: Kallymeniaceae (Rhodophyta)

The antioxidant properties of the red seaweed *Callophyllis japonica* were investigated by Kang *et al.* (2005). An ethanol extract isolated from *C. japonica* exhibited intracellular reactive oxygen species, DPPH radical scavenging activity and lipid peroxidation inhibitory acitivity. The radical scavenging activity of the extract protected the viability of Chinese hamster lung fibroblast (V79-4) cells exposed to H_2O_2. Further more, the extract reduced the proportion of apoptotic cells induced by H_2O_2, as demonstrated by decreased sub-G1 hypo-diploid cells and decreased apoptotic body formation. The extract also increased the activities of the cellular antioxidant enzymes, superoxide dismutase and catalase. Together, these findings suggest that *C. japonica* exhibits antioxidant properties.

Caulerpa racemosa (Forrsk.) Agardh. var. *macrophysa* Family: Caulerpaceae

Gracilaria tenuispitata C.F. Chang and B. M. Xia var. *tenuistipitata* Family: Gracilariaceae (Rhodophyta)

Sargassum sp. Family: Sargassaceae

Ulva lactuca L. Family: Ulvaceae (Chlorophyta)

The aqueous extract of four marine algae *Caulerpa racemosa* var. *macrophysa, Gracilaria tenuispitata* var. *tenuispitata, Sargassum* sp. and *Ulva lactuca* were prepared by boiling dried seaweed powder in water for 3 h, and by autoclaving each sample at 120°C for 3 h. They were then freeze dried and evaluated for their antioxidant activities using DPPH, hydroxyl radical and superoxide anion scavenging assays. Boiling extracts of the seaweeds except *C. racemosa* were found to have higher total phenolic contents than those obtained from the autoclaved method. The antioxidant results also showed that O_2 scavenging activity existed only in the boiling extracts of *C. racemosa*, *G. tenuispitata* and *U. lactuca*. In DPPH and OH assays, however, almost all the boiling extracts were less active than the autoclaved ones. Among the four alga species, *Sargassum* sp. was the most active. Both extracts of this seaweed had the highest TPC and also displayed the strongest DPPH and OH inhibitory activities. A strong positive-correlation between the antioxidant potency and TPC of the autoclave extracts was found while for the boiling extracts such relation was very weak (Yangthong *et al.*, 2009).

Chlamydomonas nivalis **(F. A. Bauer) Wille Family: Chlamydomonadaceae (Chlorophyta)**

Li *et al.* (2006) evaluated the antioxidant capacity and total phenolic content of different fractions of 23 microalgae includig *Chlamydomonas nivalis*, using TEAC and Folin-Ciocalteu method, respectively. *Chlamydomonas nivalis* possessed the highest antioxidant capacity. The correlation coefficients between the antioxidant capacities and the phenolic contents were very small in hexane, ethylacetate and water fractions. Thus, phenolic compounds were not a major contributor to the antioxidant capacities of these microalgae.

Chlorella stigmatophora **Butcher Family: Chlorellaceae (Chlorophyta)**

A pharmacological study of hydrosoluble and liposoluble extracts of the marine microalga *Chlorella stigmatophora* indicated that hydrosoluble components show significant anti-inflammatory and free radical scavenging activity. These activities were not detected in the liposoluble fractions (Guzman *et al.*, 2001).

Chlorella vulgaris **Beijerinck Family: Chlorellaceae (Chlorophyta)**

The antioxidant activity of the microalgal ethanolic extracts of *Chlorella vulgaris* was determined by Rodriguez-Garcia and Guil-Guerrero (2007) by means of the β-carotene-lineolate model system. The results show that the extract showed good antioxidant activity.

Cytoseira tamariscifolia **(Hads.) Papenus Family: Fucaceae**

Zubia *et al.* (2009) assessed the antioxidant activity of the crude extracts of *Cytoseira tamariscifolia* by three methods (DPPH, reducing activity and β-carotene-linoleic acid system) and their phenolic contents. The extract exhibited antioxidant activity. Further to the fractionation of these crude extracts, a significant correlation was found in most fractions between the high antioxidant activity and the high phenolic content.

Desmarestia viridis **(O. F. Muller) J.V. Lamouroux Family: Desmarestiaceae**

Free radical scavenging activity by DPPH assay and deoxyribose assay were ascertained among 27 species of common seaweeds using sequence extraction method (Xiaojun *et al.*, 1996). The results indicated that DPPH scavenging activity existed in

15 species including *Desmarestia viridis*. On the other hand the deoxyribose results showed that almost all the seaweed species had good ability to scavenge hydroxyl radicals.

Dunaliella salina **Teodoresco Family: Dunaliellaceae (Chlorophyta)**

Dunaliella salina, a green marine algae has the ability to protect against oxidative stress *in vivo* using animal models (Murthy *et al.*, 2005).

Ecklonia cava **Kiellman Family: Lessoniaceae**

The potential antiproliferative and antiradical activities of an enzymatic extract of *Ecklonia cava* together with its crude polysaccharide (CphF) and crude polyphenol fractions were evaluated *in vitro* by Athukorala *et al.* (2006). The CphF exhibited interesting antiradical properties, expressed by its capacity to scavenge superoxide anion, hydrogen peroxide and hydroxyl radical. In reducing power assay, CphF extract at 5 µg/ml was found to be as high as that of BHT at the same concentration. Also in total antioxidant assays the effect of CphF at 50µg/ml was equivalent or slightly higher than those of commercial counter parts at 5µg/ml concentration.

Ecklonia stolonifera **Okamura Family: Lessoniaceae**

The antioxidant activity of *Ecklonia stolonifera* was determined by Lee *et al.* (1996) by measuring lipid peroxide produced when a mouse liver homogenate was exposed to the air at 37°C, using 2-thiobarbituric acid (TBA) and the radical scavenging effect on DPPH radical. The methanol extract of *Ecklonia stolonifera* showed strong antioxidant activity. The methanol extract was fractionated with several solvents. With regard their fractions, the antioxidative activity were in the order of ethyl acetate>dichloromethane insoluble intermediated phase>dichloromethane>*n*-butanol>water fraction. The ethyl acetate soluble fraction exhibiting the strongest antioxidant activity was further purified. Antioxidant phloroglucinol was isolated and identified. Its antioxidant activity was similar to that of L-ascorbic acid.

Fucus ceranoides **L. Family: Fucaceae**

Zubia *et al.* (2009) assessed the antioxidant activity of the crude extracts of *Fucus ceranoides* by three methods (DPPH, reducing activity and β-carotene-linoleic acid system) and their phenolic contents. The extract exhibited antioxidant activity. Further to the fractionation of these crude extracts, a significant correlation was found in most fractions between the high antioxidant activity and the high phenolic content.

Fucus vesiculosus **L. Family: Fucaceae**

Two extracts from the brown seaweed *Fucus vesiculosus* containing 28.8 per cent polyphenols or 18 per cent polyphenols plus 0.0012 per cent fucoxanthin have been obtained and studied to determine their toxicity in mice and rats and also their antioxidant activity (Zaragozá *et al.*, 2008). Both extracts were shown to lack any relevant toxic effects in an acute toxicity test following a 4 week daily treatment in rats. The extracts exhibited antioxidant activity in noncellular systems and in activated RAW 264.7 macrophages, as well as in *ex vivo* assays in plasma and erythrocytes, after the 4 week treatment in rats. Their *ex vivo* results indicated that compounds from extract 2 may be more easily absorbed and that the antioxidants in their parent or metabolized form are more active. These findings support the view that the daily consumption of *F. vesiculosus* extract 2 (Healsea) would have potential benefits to humans.

Gelidium amansii (J.V. Lamous.) J.V. Lamour. Family: Gelidiaceae

Free radical scavenging activity by DPPH assay and deoxyribose assay were ascertained among 27 species of common seaweeds using sequence extraction method (Xiaojun *et al.*, 1996). The results indicated that DPPH scavenging activity existed in 15 species including *Gelidium amansii*. On the other hand the deoxyribose results showed that almost all the seaweed species had good ability to scavenge hydroxyl radicals.

Gloisiphonia capillaris (Hudson) Carmichael Family: Ceramiaceae

Free radical scavenging activity by DPPH assay and deoxyribose assay were ascertained among 27 species of common seaweeds using sequence extraction method (Xiaojun *et al.*, 1996). The results indicated that DPPH scavenging activity existed in 15 species including *Gloiosiphonia capillaris*. On the other hand the deoxyribose results showed that almost all the seaweed species had good ability to scavenge hydroxyl radicals.

Gracilaria verrucosa L. Gracilariaceae

Heo *et al.* (2006) compared the antioxidant activity and the total phenolic content of red sea weed (*Gracilaria verrucosa* L.) from extract of methanol, ethanol, acetone, chloroform and hexane; and the correlation between total phenolic content and the antioxidant activity of each extract; then to determine the chlorophyll a, chlorophyll b, and carotene content of each extract and their correlation with the free radical scavenging activity as well. The antioxidant activity were measured by free radical scavenging method with DPPH and by reducing power method with K4Fe(CN)6 as standard, whereas the total phenolic content was measured by Folin Ciocalteu method with gallic acid as standard. The results showed that the highest of the antioxidant activity by free radical scavenging method was found in acetone extract : 43.43 per cent (BHT: 84.15 per cent); whereas by reducing power method was found in chloroform extract : 0.1756 meq K4Fe(CN)6/g extract (BHT : 6.1767 meq K4Fe(CN)6/g extract); and the highest of the total phenolic content was also found in acetone extract : 45.29 mg/g extract. There were close correlation between phenolic content and antioxidant activity both by free radical scavenging method and by reducing power method. Chlorophyll a and carotene had also close correlation with the free radical scavenging activity, but not for chlorophyll b.

Halidrys siliquosa (L.) Lyngbye Family: Fucaceae

Zubia *et al.* (2009) assessed the antioxidant activity of the crude extracts of *Halidrys siliquosa* by three methods (DPPH, reducing activity and β-carotene-linoleic acid system) and their phenolic contents. The extract exhibited antioxidant activity. Further to the fractionation of these crude extracts, a significant correlation was found in most fractions between the high antioxidant activity and the high phenolic content.

Laminaria japonica Aiesch. Family: Laminariaceae (Phaeophyceae)

A low molecular weight sulfate polysaccharide (LMWF) was prepared from *Laminaria japonica* by Zhao *et al.* (2004). The antioxidant activity of LMWF *in vitro* was studied using three kinds of oxygen free radical systems. LMWF had effective scavenging abilities on superoxide radical, hydroxyl radical and hypochlorous acid directly *in vitro.* Addition of LMWF significantly lowered the content of serum malonaldehyde and markedly increased the activities of superoxide dismutase and glutathione peroxidase, compared with model groups.

Antioxidant capacities of 56 selected Chinese medicinal plants, including *Laminaria japonica* were evaluated by Song *et al.* (2010) using the TEAC and FRAP assays, and their total phenic content was measured by the Folin-Ciocalteu method. *Laminaria japonica* showed low antioxidant activity with TEAC value of 6.86μmol Trolox/g, FRAP value of 0.33μmol Fe^{2+}/g and phenolic content was 0.36mg GAE/g.

Neorhodomela aculeata Masuda Family: Rhodomelaceae

The antioxidant and anti-inflammatory properties of marine red alga *Neorhodomela aculeata* were investigated with neuronal and microglial cells by Lim *et al.* (2006). Extracts of *N. aculeata* had potent neuroprotective effects on glutamate-induced neurotoxicity and inhibited reactive oxygen species (ROS) generation in the murine hippocampal HT22 cell line. Also extracts inhibited H_2O_2 induced lipid peroxidation in rat brain homogenates.

Nostoc ellipsosporum (Desmazieres) Rabenhorst. Family: Nostocaceae (Cyanophyceae)

Li *et al.* (2006) evaluated the antioxidant capacity and total phenolic content of different fractions of 23 microalgae includig *Nostoc ellipsosporum*, using TEAC and Folin-Ciocalteu method, respectively. *Nostoc ellipsosporum* possessed the highest antioxidant capacity. The correlation coefficients between the antioxidant capacities and the phenolic contents were very small in hexane, ethylacetate and water fractions. Thus, phenolic compounds were not a major contributor to the antioxidant capacities of these microalgae.

Nostoc sphaeroides Kutz. Family: Nostocaceae (Cyanophyceae)

Gini *et al.* (2006) reported antioxidant property of *Nostoc sphaeroides* on CCl_4 induced hepatic damage.

Phaeodactylum tricornutum Bohlin Family: Phaeodactylaceae (Bacillariophyceae)

A pharmacological study of hydrosoluble and liposoluble extracts of the marine microalga *Phaeodactylum tricornutum* indicated that hydrosoluble components show significant anti-inflammatory and free radical scavenging activity. These activities were not detected in the liposoluble fractions (Guzman *et al.*, 2001).

The antioxidant activity of the microalgal ethanolic extracts of *Phaeodactylum tricornutum* was determined by Rodriguez-Garcia and Guil-Guerrero (2007) by means of the β-carotene-lineolate model system. The results show that the extract showed good antioxidant activity.

Polysiphonia urceolata (Lightboot ex Dillwyn) Greville Family: Polysiphoniaceae (Rhodophyta)

Free radical scavenging activity by DPPH assay and deoxyribose assay were ascertained among 27 species of common seaweeds using sequence extraction method (Xiaojun *et al.*, 1996). The results indicated that DPPH scavenging activity existed in 15 species including *Polysiphonia urceolata*. On the other hand the deoxyribose results showed that almost all the seaweed species had good ability to scavenge hydroxyl radicals.

Antioxidant activity (AA), total phenolic content and reducing power of the crude extract, fractions and subfractions derived from *Polysiphonia urceolata*, were evaluated and determined by Duan *et al.* (2006). The antioxidant activity was measured using DPPH, radical scavenging assay, and β-carotene-lineolate bleaching assay systems, and compared with that of BHT, gallic acid and ascorbic acid. The results showed that the crude extract and ethyl-acetate soluble fraction exhibited higher AA than BHT in the DPPH assay model, at all of four concentration levels tested (from 0.4 to 50 μg/ml), while, in the β-carotene-lineolate assay system, the crude extract and the ethyl acetate-soluble fraction exhibited similar or in most cases, higher AA than gallic acid and Ascorbic acid at the same concentrations from 10 to 200 μg/ml). Significant associations between the antioxidant potency and the total phenolic content, as well as between the antioxidant potency and the reducing power, were found.

Porphyra haitanensis T.J. Chang et B.F. Zheng Family: Bangiaceae (Rhodophyta)

Sulfated polysaccharide fraction F2 from *Porphyra haitanensis* showed inhibitory effect on the *in vitro* lipid peroxidation. F2 treatment increased TAOC and the activity of superoxide dismutase (SOD) and glutathione peroxidase (GSH-Px) in all the organs tested in ageing mice.

Porphyridium cruentum (Agardh) Naeg Family: Porphyridaceae (Rhodophyta)

The antioxidant activity of the microalgal ethanolic extracts of *Porphyridium cruentum* was determined by Rodriguez-Garcia and Guil-Guerrero (2007) by means of the β-carotene-lineolate model system. The results show that the extract showed good antioxidant activity.

Rhodomela confervoides (Hudson) P.C. Silva Family: Rhodomelaceae (Rhodophyta)

The methanol-chloroform extract of the marine red alga, *Rhodomela confervoides*, was measured for antioxidant activity, using the DPPH radical scavenging assay and the β-carotene-lineolate bleaching assay systems, and compared with those of the positive controls of butylated hydroxytoulene, gallic acid and ascorbic acid. The active extract was further purified by liquid-liquid partition to afford four fractions, of which the ethyl acetate-soluble (EA) fraction exhibited the strongest antioxidant activity in both the assay systems. Significant associations between the antioxidant potency and the total phenolic content, as well as between the antioxidant potency and the reducing power, were found for the tested fractions and subsfractions (Wang *et al.*, 2009).

Rhodomela teres (Perestenko) Masuda Family: Rhodomelaceae (Rhodophyta)

Free radical scavenging activity by DPPH assay and deoxyribose assay were ascertained among 27 species of common seaweeds using sequence extraction method (Xiaojun *et al.*, 1996). The results indicated that DPPH scavenging activity existed in 15 species, *Rhodomela teres* showed strongest activity. On the other hand the deoxyribose results showed that almost all the seaweed species had good ability to scavenge hydroxyl radicals. The most active species were *Rhodomela teres* and *Chorda filum*.

Sargassum filipendula Agardh Family: Sargassaceae

Fucans isolatd from *Sargassum filipendula* was evaluated by Costa *et al.* (2011). All heterofucans displayed considerable activity especially SF-1.ov which showed the most significant antioxidant potential with 90.7ascorbic acid equivalents in a total antioxidant capacity test and similar activity when campared with vitamin C in a reducing power assay.

Sargassum fulvellum (Turner) Agardh Family: Sargassaceae (Phaeophyceae)

The polysaccharide fraction of the alga *Sargassum fulvellum* was separated and investigated for its radical scavenging activities. The polysaccharide fraction of the alga showed a promising DPPH radical scavenging activity than did other fucoidans. Moreover, the sample exhibited a dose-dependent activity on hydrogen peroxide scavenging activity in the V 79-4 cell line. Interestingly, all the tested polysaccharide counterparts were more potent NO scavengers than were the commercial antioxidants BHA and α-tocopherol (Kim *et al.*, 2007).

Sargassum fusiforme (Harvey) Setchell Family: Sargassaceae (Phaeophyceae)

Antioxidant capacities of 56 selected Chinese medicinal plants, including *Sargassum fusiforme* were evaluated by Song *et al.* (2010) using the TEAC and FRAP assays, and their total phenolic content was measured by the Folin-Ciocalteu method. *Sargassum fusiforme* showed low antioxidant activity with TEAC value of 3.89μmol Trolox/g, FRAP value of 0.15μmol Fe^{2+}/g and phenolic content was 0.18mg GAE/g.

Sargassum kjellmanianum Yendo Family: Sargassaceae (Phaeophyceae)

Free radical scavenging activity by DPPH assay and deoxyribose assay were ascertained among 27 species of common seaweeds using sequence extraction method (Xiaojun *et al.*, 1996). The results indicated that DPPH scavenging activity existed in 15 species including *Sargassum kjellmanianum*. On the other hand the deoxyribose results showed that almost all the seaweed species had good ability to scavenge hydroxyl radicals.

Sargassum micranthum Yendo Family: Sargassaceae (Phaeophyceae)

Four plastoquinones isolated from the methanolic extract of the brown alga, *Sargassum micranthum* exhibited significant antioxidant activities such as an inhibitory effect on lipid peroxidation and DPPH radical scavenging activity (Mori *et al.*, 2005).

Sargassum myriocystum J. Agardh. Family: Sargassaceae

Badrinathan *et al.* (2011) evaluated the reactive oxygen radical scavenging capacities and antioxidant activities of *Sargassum myriocystum* using commonly accepted assays. It was found that phenolic contents constituted 22 mg gallic acid equivalents/gm of dry plant materials. Methanol/chloroform extracts of soxhlet showed highest reducing power, total antioxidant and hydroxyl scavenging activity. 80 to 120 per cent inhibition of hydroxyl radical quenching activity was observed and demonstrated by all soxhlet derived extracts, except petroleum ether.

Sargassum thunbergii (Mestens ex Roth) Kuntze Family: Sargassaceae (Phaeophyceae)

The free radical scavenging activity of water soluble natural antioxidants from *Sargassum thunbergii* was evaluated by Park *et al.* (2005) by examining the radical scavenging activities of the extracts of hydrylyzates from *S. thunbergii* on DPPH and alkyl radicals. The scavenging activity of the radicals increased with increased concentrations of the extracts. The scavenging results were higher for hydroxyl and alkyl radicals and lower for DPPH radical as compared with vitamin C as a reference. The hydrogen peroxide scavenging activity of the extracts was also investigated; the Alcalase extract showed the highest scavenging activity among the extracts prepared with the five proteases and five carbohydrates.

Spirulina platensis (Gomont) Geitler Family: (Cyanophyceae)

C-phycocyanin from *Spirulina platensis* effectively inhibited PCM induced lipid peroxidation in rat liver *in vivo* (Pinero Estrada *et al.*, 2001).

Wojcikowski *et al.* (2007) evaluated the antioxidant activity of aerial parts of *Spirulina platensis.* Oxygen radical absorbance capacity of the ethyl extract, methanol extract and aqueous extracts were 1.07, 37.51 and 7.03μmol Trolox equivalent per g of dried starting material, respectively.

Synechococcus sp. Family: Synechococcaceae (Cyanophyceae)

Li *et al.* (2006) evaluated the antioxidant capacity and total phenolic content of different fractions of 23 microalgae includig *Synechococcus*, using TEAC and Folin-Ciocalteu method, respectively. *Synechococcus* sp. possessed the highest antioxidant capacity. The correlation coefficients between the antioxidant capacities and the phenolic contents were very small in hexane, ethylacetate and water fractions. Thus, phenolic compounds were not a major contributor to the antioxidant capacities of these microalgae.

Turbinaria conoides J. Agardh. Family: Sargassaceae (Phaeophyceae)

Chattopadhyay *et al.* (2010) isolated a fucoidan, one alginic acid and neutral glucan from *Turbinaria conoides*. The *in vitro* antioxidant capacity of the fractions was studied by FRAP and DPPH assays. The fucoidan showed the highest potential followed by the alginate and laminarin.

Ulva lactuca L. Family: Chlorophyceae

Antioxidant activities of *Ulva lactuca* crude organic extracts were determined by Abd El-Baky *et al.* (2008). The extract exhibite remarkable antioxidant activity, compared to a synthetic antioxidant BHA and BHT, as evaluated by a rapid TLC screening, β-carotene bleaching, DPPH radical scavenging methods. The positively correlation was apparent between the antioxidant activity and present carotenoids, phenolics and chlorophylls in their extracts.

Ulva pertusa Kjllm. Family: Chlorophyceae

Polysaccharide obtained from *Ulva pertusa* (Chlorophytaa) contained a group of sulfated heteropolysaccharide referred to as Ulvan. The natural ulvan and its derivatives exhibited much higher scavenging activity on superoxide radical than vitamin C (Qi *et al.*, 2005a, b, 2006, 2010).

Undaria pinnatifida (Harvey) Suringer Family: Alariaceae

Antioxidant activity of the extracts of *Undaria pinnatifida* was investigated by Je *et al.* (2009) using ESR spectroscopy on DPPH, hydroxyl and superoxide radicals. Extracts exhibited strong radicals scavenging activity on DPPH and hydroxyl radical, and activity increased with increasing extract concentration. In contrast to DPPH and hydroxyl radical scavenging activity, extracts showed weak scavenging activity on superoxide radical. Extracts exhibited weak radical scavenging activity compared to vitamin C as a reference compound.

Fungi

Endophytic Fungi from Medicinal Plants: A potential Antioxidant Resource

Huan *et al.* (2007) investigated the relationship between Total antioxidant capacity (TAC) and Total phenolic content (TPC) for 292 morphologically distinct endophytic fungi isolated from 29 traditional Chinese medicinal plants. The antioxidant capacities of the endophytic fungal cultures were significantly correlated with their total phenolic contents, suggesting that phenolics were also the major antioxidant constituents of the endophytes. Some of the endophytes were found to produce metabolites possessing strong antioxidant activities. Several bioactive constituents from the fungal cultures and host plant extracts were identified. This investigation revealed that the metabolites produced by a wide diversity of endophytic fungi in culture can be a potential source of novel natural antioxidants.

Agaricus arvensis Schaeff. Family: Agaricaceae

The antioxidant activities of three Portuguese wild edible mushroom species were evaluated by Barros *et al.* (2007). Methanolic extracts of *Agaricus arvensis* were screened for their redcuing power, DPPH radical-scavenging acapacity, inhibition of erythrocytes hemolysis and antioxidant activity using the β-carotene lineolate model systems. The amounts of ascorbic acid, β-carotene and lycopene found in the mushroom extracts were very low. Otherwise, the high contents of phenolic compounds might account for the good antioxidant properties found in this species.

Agaricus bisporus (Lge) Sing. Family: Agaricaceae

Methanol extract of edible mushroom *Agaricus bisporus* (approx. 2 mg/ml) showed a high DPPH scavenging activity (Aguiano *et al.*, 2007).

Agaricus blazei Murril (=_Agaricus braziliensis_) Family: Agaricaceae

The ethanolic extract of *Agaricus blazei* and ethyl acetate and hydroalcoholic fractions were evaluated for their antioxidant activity (Oliveira *et al.*, 2007).

Agrocybe aegerita (V.Brig.) Singer Family: Strophariaceae

Zhang *et al.* (2004) evaluated the cyclooxygenase inhibitory and antioxidant compounds from the fruiting body of an edible mushroom *Agrocybe aegerita.* The inhibition values of liposome peroxidation by fatty acid fraction (FAF), palmitic acid and ergosterol at 100µg/ml were 75, 45 and 43 per cent respectively. The inhibition values of cyclooxygenase-I enzyme by FAF, palmitic acid and ergosterol at 100µg/ml with values of 88, 45 and 28 per cent respectively.

Amanita caesarea (Scop.: Fr.) Pess. Family: Amanitaceae (Basidiomycota)

Methanol extract of edible mushroom *Amanita caesaria* (approx. 2 mg/ml) showed a high DPPH scavenging activity close to 90 per cent (Aguiano *et al.*, 2007). It showed highest ABTS scavenging activity at approx. 0.14mg/ml. Oxidative enzymes (peroxidases and polyphenol oxidases) present in the water fractions reduced their antioxidant activity by different extents since the phenols responsible for the antioxidant activity were not only those substrates of the oxidative enzymes. Other phenolic compound and low molecular weight compounds were also involved in the antioxidant activity.

Sarikurkcu *et al.* (2010) evaluated the antioxidant activity of three edible mushrooms from Turkey including *Amanita caesaria*. The strongest super-oxide anion scavenger was *Amanita caesaria*. In the DPPH system the extract showed 79.4 per cent scavenging ability. Additionally, reducing power and chelating capacity of the mushrooms increased with concentration.

Antrodia camphorata Wu _et al._, Family: Polyporaceae

Hsiao *et al.* (2003) evaluated the ability of *Antrodia camphorata* extracts to protect against oxidative stress *in vitro* and against CCl_4-induced hepatic injury *in vivo*. An extract of *A. camphorata* inhibited nonenzymatic iron-induced lipid peroxidation in rat brain homogenates with an IC_{50} value about 3.1 mg/ml. It also scavenged the stable free radical DPPH. The dose of *A. camphorata* extract resulting in a decrease of 0.20 in the absorbance of DPPH was about 31µg/ml. Furthermore, an *A. camphorata* extract dose-dependently (250-1250mg/kg) ameliorated the increase in plasma aspartate aminotransferase (GOT) and alanine transferase (GPT) levels caused by chronic repeated CCl_4 intoxication in mice.

Antrodia salmonea T.T. Chang et W.N. Chon Family: Polyporaceae

Four new compounds were isolated by Shen *et al.* (2006) from the basidiomata of the fungus *Antrodia salmonea*. An *in vitro* cellular functional assay was performed to evaluate their antioxidative burst activity in human leucotytes. The results indicated that three of these four new compounds displayed antioxidative effect in human leucocytes with different potency and might confer anti-inflammatory activity to these drugs.

Auricularia auricula-Judae (Bullock) J. Schrot. Family: Auriculariaceae

In vitro evaluation of antioxidant activities of *Auricularia auricula-Judae* showed a significant inhibition of lipid peroxidation and potent hydroxyl scavenging activity when compared with standard drug catechin (Acharya *et al.*, 2004). IC_{50} value of crude, boiled and ethanolic extracts of *Auricularia auricula* represented 403, 510 and 373µg/ml respectively in case of hydroxyl scavenging activity and 310, 572 and 398µg/ml, respectively in case of lipid peroxidation. Furthermore, crudee, boiled and ethanolic extracts also increase significantly nitric oxide production (664, 191 and 850 pmole/mg dry wt/hr respectively) over the control.

Boletopsis grisea (Peck) Bondartsev et Singer Family: Boletaceae

The antioxidant activities of ten natural *p*-terphenyl derivatives, 1-10, obtained from the fruiting bodies of *Boletopsis grisea* indigenous to China were evaluated by Liu *et al.* (2004) in comparison with BHA and DPPH radical-scavenging method. The compounds 1-3 showed significant antioxidant activity. The compound 2 exhibited the strongest radical-scavenging activity with an EC_{50} value of 0.07.

Boletus edulis Bull. ex Fr. Family: Boletaceae (Basidiomycota)

Methanol extract of edible mushroom *Boletus edulis* (approx. 2 mg/ml) showed a high DPPH scavenging activity close to 90 per cent (Aguiano *et al.*, 2007). It showed highest ABTS scavenging activity at approx. 0.14mg/ml. Oxidative enzymes (peroxidases and polyphenol oxidases) present in the water fractions reduced their antioxidant activity by different extents since the phenols responsible for the antioxidant activity were not only those substrates of the oxidative enzymes. Other phenolic compound and low molecular weight compounds were also involved in the antioxidant activity and differed depending on mushroom species.

The methanolic extract of *Boletus edulis* was analyzed for its antioxidant activities in different test systems namely β-carotene/linoleic acid, DPPH free radical scavenging, reducing power and metal chelating activities in addition to their total phenolic and flavonoid contents. In β-carotene/linoleic acid and DPPH systems the extract showed strongest activity patterns. Its activity was as strong as the positive controls. The reducing power of the species was excellent. Chelating capacity of the extracts increased with the increasing concentration. The extract was found to have high phenolic content.

Clitocybe geotropa Family:Tricholomataceae (Basidiomycota)

Sarikurkcu *et al.* (2010) evaluated the antioxidant activity of three edible mushrooms from Turkey including *Clitocybe geotropa*. Reducing power and chelating capacity of the mushrooms increased with concentration. In DPPH system the extract showed good scavenging ability.

Cordyceps sinensis (Berk.) Sacch. Family: Clavicipitaceae (Ascomycota)

Cordyceps, one of the well-known traditional Chinese medicines, consists of the dried fungus, *Cordyceps sinensis* growing on the larva of the caterpillar. It is commonly used for the replenishment of body health. One of the known pharmacological effects is its anti-oxidation activity. However, there is a great variation of the quality in different sources of *Cordyceps*. Li *et al.* (2001) analyzed the antioxidant properties of the water extracts of various sources of natural *C. sinensis* and cultured *Cordyceps* mycelia by using three different assay methods such as the xanthine oxidase assay, the induction of hemolysis assay and the lipid peroxidation assay. The results showed that *Cordyceps*, in general, possesses a strong anti-oxidation activity in all assays tested. However, both natural and cultured *Cordyceps* showed the lowest inhibition in the lipid peroxidation when compared with the other two assay methods. The cultured *Cordyceps* mycelia had equally strong anti-oxidation activity as compared to the natural *Cordyceps*. Besides, the anti-oxidation activities were increased to 1-30 folds in the partially purified polysaccharide fractions from the cultured *Cordyceps* mycelia, which suggested that the activity could be derived partly from *Cordyceps* polysaccharides.

Wojcikowski *et al.* (2007) evaluated the antioxidant activity of aerial parts of *Cordyceps sinensis.* Oxygen radical absorbance capacity of the methanol extract and aqueous extracts were 83.49 and 33.79μmol Trolox equivalent per g of dried starting material, respectively.

Eurotium herbariorum **(F.H. Wigg) Link Family: Trichocomaceae (Ascomycota)**

The extracts prepared by culturing *Eurotium herbariorum* were shown to have DPPH radical scavenging activity (Miyake *et al.*, 2009). Five antioxidants were isolated from the extracts and identified as isodihydroauroglaucin, auroglaucin, dihydroauroglaucin, tetrahydroauroglaucin and flavoglaucin. Compared to α-tocopherol, the isolated antioxidants exhibited high antioxidative activity for the radical scavenging capacity of DPPH and superoxide, but low activity for inhibiting the autooxidation of docosahexaenoic acid (Miyake *et al.*, 2009).

Fistulina hepatica **Schaeffer Family: Schizophyllaceae**

English Name: Beefsteak fungus.

Fistulina hepatica was investigated for its capacity to act as a scavenger of DPPH radical and reactive oxygen species (superoxide radical, hydroxyl radical and hypochlorous acid). Good results were obtained against DPPH in a concentration-dependent manner. Beefsteak fungus also displayed good activity against superoxide radical, achieved by its capacity to act as both scavenger and xanthine oxidase inhibitor. A prooxidant effect was noticed for hydroxyl radical, which may be due to its capacity for iron ions reduction. Little ability for iron chelation was also observed. Beefsteak fungus showed a weak protective effect against hypochlorous acid (Ribeiro *et al.*, 2007).

Flammulina velutipes **(Curt.: Fries) Singer Family: Physalacriaceae**

Bao *et al.* (2008) evaluated the antioxidative activities of edible mushroom *Flammulina velutipes.* The addition of 5 ml of mushroom extract to 100 g of minced bigeye tuna meat was more effective than ascorbic acid sodium salt (500 ppm) or α-tocopherol (500ppm) with regard to oxidation of lipid in the tuna meat.

Bao *et al.* (2010) evaluated antioxidative properties of hydrophilic extracts prepared from the fruiting body and spent culture medium of *Flammulina velutipes* by monitoring the total reducing power ability and DPPH free radical scavenging activity, together with antioxidative activities against lipid oxidation in homogenates of yellow tail dark muscle and autoxidation of oxymyoglobin purified from yellow tail dark muscle. Generally all the extract had reducing power ability, DPPH radical scavenging activity and antioxidative activity against lipid oxidation and autoxidation of oxymyoglobin.

Ganoderma applanatum (Pers.) Patouillard Family: Ganodermataceae

In vitro evaluation of antioxidant activities of *Ganoderma applanatum* showed significant inhibition of lipid peroxidation and potent hydroxyl radical scavenging activity when compared with standard drug catechin. IC_{50} values of crude, boiled and ethanolic extract were 604.8, 624 and 627 µg/ml, respectively in case of hydroxyl radical scavenging activity, and 441, 520.5 and 166.16µg/ml respectively in casse of lipid peroxidation. Furthermore, crude boiled and ethanolic extracts also increased significantly nitric oxide production (156.67, 121.88 and 742 pmole/mg dry wt/hr, respectively) over the control (Acharya *et al.*, 2005).

Ganoderma boninense Pat Family: Ganodermataceae

Antioxidant properties of *Ganoderma boninense* was evaluated by Mitchell *et al.* (2009) by DPPH radical scavenging activity.The antioxidant activity of the extracts on DPPH scavenging activity was 24.63 per cent at 1 mg/ml with EC_{50} value of 2.08 mg/ml. The phenol contents were the major antioxidant components found in the extracts (25.02 mg/g), followed by flavonoid (16.89mg/g), ascorbic acid was found in small amounts (3.5mg/g) and β-corotene and lycopene were only found in vestigial amounts (<0.005 mg/g).

Ganoderma lucidum (Curtis) P.Karst. Family:Ganodermataceae (Basidiomycota)

The hot water extract of the mushroom *Ganoderma lucidum* was shown to have antioxidative effect against heart toxicity (Wong *et al.*, 2004). Investigations into the mechanisms of action, level of lipid peroxidation level *in vivo*, and superoxide scavenging activity were also conducted. The results of this study showed that *Ganoderma lucidum* exhibited a dose-dependent antioxidant effect on lipid peroxidation and superoxide scavenging activity in mouse heart homogenate. Wong *et el.* (2004) concluded that the antioxidative activity may therefore contribute to the cardioprotective effect of *Ganoderma lucidum*, and may therefore protect the heart from superoxide induced damage.

Wojcikowski *et al.* (2007) evaluated the antioxidant activity of aerial parts of *Ganoderma lucidum*. Oxygen radical absorbance capacity of the ethyl extract, methanol extract and aqueous extracts were 7.46, 30.29 and 54.68µmol Trolox equivalent per g of dried starting material, respectively.

Antioxidative activity of *Ganoderma lucidum* from tropical South India were investigated by Joseph *et al.* (2008) using DPPH, superoxide, nitric oxide scavenging and lipid peroxidation. The chloroform extract of the mushroom showed marked free radical scavenging activities.

Ng *et al.* (20.) evaluated the antioxidant activity of *Ganoderma lucidum*. The results suggest that the hepatoprotection afforded by *Ganoderma lucidum* pretreatment may be attributed to its antioxidant activity.

Grifola frondosa (Diskson: Fries) Gray Family: Meripilaceae

Morchella esculenta (Linn.: Fries) Persoon Family: Morchellaceae (Ascomycota)

Termitomyces albuminosus (Berkeley and Broome) Heim Family: Lyophyllaceae (Basidiomycota)

Mushrooms have been used as food and food-flavouring material in soups and sauces for centuries, due to their unique and subtle flavour. Mushrooms have recently become attractive as functional foods and as a source of physiologically beneficial medicine. Mau *et al.* (2004) conducted studies to evaluate the antioxidant properties of methanolic extracts from the above three mushroom mycelia, including antioxidant activity, reducing power, scavenging effects on radicals and chelating effects on ferrous ions. The contents of potential antioxidant components in methanolic extracts from these mycelia were also determined. Methanolic extracts from the three mycelia showed high antioxidant activities (85.4-94.7 per cent) at 25 mg/ml. Reducing powers of the three methanolic extracts were 0.97-1.02 at 25 mg/ml. Scavenging effects on DPPH radicals were 78.8-94.1 per cent at 10mg/ml. These three mycelia showed no scavenging effect on hydroxyl radicals. Chelating effects on ferrous ions were high (90.3-94.4 per cent) at 10mg/ml. Total phenols were the major naturally occurring antioxidant components found in methanolic extracts. Contents of ascorbic acid and tocopherols were similar for these three mycelia. All EC_{50} values were below 10mhg/ml, indicating that the three mycelia had good antioxidant properties except for the scavenging effect on hydroxyl radicals. On the basis of the results obtained, these three mycelia might be somewhat beneficial to the antioxidant protection system of the human body.

Hydnellum geogerirum Family: Thelephoraceae

Fractionation of methanol extracts of the fruit bodies of inedible mushroom *Hydnellum geogerirum* resulted in isolation of two new *p*-terphenyl derivatives named hydnellins A and B and sarcodonin delta possessing N-oxide. The terphenyls showed moderate antioxidant activity. (Hashimoto *et al.*, 2006).

Hydnellum suaveolens (Scop.) P. Karst. Family: Thelephoraceae

Fractionation of methanol extracts of the fruit bodies of inedible mushroom *Hydnellum suaveolens* resulted in isolation of two new *p*-terphenyl derivatives named hydnellins A and B and sarcodonin delta possessing N-oxide. The terphenyls showed moderate antioxidant activity. (Hashimoto *et al.*,.2006).

Inonotus obliquus (Fr.) Pilat Family: Hymenochaetaceae

English Name: Chaga mushroom.

Antioxidant activity of hot water extract of chaga fungus *Inonotus obliquus* used as a folk medicine in Russia was evaluated and compared with medicinal fungi. Chaga fungus showed strongest activity among fungi examined in terms of both superoxide and hydroxide radicals scavenging activities. Further determination of the antioxidant potential of isolated fruiting body and sclerotium revealed that the 80 per cent MeOH extract of fruiting body had the highest potential as high as that of Chaga fungus decoction.

Ju *et al.* (2010) investigated the effect of steam treatment on free phenolic acids in Chaga mushrooms. After the steam treatment, the soluble phenolic content determined by modified Folin-Ciocalteu method was increased and antioxidant activity was enhanced, as confirmed by a DPPH radical scavenging activity assay. The amounts of vanillic acid, protocatechuic acid, syringic acid, and 2,5-dihydroxyterephthalic acid were increased significantly as the result of steam treatment, suggesting that the liberation of low molecular weight free phenolics was enhanced by the steaming process. Consequently, the radical scavenging activity was also significantly enhanced by free phenolics produced using this method.

Inonotus xeranticus (Berk) Imaz et Aoshi Family: Hymenochaetaceae

In an effort to identify antioxidants from edible and medicinal mushrooms, three new hispidin derivatives, methylinoscavin A, inoscavin B, D and methylinoscavin B, D together with the known compounds inoscavin A, phelligridin D, F, 3,4-dihydroxybenzoic acid, Interfungins A, B and C were isolated from the methanolic extract of the fruiting bodies of *Inonotus xeranticus* (Lee *et al.*, 2006a, b; Lee and Yun, 2006, 2007). These compounds exhibited significant scavenging activity against the ABTS radical anion and DPPH radical.

Lactarius deterrimus Groger Family: Russulaceae (Basidiomycota)

The methanolic extract of *Lactarius deterrimus* was analyzed for its antioxidant activities in different test systems namely β-carotene/linoleic acid, DPPH free radical scavenging, reducing power and metal chelating activities in addition to their total phenolic and flavonoid contents. In β-carotene/linoleic acid and DPPH systems the extract showed strongest activity patterns. Its activity was as strong as the positive controls. The reducing power of the species was excellent. Chelating capacity of the extracts increased with the increasing concentration.

Lasiosphaera fenzelii Reich. Family: Lycoperdaceae

Li *et al.* (2008) evaluated the antioxidant properties of 45 medicinal plants including *Lasiosphaera fenzelii* using FRAP and TEAC assays and the total phenolic contents of these plants were measured. *Lasiosphaera fenzelii* showed low antioxidant activity with 9.08 µmol Fe(II)/g FRAP value, 16.55µmol Trolox/g of TEAC value and 5.11 mg GAE/g of Phenolic content.

Lentinula edodes (Berk) Pegler Family: Marasmiaceae (Basidiomycota)

Methanol extract of edible mushroom *Lentinula edodes* (approx. 2 mg/ml) showed a high DPPH scavenging activity close to 90 per cent (Aguiano *et al.*, 2007). It showed highest ABTS scavenging activity at approx. 0.14mg/ml. Oxidative enzymes (peroxidases and polyphenol oxidases) present in the water fractions reduced their antioxidant activity by different extents since the phenols responsible for the antioxidant activity were not only those substrates of the oxidative enzymes. Other phenolic compound and low molecular weight compounds were also involved in the antioxidant activity.

Sasidharan *et al.* (2010) investigated the antioxidant and hepatoprotective effects of methanolic extracts of *Lentinula edodes* and the determination of their total phenolics content. The amount of total phenolics was estimated to be 70.83 GAE per gram of dry extract. The antioxidant activity of *Lentinula edodes* extract was 39.0 per cent at a concentration of 1mg/ml and was also concentration dependent, with EC_{50} value of 4.4 mg/ml.

Lepista nuda (Bull.) Cooke Family: Tricholomataceae

Antioxidant capacity of *Lepista nuda* extracts obtained with ethanol were investigated by Mercan *et al.* (2006). Four complementary test systems, namely DPPH free radical scavenging, β-carotene/linoleic acid systems, total phenolic compounds and total flavonoid concentration have been used. Linoleic acid inhibition values of *Lepista nuda* ethanol extract, BHA and α-tocopherol standards were found to be 84.3 per cent, 98.9 per cent and 99.2. per cent, respectively in the concentration of 160µg/ml. Total flavonoid amount was 8.21µg/ml quercetin equivalent while phenol compound amount was 48.01µg/ml pyrocatechol equivalent in the extract.

Leucoagaricus pudicus (Bull.) Bon Family: Agaricaceae (Basidiomycota)

Sarikurkcu *et al.* (2010) evaluated the antioxidant activity of three edible mushrooms from Turkey including *Leucoagaricus pudicus*. Reducing power and chelating capacity of the mushrooms increased with concentration. In DPPH system the extract showed good scavenging ability. In the case of total phenolics *L. pudicus* was found to have the highest content.

Leucopaxillus giganteus (Sowerby) Singer Family: Tricholomataceae

The antioxidant activities of three Portuguese wild edible mushroom species were evaluated by Barros *et al.* (2007). Methanolic extracts of *Leucopaxillus gigantus* were screened for their redcuing power, DPPH radical-scavenging capacity, inhibition of erythrocytes hemolysis and antioxidant activity using the β-carotene lineolatee model systems. The amounts of ascorbic acid, β-carotene and lycopene found in the mushroom extracts were very low. Otherwise, the high contents of phenolic compounds might account for the good antioxidant properties found in this species.

Morchella angusticeps Peck Family: Morchellaceae (Ascomycota)

Seven *Morchella* species were analyzed by Gursoy *et al.* (2009) for their antioxidant activities in different test systems namely β-carotene/linoleic acid, DPPH, reducing power, chelating effect and scavenging effect (per cent) on the stable ABTS. In β-carotene/linoleic acid system, the most active mushroom was *M. angusticeps*. The reducing power and chelating capacity of the methanol extracts of mushrooms increased with concentration.

Morchella conica Pers. Family: Morchellaceae (Ascomycota)

Seven *Morchella* species were analyzed by Gursoy *et al.* (2009) for their antioxidant activities in different test systems namely β-carotene/linoleic acid, DPPH, reducing power, chelating effect and scavenging effect (per cent) on the stable ABTS. In DPPH assay, methanol extract of *M. conica* showed high antioxidant activity. The reducing power and chelating capacity of the methanol extracts of mushrooms increased with concentration. In 40μg/ml concentration, methanol extract of *M. conica*, exhibited the highest radical scavenging activity (78.66 per cent) when reacted with ABTS radical. *M. conica* was found to have the highest phenolic content among the samples.

Morchella esculenta (Linn.: Fries) Persoon Family: Morchellaceae (Ascomycota)

Methanol extract of *Morchella esculenta* (approx. 2mg/ml) showed a high DPPH scavenging activity close to 90 per cent (Anguiana *et al.*, 2007).

Seven *Morchella* species were analyzed by Gursoy *et al.* (2009) for their antioxidant activities in different test systems namely β-carotene/linoleic acid, DPPH, reducing power, chelating effect and scavenging effect (per cent) on the stable ABTS. In β-carotene/linoleic acid system, the most active mushroom was *M. esculenta* var. *umbrina*. The reducing power and chelating capacity of the methanol extracts of mushrooms increased with concentration.

Morchella rotunda (Fr.) Boud. Family: Morchellaceae (Ascomycota)

Seven *Morchella* species were analyzed by Gursoy *et al.* (2009) for their antioxidant activities in different test systems namely β-carotene/linoleic acid, DPPH, reducing power, chelating effect and scavenging effect (per cent) on the stable ABTS. The reducing power and chelating capacity of the methanol extracts of mushrooms increased with concentration. Flavonoid content of *M. rotunda* was found superior (0.59μg QEs/mg extract).

Phellinus merrillii (Murr.) Ryv. Family: Hymenochaetaceae (Basidiomycota)

Phellinus merrillii displayed antioxidant activities in a series of *in vitro* tests such as total antioxidant activity, determination of total penolic compound, ferric reducing power assay, dot blot DPPH staining, DPPH radical scavenging and reducing power methods (Chang *et al.*, 2007). Many phenolic compounds were found in the fungus such as hispolon. The ethyl acetate extract of *Phellinus merrillii* also showed antihepatotoxic activity against CCl_4-induced actute toxicity in rat liver.

Phellinus rimosus (Berkeley) Pilat Family: Hymenochaetaceae (Basidiomycota)

The methanolic extract of *Phellinus rimosus* possessed significant *in vitro* superoxide anion, hydroxyl radical and nitric oxide scavenging and lipid peroxidation inhibiting activities (Ajith and Janardhanan, 2001). They attributed the anti-inflammatory activity of the methanolic extract to it's free radical scavenging properties.

Ajith and Janardhanan (2002) studied the antioxidant and antihepatotoxic activities of *Phellinus rimosus*. The superoxide anion scavenging, Fe^{2+}-ascorbate induced lipid peroxidation inhibiting, hydroxyl radical scavenging and nitric oxide scavenging activities of the ethyl acetate extract were determined. The results indicated that ethyl acetate extract exhibited significant *in vitro* antioxidant activity.

Antioxidant, cytotoxic and antitumor activities of ethanol extract of a wood inhabiting polypore macrofungus, *Phellinus rimosus* were studied by Rony *et al.* (2008). The ethanol extract showed marked superoxide anion scavenging, Fe^{2+}/ascorbate induced lipid peroxidation inhibiting and DPPH radical scavenging activities.

***Phellinus* sp. PL3 Family: Hymenochaetaceae (Basidiomycota)**

Phellisin A, which was isolated from the culture broth of fungus *Phellinus* sp. PL3, exhibited significant low-density lipoproteins (LDL)-antioxidant activity. It inhibited the Cu^{2+}-mediated oxidation of LDL (IC_{50}: 5,3μM) and AAPH-mediated oxidation of LDL (IC_{50} 2.8 μM) in the TBARS assay as well as the macrophage-mediated LDL oxidation (73 per cent inhibition at 5μM). In addition it delayed LDL oxidation with a prolonged lag time (192 min at 2μM, control: 44 min). This compound also showed a 10-fold more potent DPPH free radical scavenging activity (IC_{50} 1.7μM) than trolox (IC_{50} 18.6μM), a known DPPH inhibitor. In addition, phellinsin A inhibited xanthine oxidase activity with an IC_{50} value of 31.0μM, whereas allopurinol, a xanthine oxidase inhibitor showed and IC_{50} value of 40.7μM.

***Pleurotus florida* (Mont.) Singer Family: Pleurotaceae (Basidiomycota)**

English Name: Oyster mushroom

The ethanolic extract of Oyster mushroom *Pleurotus florida* was studied for its free radical scavenging property on different *in vitro* models like DPPH assay, FRAP assay and *in vitro* antilipidperoxidative assay using goat liver homogenate and RBC Ghost model. The *in vitro* lipid peroxidation was inhibited to a good extent by the *P. florida* ethanolic extract and the extent of inhibition being higher in the RBC membrane model than the liver homogenate model. The mushroom extract showed good dose-dependent free radical scavenging property in both the models (Selvi and Chinnaswamy, 2007).

***Pleurotus ostreatus* Jacq. ex Fries Family: Pleurotaceae (Basidiomycota)**

Jayakumar *et al.* (2006) investigated the putative antioxidant activity of the oyster mushroom *Pleurotus ostreatus* on CCl_4-induced liver damage in male wistar rats. When rats with CCl_4-induced hepatotoxicity were treated with the extract of *P.ostreatus*, the serum SGOT, SGPT and SLAP levels reverted to near normal, while the hepatic concentration of GSH, CAT, SOD and GPx were significantly increased and that of MDA significantly lowered, when compared to CCl_4-exposed untreated rats.

Jayakumar *et al.* (2007) investigated the antioxidant activity of the oyster mushroom, *Pleurotus ostreatus*, on lipid peroxidation and antioxidant status of major organs of aged rats (24 months old) rats when compared to young (4 month old) rats.

Elevated levels of MDA and significantly lowered levels of reduced GSH and of vitamin C and E were observed in the liver, kidneys, heart and brain of aged rats, when compared to values in young rats. Quantitative analysis of the activities of the antioxidant enzymes catalase (CAT), SOD and GPx revealed significantly lower values in the liver, kidneys, heart and brain of aged rats. Administration of the extract of *P. ostreatus* to aged rats resulted in elevated levels of reduced glutathione and vitamins C and E and in increased activities of CAT, SOD and GPx so that the values in most of these parameters did not differ significantly from those in young rats. In addition, the levels of MDA was lowered on administration of mushroom extract to aged rats.

Jayakumar *et al.* (2008) also reported that *Pleurotus ostreatus* decreases the oxidative stress induced by carbon tetrachloride in rat kidneys, heart and brain.

Polyporus umbellatus (Pers) Fr. Family: Polyporaceae

Gan *et al.* (2010) evaluated the antioxidant activity of *Polyporus umbellatus* using the FRAP and TEAC assays, and its total phenolic content was measured by the Folin-Ciocalteu method. For FRAP assay the value was 7.93 µmol Fe(II)/g dry weight. For TEAC assay, the value was 4.18µmol Trolox/g dry weight. Total phenolic content was 0.38 mg gallic acid equivalent/g dry weight of plant material.

Polyporus valvatus Family: Polyporaceae (Basidiomycota)

Orhan and Ustun (2010) investigated the antioxidative activity of ethanolic extracts of a number of mushroom species growing in Turkey mainly *Polyporus* species (*Polyporus gilvus, Polyporus sulphureus, Polyporus annosus, Polyporus radiatus, Polyporus pinicola, Polyporus valvatus, Polyporus fomentarius, Polyporus stevenii, Polyporus badius*) as well as *Cantharellus cibatarius, Lactarius deliciosus* and *Trametes versicolor*. DPPH radical scavenging activity, ferrous iron-chelating effect, ferric-reducing antioxidant power (FRAP) and beta-carotene bleaching tests were followed for antioxidant studies. Total phenol contents of the extracts were also determined by Folin-Ciocalteu reagent. *P. volvatus* showed the best scavenging effect against DPPH, while all of the extracts exerted low capacity in chelating ferrous ions and reducing ferric ion.

Poria cocos F.A.Wolf Family: Polyporaceae (Basidiomycota)

The antioxidant properties of twenty medicinal herbs used in traditional Mediterranean and Chinese medicine including *Poria cocos* were studied by Schinella *et al.* (2002). Most of the extracts were weak scavengers of the hydroxyl radical and *Poria cocos* exhibited the highest scavenging activity.

Wu *et al.* (2005) evaluated the antioxidant activities of *Poria cocos*, popular ingredient of Chinese medicine. The results showed that aqueous extracts of this drug exhibited antioxidant activities in a concentration-dependent manner. The extract displayed an inhibitory effect on $FeCl_2$-ascorbic acid and lipid peroxidation in rat liver homogenate *in vitro*.

Gan *et al.* (2010) evaluated the antioxidant activity of *Poria cocos* using the FRAP and TEAC assays, and its total phenolic content was measured by the Folin-Ciocalteu method. For FRAP assay the value was 3.88 µmol Fe(II)/g dry weight. For TEAC assay, the value was 1.31μmol Trolox/g dry weight. Total phenolic content was 0.10 mg gallic acid equivalent/g dry weight of plant material.

Ramaria flava (Schaeff.) Quel Family: Gomphaceae (Basidiomycota)

Gursoy *et al.* (2010) evaluated the antioxidant activities of three edible mushrooms including *Ramaria flava* in different test systems – β-carotene/linoleic acid, DPPH free radical scavenging, reducing power and metal chelating activities in addition to their total phenolic and flavonoid contents. In DPPH system, the scavenging effects increased with the concentration. The reducing power of the mushroom also increased with concentration. Chelating effect was 96.75 per cent for *R. flava*. In the case of total phenolic and flavonoid assays, *R. flava* was found to have the highest phenolic and flavonoid content.

Rhizopogon roseolus (Corda) Th. Fr. Family: Rhizopogonaceae (Basidiomycota)

Gursoy *et al.* (2010) evaluated the antioxidant activities of three edible mushrooms including *Rhizopogon roseolus* in different test systems – β-carotene/linoleic acid, DPPH free radical scavenging, reducing power and metal chelating activities in addition to their total phenolic and flavonoid contents. In the first case, methanol extract of *Rhizopogon roseolus* showed the strongest activity. In DPPH system, the scavenging effects increased with the concentration. The reducing power of the mushroom also increased with concentration.

Russula delica Fr. Family: Russulaceae (Basidiomycota)

Antioxidant activities of the ethanolic extract of *Russula delica* was evaluated by Yaltirak *et al.* (2009) by using DPPH radical scavenging and chelating ability on ferrous ion assay. Scavenging effect of DPPH radicals was 26 per cent at 10mg/ml and chelating effects on ferrous ions was 58 per cent at 5 mg/ml. In addition, the amounts of total phenolic content (6.23 mg/g), ascorbic acid (2.93 mg/g), β-carotene (0.11 mg/g) and lycopene (0.03 mg/g) in the macrofungi ethanolic extract were determined.

Gursoy *et al.* (2010) evaluated the antioxidant activities of three edible mushrooms including *Russula delica* in different test systems – β-carotene/linoleic acid, DPPH free radical scavenging, reducing power and metal chelating activities in addition to their total phenolic and flavonoid contents. In DPPH system, the scavenging effects increased with the concentration. The reducing power of the mushroom also increased with concentration.

Saccharomyces cerevisiae Meyen ex E.C. Hansen Family: Saccharomycetaceae (Ascomycota)

English Name: Yeast.

In vitro potential antioxidant activity of (1-3), (1-6)-β-D-glucan and protein fractions from *Saccharomyces cerevisiae* cell walls was investigated by Jaehrig *et al.* (2007). The results showed that the antioxidant activity of yeast cell wall proteins exceeds that of β-glucan greatly. Especially aromatic side chains and free thiols from denatured proteins seem to work as antioxidants.

Sarcodon imbricatus (L.) Karsten Family: Bankeraceae

The antioxidant activities of three Portuguese wild edible mushroom species were evaluated by Barros *et al.* (2007). Methanolic extracts of *Sarcodon imbricatus* were screened for their reducing power, DPPH radical-scavenging capacity, inhibition of erythrocytes hemolysis and antioxidant activity using the β-carotene lineolatee model systems. The amounts of ascorbic acid, β-carotene and lycopene found in the mushroom extracts were very low. Otherwise, the high contents of phenolic compounds might account for the good antioxidant properties found in this species.

Suillus collitinus (Fr.) Kuntze Family: Suillaceae (Basidiomycota)

The methanolic extract of *Suillus collitinus* was analyzed for its antioxidant activities in different test systems namely β-carotene/linoleic acid, DPPH free radical scavenging, reducing power and metal chelating activities in addition to their total phenolic and flavonoid contents. The reducing power of the species was excellent. Chelating capacity of the extracts increased with the increasing concentration. Total flavonoid content was found to be high.

Thelephora aurantiotincta Corner Family: Thelephoraceae

The antioxidant activities of ten natural *p*-terphenyl derivatives, 1-10, obtained from the fruiting bodies of *Thelephora aurantiotincta* indigenous to China were evaluated by Liu *et al.* (2004) in comparison with BHA and DPPH radical-scavenging method. The compounds 1-3 showed significant antioxidant activity. The compound 2 exhibited the strongest radical-scavenging activity with an EC_{50} value of 0.07.

Thelephora ganbajun Zang Family: Thelephoraceae

The antioxidant activity of three poly(phenyacetyloxy)-substituted 1,1′,4′,1″-terphenyl compounds from the edible mushroom *Thelephora ganbajun* were investigated by Yang *et al.* (2004). The IC_{50} values of compounds 1-3 for lipid peroxidation in rat liver homogenate were 400, 48, 54μM, respectively. Compounds 1-3 increased SOD activity with EC_{50} values of 182, 74, 204μM. They were also assessed on the DPPH radical scavenging activity with EC_{50} values of 49, 1233 and 55μM.

The antioxidant activities of ten natural *p*-terphenyl derivatives, 1-10, obtained from the fruiting bodies of *Thelephora ganbajun* indigenous to China were evaluated by Liu *et al.* (2004) in comparison with BHA and DPPH radical-scavenging method. The compounds 1-3 showed significant antioxidant activity. The compound 2 exhibited the strongest radical-scavenging activity with an EC_{50} value of 0.07.

Thelephora vialis Schwein Family: Thelephoraceae

A new potent antioxidant, vialinin A (1), together with a known compound, ganbajunin (2) and a mixture of ganbajunins D (3) and E (4), were isolated from the dry fruiting bodies of *Thelephora vialis* (Xie *et al.*, 2005). Vialinin A had strong DPPH free radical-scavenging activity with EC_{50} value of 14.0μM, nearly equal to that of BHT (EC_{50} value of 14.0μM).

Xerocomus chrysenteron (Bull.) Quelet Family: Boletaceae (Basidiomycota)

The methanolic extract of *Xerocomus chrysenteron* was analyzed for its antioxidant activities in different test systems namely β-carotene/linoleic acid, DPPH free radical scavenging, reducing power and metal chelating activities in addition to their total phenolic and flavonoid contents. The reducing power of the species was excellent. Chelating capacity of the extracts increased with the increasing concentration.

Xylaria nigripes (Klotzsch.) Sacc. Family: Xylariaceae

Ko *et al.* (2009) investigated the antioxidant and antiradical activities of Wu Ling Shen, a popular medicinal fungus (*Xylaria nigripes*) used in traditional Chinese medicine preparations. Two different *X. nigripes* materials, the cultivated *X. nigripes* mycelia (XN) and a commercial *X. nigripes* product (XNP), were used to prepare the aqueous (XN-H vs. XNP-H) and ethanol (XN-E vss XNP-E) extracts for this study. Results showed that extracts of XN possessed stronger antioxidant and antiradical activities than XNP in all tested model systems. However, all extracts exhibited a weak activity in metal chelation and reducing power. Total antioxidant activity of XN extracts (IC_{50} = 6.20µg/ml for XN-H and 5.41µg/ml for XN-E), but not XNP extracts (IC_{50} = 128.13µg/ml for XNP-H and 96.16 µg/ml for XNP-E), was more potent than Trolox (IC_{50} = 19.64 µg/ml) and vitamin C (IC_{50} = 26.39µg/ml). XN-E (IC_{50} = 5.12µg/ml and XNP-E (IC_{50} = 8.89µg/ml) possessed a relatively similar potency as that of positive controls (IC_{50} = 6.94µg/ml for Trolox and 4.25µg/ml for vitamin C) in the superoxide radical scavenging activity. Although the DPPH radical scavenging activity of XN extracts was weaker than that of Trolox and vitamin C, it was about eight times more potent than that of XNP extracts. Among the different extracts, XN-E showed the highest total flavonoid (32.69 mg/g) and phenol (59.75 mg/g) contents, while XNP-H (7.50 per cent w/w) had the highest level in total polysaccharide content.

LICHENS

Cetraria islandica (L.) Acherius Fruticose Lichen

Gulcin *et al.* (2002) reported that aqueous extract of *Cetraria islandica* in 50, 100, 250 and 500 µg/ml concentration showed 96, 99, 100 and 100 per cent inhibition on peroxidation of linoleic acid, respectively which was more than that of 500µg/ml of α-tocopherol (77 per cent). Gulcin *et al.* (2002) reported that superoxide anion radical scavenging activity of *C.islandica* was concentration dependent and increased with increasing amount of sample.

Dermatocarpon miniatum Family: Verrucariaceae (Ascomycota)

Aslan *et al.* (2006) evaluated the antioxidant activity of the lichen *Dermatocarpon miniatum*. Extracts provided 50 per cent inhibition at 396.1µg/ml concentration on DPPH assay and gave 49 per cent inhibition of linoleic acid oxidation.

Lethariella cashmeriana Krog Family: Usneaceae

Lethariella sernanderi (Motyka) Obermayer Family: Usneaceae

Lethariella sinensis Wei et Jiang Family: Usneaceae

A yellow and new dark pigments were isolated by Kinoshita *et al.* (2009) from *Lethariella cashmeriana, L. sernanderi* and *L. sinensis* as antioxidant components. The yellow pigment was identified as canarione, and the others were determined to be 1,2-quinone derivatives, rubrocashmeriquinone and 7-chlororubrocashmeriqunone and 7-chlorocanarione by analysis of their spectroscopic data.

Peltigera rufescens (Weiss.) Humb. Family: Peltigeraceae

Tanas *et al.* (2010) evaluated the anti-inflammatory and antioxidant activities of methanol extract of *Peltigera rufescens*. The extract reduced the myeloperoxidase and inducible nitric oxide synthase activities, which were increased by carrageena injection. Carragenan injection also increased the lipid peroxidation as compared with untreated paw tissues. The administration of extract, diclofenac, and indomethacin reduced the LPO in paw tissues through amelioration of the antioxidant defense systems.

Rimelia reticulata (Taylor) Hale et Fletcher Family: Parmeliaceae

Meera *et al.* (2008) investigated the anti-inflammatory activity and antioxidant activity of *Rimelia reticulata*. The *in vitro* antioxidant activity has been investigated by Hydrogen peroxide scavenging method. The chloroform, benzene, acetone, ethyl acetate and alcohol extracts showed significant antioxidant activity in the above mentioned method.

Usnea longissima Ach. Family: Usneaceae

Odabasoglu *et al.* (2006) reported gastroprotective and antioxidant effects of usnic acid from *Usnea longissima* on indomethacin induced gastric ulcer in rats. The work of Bayir *et al.* (2006) suggest that the gastroprotective effect of diffractaic acid, isolated from *Usnea longissima* can be attributed to its enhancing effects on antioxidant defense systems as well as reducing effects of neutrophil infiltration.

Usnea undulata Stirt. Family: Usneaceae

Susithra *et al.* (2011) assessed the *In-vitro* potential anti-oxidant and anti-inflammatory activities of isolated compounds (SU-I, SU-II, SU-III) of lichen, *Usnea undulata*, by different methods *viz.* DPPH radical scavenging assay, Hydroxyl radical scavenging assay Nitric oxide scavenging assay and Hydrogen peroxide assay. Three compounds were isolated and characterized. The free radical scavenging activity of the isolated compounds was estimated by IC_{50} values at various concentrations of 20 to 80µg/ml. At 80µg/ml, DPPH radical scavenging assay, Hydroxyl radical scavenging assay, Nitric oxide assay and Hydrogen peroxide assay showed maximum inhibition of 80.31 per cent, 81.20 per cent, 83.83 per cent and 84.31 per cent respectively for the compound SU-I.

Bryophtes

Ceratodon purpureus (Hedw.) Bird. Family: Ditrichaceae

The phenolic content, reducing power, antioxidant and free radical scavenging activities of ethanol extract was determined *in vitro* by Chobot *et al.* (2006). Ethanol extract showeed scavenging and antioxidant activity, which was lower in comparison with caffeic acid. No correlation between the total phenolic content and antioxidant or scavenging activities was found.

Dicranum polysetum Sw. Family: Dicranaceae

The phenolic content, reducing power, antioxidant and free radical scavenging activities of ethanol extract of *Ceratodon purpurens* was determined *in vitro* by Chobot *et al.* (2006). Ethanol extract showed scavenging and antioxidant activity, which was lower in comparison with caffeic acid. No correlation between the total phenolic content and antioxidant or scavenging activities was found.

Dicranum scoparium Hedw. Family: Dicranaceae

The phenolic content, reducing power, antioxidant and free radical scavenging activities of ethanol extract of *Dicranum scoparium* was determined *in vitro* by Chobot *et al.* (2006). Ethanol extract showed scavenging and antioxidant activity, which was lower in comparison with caffeic acid. No correlation between the total phenolic content and antioxidant or scavenging activities was found.

Leucobryum glaucum (Hedw.) Angstr. Family: Leucobryaceae

The phenolic content, reducing power, antioxidant and free radical scavenging activities of ethanol extract was determined *in vitro* by Chobot *et al.* (2006). Ethanol extract showed scavenging and antioxidant activity, which was lower in comparison with caffeic acid. No correlation between the total phenolic content and antioxidant or scavenging activities was found.

Mnium marginatum (With.) P. Beauv. Family: Mniaceae

The phenolic content, reducing power, antioxidant and free radical scavenging activities of ethanol extract was determined *in vitro* by Chobot *et al.* (2006). Ethanol extract showed scavenging and antioxidant activity, which was lower in comparison with caffeic acid. No correlation between the total phenolic content and antioxidant or scavenging activities was found.

Pteridophytes

Abacopteris penangiana (Hook.) Ching Family: Thelypteridaceae

Zhongxiang *et al.* (207) reported about the presence of five new flavan-4-ol glycosides abactopterins E-I and seven known flavonoid glycosides were isolated from the aerial parts of the fern *Abactoperis penangiana*. The isolated glycosides were evaluated for their antioxidant activity using the TEAC assay and compounds showed TEAC values of 1.03-1.91mM.

Actiniopteris semiflbellata Pichi Sermolli Family: Actiniopteridaceae

Methanolic extracts of *Actiniopteris semiflabellata* showed high free radical scavenging activity (Mothana *et al.*, 2010).

Adiantum capillus-veneris L. Family: Adiantaceae

Pourmorad *et al.* (2006) carried out a systematic record of the relative antioxidant activity in selected Iranian medicinal plant species. The total phenol content of *Adiantum capillus-veneris*, measured by Folin Ciocalteu method was 22.3 mg/g GAE. Flavonoid contents in terms of quercetin equivalent was78.3 mg/g. DPPH radical scavenging effect of the extracts was detemined spectrophotometrically. The radical scavenging activity of *Adiantum capillus-veneris* at 4mg/ml concentration was 44 per cent.

Kumar (2009) evaluated the antioxidant potential of leaf extract of *Adiantum capillus-veneris* against hydrogen peroxide induced oxidative damage in peripheral blood lymphocytes. Incubation of peripheral blood lymphocytes with 100 µM H_2O_2 for 2 hours significantly increased lipid peroxidation and decreased the levels of glutathione and antioxidant enzymes (SOD, CAT, GPx). Pretreatment with plant leaves extract for 18 hours could effectively inhibit lipid peroxidation and enhance the activities of antioxidant enzymes and glutathione content significantly.

Adiantum pedatum L. Family: Adiantaceae

The antioxidant activity of four organic solvent extracts of *Adiantum pedatum* was evaluated by Chandrappa *et al.* (2011) by DPPH and FRAP methods. The extracts exhibited antioxidant activity.

Adiantum trapeziforme L. Family: Adiantaceae

Limaye *et al.* (2010) investigated the antioxidant activity of fronds of *Adiantum trapeziforme* by free radical scavenging DPPH method, by FRAP and hydroxyl mediated DNA damage. The results of their investigation showed that frond extracts contain high amount of ascorbic acid, *i.e.*, 1520.24 mg/g and 1770.80mg/g at vegetative and reproductive stages respectively. A linear correlation was observed between phenolic content and the reducing power. The extract also showed good antioxidant activity in DPPH assay and hydroxyl mediated DNA damage assay.

Allantodia himalayensis (C. B. clarke) Ching Family: Woodsiaceae

Ding *et al.* (2008) investigated the phenolic contents and radical scavenging capacity of *Allantodia himalayensis*. Total phenolic contents were determined by the Folin-Ciocalteu colorimetric method using gallic acid as a standard. Antioxidant activity was assessed in DPPH radical scavenging system using lutin as a positive control. The total phenolic content in *Allantodia himalayensis* was 0.16 and IC_{50} (µg/ml) values for antioxidant activity was 133.2.

Alsophila costularis Baker Family: Cyatheaceae

Ding *et al.* (2008) investigated the phenolic contents and radical scavenging capacity of *Alsophila costularis*. Total phenolic contents were determined by the Folin-Ciocalteu colorimetric method using gallic acid as a standard. Antioxidant activity was assessed in DPPH radical scavenging system using lutin as a positive control. The total phenolic content in *Alsophila costularis* was 0.79 and IC_{50} (µg/ml) values for antioxidant activity was 130

Angiopteris evecta (J.R. Forst.) Hoffmann Family: Angiopteridaceae

Povichit *et al.* (2010) investigated phenolic content and free radical scavenging effect of *Angiopteris evecta* by DPPH and ABTS assays, anti-lipid peroxidation activity by TBARS and for antiglycation activity. The results revealed that the total phenolic content showed good correlation with free radical scavenging by ABTS and anti-lipid peroxidation by TBARS, but showed no correlation with antiglycation. The extract of *Angiopteris evecta* moderate antioxidant effect and also showed a promising antiglycation effect. The IC_{50} were 0.42mg/ml for the DPPH method, TEAC value of

0.36 mg/Trolox/mg sample for the ABTS method, IC_{50} of 7.30 mg/ml for the TBARS method and IC_{50} of 0.28 for the antiglycation method.

Arachniodes exilis (Hance) Ching Family: Dryopteridaceae

Daonian *et al.* (2010) studied *Arachniodes exilis* for the antioxidant and hepatoprotective activity by different assays, including reducing power, lipid peroxidation, DPPH, ABTS, superoxide anion, hydroxyl radicals and hydrogen peroxide scavenging.

Arthromeris wallichiana (Spreng.) Ching Family: Polypodiaceae

Ding *et al.* (2008) investigated the phenolic contents and radical scavenging capacity of *Arthromeris wallichiana*. Total phenolic contents were determined by the Folin-Ciocalteu colorimetric method using gallic acid as a standard. Antioxidant activity was assessed in DPPH radical scavenging system using lutin as a positive control. The total phenolic content in *Arthromeris wallichiana* was 1.57 and IC_{50} (μg/ml) values for antioxidant activity was 66.2.

Asplenium nudus L. Family: Aspleniaceae

English Name: Bird's nest fern; Japanese: Shimaoutaniwatari.

Maeda *et al.* (2006) examined the phenolic contents of young leaf of *Asplenium nudus* and its radical scavenging activities for DPPH radical. The polyphenol content was 55 mg-gallic acid equivalent/100 g and DPPH scavenging activity was 177.6 μmol-Trolox eq./100 g).

Blechnum orientale L. Family: Blechnaceae

Leaf extract of *Blechnum orientale* was investigated for its total phenolic content (TPC), and antioxidative activities by Lai *et al.* (2009). The antioxidative activity was measured by assays for radical scavenging against DPPH, ferric ion reducing power (FRP), β-carotene bleaching (BCB) and ferrous ion chelating (FIC). The results revealed *B. orientale* to possess the highest amount of total polyphenols and strongest potential as a natural antioxidative agent.

Brainea insignis (Hook.) J.Sm. Family: Blechnaceae

Ding *et al.* (2008) investigated the phenolic contents and radical scavenging capacity of *Brainea insiginis*. Total phenolic contents were determined by the Folin-Ciocalteu colorimetric method using gallic acid as a standard. Antioxidant activity was assessed in DPPH radical scavenging system using lutin as a positive control. The total phenolic content in *Brainea insignis* was 2.56 and IC_{50} (μg/ml) values for antioxidant activity was 6.1.

Camptosorus sibiricus Rupr. Family: Aspleniaceae

Flavonoids and organic acids were recommended in the literature as the main active constituents of *Camptosorus sibiricus*. Activity-guided fractionation led to the isolation of 9 flavonoids and 8 phenolic acids. All compounds were tested for DPPH scavenging activity, SOD-like and aldose reductase inhibition. Among them 1,2,3,5,6,7,8,9,11, 15 showed activities (Li *et al.*, 2008). The most active free radical scavenger and antioxidant was compound 8, while compound 1 exhibited strong inhibiting activity of aldose reductase.

Cheilanthus glauca (Cav.) Mett. Family: Sinopteridaceae

English Name: Doradilla.

Antioxidant capacity of freeze dried extract of *Cheilanthus glauca* was investigated by Pastene *et al.* (2007) through different tests, *i.e.*, DPPH, ABTS radical scavenging activities, FRAP, ORA, lipid peroxidation in rat liver microsomes. Doradilla extracts exhibited a strong antioxidant capacity, which was concentration dependent in all assays. To ascertain the possible explanation for this potent activity, the percentages of rutin and total flavonoids were estimated, found values of 3.50 per cent and 11.70 per cent respectively.

Cibotium barometz (L.) J.Sm. Family: Dicksoniaceae

Leaf extract of *Cibotium barometz* was investigated for its total phenolic content (TPC), and antioxidative activities by Lai *et al.* (2009). The antioxidative activity was measured by assays for radical scavenging against DPPH, ferric ion reducing power (FRP), β-carotene bleaching (BCB) and ferrous ion chelating (FIC). The results revealed *C. barometz* to possess antioxidative activity.

Cyathea phalerata Mart. Family: Cyatheaceae

Hort *et al.* (2008) investigated the antioxidant properties of *Cyathea phalerata* using *in vitro* and *in vivo* assays. The *in vitro* antioxidant potential of the crude extract (CE), precipitate (PPT), aqueous fraction (AQF), n-Butanolic fraction (BUF) and ethyl acetate fraction (EAF) was evaluated through the scavenging of DPPH, superoxide anion and OH radicals, and lipid peroxidation in rat liver homogenate. In these assays, it was observed that EAF had marked antioxidant potential, especially as a scavenger of the OH radical and in inhibiting lipid peroxidation.

Cyclogramma neoauriculata (Ching) Tagawa Family: Thelypteridaceae

Ding *et al.* (2008) investigated the phenolic contents and radical scavenging capacity of *Cyclogramma neoauriculata*. Total phenolic contents were determined by the Folin-Ciocalteu colorimetric method using gallic acid as a standard. Antioxidant activity was assessed in DPPH radical scavenging system using lutin as a positive control. The total phenolic content in *Cyclogramma neoauriculata* was 0.95 and IC_{50} (μg/ml) values for antioxidant activity was 20.2.

Cyrtomium hookerianum (C. Presl) C. Chr. Family: Dryopteridaceae

Ding *et al.* (2008) investigated the phenolic contents and radical scavenging capacity of *Cyrtomium hookerianum*. Total phenolic contents were determined by the Folin-Ciocalteu colorimetric method using gallic acid as a standard. Antioxidant activity was assessed in DPPH radical scavenging system using lutin as a positive control. The total phenolic content in *Cyrtomium hookerianum* was 0.82 and IC_{50} (μg/ml) values for antioxidant activity was 30.1.

Davallia divaricata Bl. Family: Davalliaceae

Chang *et al.* (2007) evaluated the antioxidant activities and polyphenol contents of six folk medicinal ferns used as "Gusuibu". Ethanol and aqueous extracts of these six sources were characterized for their antioxidant, scavenging activities, reducing power, total polyphenols, flavonols, flavonoids, condensed tannins, and proanthocyanidin contents. Most samples in aqueous extracts had higher antioxidant potencies and polyphenol contents than the ethanol extracts. Extracts of *Davillea divaricata* showed moderate radical scavenging activity.

Davallia mariesii Moore ex Bak. Family: Davalliaceae

The folk remedy Gusuibu, Chinese medicine is obtained from six different ferns. Ethanol and aqueous extracts of six ferns were evaluated and characterized by Chang *et al.* (2007) for their various activities and components. Results showed that most samples in aqueous extracts had higher antioxidant potencies and polyphenol contens than the ethanol extracts, indicating that the aqueous preparation of Gusuibu is more potent than the ethanol. EC_{50} values of reducing capacities, and scavenging activities against DPPH radicals showed significant variation among the six ferns within ethanol or aqueous extracts and between the two solvents. The maximum (1.27) Trolox Equivalent Antioxidant Capacity was recorded in aqueous extract of fern *Davallia mariesii. Davallia mariesii* had a higher degree of reducing power (aqueous extract $EC_{50} = 26.58$ µg/ml).

Davallia solida (Forst.) Sw. Family: Davalliaceae

Chang *et al.* (2007) evaluated the antioxidant activities and polyphenol contents of six folk medicinal ferns used as "Gusuibu". Ethanol and aqueous extracts of these six sources were characterized for their antioxidant scavenging activities, reducing power, total polyphenols, flavonols, flavonoids, condensed tannins, and proanthocyanidin contents. Most samples in aqueous extracts had higher antioxidant potencies and polyphenol contents than the ethanol extracts. Extracts of *Davallia solida* had highest radical scavenging activity (expressed as µg extract/ml (26.89).

Davallia solida rhizome contains a high content of phenolic compound and showed a strong DPPH radical scavenging activity (Y.-H. Chen *et al.*, 2008)

Dicranopteris linearis (Burm.f.) Underw. Family: Gleicheniaceae

Leaf extract of *Dicranopteris linearis* was investigated for its total phenolic content (TPC), and antioxidative activities by Lai *et al.* (2009). The antioxidative activity was measured by assays for radical scavenging against DPPH, ferric ion reducing power (FRP), β-carotene bleaching (BCB) and ferrous ion chelating (FIC). The results revealed *D. linearis* to possess antioxidative activity.

Zakaria (2007) evaluated the free radical scavenging properties of *Dicranopteris linearis* using DPPH and superoxide anion radical scavenging assays. The extract was found to show remarkable antioxidant activity in both assays. The DPPH radical scavenging activity was 96.60 per cent and superoxide scavenging activity was 89.40 per cent. Phytochemicals screening of this plant demonstrated the presence of flavonols, triterpenes, tannins, saponins and steroids.

Diplazium esculentum **(Retz.) Sw. Family: Polypodiaceae**

Wong *et al*. (2006) investigated the antioxidant properties of 25 edible tropical plants including *Diplazium esculentum* using DPPH and FRAP assays. Total phenolic content was also estimated. *Diplazium esculentum* showed moderate antioxidant activity.

Drynaria fortunei **(Kunze) J.Sm. Family: Polypodiaceae**

Cai *et al*. (2004) evaluated the antioxidant activity and phenolic compounds in traditional Chinese medicinal plants associated with anticancer. The improved ABTS.+ method was used to systematically assess the total antioxidant capacity (Trolox equivalent antioxidant capacity, TEAC) of the medicinal extracts. The TEAC values and total phenolic content for methanolic extract of root of *Drynaria fortunei* were 128.7 µmol Trolox equivalent/100g dry weight (DW), and 0.34 g of gallic acid equivalent/100g DW respectively. A positive significant linear relationship between antioxidant activity and total phenolic content showed that phenolic compounds were the dominant antioxidant components in the tested herbs. Major types of phenolic compounds in *Drynaria fortunei* include flavonones (dihydroflavononе, naringin).

Chang *et al*. (2007) evaluated the antioxidant activities and polyphenol contents of six folk medicinal ferns used as "Gusuibu". Ethanol and aqueous extracts of these six sources were characterized for their antioxidant, scavenging activities, reducing power, total polyphenols, flavonols, flavonoids, condensed tannins, and proanthocyanidin contents. Extracts of *Drynaria fortunei* had lower radical scavenging activity.

Dryopsis heterolaena **(C. Chr.) Holttum et P.J. Edwards Family: Dryopteridaceae**

Ding *et al*. (2008) investigated the phenolic contents and radical scavenging capacity of *Dryopsis heterolaena*. Total phenolic contents were determined by the Folin-Ciocalteu colorimetric method using gallic acid as a standard. Antioxidant activity was assessed in DPPH radical scavenging system using lutin as a positive control. The total phenolic content in *Dryopsis heterolaena* was 1.90 and IC_{50} (µg/ml) values for antioxidant activity was 12.5.

Dryopsis mariformis **(Rosenst.) Holttum et P.J. Edwards Family: Dryopteridaceae**

Ding *et al*. (2008) investigated the phenolic contents and radical scavenging capacity of *Dryopsis mariformis*. Total phenolic contents were determined by the Folin-Ciocalteu colorimetric method using gallic acid as a standard. Antioxidant activity was assessed in DPPH radical scavenging system using lutin as a positive control. The total phenolic content in *Dryopsis mariformis* was 1.34 and IC_{50} (µg/ml) values for antioxidant activity was 17.8.

Dryopteris conjugata **Ching Family: Dryopteridaceae**

Ding *et al*. (2008) investigated the phenolic contents and radical scavenging capacity of *Dryopteris conjugata*. Total phenolic contents were determined by the Folin-Ciocalteu colorimetric method using gallic acid as a standard. Antioxidant activity was assessed in DPPH radical scavenging system using lutin as a positive control. The total phenolic content in *Dryopteris conjugata* was 1.17 and IC_{50} (μg/ml) values for antioxidant activity was 15.5.

Dryopteris crassirhizoma Nakai Family: Dryopteridaceae

The rhizome of *Dryopteris crassirhizoma* exhibited significant antioxidant activity, as assessed by the DPPH radical scavenging activity *in vitro* by Lee *et al.* (2003). Two phloroglucinol derivatives, flavaspidic acids PB and AB, isolated from the rhizome were tested for their antioxidant activities, such as their DPPH radical scavenging, superoxide radical scavenging and lipid peroxidation inhibitory activities. These compounds exhibited potent antioxidant activity against the LPO inhibitory test with IC_{50} values of 12.9 and 13.1 μM, respectively, compared with α-tocopherol (IC_{50} 15.6 μM) and BHA (IC_{50} 10.8 μM) while the two compounds had a moderate effect on the DPPH radical scavenging activity (IC_{50} 71.7, 76.3 μM) as well as superoxide radical scavenging activity (IC_{50} 58.6, 64.4 μM). The potent activity of the flavaspidic acids on inhibiting LPO might be due to possible stabilization as a result of chelating iron (Lee *et al.*, 2003).

Dryopteris neorosthornii Ching Family: Dryopteridaceae

Ding *et al.* (2008) investigated the phenolic contents and radical scavenging capacity of *Dryopteris neorosthornii*. Total phenolic contents were determined by the Folin-Ciocalteu colorimetric method using gallic acid as a standard. Antioxidant activity was assessed in DPPH radical scavenging system using lutin as a positive control. The total phenolic content in *Dryopteris neorosthornii* was 1.92 and IC_{50} (μg/ml) values for antioxidant activity was 18.1.

Equisetum arvense L. Family: Equisetaceae

English Name: Field horsetail, Horsetail.

Hepatoprotective and free radical scavenging activities of phenolic petrosins and flavonoids isolated from *Equisetum arvense* were investigated by Oh *et al.* (2004). Petrosins Onitin (1), onitin-9-O-glucoside (2), flavonoids apigenin (3), luteolin (4) isolated from the MeOH extracts showed superoxide scavenging effects (IC_{50} = 35.3 μM and 5.9μM, respectively) and DPPH free radical scavenging effect (IC_{50} of 35.8 and 22.7μM, respectively).

Stajner *et al.* (2006) evaluated the scavenger activities of above ground parts of *Helminthostachys zeylamica* in phosphate buffer using three different methods: DPPH assay, ESR and NO radical inhibition assay. Total reducing power was determined by FRAP assay. DPPH scavenger capacity was 51.4 per cent, NO inhibition was 9.13, Ferric reducing antioxidant power was 2.85 and scavenger effect was 73.5 per cent.

Mimica-Dukic *et al.* (2008) presented the antioxidant activity and phenolic composition of three different extracts (EtOH, n-BuOH and H_2O) of field horsetail. The antioxidant activity has been evaluated measuring the total reducing power, inhibition of lipid peroxidation and free radical scavenging capacity towards DPPH radical and nitric oxide, respectively. The results obtained show that the highest radical scavenging capacity regarding both DPPH and NO radicals is expressed by

EtOAc extract (EC_{50} = 2.37µg/ml and EC_{50} = 90.07µg/ml, respectively) and the lowest by H_2O extract. N-BuOH extract showed the highest total reducing power (AEAC = 13.40µg/ml). Isoquercetin was the main flavonoid. In H_2O extract high content of phenolic acids and low percentage of flavonoids were detected.

Cetojevic-Simon *et al.* (2010) investigated antioxidative and antiproliferative activity of different horsetail (*Equisetum arvense*) extracts. The antioxidative activity was measured by the electron spin resonance (ESR) spectroscopy-spin trapping method. The influence of different horsetail extracts during lipid peroxidation of (1) sunflower oil induced by the lipophilic azo-initiator 4,4'-azobis(4-cyanovaleric acid) and (2) soybean phosphatidylcholine liposomes induced by the hydrophilic azo-initiator 2,2'-azobis(2-amidinopropane) dihydrochloride was studied. The results of ESR analysis confirmed that the extracts investigated suppressed the formation of lipid peroxyl radicals in both systems investigated in a dose-dependent manner. The results indicate that n-butanol, methanol, ethyl acetate, and water extracts had significant peroxyl radical scavenging activity. Extracts inhibited cell growth that was dependent on cell line, type of extract, and extract concentration.

Equisetum hiema L. Family: Equisetaceae

Antioxidant capacities of 56 selected Chinese medicinal plants, including *Equisetum hiema* were evaluated by Song *et al.* (2010) using the TEAC and FRAP assays, and their total phenolic content was measured by the Folin-Ciocaleteu method. *Equisetum hiema* showed low antioxidant activity with TEAC value of 10.66µmol Trolox/g, FRAP value of 13.79µmol Fe^{2+}/g and phenolic content was 52.68mg GAE/g.

Equisetum maximum Lam. Family: Equisetaceae

Pourmorad *et al.* (2006) carried out a systematic record of the relative antioxidant activity in selected Iranian medicinal plant species. The total phenol content of *Equisetum maximum*, measured by Folin Ciocalteu method was 54.5 mg/g GAE. Flavonoid contents in terms of quercetin equivalent was 36.2 mg/g. DPPH radical scavenging effect of the extracts was detemined spectrophotometrically. The radical scavenging activity of *Equisetum maximum* at 0.8mg/ml concentration was 89.6 per cent.

Equisetum ramosissimum Desf. Family: Equisetaceae

Stajner *et al.* (2006) evaluated the scavenger activities of above ground parts of *Equisetum ramosisimum* in phosphate buffer using three different methods: DPPH assay, ESR and NO radical inhibition assay. Total reducing power was determined by FRAP assay. DPPH scavenger capacity was 63.2 per cent, NO inhibition was 3.48, Ferric reducing antioxidant power was 5.44 and scavenger effect was 97.8 per cent.

Equisetum telmateica Ehrh. Family: Equisetaceae

Stajner *et al.* (2006) evaluated the scavenger activities of above ground parts in phosphate buffer using three different methods: DPPH assay, ESR and NO radical inhibition assay. Total reducing power was determined by FRAP assay. DPPH scavenger capacity was 90.6 per cent, NO inhibition was 34.56, Ferric reducing antioxidant power was 44.1 and scavenger effect was 98.9 per cent. The extract showed highest antioxidant activity.

Helminthostachys zeylanica Hook.f. Family: Ophioglossaceae

Three new cylclized geranyl stilbenes, ugonstilbene A, ugonstilbene B, ugonstilbene C and one known compound 3-hydroxyacetophenone, were isolated from the dried rhizome of *Helminthostachys zeylanica* by Chen *et al.* (2003). All the compounds exhibited moderate antioxidant activity by the DPPH assay with IC_{20} of 11.31, 38.72 and 30.80μM, respectively, while the IC_{20} of Trolox was 8.70 μM.

Humata griffithiana (Hk.) C. Chr. Family: Davalliaceae

Chang *et al.* (2007) evaluated the antioxidant activities and polyphenol contents of six folk medicinal ferns used as "Gusuibu". Ethanol and aqueous extracts of these six sources were characterized for their antioxidant, scavenging activities, reducing power, total polyphenols, flavonols, flavonoids, condensed tannins, and proanthocyanidin contents. Most samples in aqueous extracts had higher antioxidant potencies and polyphenol contents than the ethanol extracts. Extracts of *Humata griffithiana* had higher radical scavenging activity.

Humata platylepis (Bak.) Ching Family: Davalliaceae

Ding *et al.* (2008) investigated the phenolic contents and radical scavenging capacity of *Humata platylepis*. Total phenolic contents were determined by the Folin-Ciocalteu colorimetric method using gallic acid as a standard. Antioxidant activity was assessed in DPPH radical scavenging system using lutin as a positive control. The total phenolic content in *Humata platylepis* was 2.36 and IC_{50} (μg/ml) values for antioxidant activity was 23.7.

Lindsaea odorata Roxb. Family: Lindseaceae

Ding *et al.* (2008) investigated the phenolic contents and radical scavenging capacity of *Lindsaea odorata*. Total phenolic contents were determined by the Folin-Ciocalteu colorimetric method using gallic acid as a standard. Antioxidant activity was assessed in DPPH radical scavenging system using lutin as a positive control. The total phenolic content in *Lindsaea odorata* was 0.34 and IC_{50} (μg/ml) values for antioxidant activity was 47.3.

Lygodium flexuosum (L.) Sw. Family: Lygodiaceae

In vitro antioxidant activity of different extracts of plant *Lygodium flexuosum* were evaluated by Jeetedra and Manish (2011) using models of DPPH radical scavenging activity, total antioxidant capacity, nitric oxide scavenging activity, iron chelating activity and hydroxyl radical scavenging activity. Phenolic contents of extracts were estimated as GAE by Folin-Ciocalteu method and correlated with antioxidant activity. Highest phenolic content (6.24 per cent) was found in methanolic extract with highest antioxidant activity. The fractionation and characterization of methanolic extract showed presence of quercetin.

Lygodium japonicum (Thunb. ex Murr.) Sw. Family: Lygodiaceae

Li *et al.* (2006) evaluated the antioxidant activities of purification polysaccharide from *Lygodium japonicum* by various antioxidant assays. The results indicated that the purification polysaccharides showed strong antioxidant activity against liposome peroxidation, DPPH radical scavenging, superoxide anion radical scavenging and metal chelating activities. This antioxidant activity was concentration dependent.

Matteucia orientalis **(L.) Todaro Family:Onocleaceae**

Cai *et al.* (2004) evaluated the antioxidant activity and phenolic compounds in traditional Chinese medicinal plants associated with anticancer. The improved ABTS.+ method was used to systematically assess the total antioxidant capacity (Trolox equivalent antioxidant capacity, TEAC) of the medicinal extracts. The TEAC values and total phenolic content for methanolic extract of rhizome of *Matteucia orientalis* were 2176.4 µmol Trolox equivalent/100g dry weight (DW), and 10.08 g of gallic acid equivalent/100g DW respectively. A positive significant linear relationship between antioxidant activity and total phenolic content showed that phenolic compounds were the dominant antioxidant components in the tested herbs. Major types of phenolic compounds in *Matteucia orientalis* include m-benzo-triphenol derivatives (filicic acids), tannins.

Ding *et al.* (2008) investigated the phenolic contents and radical scavenging capacity of *Matteucia orientalis*. Total phenolic contents were determined by the Folin-Ciocalteu colorimetric method using gallic acid as a standard. Antioxidant activity was assessed in DPPH radical scavenging system using lutin as a positive control. The total phenolic content in *Matteucia orientalis* was 1.37 and IC_{50} (µg/ml) values for antioxidant activity was 105.1.

Matteuccia struthiopteris **(L.) Todaro Family: Onocleaceae**

English Name: Ostrich fern.

Kimura *et al.* (2004) isolated a caffeic acid derivative from *Matteuccia struthiopteris* as a major radical scavenger.

Nothoperanema hendersonii **(Bedd.) Nakaika Family: Dryopteridaceae**

Ding *et al.* (2008) investigated the phenolic contents and radical scavenging capacity of *Nothoperanema hendersonii*. Total phenolic contents were determined by the Folin-Ciocalteu colorimetric method using gallic acid as a standard. Antioxidant activity was assessed in DPPH radical scavenging system using lutin as a positive control. The total phenolic content in *Nothoperanema hendersonii* was 2.35 and IC_{50} (µg/ml) values for antioxidant activity was 5.9.

Osmunda cinnamomea **L. Family: Osmundaceae**

Ding *et al.* (2008) investigated the phenolic contents and radical scavenging capacity of *Osmunda cinnamomea*. Total phenolic contents were determined by the Folin-Ciocalteu colorimetric method using gallic acid as a standard. Antioxidant activity was assessed in DPPH radical scavenging system using lutin as a positive control. The total phenolic content in *Osmunda cinnamomea* was 1.02 and IC_{50} (µg/ml) values for antioxidant activity was 11.1.

Osmunda claytoniana **L. Family: Osmundaceae**

Ding *et al.* (2008) investigated the phenolic contents and radical scavenging capacity of *Osmunda claytoniana*. Total phenolic contents were determined by the Folin-Ciocalteu colorimetric method using gallic acid as a standard. Antioxidant activity was assessed in DPPH radical scavenging system using lutin as a positive control. The total phenolic content in *Osmunda claytoniana* was 0.66 and IC_{50} (μg/ml) values for antioxidant activity was 110.

Parathelypteris nipponica (Franch. et Sav.) Ching Family: Thelypteridaceae

Fu *et al.* (2010) evaluated the antioxidant, free radical scavenging, hepatoprotective and anti-inflammatory potential of *Parathelypteris nipponica*. Total flavonoids content was 262 mg/g (w/w). The methanolic extract exhibited strong antioxidant activity with EC_{50} values in reductive ability (0.18 mg/ml) and FTC assay (0.10 mg/ml), strong free radical scavenging activity as evidenced by the low EC_{50} values in DPPH (2.00 mg/ml), superoxide anion (0.60 mg/ml), OH radicals (0.26 mg/ml), and hydrogen peroxide (0.45 mg/ml) methods.

Phlegmariurus henryi (Baker) Ching Family: Lycopodiaceae

Ding *et al.* (2008) investigated the phenolic contents and radical scavenging capacity of *Phlegmariurus henryi*. Total phenolic contents were determined by the Folin-Ciocalteu colorimetric method using gallic acid as a standard. Antioxidant activity was assessed in DPPH radical scavenging system using lutin as a positive control. The total phenolic content in *Phlegmariurus henryi* was 1.15 and IC_{50} (μg/ml) values for antioxidant activity was 25.4.

Phymatosorus cuspidatus (Don) Pich. Serm. Family: Polypodiaceae

Ding *et al.* (2008) investigated the phenolic contents and radical scavenging capacity of *Phymatosorus cuspidatus*. Total phenolic contents were determined by the Folin-Ciocalteu colorimetric method using gallic acid as a standard. Antioxidant activity was assessed in DPPH radical scavenging system using lutin as a positive control. The total phenolic content in *Phymatosorus cuspidatus* was 0.93 and IC_{50} (μg/ml) values for antioxidant activity was 43.8.

Plagiogyria maxima C.Chr. Family: Plagiogyriaceae

Ding *et al.* (2008) investigated the phenolic contents and radical scavenging capacity of *Plagiogyria maxima*. Total phenolic contents were determined by the Folin-Ciocalteu colorimetric method using gallic acid as a standard. Antioxidant activity was assessed in DPPH radical scavenging system using lutin as a positive control. The total phenolic content in *Plagiogyrea maxima* was 0.62 and IC_{50} (μg/ml) values for antioxidant activity was 22.1.

Plagiogyria distinctissima Ching Family: Plagiogyriaceae

Ding *et al.* (2008) investigated the phenolic contents and radical scavenging capacity of *Plagiogyria distinctissima*. Total phenolic contents were determined by the Folin-Ciocalteu colorimetric method using gallic acid as a standard. Antioxidant activity was assessed in DPPH radical scavenging system using lutin as a positive control. The total phenolic content in *Plagiogyrea distinctissima* was 0.62 and IC_{50} (μg/ml) values for antioxidant activity was 22.1.

Polypodiodes amoena (Wall. ex Mett.) Ching Family: Polypodiaceae

Ding *et al.* (2008) investigated the phenolic contents and radical scavenging capacity of *Polypodiodes amoena.* Total phenolic contents were determined by the Folin-Ciocalteu colorimetric method using gallic acid as a standard. Antioxidant activity was assessed in DPPH radical scavenging system using lutin as a positive control. The total phenolic content in *Polypodiodes amoena* was 1.05 and IC_{50} (µg/ml) values for antioxidant activity was 86.8.

Polypodiodes mengtzeense (H. Christ) Ching Family: Polypodiaceae

Ding *et al.* (2008) investigated the phenolic contents and radical scavenging capacity of *Polypodiodes mengtzeense.* Total phenolic contents were determined by the Folin-Ciocalteu colorimetric method using gallic acid as a standard. Antioxidant activity was assessed in DPPH radical scavenging system using lutin as a positive control. The total phenolic content in *Polypodiodes mengtzeense* was 2.01 and IC_{50} (µg/ml) values for antioxidant activity was 23.8.

Polypodium leucotomos L. Family: Polypodiaceae

Antioxidant activity of *Polypodium leucotomus* was assessed by Gombau *et al.* (2005). The antioxidant capacity increased in a concentration-dependent manner, with ferulic and caffeic acids as the most powerful antioxidants.

Fernblock, an aqueous extract of the aerial parts of the fern *Polypodium leucotomos,* used as a raw material for topical and oral photoprotective formulations, was fractioned by HPLC and the main components with antioxidant capability were identified by means of UV spectra, electrochemical detection and MSn. Phenolic compounds were identified as 3,4-dihydroxybenzoic acid, 4-hydroxybenzoic acid, vanillic acid, caffeic acid, 4-hydroxycinnamic acid, 4-hydroxycinnamoyl-quinic acid, ferulic acid, and five chlorogenic acid isomers. Total ferric antioxidant capacity (FRAP) of HPLC eluted fractions was measured. The results suggest that the herein identified compounds support, atleast partially, the antioxidant and radical scavenging capacities of Fernblock (Garcia *et al.*, 2006).

Polypodium vulgare L. Family: Polypodiaceae

English Name: Yarrow.

Souri *et al.* (2008) investigated the antioxidant activity and radical scavenging activity of methanolic extract. The antioxidant activity of this plant extract against linoleic acid peroxidation expressed as IC_{50} was 63.48ng/ml and radical scavenging activity against DPPH expressed as IC_{50} was 2.14 µg/ml.

Polystichum pycnopterum (H.Christ.) Ching ex W.M.Chu et Z.R.He Family: Dryopteridaceae

Ding *et al.* (2008) investigated the phenolic contents and radical scavenging capacity of *Polystichum pycnopterum.* Total phenolic contents were determined by the Folin-Ciocalteu colorimetric method using gallic acid as a standard. Antioxidant activity was assessed in DPPH radical scavenging system using lutin as a positive control. The total phenolic content in *Polystichum pycnopterum* was 1.08 and IC_{50} (µg/ml) values for antioxidant activity was 14.2.

Polystichum semifertile **(C.B. Clarke) Ching Family: Dryopteridaceae**

Ding *et al.* (2008) investigated the phenolic contents and radical scavenging capacity of *Polystichum semifertile*. Total phenolic contents were determined by the Folin-Ciocalteu colorimetric method using gallic acid as a standard. Antioxidant activity was assessed in DPPH radical scavenging system using lutin as a positive control. The total phenolic content in *Polystichum semifertile* was 1.37 and IC_{50} (µg/ml) values for antioxidant activity was 5.44.

Pseudodrynaria coronans **(Wall. ex Mett.) Ching Family: Polypodiaceae**

Chang *et al.* (2007) evaluated the antioxidant activities and polyphenol contents of six folk medicinal ferns used as "Gusuibu". Ethanol and aqueous extracts of these six sources were characterized for their antioxidant, scavenging activities, reducing power, total polyphenols, flavonols, flavonoids, condensed tannins, and proanthocyanidin contents. Most samples in aqueous extracts had higher antioxidant potencies and polyphenol contents than the ethanol extracts. Extracts of *Pseudodrynaria coronans* had lower radical scavenging activity.

Pteridium aquilinum **(L.) Kuhn Family: Dennstaedtiaceae**

Xu *et al.* (2009) evaluated the antioxidant activity of a water-soluble polysaccharide purified from *Pteridium aquilinum*. The purified polysaccharide was demonstrated to have strong reductive power (FRAP value: 827.6 micromol/L), moderate scavenging activities against DPPH radicals (83.1 per cent) and superoxide radicals (60.5 per cent), and moderate inhibiting power for self-oxidation of 1,2,3-phentriol (52.4 per cent).

Pteris multifida **Poir. Family: Pteridaceae**

Cai *et al.* (2004) evaluated the antioxidant activity and phenolic compounds in traditional Chinese medicinal plants associated with anticancer. The improved ABTS.+ method was used to systematically assess the total antioxidant capacity (Trolox equivalent antioxidant capacity, TEAC) of the medicinal extracts. The TEAC values and total phenolic content for methanolic extract of whole plant of *Pteris multifida* were 103 µmol Trolox equivalent/100g dry weight (DW), and 0.72 g of gallic acid equivalent/100g DW respectively. A positive significant linear relationship between antioxidant activity and total phenolic content showed that phenolic compounds were the dominant antioxidant components in the tested herbs. Major types of phenolic compounds in *Pteris multifida* include flavonols (quercetin), tannins.

Pteris yunnanensis **H. Christ. Family: Pteridaceae**

Ding *et al.* (2008) investigated the phenolic contents and radical scavenging capacity of *Pteris yunnanensis*. Total phenolic contents were determined by the Folin-Ciocalteu colorimetric method using gallic acid as a standard. Antioxidant activity was assessed in DPPH radical scavenging system using lutin as a positive control. The total phenolic content in *Pteris yunnanensis* was 2.40 and IC_{50} (µg/ml) values for antioxidant activity was 86.9.

Pyrrosia lingua (Thunb.) Farw. Family: Polypodiaceae

Ding *et al.* (2008) investigated the phenolic contents and radical scavenging capacity of *Pyrrosia lingua*. Total phenolic contents were determined by the Folin-Ciocalteu colorimetric method using gallic acid as a standard. Antioxidant activity was assessed in DPPH radical scavenging system using lutin as a positive control. The total phenolic content in *Pyrrosia lingua* was 1.75 and IC_{50} (μg/ml) values for antioxidant activity was 43.0.

Cai *et al.* (2004) evaluated the antioxidant activity and phenolic compounds in traditional Chinese medicinal plants associated with anticancer. The improved ABTS.+ method was used to systematically assess the total antioxidant capacity (Trolox equivalent antioxidant capacity, TEAC) of the medicinal extracts. The TEAC values and total phenolic content for methanolic extract of leaf of *Pyrrosia lingua* were 233 μmol Trolox equivalent/100g dry weight (DW), and 1.65 g of gallic acid equivalent/100g DW respectively. A positive significant linear relationship between antioxidant activity and total phenolic content showed that phenolic compounds were the dominant antioxidant components in the tested herbs. Major types of phenolic compounds in *Pyrrosia lingua* include phenolic acids (caffeic acid), anthraquinones and tannins.

Pyrrosia petiolosa (Christ.) Ching Family: Polypodiaceae

The ethanol extract of *Pyrrosia petiolosa* showed an antioxidant activity as determined by free radical scavenging, superoxide radical scavenging, lipid peroxidation, and hydroxyl radical-induced DNA strand scission assays (Hsu, 2008).

Pyrrosia sheaeri (Bak.) Ching. Family: Polypodiaceae

Ding *et al.* (2008) investigated the phenolic contents and radical scavenging capacity of *Pyrrosia sheaeri*. Total phenolic contents were determined by the Folin-Ciocalteu colorimetric method using gallic acid as a standard. Antioxidant activity was assessed in DPPH radical scavenging system using lutin as a positive control. The total phenolic content in *Pyrrosia sheaeri* was 1.87 and IC_{50} (μg/ml) values for antioxidant activity was 31.0.

Gan *et al.* (2010) evaluated the antioxidant activity of *Pyrrosia sheari* using the FRAP and TEAC assays, and its total phenolic content was measured by the Folin-Ciocalteu method. For FRAP assay the value was 316.72 μmol Fe(II)/g dry weight. For TEAC assay, the value was 147.26μmol Trolox/g dry weight. Total phenolic content was 20.29 mg gallic acid equivalent/g dry weight of plant material.

Pyrrosia subfuracea Family: Polypodiaceae

Ding *et al.* (2008) investigated the phenolic contents and radical scavenging capacity of *Pyrrosia subfuracea*. Total phenolic contents were determined by the Folin-Ciocalteu colorimetric method using gallic acid as a standard. Antioxidant activity was assessed in DPPH radical scavenging system using lutin as a positive control. The total phenolic content in *Pyrrosia subfuracea* was 2.54 and IC_{50} (μg/ml) values for antioxidant activity was 14.5.

Sphaeropteris brunoniana (Hook.) Tryon Family: Cyatheaceae

Ding *et al.* (2008) investigated the phenolic contents and radical scavenging capacity of *Sphaeroteris brunoniana*. Total phenolic contents were determined by the Folin-Ciocalteu colorimetric method using gallic acid as a standard. Antioxidant activity was assessed in DPPH radical scavenging system using lutin as a positive control. The total phenolic content in *Sphaeropteriss brunoniana* was 2.92 and IC_{50} (µg/ml) values for antioxidant activity was 33.1.

Woodwardia japonica (Hook.) Tryon Family: Blechnaceae

Ding *et al.* (2008) investigated the phenolic contents and radical scavenging capacity of *Woodwardia japonica*. Total phenolic contents were determined by the Folin-Ciocalteu colorimetric method using gallic acid as a standard. Antioxidant activity was assessed in DPPH radical scavenging system using lutin as a positive control. The total phenolic content in *Woodwardia japoinica* was 3.14 and IC_{50} (µg/ml) values for antioxidant activity was 7.5.

Woodwardia unigemmata (Makino) Nakai Family: Blechnaceae

Ding *et al.* (2008) investigated the phenolic contents and radical scavenging capacity of *Woodwardia unigemmata*. Total phenolic contents were determined by the Folin-Ciocalteu colorimetric method using gallic acid as a standard. Antioxidant activity was assessed in DPPH radical scavenging system using lutin as a positive control. The total phenolic content in *Woodwardia unigemmata* was 3.05nd IC_{50} (µg/ml) values for antioxidant activity was 7.2

GYMNOSPERMS

Abies alba Mill. Family: Pinaceae

The antiradical capacity was evaluated by measuring the scavenging activity by Yang *et al.* (2009) of the essential oil of *Abies alba* on the DPPH and ABTS radicals. The oil was able to reduce the both radicals dose-dependently, and the concentration required for 50 per cent reduction (RC_{50}) against DPPH radicals (2.7 per cent) was lower than ABTS radicals (8.5 per cent).

Abies pindrow Royle Family: Pinaceae

Kalim *et al.* (2010) evaluated the antioxidant activity of Methanol (50 per cent) extract of *Abies pindrow*. IC_{50} values for DPPH, ABTS, NO, OH, O_2 and ONOO were 84.23, 19.10, 31.43, 286.59 74.54 and 987.42µg/ml. The extract contained significant amount of polyphenols.

Dichloromethane, methanol and acetone extracts of *Abies pindrow* leaves were analysed by Gupta *et a!.* (2011) for their phytochemical content and *in vitro* antioxidant activities. The methanol extract exhibited highest antioxidant activity while acetone extract showed presence of relatively high total phenol and flavonoids contents.

Araucaria angustifolia (Bertol) Kuntze Family: Araucariaceae

Diverse flavonoids such as quercetin, rutin, gingettin, amento-flavone and tetra-O-methylamentoflavone isolated from needles of *Aruacaria angustifolia* decreases, in a dose-dependent form, lipid peroxidation levels in phosphatidylcholine liposomes (Yamaguchi *et al.*, 2009).

The antioxidant activities of the extract and its relevant fraction as well as isolated compounds from the dead bark of *Araucaria angustifolia* were investigated by Secon *et al.* (2010). Using several *in vitro* models Secon *et al.* (2010) described the antioxidant potential of the crude extract and its component ethyl acetate fraction and also of some compounds isolated from the ethyl acetate fraction. They provided the first description of the isolation of two natural product afzelechin derivatives. The extract and isolated compounds, particularly epiafzelechin protocatechuate (5), displayed very high antioxidant activity, as did compound 4, quercetin, which is well known for its antioxidant properties. In a DPPH assay, compound 5 exhibited an IC(50) of 0.7 μM; in a lipid peroxidation assay; IC(50) values of 21 μM and 35 μM were obtained when the oxidation was induced by UV and ascorbate free radicals, respectively.

Araucaria excelsa Lamb. Family: Araucariaceae

English Name: Norfolk pine.

Chemical investigations by Michael *et al.* (2010) from a foliar extract of *Araucaria excelsa* resulted in the identification of seven phenolic metabolites including one flavananol, one flavananol 3-O-glycoside, four C-glycoside flavonoids, and one phenolic acid. Moderate antioxidant activity was observed with a DPPH free radical scavenging assay in comparison with the reference antioxidant ascorbic acid.

Biota orientalis (L.) Endl. Family: Cupressaceae

Gan *et al.* (2010) evaluated the antioxidant activities of 40 medicinal plants associated with prevention and treatment of cardiovascular and cerebrovascular diseases using FRAP and TEAC assays and their total phenolic contents were measured by the Folin-Ciocalteu methods. The antioxidant activity of *Biota orientalis* was 181.64μmol Fe(II)/g in FRAP assay, 135.95μmol Trolox/g and Total phenolic content was 9.12 mg GAE/g. A high correlation between antioxidant capacity and total phenolic content indicated that phenolic compounds could be the main contributor of the antioxidant activity of these plants.

Calocedrus formosana Florin Family: Cupressaceae

Wang *et al.* (2004) evaluated the antioxidant activity of heartwood, bark and leaf of *Calocedrus formosana* using DPPH radical and superoxide anion scavenging activitiess. Similar IC_{50} values against the DPPH radical were found for the heartwood and bark extracts at approximately 23 μg/ml. Moreover, the heartwood extract exhibited the highest inhibitory activity against superoxide radicals among the test samples; a 2.3-fold lower value of IC_{50} for superoxide radical inhibition was found in the heartwood extract relative to that of (+)-catechin. Much less effect on inhibition of DPPH and superoxide radicals was found from the leaf extract of *C. formosana*. More than 70 per cent of superoxide radicals were inhibited in the presence of 10 μg/ml heartwood extract, whereas only 15 per cent inhibition was obtained from the leaf extract.

Calocedrus macrolepis Kurz. Family: Cupressaceae

Chen *et al.* (2009) studied the antioxidant activity of leaves extract of *Calocedruss macrolepis* towards xanthine oxidase, lipoxygenase and tyrosinase activities. The inhibitory effect of acetone extract on xanthine oxidase activity was nil, tyrosine activity was about 40 per cent and lipoxygenase activity was about 60 per cent.

Cedrus deodora (Roxb.) Loud. Family: Pinaceae

An activity-directed fractionation and purification process was used by Tiwari *et al.* (2001) to identify the antioxidant components of *Cedrus deodara*. The chloroform extract of dried heartwood showed strong antioxidant activity on DPPH free radical. Three compounds with potent antioxidant activity were isolated in significant yields. They were identified as (-)-matairesinol, (-)-nortrachelogenin, and a dibenzylbutyrolactollignan (4,4',9-trihydroxy-3,3'-dimethoxy-9,9'-epoxylignan).

Chamaecyparis formosensis (Matsumura) A.Henry Family: Cupressaceae

Chen *et al.* (2009) studied the antioxidant activity of leaves extract towards xanthine oxidase, lipoxygenase and tyrosinase activities. The inhibitory effect of acetone extract on xanthine oxidase activity was less than 40 per cent, tyrosinase activity was about 60 per cent and lipoxygenase activity was more than 70 per cent.

Chamaecyparis obtusa (Hinoki) Mattress var. _formosana_ (Hayata) Rehder Family: Cupressaceae

Chen *et al.* (2009) studied the antioxidant activity of leaves extract towards xanthine oxidase, lipoxygenase and tyrosinase activities. The inhibitory effect of acetone extract on xanthine oxidase activity was nil, tyrosine activity was more than 30 per cent and lipoxygenase activity was about 50 per cent.

Cupressus arizonica Greene Family: Cupressaceae

The essential oils of different parts of *Cupressus arizonica* were examined by Eamami *et al.* (2010) for their potential radical scavenging activity. Fresh leaves and fruits of *C. arizonica* were steam distilled and the oils were analyzed qualitatively and quantitatively by GC and GC/MS. Different methods of rapid TLC screening, DPPH assay, desoxyribose assay, assay for site-specific reaction and non-enzymatic lipid peroxidation were employed to evaluate possible antioxidant activity of the oils and their main components, as well as the positive controls. The main components of these essential oils were α-pinene, cymene, myrcene, limonene, sabinene hydrate, cedrane and germacrene D. In rapid TLC screening, the tested compounds mainly showed antioxidant activity. In DPPH assay the stronger antioxidant effects were shown by the fruit oil of *C. arizonia*. In deoxyribose degradation assay test, most of the tested compounds showed some antioxidant effects. In non-enzymatic lipid peroxidation test, most antioxidant activity was shown by the fruit oil.

Cupressus macrocarpa Hartw. ex Gordon Family: Cupressaceae

Volatile oils, EtOH and DCM ecxtracts remarkably reduce DPPH radicals compared with BHA (El Ghorab *et al.*, 2007).

Cupressus sempervirens L. Family: Cupressaceeae

Emami *et al.* (2007) evaluated the antioxidant activity of the antioxidant acivity of the methanol extract of leaves and fruits of *Cupressus sempervirens* using FTC and TBA methods. The extracts exhibited strong antioxidant activity. Methanol extract of fruits of *C. semipervirens* cv. *cereifeormis* showed the highest antioxidant actvity while the methanol extract of leaves of *C. sempervirens* var. *sempervirens* possessed the lowest antioxidant activity.

Ephedra americana Bonpl. ex Willd. Family: Ephedraceae

Rivera and Pareja (2001) evaluated the antioxidant activity of ethanolic extracts and isolates of *Ephedra americana*. Ethanolic extract exhibited strong antioxidant activity. Five flavonoids – hesperidin, chrisyn, 3′,4′,5,7-tetrahydroxy-6,8-di R flavone, 3′,5,7-trihydroxy-4′,6-dimethoxy flavonol and 4′-hydroxy-5,7-dimethoxyflvaone - were isolated from the extract. All these compounds showed good antioxidant activity.

Ephedra gerardiana Wall. Family: Ephedraceae

Kumar and Singh (2011) evaluated the antioxidant activities of ethanol extracts of Trans Himalayan plants. *Ephedra gerardiana* leaf showed maximum DPPH radical scavenging activity with IC_{50} value of 13.30µg/ml.

Ephedra nebrodensis Tineo Family: Ephedraceae

Ballero *et al.* (2010) reported that the ethanol : acetone (1:1) extract of *Ephedra nebrodensis* exhibited dose related antioxidant activity.

Ephedra pachyclada Boiss. Family: Ephedraceae

Parsaeimehr *et al.* (2010) evaluated the antioxidant activity of three endemic species of *Ephedra* using FRAP test. *Ephedra pachyclada* showed an antioxidant activity of 1.56 m mol eq. quercetin/g extract and 0.275 m mol eq quercetin/g extract for the wild plant and callus respectively. *Ephedra platyclada* had total phenol content of 454.5 µmol eq. Catechin/g extracts and 108.77 µmol eq. Catechin/g extracts for the wild plants and callus respectively.

Ephedra procera Fisch. et Mey Family: Ephedraceae

Parsaeimehr *et al.* (2010) evaluated the antioxidant activity of three endemic species of *Ephedra* using FRAP test. *Ephedra procera* showed an antioxidant activity of 1.6 m mol eq. quercetin/g extract and 0.248 m mol eq quercetin/g extract for the wild plant and callus respectively. *Ephedra procera* had total phenol content of 436.44µmol eq. Catechin/g extracts and 105.56 µmol eq. Catechin/g extracts for the wild plants and callus resspectiveely.

Ephedra sinica Seapf. Family: Ephedraceae

Antioxidant capacities of 56 selected Chinese medicinal plants, including *Ephedra sinica* were evaluated by Song *et al.* (2010) using the TEAC and FRAP assays, and their total phenolic content was measured by the Folin-Ciocaleteu method. *Ephedra sinica* showed high antioxidant activity with TEAC value of 197.69µmol Trolox/g, FRAP value of 388.68µmol Fe^{2+}/g and phenolic content was 27.70mg GAE/g. The strong correlation between TEAC value and FRAP value suggested that the antioxidants in these plants possess free radical scavenging activity and oxidant reducing power, and the high positive correlation between antioxidant capacities and total phenolic content implied that phenolic compounds are a major contributor to the antioxidant activity of these plants.

Ephedra strobilacea (Ghahr.) Freitag et M. Maier Family: Ephedraceae

Parsaeimehr *et al.* (2010) evaluated the antioxidant activity of three endemic species of *Ephedra* using FRAP test. *Ephedra strobilacea* showed an antioxidant activity of 1.61 m mol eq. quercetin/g extract and 0.278 m mol eq quercetin/g extract for the wild plant and callus respectively. *Ephedra strobilacea* had total phenol content of 504.9 µmol eq. Catechin/g extracts and 114.61 µmol eq. Catechin/g extracts for the wild plants and callus respectively.

Ephedra viridis Coville Family: Ephedraceae

Phytochemical investigation of *Ephedra viridis* by Pullela *et al.* (2005) led to the isolation of four lignans including (+)-9-acetoxyisolariciresinol, lariciresinol, 9-acetoxylariciresinol and isolariciresinol. All the isolates were tested for their antioxidant activity and cytotoxicity against a panel of solid tumor and human leukemia cells. Most of the lignans exhibited moderate antioxidant activity without any cytotoxicity.

Ginkgo biloba L. Family: Ginkgoaceae

The antioxidant potential of *Ginkgo biloba* extract (EGb 761) on healthy human erythrocyte membranes was investigated by Kose and Dogan (1995). Lipoperoxidation was induced in erythrocyte suspensions using hydrogen peroxide, in the presence of EGb761 at 37 degrees C; malondialdehyde production was determined as the indicator of lipoperoxidation during the incubation period. The results for EGb 761 at concentrations of 20, 25, 50, 125, 250 and 500 µg/ml suggest that the antioxidant potential of EGb 761 in erythrocyte membranes increases with dose. Similarly, using different incubation periods (0, 15, 30, 45 or 60 min) indicated that the antioxidant effect of EGb 761 increased by the incubation time.

The findings of Wojcicki and Samochowiec (1996) sugget that *Ginkgo biloba,* in addition to lowering serum lipid level and possessing antioxidant properties in rabbits on an experimental diet, also affects metabolic processes in the liver and may modify lipid deposition in major arteries.

Extract from *Ginkgo biloba* leaves, the EGb 761 has a great number of therapeutical properties and anti-stress effects, for example the antioxidative one. Butnaru *et al.* (1997) estimated the influence of achronic treatment with EGb 761 on the concentration of the main product of lipidic degradation-malondialdehyde (MDA), in cerebral cortex, in diencephalons, thymus and gastric mucosa. The experiment was performed on wistar rats, male, young. The MDA estimation was performed by the spectrocalorimetry. They have determined that the stress raised the MDA concentration in all examined structures with approximately 100 per cent and 80 per cent of animals manifested stress ulcer. The treatment with EGb 761 before a stress inhibits the post-stress growth of MDA concentration and the process of stress ulcer formation (Butnaru *et al.*, 1997).

The antioxidant effect of *Ginkgo biloba* extract (EGb761) at 25 and 250 µg/ml concentrations on lipoperoxidation induced by hydrogen peroxide in erythrocyte obtained from Behcets disease (BD) patients was examined *in vitro* conditions. When compared to healthy controls, basal erythrocyte MDA levels were found to be higher in BD patients. In the *in vitro* study, there was also a significant increase in H_2O_2-induced MDA production in the medium containing no EGb 761 in the patient group, whereas significant decreases in MDA levels were observed in the medium containing EGb 761 both in the patient and control groups. The decrease in MDA production was found to be related to EGb 761 concentration (Kose *et al.*, 1997).

Free radicals are involved in numerous skin diseases, especially inflammatory reactions and photosenescence. To identify possible free-radical scavenging by an original terpene-free *Ginkgo biloba* extract containing 33 per cent Ginkgo flavone glycosides, mostly quercetin and kaempferol derivatives, Hibatallah *et al.* (1999) studied its activity by means of *in vitro* and *in vivo* experiments, using superoxide dismutase as a positive control. The data confirmed the free-radical scavenging property of both Gingkgo extract and SOD.

Yoshikawa *et al.* (1999) reviewed the biological actions and clinical applications of *Ginkgo biloba* leaf extract (GBLE). GBLE contains many different flavone glycosides and terpenoids. GBLE has an antioxidant action as a free radical scavenger, a relaxing effect on vascular walls, an antagonistic action on platelet-activating factor, an improving effect on blood flow or microcirculation, and a stimulating effect on neurotransmitters. Besides a direct scavenging action on active oxygen species, GBLE exerts an anti-inflammatory effect on inflammatory cells by suppressing the production of active oxygen and nitrogen species. GBLE inhibited the increase in the products of the oxidative decomposition low-density lipoprotein (LDL), reduced the cell death in various types of neuropathy, and prevented the oxidative damage to mitochondria, suggesting that GBLE exhibits beneficial effects on neuron degenerative diseases by preventing chronic oxidative damage. The study using a model of ischemia-reperfusion injury has also demonstrated the protective effect of GBLE on cardiac muscle and its antioxidative action *in vivo*. GBLE showed a very strong scavenging action on free radicals and is thus considered to be useful for the treatment of diseases related to the production of free radicals, such as ischemic heart disease, cerebral infarcion, chronic inflammation and aging.

Bridi *et al.* (2001) evaluated the lipid peroxidation (LPO) and the activity of the antioxidant enzymes, catalase (CAT) and superoxide dismutase (SOD) were evaluated in the hipocampus, striatum and substatia nigra (SN) of rats treated with EGb761. An increase in the CAT and SOD activities in the hippocampus, striatum and SN and a decrease of the LPO in the hippocampus were observed.

Antioxidant effect of *Ginkgo biloba* extract on hydrogen peroxide-induced lipoperoxidation of G-6-PD deficient erythrocytes was investigated by Sarikcioglu *et al.* (2004). The findings demonstrate that irrespective of administration time, EGb 761 significantly reduced TBARS levels in the erythrocytes of control and G-6-PD deficient patients subjected to oxidative stress.

The extract of the plant improves blood circulation, prevents free radical damage and prevents premature ageing. EGB 761 and ginkgolide are the chief antioxidative constituents (Mishra *et al.*, 2007).

Feng *et al.* (2009) have investigated the anticancer effects of three analogues of EGb 761 samples on sarcoma 108 (S180)-bearing mice and leukemic 1210 (L1210) cell lines. They have also evaluated the changes of endogenous antioxidant scavenging enzymes, including superoxide dismutase (SOD), glutathione(GST), lipid peroxidation (LPx) and catalase (CAT), in the blood of the S-180-bearing mice. From the study it was concluded that the anticancer activity of EGb 761 can be improved by increasing the concentration of the aglycone from of the flavonoid. Terpene trilactones cannot exert the anticancer effects of flavonoids *in vivo*. Raising the levels of the free radical scavenger enzymes GST, SOD and CAT may be one of the involved anticancer mechanisms.

Antioxidant capacities of 56 selected Chinese medicinal plants, including *Ginkgo biloba* were evaluated by Song *et al.* (2010) using the TEAC and FRAP assays, and their total phenolic content was measured by the Folin-Ciocalteu method. *Ginkgo biloba* showed high antioxidant activity with TEAC value of 82.89 (Leaf), 11.63 (fruit) μmol Trolox/g, FRAP value of 88.76μmol Fe^{2+}/g (leaf), 11.67μmol Fe^{2+}/g (fruit) and phenolic content was 11.55 (leaf) and 2.14 (fruit) mg GAE/g. The strong correlation between TEAC value and FRAP value suggested that the antioxidants in these plants possess free radical scavenging activity and oxidant reducing power, and the high positive correlation between antioxidant capacities and total phenolic content implied that phenolic compounds are a major contributor to the antioxidant activity of these plants.

Gnetum bucholzianum Welw. Family: Gnetaceae

Odukoya *et al.* (2007) evaluated the antioxidant activity and total phenolic contents of Nigerian green leafy vegetables including *Gnetum bucholzianum*. Phenol content was 546.97 mg/100 g dry weight and Ascorbic content was 187.11 mg/100 g dry weight. There was low correlation between AA/TAE, AA/ASC and TAE/ASC.

Gnetum gnemon L. Family: Gnetaceae

Four stilbene derivatives, gnemonols K and L (resveratrol trimers), M (isorhapontigenin dimer), and gnemonoside K (glucoside of resveratrol trimer) together with eleven known stilbenoids and a lignan were isolated by Iliya *et al.* (2003) from the acetone, methanol and 70 per cent methanol soluble parts of the root of *Gnetum gnemon*. The antioxidant activity of the stilbenoids on lipid peroxide inhibition and super oxide scavenging activity were also investigated

Wong *et al.* (2006) investigated the antioxidant properties of 25 edible tropical plants including *Gnetum gnemon* using DPPH and FRAP assays. Total phenolic content was also estimated. *Gnetum gnemon* showed moderate antioxidant activity.

Santoso *et al.* (2010) reported on the antioxidant components and activities of the edible parts of *Gnetum gnemon* (young and mature leaves, seed skin and endosperm), as well as their ability to inhibit DNA damage. Since *Gnetum gnemon* is usually consumed daily as a boiled vegetable, they also measured the effect of boiling on the antioxidant components and activities. The mature leaves contain the greatest levels of antioxidant components, followed by the young leaves. Accordingly, the mature and young leaves possess the highest antioxidant component and DNA damage prevention activities. However, the endosperm also has significant activity in ORAC, peroxyl radical-scavenging activity and DNA damage prevention activity at higher concentrations. Boiling decreased the amount of antioxidant components as well as the antioxidant activities in all edible parts examined, except for the seed skin.

The protein from the seeds of melinjo (*Gnetum gnemon*) was purified by Siswoyo *et al.* (2011) to identify the potent antioxidant and free radical scavenging activities. Two antioxidant protein fractions were isolated from *G. gnemon* seed with molecular weights of approximately 30 kDa (Gg-AOPI) and 12 kDa (Gg-AOPII). The antioxidant or free radical scavenging activities of Gg-AOPs were investigated by employing *in vitro* assay systems including the inhibition of linoleic acid autoxidation, scavenging effect on DPPH, ABTS, reducing power, chelating abilities of metal ions Cu^{2+} and Fe^{2+}, and protections against hydroxyl radical-mediated DNA damages. The result showed that two protein fractions exhibited significant antioxidant activities against free radicals such as DPPH, ABTS, and superoxide anion and showed activities similar to those of glutathione (G-SH) and BHT in a linoleic acid emulsion assay system. Moreover, Gg-AOPI and Gg-AOPII also exhibited notable reducing power and strong chelating effect on Fe^{2+} and protected hydroxyl radical induced oxidative DNA damage.

Analyses on biological activities of *Gnetum gnemon* were done by Wazir *et al.* (2011) to determine the total phenolic and antioxidants of the plant. Four parts of *G. gnemon* were used in this study, which were leaf, bark, twig, and seeds of the plant. All parts were extracted in methanol, ethanol, hexane, chloroform and hot water using reflux. The total phenolic content of the plant extracts were determined by using Folin-Ciocalteu method. The results demonstrated that the bark from hot water extract showed the highest total phenolic at 10.71mg GAE/FDW, while the lowest was chloroform extract of seed at 2.15mg GAE/FDW. The antioxidant activity of the plant extracts were determined by using DPPH and FRAP assays, respectively. The DPPH results showed that all plant extracts demonstrated weak free radical scavenging activity tested at the final concentration of 300 µg/ml. In contrast, the methanolic twig extract showed strong reducing power activity (FRAP) at 83.55 per cent, while the hot water seed extract showed the least activity at 41.86 per cent tested at the final concentration of 300 µg/ml. However, there were no correlation between total phenolics and both antioxidant assays tested.

Gnetum macrostachyum Hook.f. Family: Gnetaceae

A new flavonoid, named 5,7,2′-trihydroxy-5′-methoxyflavone (1), was isolated by Saisin *et al.* (2009) from the acetone extract of the lianas of *Gnetum macrostachyum* together with 5,7,4-trihydroxy-3′-methoxyflavanone (2) and seven known stilbenoids (3-7). The isolated compounds showed radical scavenging activity against DPPH with IC_{50} values in the range of 0.21-19.90 mM.

Gnetum montanum Markgraf Family: Gnetaceae

Extracts of *Gnetum montanum* stem showed remarkable free radical scavenging activity with IC_{50} value of 14.5µg/ml and significant inhibitory effects on lipid peroxidation with 83.8 per cent inhibition at the concentration of 50ng/ml (Thoung *et al.*, 2006).

Juniperus chinensis L. Family: Cupressaceeaae

The methanol extract of *Juniperus chinensis* heartwood revealed the strong antioxidant activity (Lim *et al.*, 2002). The antioxidant activity of n-BuOH soluble fraction was stronger than that of the others. Quercetin, naringenin, taxifolin, aomadendrin and isoquercetin were isolated from the n-BuOH fraction.

Juniperus communis L. Family: Cupressaceae

English Name: Juniper

Hydrodistillated extracts from juniper were assessed for their total phenol content, and antioxidant (iron(III) reduction, inhibition of linoleic acid peroxidation, iron (II) chelation, DPPH radical-scavenging and inhibition of hydroxyl radical-mediated 2-deoxy-D-ribose degradation, site and nonsite-specific) activities by Hinneburg *et al.* (2005). The extracts from juniper possessed the moderate antioxidant activities. Juniper performed better in the site-specific assay; thus they have a higher iron chelation ability than scavenging activity for hydroxyl radicals.

Wojcikowski *et al.* (2007) evaluated the antioxidant activity of aerial parts of *Juniperus communis*. Oxygen radical absorbance capacity of the ethyl extract, methanol extract and aqueous extracts were 37.79, 55.55 and 13.27µmol Trolox equivalent per g of dried starting material, respectively.

Emami *et al.* (2007) evaluated the antioxidant activity of the antioxidant acivity of the methanol extract of leaves and fruits of *Juniperus communis* using FTC and TBA methods. The extracts exhibited strong antioxidant activity.

Al-Mustafa and Al-Thunibat (2008) evaluated the antioxidant activity *Juniperus communis* leaves. The level of antioxidant activity was determined by DPPH and ABTS assay in relation to total phenolic contents of the medicinally used part. They concluded that *Juniperus communis* leaves have high antioxidant capacity (DPPH-TEAC> or = 80mg/g).

Juniperus excelsa M.B. Family: Cupressaceeae

Emami *et al.* (2007) evaluated the antioxidant activity of the antioxidant acivity of the methanol extract of leaves and fruits of *Juniperus excelsa* using FTC and TBA methods. The extracts exhibited strong antioxidant activity.

Öztürk *et al.* (2011) evaluated the antioxidant activity of *Juniperus excelsa*. In the β-carotene-linoleic acid assay, acetone extracts effectively inhibited oxidation of linoleic acid.

Juniperus foetidissima **Willd. Family: Cupressaceeae**

Emami *et al.* (2007) evaluated the antioxidant activity of the methanol extract of leaves and fruits of *Juniperus foetidissima* using FTC and TBA methods. The extracts exhibited strong antioxidant activity.

Emami *et al.* (2011) evaluated the antioxidant activity of essential oils obtained from branchlets of male and female trees as well as fruits of *Juniperus foetidissima*, from Iran. For this purpose, essential oils of *J. foetidissima* were phytochemically analyzed and different concentrations of them were tested in five oxidative systems: 1) low-density lipoprotein oxidation; 2) linoleic acid peroxidation; 3) red blood cell hemolysis; 4) hemoglobin glycation; and 5) insulin glycation assays. In all employed systems, antioxidant effects were observed from the three tested oils though in varying degrees. The most promising activities of the oils were observed against hemoglobin and insulin glycation. Antioxidant activities of the oils did not appear to be dose-dependent. In addition, no consistent superiority in antioxidant effects was observed from a single oil in different assays.

Juniperus oblonga **M.B. Family: Cupressaceeae**

Emami *et al.* (2007) evaluated the antioxidant activity of the methanol extract of leaves and fruits of *Juniperus oblonga* using FTC and TBA methods. The extracts exhibited strong antioxidant activity.

Juniperus oxycedrus **subsp.** ***oxycedrus*** **Family: Cupressaceae**

Öztürk *et al.* (2011) evaluated the antioxidant activity of *Juniperus oxycedrus oxycedrus*. In the β-carotene-linoleic acid assay, acetone extracts effectively inhibited oxidation of linoleic acid.

Juniperus phoenicea **L. Family: Cupressaceae**

Extracts of *Juniperus phoenicea* were obtained by different extraction solvents: methanol, ethanol, ethyl acetate and dichloromethane and evaluated composition for polyphenols (GAE 52 to 217g/kg), tannins (catechin equivalent 6.5 to 60.2 g/kg) anthocyanins (cyanidin equivalent 84 to 373 mg/kg), and flavonoids (quercetin equivalent 6.4 to 29.3 g/kg). The samples (essential oils and extracts) were subjected to a screening for their antioxidant activity by using DPPH and ABTS assays. The strongest antioxidant activity was obtained by the methanolic extract (IC_{50} = 6.5 mg/l) (Ennajar *et al.*, 2009, 2010).

Öztürk *et al.* (2011) evaluated the antioxidant activity of *Juniperus phoenicea*. In the β-carotene-linoleic acid assay, methanol extracts effectively inhibited oxidation of linoleic acid.

Juniperus procera **Hoechst ex Endl. Family: Cupressaceae**

The essential oil of *Juniperus procera* was examined for its potential radical scavenging activities (Burits *et al.*, 2001) in DPPH assay. The oil of *Juniperus procera* was effective hydroxyl radical scavenging agent when assessed in the deoxyribose degradation assay. In the *in vitro* assay for non-enzymatic lipid peroxidation in liposomes, the oil of *J. procera* also displayed antioxidant potential. The compounds that contribute to the radical scavenging activities were identified and then assessed for their effects in various test systems.

***Juniperus sabina* L. Family: Cupressaceeae**

Emami *et al.* (2007) evaluated the antioxidant activity of the methanol extract of leaves and fruits of *Juniperus sabina* using FTC and TBA methods. The extracts exhibited strong antioxidant activity.

Öztürk *et al.* (2011) evaluated the antioxidant activity of *Juniperus sabina*. In the β-carotene-linoleic acid assay, acetone and methanol extracts effectively inhibited oxidation of linoleic acid.

***Larix gmelinii* (Rupr.) Rupr. Family: Pinaceae**

Kolhir *et al.* (1996) demonstrated that dihydroquercetin is an antioxidant and possesses the properties of protecting capillaries, which is consistent with its action as an antiinflammatory agent, gastro- and hepatoprotecting agent, and its diuretic and hypolipidaemic activity. As a capillary protector, dihydroquercetin is 3.4 times more effective at a dose of 100mg/kg and 4.9 times at a dose of 300mg/kg than the well known bioflavonoid, quercetin. Dihydroquercetin shows a dose dependent suppression of lipid peroxidation in the rat liver. The hypolipidaemic activity of dihydroquercetin at 250mg/kg is comparable to that of polysponin.

***Larix sibirica* Ledeb. Family: Pinaceae**

The antioxidant properties of the ethylacetate extract of *Larix sibirica* bark and the flavonoids, quercetin and dihydroquercetin found in it were studied by Khairullina *et al.* (2006) using a model radical-chain oxidation of propan-2-ol in the kinetic regime. It was found that dihydroquercetin had the highest antioxidant activity among the studied natural compounds.

***Metasequoia glyptostroboides* Miki ex Hu Family: Taxodiaceae**

Bajpai *et al.* (2009a,b) assayed the antioxidant potential of the essential oil and extracts (hexane, chloroform, ethyl acetate and methanol) of *Metasequoia glyptostroboides* employing DPPH assay. The free radical scavenging activities of the oil and ethyl acetate extract were found to be superior (IC_{50} = 9.1 and 14.24μg/ml respectively) as compared to BHA (IC_{50} = 18.27μg/ml). Also the ethyl acetate extract revealed the highest phenolic contents (93.26 mg/g of dry wt) as compared to the other extracts.

***Picea abies* (L.) Karst. Family: Pinaceae**

English Name: Norway Spruce, Spruce

The antioxidative activity of phenolic extracts from Norway Spruce needles was examined by autoxidation of methyl linoleate by Kahkonen *et al.* (1999). Remarkable high antioxidant activity and high total phenolic content were found in the extracts of Norway spruce. Several flavonoids have been isolated from the needles of Norway spruce. The flavan-3-ols (+)-catechin, (+)-gallocatechin, dilignols and their glycosides have been found in spruce needles (Lundgren, 1987; Strack *et al.*, 1989). Lee *et al.*

(2006) evaluated the antioxidant activity of the extractives from the needles of *Picea abies*.

Co *et al.* (2011) developed environmentally sustainable extraction technique for the extraction of antioxidants from the bark of spruce. The antioxidant activity of the extracts was determined with DPPH assay. Pressurised fluid extraction using water and ethanol as solvent at 160 and 180°C, respectively, gave extracts of the highest antioxidant capacity. Stilbene glucosides such as isorhapontin, piceid and astringin were identified in the extracts.

Pinus sp. Family: Pinaceae

To determine the medicinal properties of pine pollen, the antioxidant and anti-inflammatory activities of the ethanol extract of pine pollen (PPE) were investigated by Lee *et al.* (2008). PPE displayed a strong free radical scavenging activity on DPPH radical and hydrogen peroxide. It was observed also that the antioxidant activity, measured by the ferric thiocyanate method, increased with the addition of PPE to the linoleic acid emulsion system. PPE was also found to inhibit significantly the amount of malondialdehyde protein carbonyls formed from liver homogenate. Like the antioxidant activity, the reducing power of PPE was excellent.

Pinus armandii Franch. Family: Pinaceae

Yang *et al.* (2010) investigated the antioxidant activity of the essential oil of *Pinus armandii* using DPPH assay. The DPPH scavenging activity of the essential oil was lower than BHT and ascorbic acid.

Pinus densiflora Sieb et Zucc. Family: Pinaceae

Antioxidant activity of *Pinus densiflora* was evaluated by Jung *et al.* (2003) for potential to inhibit hydroxyl radicals, inhibit total reactive oxygen species generation in kidney homogenates using 2′,7′-dichlorodihydrofluorescein diacetate (DCHF-DA) and scavenge authentic peroxynitrites. The methanolic extract showed strong antioxidant activity in the tested model systems and thus fractionated with several solvents. The antioxidant activity potential of the individual fraction was in the order of ethyl acetate>n-butanol>water>dichloromethane fraction. The ethyl acetate soluble fraction exhibiting strong antioxidant activity was further purified by repeated silica gel and Sephadex LH-20 column chromatographies. An active lignan (+)-isolarisiresinol xylopyranoside, as well as two active flavonoids (kaempferol 3-O-beta-galactopyranoside and its 6″acetyl derivative), were isolated (Jung *et al.*, 2003).

Pinus koraiensis S. et Z. Family: Pinaceae

A polyphenolics extract from *Pinus koraiensis* seed (PKS) has been investigated for its *in vivo* and *in vitro* antioxidant activity. The total phenolic content of PKS extract was 264 mg of gallic acid equivalents/g dry material. The extract was found to show remarkable scavenging activity on DPPH (EC_{50} 0.023 mg/ml), OH· (EC_{50} 0.391 mg/ml), O_2^- (EC_{50} 4.37 mg/ml), as well as a high and dose-dependent reducing power. The extract also showed strong suppressive effect on rat liver lipid peroxidation and erythrocyte haemolysis. Ageing and radiation models were used to evaluate the antioxidant effect *in vivo*. PKS extract highly inhibited oxidative damage induced by δ-galactose and γ-ray. The results have indicated that intragastric administration of the extract can increase levels of superoxide dismutase (SOD) and glutathione (GSH), and decrease malondialdehyde (MDA) content.

Pinus maritima Lam. Family: Pinaceae

English Name: French Maritime pine.

Pycnogenol (PYC) is a standardized extract composed of a mixture of flavonoids, mainly procyanidins and phenolic acids from *Pinus maritima*. PYC has strong free radical-scavenging activity against reactive oxygen and nitrogen species. The oligomeric components of PYC contribute significantly to the ESR free radical signal. PYC also participates in the cellular antioxidant network as indicated by its ability to regenerate the ascorbyl radical and to protect endogenous vitamin E and glutathione from oxidative stress. PYC modulated NO metabolism in activated macrophages by quenching the NO radical and inhibiting both iNOS mRNA expression and iNOS activity. The spectrum of different effects of NO in the circulation and the nervous system suggest the potential applications of PYC in immune and circulatory disorders as well as in neurodegenerative disease. PYC effects redox-sensitive signal transduction pathways and alters gene expression (Packer *et al.*, 1999).

The effects of pycnogenol on DNA damage *in vitro* and expression of superoxide dismutase and HP1 in *Escherichia coli* SOD and catalase deficient mutant cells was investigated by Kam and Park (2004). Thes results showed that pycnogenol (50µg/ml) did not inhibit DNA damage from Fenton reaction-generated oxyradicals, when the Fenton reaction was allowed to proceed. However, pycnogenol did protect against DNA damage when added prior to the initiation of the Fenton reaction, suggesting that protection resulted from chelation of Fe rather than from the scavenging of radicals.

Pinus massoniana Lamb Family: Pinaceae

Cui *et al.* (2005) examined the antioxidant activity, superoxide anion radical scavenging and free radical scavenging activities. The antioxidant activity in a linoleic acid emulsion increased with increasing amounts of the bark extract, with 100, 300 and 500 microg extract inhibiting 64.7 per cent, 68.6 per cent and 76.6 per cent or peroxidation, respectively. The radical scavenging activity increased in a dose-dependent manner.

The antioxidant activities of phenolic extract from *Pinus massoniana* bark (PEPB) were investigated by Yu *et al.* (2008) employing *in vitro* established systems which included the total antioxidant activity in ferric thiocyanate (FTC) system, DPPH radical scavenging activity, reducing power, metal chelating ability and nitric-oxide radical scavenging activity. It was found that PEPB had stronger antioxidant activities to inhibit the oxidation of linoleic acid in the FTC system than the standards, vitamin C. PRPB had the highest inhibition (90.38 per cent), followed by vitamin E (88.61 per cent) and C (85.68 per cent) at 100 µg/ml in the FTC system. Further more, PEPB showed strong DPPH radical scavenging activity, reducing power, nitric-oxide scavenging activity and ferrous-ion chelating ability.

The antioxidant activities of both water and ethanol extracts from hot air dried and freeze-dried *Pinus massoniana* needles were evaluated by Pan *et al.* (2009) using different antioxidation tests, reductive potential and free radical scavenging. Extracts of *Pinus massoniana* needles exhibited strong free radical scavenging activity.

Pinus morrisonicola Hay Family: Pinaceae

The free radical scavenging and anti-cancer activities of *Pinus morrisonicola* were studied by Hsu *et al.* (2005) using different parts of the pine, namely, needle, bark and cone. Results showed that pine needle water extract had the highest scavenging superoxide anion activity and the lowest IC_{50} value in inhibiting superoxide anion formation; however, the bark water extract showed the best anti-lipid peroxidation activity.

Pinus nigra Arn. subsp. _pallisiana_ (Lamb.) Holmboe. Family: Pinaceae

Gulcin *et al.* (2004) studied antioxidant and analgesic activities of turpentine of *Pinus nigra* subsp. *pallisiana*. Total antioxidant activity, reducing power, superoxide anion radical scavenging, free radical scavenging, metal chelating and hydrogen peroxide scavenging activities were increased with the increasing amount of extracts (100, 300 and 500 microg) added to linoleic acid emulsion. All the doses of turpentine exude from *Pinus nigra* subsp. *pallisiana* showed higher antioxidant activity than α-tocopherol. The sample showed 49, 70, and 91 per cent inbhibition on peroxidation of linoleic acid emulsion respectively. On the other hand, the 300 microg of α-tocopherol showed 40 per cent inhibition on peroxidation of linoleic acid emulsion.

Pinus pinaster Ait. and _Pinus radiata_ D.Don Family: Pinaceae

The composition and antiradical activity of procyanidins from the bark of two kinds of pine, *Pinus pinaster* and *Pinus radiata*, were compared by Jerez *et al.* (2007). Both the total bark extract and the fraction soluble in both water and ethyl acetate (OW) were evaluated, because of their promising results in previous experiments. Results showed that *P. radiata* bark was richer in total phenols and also in procyanidins, catechin always being the main unit, so terminal as extensional. For *P. pinaster*, epicatechin was the predominant extension unit. The mean degree of polymerization (mDP) was higher for the latter. Interestingly, opposite results were encountered for the corresponding OW fractions, where *P. radiata* showed a mDP of 2.9 vs. 2.3 of *P. pinaster*. It was also found that the higher the mDP the higher was the specific antiradical activity. The different procyanidin composition and specific antiradical activity of the two kinds of barks, and particularly their OW fractions, may lead to the design of efficient natural antioxidants with application in the food industry.

Pinus radiata D.Don and _Pinus rigida_ Mill. Family: Pinaceae

The phenolic composition of pine bark from a variety of *Pinus* species was estimated by Ku *et al.* (2007) by measuring Klason lignin, acid-soluble lignin, and a 1 per cent NaOH extract. Polyphenol contents of hot water extracts from pine bark were determined by the Folin-Ciocalteu assay and the vanillin-H_2SO_4 assay. Among the pine bark varieties investigated, *Pinus radiata* bark showed the highest polyphenol content and potent antioxidant activity. *Pinus rigida* bark was also a usable polyphenol-rich source, whereas *Pinus densiflora* bark had a low yield (5.1 per cent) of hot water extract, although it showed potent antioxidant activity. Correlations between proanthocyanidin content in pine bark and antioxidant activity were observed. The results suggested that proanthocyanidin was the crucial contributor to potent antioxidant activity in pine bark.

Pinus roxburghii Sarg. Family: Pinaceae

The needles of *Pinus roxburghii* were investigated by Puri *et al.* (2010) for antidyslipidemic activity in high fat diet (HFD)-fed hyperlipidemic Golden-Syrian hamsters. The alcoholic extracts of *P. roxburghii* needle significantly decreased the plasma triglyceride (TG) by 38 per cent, total cholesterol (TC) 23 per cent, glycerol (Gly) 44 per cent, and low density lipoproteins-cholesterol (LDL-C) 27 per cent accompanied with increase in high density lipoproteins-cholesterol and its ratio with total cholesterol (HDL–C/TC ratio) by 28 per cent at the dose of 100 mg/kg body weight (b.w.). Further extraction with several solvents exhibited significant lipid lowering in hexane soluble and *n*-butanol soluble fractions. These fractions lowered plasma TG by 56 and 53 per cent, TC 41 and 36 per cent, Gly 51 and 29 per cent, LDL-C 51 and 44 per cent, and increased the HDL–C/TC ratio by 75 and 57 per cent, respectively. Antioxidant activity of all the four fractions and alcoholic extract was assessed by Trolox equivalent antioxidant capacity (TEAC) assay, and activity was found to be significant in alcoholic extract and *n*-butanol insoluble fractions as capacity to scavenge the $ABTS^{\bullet+}$ radical cation *in vitro*.

Pinus sylvestris L. Family: Pinaceae

English Name: Scotch Pine, Finnish pine

The antioxidative activity of phenolic extracts from fruits of Scotch pine was examined by autoxidation of methyl linoleate by Kahkonen *et al.* (1999). Remarkable high antioxidant activity and high total phenolic content were found in extracts of bark, cork and needles. Several flavonoids have been isolated from the needles of scotch pine. The flavan-3-ols (+)-catechin, (+)-gallocatechin, dimeric and trimeric anthocyanidins, dilignols and their glycosides have been found in scotch pine needles. Several glycosides of quercetin, isorhamnetin, and kaempferol have also been found in this species, but the predominant occurrence of dihydroflavonol glucosides is typical in pine leaves, dihydroquercetin 3-O-glucoside being the most abundant (2 per cent of the needle dry weight).

A modified *in vitro* lipid peroxidation inhibition assay was used by Saleem *et al.* (2003) to guide the fractionation and the isolation of antioxidative principles of Finnish pine bark extract. This approach yielded 3,4-dihydroxybenzoic acid (protocatechuic acid) and taxifolin-3-O-glucopyranoside as major antioxidative compounds from the plant material.Their IC_{50} values, in comparison to trolox (6-hydroxy-2,5,7,8 tetramethylchroman-2-carboxylic acid), were: trolox (1.78 d 0.56 μm) < protocatechuic acid (5.77 d 1.63 μm) < taxifolin-3-O-glucopyranoside (16.30 d 1.98 μm).

Pinus tabulaeformis Carr. Family: Pinaceae

Gan *et al.* (2010) evaluated the antioxidant activity of *Pinus tabulaeformis* using the FRAP and TEAC assays, and its total phenolic content was measured by the

Folin-Ciocalteu method. For FRAP assay the value was 17.72 µmol Fe(II)/g dry weight. For TEAC assay, the value was 13.30µmol Trolox/g dry weight. Total phenolic content was 1.46 mg gallic acid equivalent/g dry weight of plant material.

***Platycladus orientalis* (L.) Franco Family: Cupressaceeae**

Emami *et al.* (2007) evaluated the antioxidant activity of the methanol extract of leaves and fruits of *Platycladus orientalis* using FTC and TBA methods. The extracts exhibited strong antioxidant activity.

***Podocarpus elongatus* (Ait.) L'Herit. ex Pess. Family: Podocarpaceae**

Abdillahi *et al.* (2011) evaluated the antioxidant activity of *Podocarpus elongatus* using DPPH, FRAP and β-carotene-linoleic acid assays. The extract showed high reducing power.

***Podocarpus falcatus* (Thunb.) R. Br. ex Misb. Family: Podocarpaceae**

Abdillahi *et al.* (2011) evaluated the antioxidant activity of *Podocarpus falcatus* using DPPH, FRAP and β-carotene-linoleic acid assays. The extract showed high reducing power.

***Podocarpus henkelii* Stapf ex Dallim. & A.B. Jacks Family: Podocarpaceae**

Abdillahi *et al.* (2011) evaluated the antioxidant activity of *Podocarpus henkelii* using DPPH, FRAP and β-carotene-linoleic acid assays. The extract showed high reducing power.

***Podocarpus latifolius* (Thumb.) R. Br. ex Mirb. Family: Podocarpaceae**

Abdillahi *et al.* (2011) evaluated the antioxidant activity of *Podocarpus latifolius* using DPPH, FRAP and β-carotene-linoleic acid assays. Stems of *Podocarpus latifolius* exhibited the lowest EC50 (0.84µg/ml) inhibition against DPPH. The percentage antioxidant activity based on the bleaching rate of β-carotene ranged from 96 per cent to 99 per cent. The extract also showed high reducing power.

***Podocarpus nagi* (Thumb.) Zoll. et Morr. Family: Podocarpaceae**

Diterpenoids, totarol (1), totaridiol (2), 19-hydroxytotarol (3), totaral (4), 4 beta-carboxy-19-nortotarol (5), sugiol (6), isolated from *Podocarpus nagi* were evaluated as antioxidants by Haraguchi *et al.* (1996, 1997). These diterpenoids inhibited autoxidation of linoleic acid. Mitochondrial and microsomal lipid peroxidation induced by Fe (III)-ADP/NADH or Fe (III)-ADP/NADPH were also inhibited by Totarol. They inhibited linoleic acid autoxidation but not generation of superoxide anion. Nagilactone E (2), a norditerpene lactone isolated from the same source, had no antioxidative activity. Furthermore, totarol protected red cells against oxidative hemolysis.

***Sabina przewalski* (Kom.) W.C.Cheng et L.K.Fu Family: Cupressaceae**

Ip *et al.* (2004) reported that the protective action of the *Sabina przewalski* extract against menadione-induced hepatotoxicity is associated with its antioxidant activities including the scavenging of superoxide anion radicals and inhibition of lipid peroxidation.

Taxus baccata L. Family: Taxaceae

Emami *et al.* (2007) evaluated the antioxidant activity of the methanol extract of leaves and fruits of *Taxus baccata* using FTC and TBA methods. The extracts exhibited strong antioxidant activity.

Serteser *et al.* (2009) studied the antioxidant activity of 50 per cent aqueous methanol extracts of fruits by various antioxidant assays, including free radical scavenging, hydrogen peroxide scavenging and metal (Fe^{2+}) chelating activities. The DPPH radical scavenging effect of fruit extract was 0.503. Fe^{2+} chelating activity (per cent) of fruit extract is 28.32 and Hydrogen peroxide inhibition activity of fruit extract was 21.28 per cent.

Phytochemical investigation of *Taxus baccata* led to the isolation of the lignans; lariciresinol (1), taxiresinol (2), 3′demethylisolariciresinol-9′-hydroxyisopropylether (3), isolariciresinol (4) and 3-demethylisolariciresinol (5) as well as taxoids. These compounds were evaluated for their antioxidant activity. The compounds displayed a great scavenging activity against DPPH especially at 500 and 1000μg/ml. Besides they were found to exert noteworthy reducing antioxidant power on ferric ion. In particular, FRAP of compounds 2 (3.552), 4 (3.021) and 5 (3.533) were as high as the reference chlorogenic acid (3.618) at 1000μg/ml. None of the compounds exhibited chelating ability against ferrous ion (Kucukboyaci *et al.*, 2009).

Taxus cuspidata Sieb. et Zucc. Family: Taxaceae

English Name: Far-east yew.

Phenolic components of the wood and roots of far-east yew (*Taxus cuspidata*) have been isolated and investigated by Veselova *et al.* (2007). Four lignans [(+)-taxiresinol, (+)-isotaxiresinol, (+)-isolariciresinol, (")-secoisolariciresinol] and two catechins [(+)-catechin, (")-epicatechin] were identified using spectroscopic techniques. HPLC data showed that wood, bark, roots, stems, and needles of the plant contain different amounts of lignans and catechins. The antioxidant and radical-scavenging activities of the polyphenols were evaluated on two *in vitro* model systems.

Taxus yunnanensis W.C.Cheng et L.K.Fu Family: Taxaceae

Banskota *et al.* (2003) evaluated DPPH radical scavenging and nitric oxide inhibitory activities of the constituents from the wood of *Taxus yunnanensis*. Chemical investigation of H_2O, H_2O/MeOH (1:1) and MeOH extracts of the wood led to the isolation of nineteen compounds, *i.e.*, five lignans, two simple phenolics, and twelve taxane-type diterpenes. Isotaxiresinol and seco-isolariciresinol, two major lignans of the wood, possessed potent DPPH radical scavenging activities with IC_{50} values of 21.7 and 29.9μM, respectively. Similarly, coniferylaldehyde, taxusin, 10-deacetyltaxyunnanine C, hongdoushan A, and 2α, 5α, 10β-triacetoxy-14 β-{(S)-2-methylbutyryloxy-4(20), 1,1-taxadiene showed potent NO inhibitory activity in immune macrophage like J774.1 cells, with IC_{50} values of 18.0, 22.1, 29.5, 15.0 and 26.4μM, respectively, which were either equal or lower than the positive control.

Thuja occidentalis L. Family: Cupressaceae

Deb *et al.* (2009) evaluated the free radical scavenging activity and anti-inflammatory activity. The study showed significant free radical scavenging activity by the ethanolic extract but failed to to exhibit any anti-inflammatory activity.

Dubey and Batra (2009a,b) evaluated the antioxidant activities of and ethanol fraction of aerial part of *Thuja occidentalis* in various systems. DPPH radical, super oxide anion radical, hydroxyl radical scavenging and lipid peroxidation were carried out to evaluated the antioxidant potential of the extract. The antioxidant activity of ethanol extract were increased in a concentration dependent manner. About 100, 150, 200, 250 and 300 µg of ethanol fraction inhibited $FeSO_4$ induced lipid peroxidation in a dose dependent manner and showed IC_{50} value 195.60µg/ml. In DPPH radical scavenging assay the IC_{50} values of the extract was 202.45µg/ml. The extract also inhibited hydroxyl radical generated by Fenton's reaction with an IC_{50} value of 158.59µg. Moreover the ethanol fraction also scavenged the superoxide generated by PMS/NADH-NBT system and showed IC_{50} 124.11 µg.

References

Aaby, K., Skrede, G. and Wrolstad, R. E. **2005.** Phenolic composition and antioxidant activities in flesh and achenes of strawberries (*Fragaria ananassa*). J. Agric. Food Chem. **53 (10)**: 4032-4040.

Abbas, F., Lajis, N.H., Israf, D.A., Khozirah, S. and Kalsom, Y.U. 2006. Antioxidant and nitric oxide inhibition activities of selected Malay traditional vegetables. Food Chem. **95(4)**: 566-573.

Abdalla, A.E.M., Darwish, S.M., Ayad, E.H.E and El-Hamahmy, R.M. 2007. Egyptian mango byproduct 2: Antioxidant and antimicrobial activities of extract and oil from mango seed kernel. Food Chem. **103**: 1141-1152.

Abd El-Baky, H.H., El Baz, F.K. and El-Baroty, G.S. 2004. Production of antioxidant by the green alga *Dunaliella salina*. Int. J. Agric. Biol. **6**: 49-57.

Abd El-Baky, H.H., El Baz, F.K. and El-Baroty, G.S. 2008. Evaluation of marine alga *Ulva lactuca* L. as a source of natural preservative ingredient. Amer.-Eurasian J. Agric. Environ. Sci. **3(3)**: 434-444.

Abdelshafeek, K.A., Abou-Setta, L.M. and Nazif, N.M. 2010. Study on some chemical constituents and antioxidant activity of *Beaumontia grandiflora* Wall. grown in Egypt. Australian J. Basic Appl. Sci. **4(6)**: 1063-1069.

Abdel Wahab, M.A., Said, A. and Hefner, A. 2006. NMR and radical scavenging activities of patuletin from *Urtica urens* against aflatoxin B1. Pharm. Biol. **43(6)**: 515-525.

Abdillahi, H.S., Finnie, J.F. and Van Staden, J. 2011. Anti-inflammatory, antioxidant, antityrosinase and phenolic contents of four *Podocarpus* species used in traditional medicine in South Africa. J. Ethnopharmacol. **136(3)**: 496-503.

Abdollahi, M., Larijani, B., Rahimi, R. and Salari, P. 2005. Role of oxidative stress in osteoporosis. Therapy **2**: 787-796.

Abdulnabi, A.A., Emhemed, A.H., Hussain, G.D. and Biacs, P.A. 1997. Determination of antioxidant vitamins in tomatoes. Food. Chem. **60**: 207-212.

Abe, K. and Terada, H. 1997. Radical scavenging and antioxidative effect of jyuzentaihoto. Nat. Med. **51(6)**: 528-531.

Abeer, T. 2011. Flavonoidal content of *Vangueria infausta* extract grown in Egypt: investigation of its antioxidant activity. Int. Res. J. Pharm. **2(3)**: 157-161.

Abeysinghe, D.C., Li, X., Sun, C.D., Zhang, W.S., Zhou, C.H. and Chen, K.S. 2007. Bioactive compounds and antioxidant capacities in different edible tissues of citrus fruit of four species. Food Chem. **104**: 1338-1344.

Abeywickrama, K.R.W., Amarakoon, A.M.T. and Ratnasooriya, W.D. 2005. *In-vitro* and *in vivo* antioxidant activity of high-grown Sri Lankan black tea (*Camellia sinensis* L.) Sri Lanka J. Tea Sci. **70(2)**: 57-68.

Ablise, M., Muller, B.L., Wong, C.D., Siest, G., Loppinet, V. and Visvikis, S. 2004. Synthesis and *in vitro* antioxidant activity of glycyrrhetinic acid derivatives tested with the cytochrome P450/NADPH system. Chemical and Pharmaceutical Bull. **52(12)**: 1436-1439.

Abou-Gazar, H., Bedir, E., Takamatsu, S., Ferreira, D. and Khan, I.A. 2004. Antioxidant lignans from *Larrea tridentata*. Phytochemistry **65(17)**: 2499-2505.

Aboul-Enein, A.M., El-Baz, F.K., El-Baroty, G.S., Youssef, A.M., Abd El-Baky, H.H. 2003. Antioxidant activity of algal extracts on lipid peroxidation. J. Med. Sci. **3**: 87-88.

Abouzid, S., Elshahaat, A., Ali, S. and Choudhary, M.I. 2008. Antioxidant activity of wild plants collected in Beni Sueif governarate, Upper Egypt. Drug Discov. Ther. **2(5)**: 286-288.

Abreu, P.M, Matthew, S., Gonzalez, T., Costa, D., Segundo, M.A. and Fernandes, E. 2006. Anti-inflammatory and antioxidant activity of a medicinal tincture from *Pedilanthus tithymaloides*. Life Sci. **78**: 1578-1585.

Abreu, P.M, Matthew, S., Gonzalez, T., Vanickova, L., Costa, D., Gomes, A., Segundo, M.A. and Fernandes, E. 2008. Isolation and identification of antioxidants from *Pedilanthus tithymaloides*. J. Nat. Med. **62(1)**:

Abreu, P.M., Braham, H., Jannet, H.B., Mighri, Z. and Matthew, S. 2007. Antioxidant compounds from *Ebenus pinnata*. Fitoterapia **78**: 32-34.

Abu Backer, M.N., Mangalyarkarasi, K. and Chandra, R. 2009. Antioxidant and antibacterial activity of *Cassia tora*. J. Trop. Med. Plants **10**: 13-18.

Abu Bakar, M.F., The, A.H., Rahmat, A. and Hashim, N. 2004. Antioxidant tea from leaves of *Strobilanthes crispus*. J. Trop. Med. Plants **5**: 199-204.

Acero, N., Llinares, F., de Mera, A.G., Oltra, B. and Munoz-Mingarro, D. 2006. Apoptotic and free radical scavenging properties of the methanolic extract of *Gentianella alborosea*. Fitoterapia **77**: 475-477.

Acharya, K., Samui, K., Rai, M., Dutta, B.B. and Acharya, R. 2004. Antioxidant and nitric oxide synthase activation properties of *Auricularia auricula*. Indian J. Exp. Biol. **42(5)**: 538-40.

Acharya, K., Yonzone, P., Rai, M. and Rupa, A. 2005. Antioxidant and nitric oxide synthase activation properties of *Ganoderma applanatum*. Indian J. Exp. Biol. **43(10)**: 926-929.

Acharyya, S., Dash, G.K., Mondal, S. and Dash, S.K. 2010. Antioxidative and antimicrobial study of *Spondias mangifera* Willd. root. Int. J. Pharm. Pharmceutical Sci. **2**(Suppl.4):

Achliya, G., Deshpande, J., Mishra, M., Asati, A., Wadodkar, S. and Dorle, A. 2003. Investigation of the free radical scavenging activity of some Indian medicinal plants. National symposium on Emerging Trends in Indian Medicinal plants. Lucknow, p. 58.

Achutan, C.R., Babu, B.H. and Padikkala, J. 2003. Antioxidant and hepatoprotective effects of *Rosa damascena*. Pharm. Biol. **41(5)**: 357-361.

Acosta, O., Perez, A.M. and Vaillant, F. 2009. Chemical characterization, antioxidant properties and volatile constituents of naranjilla (*Solanum quitoense* Lam.) cultivated in Costa Rica. ALAN **59(1)**:

Actis-goretta, L., Mackenzie, G.G., Oteiza, P.I. and Fraga, C.G. 2002. Comparative study on the antioxidant capacity of wines and other plant-derived beverages. Alcohol and Wine in Health and Disease **957**: 279-283.

Acuna, U.M., Atha, D.E., Ma, J. Neea, M.H. and Kennely. 2002. Antioxidant capacities of ten edible north American plants. Phytother. Res. **16**: 63-65.

Adebajo, A.C., Iwalewa, E.O., Obuotor, E.M., Ibikunle, G.F., Omisore, N.O., Adewunmi, C.O., Obaparusi, O.O., Klaes, M., Adetogun, G.E., Schmidt, T.J. and Verspohl, E.J. 2009. Pharmacological properties of the extract and some isolated compounds of *Clausena lansium* stem bark: anti-trichomonal, antidiabetic, anti-inflammatory, hepatoprotective and antioxidant effects. J. Ethnopharmacol. **122(1)**: 10-19.

Adebayo, A.H., Tan, N.H., Akindahunsi, A.A., Zeng, G.Z. and Zhang, Y.M. 2010. Anticancer and antiradical scavenging activity of *Ageratum conyzoides* L. (Asteraceae). Phcog. Mag. **6(21)**: 62-66. DOI: 10.4103/0973-1296.59968.

Adebooye, O.C. 2007. Activity and components of antioxidants in the fruits pulp of *Trichosanthes cucumeriana*. Electronic journal of Environmental, Agricultural and Food Chemistry **6(3)**: 1912-1920.

Adebooye, O.C., Ram, V. and Vasudeva, S. 2008. Peroxidase activity, chlorophylls and antioxidant profile of two leaf vegetables (*Solanum nigrum* L. and *Amaranthus cruentus* L.) under six pretreatment methods before cooking. Int. J. Food Sci. Technol. **43(1)**: 173-178.

Adedapo, A.A., Jimoh, F.O., Afolayan, A.J. and Masika, P.J. 2008. Antioxidant activities and phenolic contents of the methanol extracts of the stems of *Acokanthera oppositifolia* and *Adenia gummifera*. BMC Complement alternat. Med. **8**: 54.

Adedapo, A.A., Jimoh, F.O., Koduru, S., Afolayan, A.J. and Masika, P.J. 2008. Antibacterial and antioxidant properties of the methanol extracts of the leaves and stems of *Calpurnia aurea*. BMC Complement Altern. Med. **8**: 53.

Adefegha, S.A. and Pbph, G. 2011. Cooking enhances the antioxidant properties of some tropical green leafy vegetables. Afr. J. Biotechnol. **10(4)**: 632-639.

Adejuwon, A.A., Oluwatoyin, S.M. and Sunday, A.O. 2011. Anti-inflammmatory and antioxidant activities of *Hunteria umbellata* seed fractions. Pharmacologia **2(6)**: 165-171.

Adelkun, O.E., Oyelade, O.J., Ade-Omowaye, B.I.O., Adeyemi, I.A., van den Venter and Koekemoer, T.C. 2009a. Influence of pre-treatment on yield chemical and antioxidant properties of Nigerian okra seed (*Abelmoschus esculentus* Moench.) flour. Food Chem. Toxicol. **46**: 657-661.

Adelkun, O.E., Oyelade, O.J., Ade-Omowaye, B.I.O., Adeyemi, I.A. and van den Venter. 2009b. Chemicaal composition and the antioxidative properties of Nigerian Okra seed (*Abelmoschus esculentus* Moench.) flour. Food Chem. Toxicol. **47(6)**: 1123-1126. doi: 10.1016/jft.2009.01.036

Adeloye, O.A., Akinpelu, A.D., Ogundaini, O.A. and Obafemi, A.C. 2007. Studies on antimicrobial, antioxidant and phytochemical analysis of *Urena lobata* leaves extract. J. Phys. Nat. Sci. **1(2)**.

Aderogba, M.A., Falade, O.S., Olajide, O.A. and Adewusi, S.R.A. 2007. Antioxidant activity of *Hibiscus sabdariffa* calyx extract and its constituents. J. Trop. Med. Plants **8(1)**.

Aderogba, M.A., McGaw, I.J.,Benzabih, M. and Abegaz, B.M. 2011. Isolation and characterisation of antioxidant constituents of *Croton zambesicus* leaf extract. Nat. Prod. Res. (in press)

Aderogba, M.A., McGaw, I.J., Ogundaini, A.O. and Eloff, J.N. 2007. Antioxidant activity and cytotoxicity study of the flavonol glycosides from *Bauhinia galpinii*. Nat. Prod. Res. **25(13)**: 1224-1233.

Aderogba, M.A., Ogundaini, A.O. and Eloff, J.N. 2006. Isolation of two flavonoids from *Bauhinia monandra* (Kurz.) leaves and their antioxidative effects. Afr. J. Trad. CAM **3(4)**: 59-65.

Aderogba, M.A., Okoh, E.K., Adelanwa, T.A. and Obuotor, E.M. 2004. Antioxidant properties of the Nigerian *Piliostigma* species. J. Biol. Sci. **4(4)**: 501-503.

Aderogba, M.A., Okoh, E.K. and Idowu, T.O. 2005. Evaluation of antioxidant activity of the secondary metabolites from *Piliostigma reticulatum* (DC.) Hochst. J. Biol. Sci. **5 (2)**: 239-242.

Adesegun, S.A., Fajana, A., Orabueze, C.I. and Cooker, H.A.B. 2007. Evaluation of antioxidant properties of *Phaulopsis fascisepala* C.B.Cl. (Acanthaceae). Evidence Based Complement Altern. Med. **6(2)**: 227-231.

Adib, S.M., Rahman, M.S., Rahman, M.Z., Ahmed, K.S. and Rashid, M.A. 2010. Free radical scavenging activities of some indigenous plants of Bangladesh. Bangladesh Pharmaceutical Journal **13(1)**: 68-70.

Adiguzel, A., Ozer, H., Sokmen, M., Gulluce, M., Sokmen, A., Kilic, H., Sahin, F. and Baris, O. 2009. Antimicrobial and antioxidant activity of the essential oil and methanol extract of *Nepeta cataria*. Pol J. Microbiol. **58(1)**: 69-76.

Afanasev, I.B. 2005. Free radical mechanisms of aging process under physiological conditions. Biogerontology **6**: 283-290.

Afaq, S.H., Tajuddin, S. and Afridi, R. 1991. Bisheri Booti (*Aerva lanata*) some lesser known uses and pharmacognosy. Ethnobotany **5**: 37-40.

Afolabi, C.A., Ibukum, E.O., Emmanuel, A., Obuotor, E.M. and Farombi, E.O. 2007. Phytochemical constituent and antioxidant activity of extract from the leaves of *Ocimum gratissimum*. Sci. Res. Essay **2(5)**: 163-166.

Aftab, N. and Vieira, A. 2009. Antioxidant activities of curcumin and combinations of this curcuminoid with other phytochemicals. Phytother. Res. **24(4)**: 500-502.

Agbor, A.G., Kuante, D. and Oben, J.E. 2007. Medicinal plants can be good source of antioxidant: case study of Cameroon. Pakistan J. Biol. Sci. **10 (4)**: 537-544.

Agbor, A.G., Oben, J.E. and Ngogang, J.Y. 2003. Antioxidant activity of *Hibiscus cannabinus* Linn. Food Africa International Working meeting. Proceeding.

Agbor, A.G., Oben, J.E., Ngogang, J.Y., Cai Xinxing and Vinson, J.A. 2005. Antioxidant capacity of herbs/spices from Cameroon. A comparative study of two methods. J. Agric. Food Chem. **53(17)**: 6819-6824.

Agbor, A.G, Vinson, J.A., Oben, J.E. and Ngogang, J.Y. 2007. *In vitro* antioxidant activity of three *Piper* species. J. Herbal and Pharmacotherapy **7 (2)**: 49.

Agnaniet, H., Makani, T., Akagah, A., Menut, C. and Bessiere, J.M. 2005. Volatile constituents and antioxidant activity of essential oils from *Lippia multiflora* Mold. growing in Gabon. Flavour and Fragrance Journ. **20(1)**: 34-38.

Agnaniet, H., Menut, C. and Bessiere, J.M. 2004. Aromatic plants of tropical Central Africa. Part XLIX: Chemical composition of essential oils of the leaf and rhizome of *Aframomum giganteum* K.Schum. from Gabon. Flavour and Fragrance Journ. **19(3)**: 205-209.

Agnieszka, T. and Bozena, B. 2002. Antioxidant activity of crude tannins of pea (*Pisum sativum* L.) seed coat and their hypocholesterolemic effect in rats. Polish J. Food Nutr. Sci. **11(3)**: 33-38.

Agnihotri, V., El Sohly, H.N., Khan, S.I., Smille, T.J., Khan, I.A. and Walker, L.A. 2008. Antioxidant constituents of *Nymphaea caerulea* flowers. Phytochem. **69(10)**: 2061-2066 doi: 10.1016/jphytochem.2008.04009

Agostini, L.R., Moron Jimenez, M.J., Ramon, A.N. and Ayala Gomez, A. 2004. Determination of the antioxidant capacity of flavonoids in fruits and fresh and thermally treated vegetables. Arch Latinoam. Nutr. **54(1)**: 89-92

Ahmad, M., Kashmiri, M.A., Mohy-ud-din, A. and Sammia, Y. 2009. Characterization, compositional studies, antioxidant and antibacterial activity of seeds of *Abutilon indicum* and *Abutilon muticum* grown wild in Pakistan. Res. Rep. **73**: 39-47.

Ahmad, N., Fazal, H., Abbasi, B.H., Inayat-Ur-Rahman, Anwar, S., Khan, M.A., Basir, A., Inayat, H., Zameer, R., Khalil, S.A. and Khan, K.Y. 2011. DPPH-scavenging antioxidant potential in regenerated tissues of *Stevia rebaudiana, Citrus sinensis* and *Saccharum officinarum*. J. Med. Plants Res. **5(14)**: 3293-3297.

Ahmad, N., Shahwar, D., Khan, M.A. and Mughal, M.S. 2010. Antioxidant potential and xanthine oxidase inhibitory activity of *Croton sparsiflorus*. Asian J. Chem. **22(1)**: 555-561.

Ahmad, R., Ali, A.M., Israf, D.A., Ismail, N.H., Shaari, K. and Lajis, N.H. 2005. Antioxidant, radical-scavenging, anti-inflammatory, cytotoxic and antibacterial activities of methanolic extracts of some *Hedyotis* species. Life Sci. **76(17)**: 1953-1964.

Ahmad, R., Mahbob, E.N.M., Noor, Z.M., Ismail, N.H., Lajis, N.H. and Shaari, K. 2010. Evaluation of antioxidant potential of medicinal plants from Malaysian Rubiaceae (subfamily Rubioideae). Afr. J. Biotech. **9(46)**: 7948-7954.

Ahmad, S., Riaz, N., Saleem, M., Jabbar, A., Nisar-Ur-Rehman and Ashraf, M. 2010. Antioxidant flavonoids from *Alhagi maurorum*. J. Asian Nat. Prod. Res. **12(2)**: 138-143.

Ahmadi, F., Kadivar, M. and Shahedi, M. 2007. Antioxidant activity of *Kelussia odoratissima* Mozaff. in model and food systems. Food Chem. **105(1)**: 57-64.

Ahmadi, F., Sadeghi, S., Modarresi, M., Abiri, R. and Mikaeli, A. 2010. Chemical composition, *in vitro* antimicrobial, antifungal and antioxidant activities of the essential oil and methanolic extract of *Hymenocrater longiflorus* Benth. of Iran. Food Chem. Toxicol. **48**: 1137-1144.

Ahmed, D., Arshad, M.A., Asghar, M.N. and Aujla, M.I. 2010. Antioxidant and free radical scavenging potential of *Otostegia limbata*. Asian J. Chem. **22(6)**: 2524-2532.

Ahmed, F., Sadashiva, M.P. and Urooj, A. 2010. Radical scavenging and anticholinesterase activities of standardized extracts of *Ficus racemosa* stem bark. National Conference on Plant Diversity and Plant Health. Univ. Mysore. Souvenir, Abstract pp. 114-115.

Ahmed, F. and Urooj, A. 2010. Hepatoprotective effects of *Ficus racemosa* stem bark against carbon tetrachloride-induced hepatic damage in albino rats. Pharm. Bull. **48(2)**: 210-216.

Ahmed, R.S., Suke, S.G., Seth, V, Chakraborti, A, Tripathi, A.K. and Banerjee, B.D. 2008. Protective effects of dietary ginger (*Zingiber officinalis* Rosc.) on lindane-induced oxidative stress in rats. Phytother. Res. **22(7)**: 902-906.

Ahmed, R.S., Seth, V. and Banerjee, B.D. 2000. Influence of dietary ginger (*Zingiber officinale* Roscoe) on antioxidant defense system in rat.: comparison with ascorbic acid. Indian J. Exp. Biol. **38(6)**: 604-606.

Ahmed, S., Rahman, A., Alam, A., Saleem, M., Athar, M. and Sultana, S. 2000. Evaluation of the efficacy of *Lawsonia alba* in the alleviation of the CCl_4 induced oxidative stress. J Ethnopharmacol. **69**: 157-164.

Ahmed, S., Rahman, A., Mathur, M., Athur, M. and Sultana, S. 2001. Antitumor promoting activity of *Asteracantha longifolia* against experimental hepatocarcinogenesis in rats. Food and Chem. Toxicol. **39**: 19-28.

Ai-Li, J. and Chang-hai, W. 2005. Antioxidant properties of natural components from *Salvia plebeia* on oxidative stability of ascidian oil. Process Biochem. **41(5)**: 1111-1116.doi: 10.1016/j.probio.2005.12.001.

Aiyegoro, O.A. and Okoh, A.I. 2010. Preliminary chemical screening and *in vitro* antioxidant activities of the aqueous extract of *Helichrysum longifolium*. BMC Complement Altern. Med. **10(1)**: 21.

Ajaikumar, K.B., Babu, B.H., Thankachar, T.T., Josha, K.G. and Padikala, J. 2002. Antioxidant and gastroprotective activity of *Hibiscus tiliaceus* Linn. Amala Res. Bull. **22**: 130-136.

Ajiboye, T.O. 2011. *In vivo* antioxidant potentials of *Piliostigma thonningii* (Schum.) leaves: Studies on hepatic marker enzyme, antioxidant system, drug detoxifying enzyme and lipid peroxidation. Hum. Exp. Toxicol. **30(1)**: 55-62. doi: 10. 1177/ 0960327110366785.

Ajila, C.M., Naidu, K.A., Bhat, U.J.S. and Rao, P. 2007. Boactive compounds and antioxidant potential of mango peel extract. Food Chemistry **105**: 982-988.

Ajith, T. A. and Janardhanan, K.K. 2001. Antioxidant and anti-inflammatory activities of methanol extract of *Phellinus rimosus* (Berk) Pilat. Indian J. Exp. Biol. **39(11)**: 1166-1169.

Ajith, T. A. and Janardhanan, K.K. 2002. Antioxidant and antihepatotoxic activities of *Phellinus rimosus* (Berk) Pilat. J. Ethnopharmacol. **81**: 387-391.

Ajith, T. A. and Janardhanan, K.K. 2006. Chemopreventive activity of a macrofungus *Phellinus rimosus* against N-nitrosodiethylamine induced hepatocellular carcinoma in rat. J. Exp. Ther. Oncol. **5(4)**: 309-21.

Ajith, T.A., Sheena, N. and Janardhanan, K.K. 2006. *Phellinus rimosus* protects carbon tetrachloride induced chronic hepatotoxicity in rats: Antioxidant defense mechanism. Pharm. Biol. **44(6)**: 467-474.

Ajith, T., Nivitha, V. and Usha, S. 2007. *Zingiber officinale* Roscoe alone and in combination with α-tocopherol protect the kidney against cisplatin induced acute renal failure. Food and Chemical Toxicol. **45**: 921-927.

Akanitapichat, P., Phraibung, K., Nuchklang, K. and Prompitakkul, S. 2010. Antioxidant and hepatoprotective activities of five egg plant varieties. Food Chem. Toxicol. **48(10)**: 3017-3021.

Akarta, O. and Odozi, T.O. 1986. Antioxidant properties of red onion skin tannin extract. Agric. Waste **18**: 299-306.

Akbulut, M., Ercisli, S. and Tosun, M. 2009. Physicochemical characteristics of some wild grown European elderberry (*Sambucus nigra* L.) genotypes. Phcog. Mag. **20**: 320-323.

Akinboro, A., Mohamed, K.B., Asmawi, M.Z., Sulaiman, S.F., Sofiman, O.A. 2011. Antioxidants in aqueous extract of *Myristica fragrans* (Houtt.) suppressed mitosis and cyclophosphamide induced chromosomal aberration of *Allium cepa* L. cell. J. Zheijiang Univ. Sci. B. (Biomed. and Biotechnol.) **12(11)**: 915-922.

Akindahunsi, A.A. and Salawu, S.O. 2005. Antioxidant indices of some green leafy vegetables. Trop. Sci. **45**: 35-

Akinmoladun, A.C., Ibukun, E.O., Afor, E., Akinrinlola, B.L., Onibon, T.R., Akinboboy, A.O., Obuotor, E.M. and Farombi, E.O. 2007. Chemical constituents and antioxidant activity of *Alstonia boonei*. Afr. J. Biotech. **6(10)**: 1197-1201.

Akinmoladun, A.C., Obuotor, E.M. and Farombi, E.O. 2010. Evaluation of antioxidant and free radical scavenging capacities of some Nigerian indigenous medicinal plants. J. Med. Food. **13(2)**: 444-451.

Akowuah, G.A., Mariam, A. and Chin, J.H. 2009. The effect of extraction temperature on total phenols and antioxidant activity of *Gynura procumbens* leaf. Phcog. Mag. **4(17)**: 81-85.

Akowuah, G.A., Zhari, I., Norhayati, I. Khamsah, S.M., Nor Amin, M.N., Karupaiah, S. and Zakri, Z. 2003. Antioxidant activity of aqueous leaf extracts of *Orthosiphon stamineus*. J.Trop. Med. Plants **4**: 163-168.

Akroum, S., Satta, D. and Lalaoni, K. 2009. Antimicrobial, antioxidant, cytotoxic activities and phytochemical screening of some Algerian plants. Eur. J. Sci. Res. **31(2)**: 289-295.

Akter, R., Hasan, S.M.R., Hossain, M.M., Jamila, M., Mazumder, M.E.H. and Rahman, S.2009. *In vitro* antioxidant and *in vivo* antidiarrhoeal activity of hydromethanolic extract of *Xanthium indicum* Koenig leaves. European J. Sci. Res. **33(2)**: 305-312.

Akter, R., Hasan, S.M.R., Siddiqua, S.A., Majumdar, M.M., Hossain, M.M., Alam, M.A., Haque, S. and Ghani, A. 2008. Evaluation of analgesic and antioxidant potential of the leaves of *Curcuma alismatifolia* Gagnep. S.J. Pharm. Sci. **1(1&2)**: 3-9.

Aladedunye, F.A., Okorie, D.A. and Ighodaro, O.M. 2008. Anti-inflammatory and antioxidant activities and constituents of *Platostoma africanum* P.Beauv. Nat. Prod. Res. **22(12)**: 1067-1073.

Alam, M.A., Ghani, A., Subhan, N., Rahman, M.M., Haque, M.S., Majumder, M.M., Mazumder, M.E.H., Akter, R.A., Nahar, L. and Sarker, S.D. 2008a. Antioxidant and membrane stabilizing properties of the flowering tops of *Anthocephalus cadamba*. Nat. Prod. Comm. 3: 65-67.

Alam, M.A., Nyeem, M.A.B., Sawal, M.A., Mostofa, M., Alam, M.S., Subhan, N. and Rahman, M.M. 2008b. Antioxidant and hepatoprotective action of the crude methanolic extract of the flowering top of *Rosa damascena*. Oriental Pharmacy and Experimental Medicine **8**: 164-170.

Alam, M.A., Rahman, M.M., Subhan, N., Majumder, M., Hasan, S.M.R., Akter, R., Mazumder, M.E.H. and Faruque, A. 2009. Antioxidant potential of the ethanol extract of the leaves of *Vitex negundo* L. Turk. J. Pharm. Sci. **6(1)**: 11-20.

Alam, M.I., Auddy, B. and Gomes, A. 1994. Isolation, purification and partial characterization of viper venom inhibiting factor from the root extract of the Indian medicinal plant sarsaparilla (*Hemidesmus indicus* R.Br.). Toxicon **32**: 1551.

Alam, M.I. and Gomes, A. 1998a. Viper-venom-induced inflammation and inhibition of free radical formation by pure compound (2-hydroxy-4-methoxy benzoic acid) isolated and purified from anantamul (*Hemidesmus indicus* R.Br.) root extract. Toxicon **36**: 207.

Alam, M.I. and Gomes, A. 1998b Adjuvant effects and antiserum action potentiation by a (herbal) compound 2-hydroxy-4-methoxy benzoic acid isolated from the root extract of the Indian medicinal plant 'sarsaparilla' (*Hemidesmus indicus* R.Br.) Toxicon **36**: 1423.

Alam, M.S., Kaur, G., Jabbar, Z., Javed, K. and Athar, M. 2006. Evaluation of antioxidant activity of *Salix capraea* flowers. Phytother. Res. **20**: 479-483.

Alam, M.S., Kaur, G., Jabbar, Z., Javed, K. and Athar, M. 2007. *Eruca sativa* seeds possess antioxidant activity and exert protective effect on mercuric chloride induced renal toxicity. Food Chem. Toxicol. **45(6)**: 910-920.

Alang, G., Kaur, R., Singh, A., Budhlakoti, P. and Singh, A. 2010. Free radical scavenging activity of methanolic root extract of the plant *Crotalaria burhia* Linn. Pharmacology online **1**: 78-85.

Alasalvar, C., Gregor, J.M., Hang, D., Quantick, P.C. and Shahidi, F. 2001. Comparison of volatiles, phenolics, sugars, antioxidant vitamins and sensory quality of different colored carrot varieties. J. Agric. Food Chem. **49**: 1410-1416.

Alasalvar, C., Karama 2006. Antioxidant and antiradical activities in extracts of hazelnut kernel (*Corylus avellana* L.) and hazelnut green leafy cover. J. Agric. Food Chem. **54(13)**: 4826-32.

Al-Bandak, G. and Oreopoulou, V. 2011. Inhibition of lipid oxidation in fried chips and cookies by *Majorana syriaca*. Int. J. Food Sci. Tech. **46(2)**: 290-296.

Al-Bandak, G., Tsironi, T, Taoukis, P. and Oreopoulou, V. 2009. Antimicrobial and antioxidant activity of *Majorana syriaca* in yellowfin tuna. Int. J. Food Sci. Tech. **44(2)**: 373-379.

Albanes, D. and Hartman, T.J. 1999. Antioxidants and cancer: evidence from human observational studies and intervention trials. In: Papas, A.M. (ed.). Antioxidant status, diet and health. CRC Press, Boca Raton, Florida, pp. 497-54.

Albayrak, S., Aksoy, A. and Hamzaoglu, E. 2008. Determination of antimicrobial and antioxidant activities of Turkish endemic *Salvia halophila* Hedge. Turk. J. Biol.**32**: 265-270.

Albayrak, S., Aksoy, A., Sagdic, O. and E. Hamzaoglu. 2010. Compositions, antioxidant and antimicrobial activities of *Helichrysum* (Asteraceae) species collected from Turkey. Food Chem. **119(1)**: 114-122.

Albayrak, S., Sagdic, O., Aksoy, A. and Hamzaoglu, E. 2008. Antimicrobial and antioxidant activities of *Helichrysum* species from the Mediterranean region of Turkey. Asian J. Chem. **20**: 3143-3152.

Al-Burtamani, S.K.S. and Fatope, M.O. 2004. Flavonoids from *Commiphora wightii*. IUPAC International Conference on Biodiversity and Natural Products Chemistry and Medicinal Applications. New Delhi, p. 29, 26-31.

Alcantara, J.M., Yamaguchi, K.K.L. and Junior, V.F.V. 2010. Composicao quimica de oleos essenciais de especies de *Aniba* e *Licaria* e suas actividades antioxidante e antiagregante plaquetaria. Quim. Nova **33**: 141-145.

Al-Dabbas, M.M., Al-Ismail, K. and Al-Qudah, Y. 2010. Antioxidant activity of different extracts from *Varthemia iphionoides*. Larivista Italiana Delle Sostane Grasse **87**: 243-249.

Al-Dabbas, M.M., Kitahara, K., Suganuma, T., Hashimoto, F. and Tadera, K. 2006a. Antioxidant and alpha-amylase inhibitory compounds from aerial parts of *Varthemia iphionoides* Boiss. Biosci. Biotechnol. Biochem. **70(9)**: 2178-2184.

Al-Dabbas, M.M., Suganuma, T., Kitahara, K., Hou, D.X. and Fuji, M. 2006b. Cytotoxic, antioxidant and antibacterial activities of *Varthemia iphionoides* Boiss. extracts. J. Ethnopharmacol. **108(2)**: 287-293.

Aldini, G., Piccoli, A., Beretta, G., Morazzoni, P., Riva, A., Marinello, C and Facino, R.M. 2006. Antioxidant activity of polyphenols from solid olive residues of c.v.Coratina. Fitoterapia **77**: 121-128.

Al-Dosani, M.S. 2010. Antioxidant and protective effects of spinach (*Spinacea oleracea*) leaves against carbon tetrachloride-induced liver injury. Clin. Exp. Med. J. **4(1)**: 129-140. doi: 10.1556/CEMED.4.2010.1.13

Alesiani, D., Canini, A., D'Abrosca, B., DellaGreca, M., Fiorentino, A., Mastellone, C., Monaco, P. and Pacifico, S. 2010. Antioxidant and antiproliferative activities of phytochemicals from quince (*Cydonia vulgaris*) peels. Food Chem. **118(2)**: 199-207.

Al-Farsi, M., Alasalvar, C., Morris, A., Baron, M. and Sahidi, F. 2005. Comparison of antioxidant activity, anthocyanins, carotenoids, and phenolics of three native fresh and sun-dried date (*Phoenix dactylifera* L.) varieties grown in Oman. J. Agric. Food Chem. **53**: 7592-7599.

Al-Fatimi, M., Wuster, M., Schroeder, G. and Lindequist, U. 2007. Antioxidant, antimicrobial and cytotoxic activities of selected medicinal plants from Yemen, J. Ethnopharmacol. **111**: 657-666.

Alfredo-Flores. 2007. Great Guava! Tropical fruits offer nutrition-along with color, taste and variety. Agric. Res. Mag. **55(9)**: 10-11.

Ali, B.H., Alqarwi, A.A., Bashir, A.K. and Tanira, M.O. 2000. Antioxidant action of extract of the traditional medicinal plant, *Rhazya stricta* Decne. in rats. Phytother. Res. **14**: 469-471.

Ali, M., Thomson, M. and Afzal, M. 2000. Garlic and onions: their effect on eicosanoid metabolism and its clinical relevance. Prostaglandins Leukor Essent. Fatty Acids **62**: 55.

Ali, M.M., Elgindi, M.R., Abd El Kader, M.A., Mahmoud, I.I and Atia, H.G. 2004. Antioxidant, anti-inflammatory, and analgesic activities of the chloroform extract of *Crotalaria aegyptiaca*. Untitled Document of Faculty of Pharmacy, Cairo Univ. www.Pharma.edu.eg./Files/BFOPCU/Vol. 42(2).

Ali, S.S., Naresh Kasauju, Abhinav Luthra, Angad Singh, Hallihosur Sharanabasava, Abhishek Sahu and Utpal Bora 2007. Indian medicinal herbs as sources of antioxidants. Food Research International **41**: 1-15.

Aligiannis, N., Mitaku, S., Tsitsa-Tsardis, E, Harvala, C., Tsaknis, I., Lalas, S. and Haroutounian, S. 2003. Methanolic extract of *Verbascum macrurum* as a source of natural preservatives against oxidative rancidity. J. Agric. Food Chem. **51**: 7308-7312.

Al-Ismail, K. and Aburjai, T. 2004. Antioxidant activity of water and alcohol extracts of chamomile flowers, anise seeds and dill seeds. J. Science of Food and Agriculture **84**: 173-178.

Aliyu, B., Ibrahim, H., Musa, A.M., Ibrahim, M.A., Oyewale, A.O. and Amupitan, J.O. 2010. *In vitro* evaluation of antioxidant activity of *Anisopus mannii* N.E.Br. Afr. J. Biotech. **9(16)**: 2437-2441.

Allais, D.P., Simon, A., Bennini, B., Chulia, A.J., Kaouadji, M. and Delage, C. 1991. Flavone and flavonol glycosides from *Calluna vulgaris*. Phytochemistry **30**: 3309-3101.

Alma, M.H., Mavi, A., Yildirim, A., Digrak, M. and Hirata. 2003. Screening chemical composition and *in vitro* antioxidant and antimicrobial activities of essential oils from *Origanum syriacum* growing in Turkey. Biol. Pharm. Bull. **26**: 1725-1729.

Al-Mamun, M., Yamaki, K., Masumizu, T., Nakai, Y., Saito, K., Sano, H. and Tamura, Y. 2007. Superoxide anion radical scavenging activities of herbs and pastures in Northern Japan determined using electron spin resonance spectrometry. Int. J. Biol. Sci. **3**: 349-355.

Almeida, I.F., Fernandes, E., Lima, J.L.F.C., Costa, P.C. and Bahia, F.M. 2008. Walnut (*Juglans regia*) leaf extracts are strong scavengers of pro-oxidant reactive species. Food Chem. 106: 1014-1020.

Almeida, I.F., Fernandes, E., Lima, J.L.F.C., Costa, P.C. and Bahia, F.M. 2008. Protective effect of *Castanea sativa* and *Quercus robur* leaf extracts against oxygen and nitrogen reactive species. J. Photochem. Photobiol. **91(2)**: 87-95.

Almela, L., Sanchez-Munoz, B., Fernandez-Lopez, J.A., Roca, M. and Rabe, V. 2006. Liquid chromatographic-mass spectrometric analysis of phenolics and free radical scavenging activity of rosemary extract from different raw material. J. Chromatogr. A. **1120**: 221-229.

Al-Mustafa, A. and Al-Thunibat, O.Y. 2008. Antioxidant activity of some Jordanian medicinal plants used traditionally for the treatment of diabetes. Pakistan J. Biol. Sci. **11**: 351-358.

Alothman, M., Bhat, R. and Karim, A.A. 2009. Antioxidant capacity and phenolic content of selected tropical fruits from Malaysia extracted with different solvents. Food Chem. **115(3)**: 785-788.

Alpsoy, L., Uzonur, I., Sakcali, M.S. and Karaman, S. 2010. Antioxidant and antimutagenic activities of *Viscum album* fruit ethanolic extract in human lymphocytes. African J. Biotech. **9(17)**: 2539-2543.

Al-Reza, M., Rahman, A., Sattar, M.A., Rahman, M.O. and Fida, H.M. 2010. Essential oil composition and antioxidant activities of *Curcuma aromatica* Salisb. Food Chem. Toxicol. **48(b)**: 1757-1760.

Al-Reza, S.M., Bajpai, V.K. and Kang, S.C. 2009. Antioxidant and antiliterial effect of seed essential oil and organic extracts from *Zizyhus jujuba*. Food Chem. Toxicol. **47(9)**: 2374-2380.

Al-Reza, S.M., Rahman, A., Cho, Y.-S., and Kang, S.C. 2010. Chemical composition and antioxidant activity of essential oil and organic extracts of *Cestrum nocturnum* L. J. Essential Oil bearing plants **13(5)**: 615-624.

Al-Saikhan, M.S., Howar, L.R. and Miller, J.C. Jr. 1995. Antioxidant activity and total phenolics in different genotypes of potato (*Solanum tuberosum* L.). J. Food Sci. **60**: 341-343.

Alscher, R.G. and J.L.Hess. 1993. Antioxidants in higher plants. CRC Press.

Alterejos, J., Salido, S., Perez-Bonilla, M., Linares-Palomino, P.J., van Beek, T.A., Nogueras, M. and Sanchez, A. 2005. Preliminary assay on the radical scavenging activity of olive wood exracts. Fitoterapia **76(3-4)**: 348-351.

Altun, M.L., Citoglu, G.S., Yilmaz, B.S. and Coban, T. 2008. Antioxidant properties of *Viburnum opulus* and *Viburnum lantana* growing in Turkey. Int. J. Food Sci. Nutr. **59(3)**: 175-180.

Alvarez, E., Leiro, J.M., Rodriguez, M. and Orallo, F. 2004. Inhibitory effects of leaf extracts of *Stachytarpheta jamaicensis* (Verbenaceae) on the respiratory burst of rat macrophages. Phytother. Res. **18(6)**: 457-462.

Alvarez-Jubete, L., Wijngaard, H., Arendt, E.K. and Gallagher, E. 2010. Polyphenol composition and *in vitro* antioxidant activity of amaranth, qunoa, buckwheat and wheat as affected by sprouting and baking. Food Chem. **119(2)**: 770-779.

Alviano, D.S. and Alviano, C.S. 2009. Plant extracts: Search for new alternatives to treat microbial diseases. Current Pharmaceutical Biotech. **10**: 106-121.

Alviano, W.S., Alviano, D.S., Diniz, C.G., Antoniolli, A.R., Alviano, C.S., Farias, L.M., Carvalho, M.A.R., Souza, M.M.G. and Bolognese, A.M. 2008. *In vitro* antioxidant potential of medicinal plant extracts and their activities against oral bacteria based on Brazilian folk medicine. Arch. Oral Biol. **56(3)**: 545-552.

Amado, L.L., Jaramillo M., Rocha, A.M., Ferreira, J.L.R., Garcia, L.M., Ramos, P.B. *et al.*, 2007. A new method to evaluate total antioxidant capacity against reactive oxygen and nitrogen species (RONS) in aquatic organisms. Comparative Biochemistry and Physiology – part A: Molecular and Integrative Physiology. **148**: S75-S76.

Amakura, Y., Umino, Y., Tsuji, S., Ito, H., Hatano, T., Yoshida, T. and Tonogai, Y. 2002. Constituents and their antioxidative effects in eucalyptus leaf extract used as a natural food additive. Food Chem. **77**: 47-56.

Amakura, Y., Yoshimura, M., Sugimoto, N., Yamazaki, T. and Yoshida, T. 2009. Marker constituents of the natural antioxidant *Eucalyptus* leaf extract for the evaluation of food additives. Biosci. Biotech. Biochem. **73(5)**: 1060-1065.

Amarowicz, R., Estrella, I., Hernandez, T., Robredo, S., Troszynska, A., Kosinska, A and Pegg, R.B. 2010. Free radical-scavenging capacity, antioxidant activity, and phenolic composition of green lentil (*Lens culinaris*). Food chem. **121(3)**: 705-711.

Amarowicz, R., Naczk, M., and Shahidi, F. 2000. Antioxidant activity of crude tannins of canola and rapeseed hulls. J. Amer.Oil Chemists Soc. **77**: 957-961.

Amarowicz, R., Naczk, M., and Shahidi, F. 2000. Antioxidant activity of various fractions of non-tannin phenolics of canola hulls. J. Agric. Food Chem. **48**: 2755-2759.

Amarowicz, R., Pegg, R. B., Rahimi–Moghaddam, P., Barl, B. and Weil, J. A. 2004. Free-radical scavenging capacity and antioxidant activity of selected plant species from the Canadian prairies. Food Chem. **84**: 551–562.

Amarowicz, R., Troszynska, A. and Pegg, R. B. 2008. Antioxidative and radical scavenging effects of phenolics from *Vicia sativa*. Fitoterapia **79((2)**: 121-122.

Ames, B.N., Shigenga, M.K. and Hagen, T.M. 1993. Oxidants, antioxidants and the degenerative diseases of ageing. Proc. Nat. Acad. Sci. USA **90**: 7915-7922.

Amin, I., Norazaidah, Y. and Hainida K.I. E. 2006. Antioxidant activity and phenolic content of raw and blanched *Amaranthus* species. Food Chem. **94 (1)**: 47-52.

Aminimoghadamfarouj, N., Nematollahi, A. and Wiart, C. 2011. Anti-bacterial, antioxidant activity and phytochemical study of *Uvaria grandiflora*: A rare species of Annonaceae. J. Pharm. Res. **4(4)**:

Amom, Z., Bahari, H., Isemaail, S., Ismail, N.A., Shah, Z.Md. and Arsyad, M.S. 2009. Nutritional composition, antioxidant ability and flavonoid content of *Tinospora crispa* stem. Advances in Natural and Applied Sciences.

Amresh, G., Venkateswara Rao, C. and Singh, P.N. 2007. Antioxidant activity of *Cissampelos pareira* on benzo(a)pyrene-induced mucosal injury in mice. Nutr. Res. **27(10)**: 625-632.

Amro, B., Aburjai, T. and Sl-Khalil, S. 2003. Antioxidative and radical scavenging effects of olive cake-extract. Fitoterapia **73**: 456-461.

An, L.J., Guan, S., Shi, G.F., Bao, Y.M., Duan, Y.L. and Jiang, B. 2005. Protocatechuic acid from *Alpinia oxyphylla* against MPP-induced neurotoxicity in PC12 ceells. Food Chem. Toxicol. **44(3)**: 436-443.

An, R.B., Oh, S.H., Jeong, G.S. and Kim, Y.C. 2006.Gomisin J with protective effect against t-BHP-induced oxidative damage in HT22 cells from *Schizandra chinensis*. Nat. Prod. Sci. **12(3)**: 134-137.

Anagnostopoulou, M.A., Kefalas, P., Papageorgiou, V.P., Assimopoulou, A.N. and Boskou, D. 2005. Radical scavenging activity of various extracts and fractions of sweet orange peel (*Citrus sinensis*). Food Chem. **94(1)**: 19-25.

Anandan, R., Deepika, R. and Devaki, T. 1999. Protective effects of *Picrorrhiza kurroa* on mitochondrial glutathione antioxidant system in D-galactosamine induced hepatitis in rats. Curr. Sci. **76**: 1542-1545.

Anandan, R. and Devaki, T. 1999. Hepatoprotective effect of *Picrorrhiza kurroa* on tissue defence system in D-galactosamine-induced hepatitis in rats. Fitoterapia **70**: 54-57.

Anandan, R., Rekha, R.D. and Devaki, T. 1999. Protective effect of *Picrorrhiza kurroa* on mitochondrial glutathione antioxidant stem in D-glactosamine-induced hepatitis rats. Curr. Sci. **76(12)**: 1543-1545.

Ananthan, R., Baskar, C., Narmatha Bai, V., Pari, L., Latha, M. and Ramkumar, K.M. 2003. Antidiabetic effect of *Gymnema montanum* leaves: Effect on lipid peroxidation induced oxidative stress in experimental diabetes. Pharmacol. Res. **48**: 551-556.

Anarthe, S.J., Bhalke, R.D., Jadhav, R.B. and Surana, S.J. 2008. *In vitro* antioxidant activities of methanol extract of *Dendrophthoe falcata* Linn. stem. Biomed. **3**: 182-189.

Anasuya, N. and Gomathi, R. 2010. *In vitro* antioxidant activity of Indian Oregano. 20th Annual Conference of IAAT and Intern. Symposium on Taxonomy, Plant Diversity and Conservation. Abstract pp. 235.

Anbazhagi, T., Kadavul, K, Suguna, G. and Petrus, A.J.A. 2009. Studies on the pharmacognostical and *in vitro* antioxidant potential of *Cleome gynandra* Linn. leaves. Natural Product Radiance **8(2)**: 151-157.

Andallu, B., Haripriya, A. and Vijayadurga, R. 2003. Antidiabetic and antioxidant role of knoll khol (*Brassica oleracea* var. *caulocarpa*) in NIDDM subjects. Nutrition Society of India, 35th Annual Meeting NIN, Hyderabad p. 41.

Andallu, B. and Varadacharyalu, N.Ch. 2003. Antioxidant role of mulberry (*Morus indica* L. cv. Anantha) leaves in streptozotocin diabetic rats. Clinica Chimica Acta **338**: 3-10.

Andarwulan, N., Batari, R., Sandrasari, D.A., Bollin, B. and Wijaya, H. 2010. Flavonoid content and antioxidant activity of vegetables from Indonesia. Food Chem. **121(4)**: 1231-1235.

Andlauer, W. and Furst, P. 1998. Antioxidative power of phytochemicals with special reference to cereals. Cereal Foods World **43**: 356-359.

Andrade, A.M., Oliveira, J.P.R., Santos, A.L.L.M., Franco, C.R.P., Antoniolli, A.R., Estevan, C.S. and Thomazzi, S.M. 2010. Preliminary study on the anti-inflammatory and antioxidant activities of the leaves extract of *Hyptis fruticosa* Salzm. ex Benth., Lamiaceae. Braz. J. Pharmacogn. **20(6)**: 962-968.

Andreeva, T.I., Komarova, E.N., Yusubov, M.S. and Korotkova, E.I. 2004. Antioxidant activity of cranberry tree (*Viburnum opulus* L.) bark. Pharmaceutical Chem. J. **38(10)**:

Andrensek, S., Simonovska, B., Vovk, I., Fyhrquist, P., Vuorela, H and Vuorela, P. 2004. Antimicrobial and antioxidative enrichment of oak (*Quercus robur*) bark by rotation planar extraction. Int. J. Food Microbiol. **92(2)**: 181-187.

Andreo, M.A., Rodriguez, K.V.B.Hiruma-Lima, C.A., Da Rocha, L.R.M., Souza Brito, A.R.M. and Vilegas, W. 2006. Effect of *Mouriri pusa* extracts on experimentally induced gastric lesions in rodents: role of endogenous sulfhydryls compounds and nitric oxide in gastroprotection. J. Ethnopharmacol. **107**: 431-441.

Andrews, P. K., Fahy, D. A. and Foyer. C. H. 2004. Relationships between fruit exocarp antioxidants in the tomato (*Lycopersicon esculentum*) high pigment-1 mutant during development. Physiologia Plantarum. **120(4)**: 519-528.

Anguiana, A.C., Ramirez, Santoyo, Reglero, G., Rivas, C. and Soler. 2007. Radical scavenging activities, endogenous oxidative enzymes and total phenols in edible mushrooms commonly consumed in Europe. J. Sci. Food Agric. **87(12)**: 2272-2278.

Anil, P., Manish, S., Garvendra, R.S., Vijay, B. and Tarachand, K. 2010. *In vitro* antioxidant studies of *Lagerstroemia speciosa*. Phcog. J. **2(10)**: 357-360.

Anilakumar, K.R., Khanum F., Sudarshanakrishna, K.R. and Santhanam, K. 2004. Effect of amaranth leaves on dimethylhydrazine-induced changes in multicomponent antioxidant system of rat liver. Indian J. Exp. Boil. **42(6)**: 595-600.

Anilakumar, K.R., Saritha, V., Khanum, F. and Bawa, S.A. 2009. Amelioratívbe effect of ajwan extract on hexachlorocyclohexane-induced lipid peroxidation in rat liver. Food Chem. Toxicol. **47**: 279-282.

Anilakumar, K.R., Sudarshanakrishna, K.R., Chandramohan, G., Ilaiyaraja, Khanum, F. and Bawa, A.S. 2010. Effect of *Aloe vera* gel extract on antioxidant enzymes and azoxymethane-induced oxidative stress in rats. Indian J. Exp. Biol. **48**: 837-842.

Aniya, Y., Miyagi, C., Nakandakari, A., Kamiya, S., Imaizumi, N. and Ichiba, T. 2002. Free radical scavenging action of the medicinal herb *Limonium wrightii* from the Okinawa islands. Phytomedicine **9(3)**: 239-244.

Aniya, Y., Shimobukuro, M., Kohatsu, M., Gyamfi, M.A., Kunii, D., Takayama, F. and Egashira, T. 2000. Antioxidant and hepatoprotective actions of the medicinal herb *Artemisia campestris* from the Okinawa islands. Biol. Pharm. Bull. **23(3)**: 309-312.

Aniya, Y., Koyama, T., Miyagi, C., Miyahira, M., Inomata, C., Kinoshita, S. and Ichiba, T. 2005. Free radical scavenging and hepatoprotective actions of the medicinal herb, *Crassocephalum crepidioides* from the Okinawa Islands. Biol. Pharm. Bull. **28(1)**: 19-23.

Annan, K. and Houghton, P.J. 2008. Antibacterial, antioxidant and fibroblast growth stimulation of aqueous extracts of *Ficus asperifolia* Miq. and *Gossypium arboreum* L. wound healing plants of Ghana. J. Ethnopharmacol. **119 (1)**: 141-144.

Annu, W., Lath, P.G., Shaji, J., Anuja, G.I., Suja, S.R., Sini, S., Shyamal, S., Shine, V.J., Shikha, P., Rajasekharan, R. 2010. Anti-inflammatory, analgesic and anti-lipid peroxidation studies on leaves of *Commiphora caudata* (Wight et Arn.) Engl. Indian J. Nat. Prod. Resources **1(1)**: 44-48.

Anonymous. 1992. Wealth of India: Raw materials. 3rd ed. Vol.3. CSIR, New Delhi.

Anonymous 2010. Ultrasonic-microwave synergistic extraction and antioxidant activity of total flavonoids from *Wickstroemia indica* (L.) C.A.Mey. Food Sci. **31(1)**: 101-105

Ansari, N.M. and Bhandari, U. 2009. Cardioprotective activity of *Zingiber officinale* Roscoe in rats. Hamdard Medicus **51(4)**: 26-29.

Ansari, N. M., Houlihan, L., Hussain, B. and Pieroni, A. 2005. Antioxidant activity of five vegetables traditionally consumed by South-Asian migrants in Bradford, Yorkshire, UK. Phytother. Res. **19**: 907–911.

Antal, D.S. 2010. Medicinal plants with antioxidant properties from Banat region (Romania): A rich pool for the discovery of multi-target phytochemicals active in free-radical related disorders. Analele Universitatii din Oradea – Fascicula Biologie **17(1)**: 14-22.

Anto, R.J., Kuttan, Babu, D.K., Rajasekharan, K.N. and Kuttan, R. 1995. Antitumor and antioxidant activity of natural curcuminoids. Cancer Letters **94**: 79-83.

Antolovich, M.; Prenzier, P. and Patsalides, E. 2002. Methods for testing antioxidant activity. Analyst **127**: 183-198.

Anupama, K., Satyanaryana, S., Mukkanti, K. and Abdulla Khan, K. 2010. *In vitro* antioxidant activity of Hab-E-Jund (Unani Medicine) prescribed in febrile conditions. Pharmacology online **1**: 943-949.

Anuradha, K., Naidu, M.M., Manohar, R.S. and Indiramma, A.R. 2010. Effect of Vanilla extract on radical scavenging activity in biscuits. Flavour and Frangrance Journ. **25(6)**: 488-492.

Anyasor, G.N., Ogunwenmo, K.O., Oyelana, O.A. and Akpofunure, B.E. 2010. Phytochemical constituents and antioxidant activities of aqueous and methanol stem extracts of *Costus afer* Ker.-Gawl. (Costaceae). Afri. J. Bioteh. **9(31)**: 4880-4884.

Ao, C., Li, A., Elzaawely, A.A., Xuan, T.D. and Tawata, S. 2008. Evaluation of antioxidant and antibacterial activities of *Ficus microcarpa* L.f. extract. Food Contro. **19**: 940-948.

Aoyagi, Y., Kasuga, A., Fujihara, S. and Sugahara, T. 1995. Isolation of antioxidative compounds from *Saxifraga stolonifera*. J. Jap. Soc. Food Sci. Tech. **42(12)**: 1027-1030.

Aoyama, S. and Yamamoto, Y. 2007. Antioxidant activity of flavonoid content of welsh onion (*Allium fistulosum*) and the effect of thermal treatment. Food Sci. Tech. Res. **13(1)**: 67-72.

Apak, R., Guclu, K., Ozyurek, M., Karademir, S.E., and Ercag, E.2006. The cupric ion reducing antioxidant capacity (CUPRAC) and polyphenolic content of some herbal teas. Int. J. Food Sci. Nutr. **57**: 292-304.

Apati, P., Szentmihalyi, K., Kristo, Sz, T., Papp, I. Vinkler, P., Szoke, E. and Kery, A. 2003. Herbal remedies of *Solidago*, correlation of phytochemical characteristics and antioxidative properties. J. Pharm. Biomed. Anal. **32**: 1045-1053.

Aqil, F., Ahmed, I. and Mehmood, Z. 2006. Antioxidant and free radical scavenging properties of twelve traditionally used Indian medicinal plants. Turk. J. Biol. **30**: 177-183.

Aquino, R., Caceras, A., Morelli, S. and Rastrelli, L. 2002. An extract of *Tagetes lucida* and its phenolic constituents as antioxidants. J. Nat. Prod. **65(12)**: 1773-1776.

Aquino, R., Morelli, S., Lauro, M.R., Abdo, S., Saija, A. and Tomaino. A. 2001. Phenolic constituents and antioxidant activity of an extract of *Anthurium versicolor* leaves. J. Nat. Prod. **64(8)**: 1019-1023.

Aquino, R., Morelli, S., Tomaino, A., Pellegrino, M., Saija, A., Grumetto, L., Puglia, C., Ventura, D., Bonina, F. and Grumetto, L. 2002. Antioxidant and photoprotective activity of a crude extract of *Culcitum reflexum* H.B.K. leaves and their major flavonoids. J. Ethnopharmacol. **79**: 183-191.

Ara, N. and Nur, H. 2009. *In vitro* antioxidant activity of methanolic leaves and flower extracts of *Lippia alba*. Res. J. Med. Med. Sci. **4**: 107-110.

Arabshahi, D.S., Visalakshi Devi, D. and Urooj, A. 2005. Evaluation of antioxidant activity of some plant extracts and their heat, pH and storage stability. Food Chem. **100 (3)**: 1100-1105.

Araghiniknam, M., Chung, S., Nelson-White, T., Eskelson, C. and Watson, R.R. 1996. Antioxidant activity of dioscorea and dehydroepiandrosterone (DHEA) in older humans. Life Sci. **59**: PL147-157.

Ara Jahan, I., Nahar, N., Mosohuzzaman, M., Rokeya, B., Ali, L., Khan, A.K.A., Makhmur, T. and Choudhary, M.I. 2009. Hypoglycemic and antioxidant activities of *Ficus racemosa* Linn. fruits. Nat. Prod. Res. **23(4)**: 399-408.

Arena, E., Fallico, B. and Maccarone, E. 2001. Evaluation of antioxidant capacity of blood orange juice as influenced by constituents, concentration process and storage. Food Chem. **74**: 423-427.

Argolo, A.C.C., Sant'Ana, A.*E.G.*, Pletsch, M. and Coelho, L.C.B.B. 2004. Antioxidant activity of leaf extracts from *Bauhinia monandra*. Bioresource Technol. **95(2)**: 229-233. doi: 10.1016/j.biotech.2003.12.014. PMID: 15246449

Arlorio, M., Coisson, J.D., Travaglia, F., Varsaldi, F., Miglio, G., Lombardi, G. and Marteelli, A. 2005. Antioxidant and biological activity of phenolic pigments from *Theobroma cacao* hulls extracted with supercritical CO_2. Food Res. Int. **38(8-9)**: 1009-1014.

Armata, M., Gabrieli, C., Termentzi, Zervou, M. and Kokkalou, E. 2008. Constituents of *Sideritis syriaca* ssp. *syriaca* (Lamiaceae) and their antioxidant activity. Food Chem. **111(1)**: 179-186.

Arora, P., Chawla, R., Chaudhary, P., Sharma, R.K., Puri, S.C., Gupta, V.K., Handa, G. and Qazi, G.N. 2004. Antioxidant activities of fractionated extract of *Podophyllum hexandrum* Royle. Chemistry Biology Interface: Synergistic New frontiers, New Delhi p25-46, November 21-26, 2004.

Arora, R., Chawla, R., Puri, S.C., Sagar, R., Singh, S., Kumar, R., Sharma, A.K., Prasad, J., Singh, S., Kaur, G., Chaudhary, P., Qazi, G.N. and Sharma, R.K. 2005. Radioprotective and antioxidant properties of low-altitude *Podophyllum hexandrum* (LAPH). J. Environ. Pathol. Toxicol. Oncol. **24(4)**: 299-314.

Arora, A., Sairam, R.K. and Srivastava, G.C. 2002. Oxidative stress and antioxidative system in plants. Curr. Sci. **82(10)**: 1127-1238.

Arrendondo, M.F., Blasina, F., Echeverry, C., Morquio, A.Ferreira, M., Abin-Carriquiry, J.A., Lafon. L. and Dajas, F. 2004. Cytoprotection by *Achyrocline satureioides* (Lam.) DC. and some of its main flavonoids against oxidative stress. J. Ethnopharmacol. **91(1)**: 13-20.

Arriaga, A.M., Lima, J.Q., e Vasconcelos, J.N., de Oliveira, M.C., Lemos, T.L., Fneska, A.M., Malcher, G.T., Santiago, G.M., Mafezoli, J. and Braz-Filho, R. 2009. Antioxidant and larvicidal activities of *Tephrosia egregia* Sandw. against *Aedes aegypti*. Nat. Prod. Commun. **4(4)**: 529-530.

Arruda, S.F., Siqueira, E.M.A. and Souza, E.M.T. 2004. Malanga (*Xanthosma sagittifolium*) and Purslane (*Portulaca oleracea*) leaves reduce oxidative stress in vitamin A-deficient rats. Nutrition and Metabolism **48(4)**: 288-295.

Arteel, G.E. 2003. Oxidants and antioxidants in alcohol-induced liver disease. Gastroenterology **124**: 778-790.

Arulmozhi, S., Mazumder, P.M., Ashok, P. and Narayan S.L. 2007. *In vitro* antioxidant and free radical scavenging activity of *Alstonia scholaris* (Linn.) R. Br. Iranian J. Pharmocol. Therapeutics. **6(2)**: 191-196.

Arulmozhi, S., Mazumder, P.M., Narayan, L.S. and Thakurdesai, P.A. 2010. *In vitro* antioxidant and free radical scavenging activity of fractions from *Alstonia scholaris* (L.) R.Br. Int. J. Pharmatech. Res. **2(1)**: 18-25.

Arumugam, P., Padmaja, M.V., Santhiya, S.S. and Ramesh, A. 2004. *In-vitro* screening of antioxidant activity in different fractions of *Mentha viridis* Linn. Chemistry Biology interface: Synergistic New frontiers, New Delhi, India P. 23-8, November 21-26, 2004.

Arunabh, B., Abhijit, C., Ghosal, S. and Battacharya, S.K. 1999. Antioxidant activity of active tannoid principles of *Emblica officinalis* (amla). Indian J. Exp. Biol,. **37**: 676-680.

Aruoma, O.I. 1998. Free radicals, oxidative stress and antioxidants in human health and disease. J. Am. Oil Chem. Soc. **75**: 199-212.

Aruoma, O.I. 2003. Methodological considerations for characterizing potential antioxidant actions of bioactive components in food plants. Mut. Res. **523-524**: 9-20.

Aruoma, A.M., Butler, M.J. and Halliwell, B. 1993. Evaluation of the antioxidant and prooxidant actions of gallic acid and its derivatives. J. Agric. Food Chem. 41: 1880-1885.

Aruoma, O.I. and Cuppett, S.L. 19994. Anioxidant methodology: *In Vivo* and *In vitro* concepts. AOC Press, Illinois.

Arya, D.S., Nandave, M.,Ojha, S.K., Kumari, S., Joshi, S. and Mohanty, I. 2006. Myocardial salvaging effects of *Ocimum sanctum* in experimental model of myocardial necrosis: A haemodynamic, biochemical and histoarchitectural assessment. Curr. Sci. **91(5)**: 667-672.

Arya, V. and Yadav, J.P 2010. Antioxidant activity and total phenols in leaves extracts of *Cassia tora* L. Pharmacology online **2**: 1030-1036.

Asgarirad, H., Pourmorad, F., Hosseinimehr, S.J., Saeidnia, S., Ebrahimzadesh, M.A. and Lotfi, F. 2010. *In vitro* antioxidant analysis of *Achillea tenuifolia*. African J. Biotechn. **9(24)**: 3536-3541.

Asamari, A.M., Addis, P.B., Epley, R.J. and Krick, T.P. 1996. Wild rice hull antioxidants. J. Agric. Food Chem. **44**: 126-130.

Asha, V.V., Akhila, S., Wills, P.J. and Subromaniam, A. 2004. Further studies on the antihepatotoxic activity of *Phyllanthus maderaspatensis* Linn. J. Ethnopharmacol. **92**: 67-70.

Ashawat, M.S. and Saraf, S. 2007. Antioxidant activity of skin care herbal cosmetic cream and lotion. Plant archives **7(2)**: 685-687.

Ashida, S., Noguchi, S.F. and Suzuki, T. 1994. Antioxidative components, xanthone derivatives, in *Swertia japonica* Makino. J. Amer. Oil Chem. Soc. **71(10)**: 1095-1099. doi: 10.1007/BF02675902.

Ashok Kumar, B.S., Lakshman, K., Jayaveera, K.N., Shekar, D.S., Arun Kumar, A. and Manoj, B. 2010. Antioxidant and antipyretic properties of methanolic extract of *Amaranthus spinosus* leaves. Asian Pacific J. Trop. Medicine **3(9)**: 702-706.

Ashok Kumar, B.S., Lakshman, K., Nandeesh, R., Arun Kumar, P.A., Manoj, B., Vinod Kumar and Shkar, D.S. 2010. *In vitro* alpha-amylase inhibition and *in vivo* antioxidant potential of *Amaranthus spinosus* in alloxan-induced oxidative stress in diabetic rats. Saudi J. Biol. Sci. **18(1)**: 1-5.

Ashokkumar, D., Mazumder, U. K., Gupta, M., Senthilkumar, G.P. and Selvan, V. T. 2008. Evaluation of Antioxidant and free radical scavenging activities of *Oxystelma esculentum* in various *In vitro* Models. J. Complement. Integr. Med. **5(1)**: Article 9. DOI: 10.2202/1553-3840.1124

Ashokkumar, D., Selvan, V.T., Mazumdar, U.K. and Gupta, M. 2009. Antineoplastic and antioxidant activities of *Oxystelma esculentum* on swiss albino mice bearing Ehrlich's ascites carcinoma. Pharmac. Biol. **47(3)**: 195-202.

Ashok Kumar, K., Uma Maheshwari, M., Shivashanmugam, A.T., Subhadra Devi, V., Subhashini, N. and Ravi, T.K. 2008. Antioxidant activities of *Erythrina stricta* Roxb. using various *in vitro* and *ex vivo* models. Int. J. Oriental Pharmacy Exp. Med. **8**: 266-278.

Ashok Kumar, K., Uma Maheshwari, M., Shivashanmugam, A.T., Subhadra Devi, V., Subhashini, N. and Ravi, T.K. 2009. Free radical scavenging and antioxidant activities of *Glinus oppositifolius* (Carpet weed) using different *in vitro* assay systems. Pharm. Biol. **47(6)**: 474-482.

Ashour, M.L., El-Readi, M., Youn, M., Mulyaningsih, S., Sporer, F., Efferth, T. and Wink, M. 2009. Chemical composition and biological activity of the essential oil obtained from *Bupluerum marginatum* (Apiaceae). J. Pharm. Pharmacol. **61(8)**: 1079-1087.

Aslan, A., Gulluce, M., Sokmen, M., Adiguzel, A., Sahin, F. and Ozkan, H. 2006. Antioxdant and antimicrobial properties of the lichens *Cladonia foliacea, Dermatocarpon miniatum, Evarinia divaricata, Evernia prunastri* and *Neofuscella pulla*. Pharm. Biol. **44(4)**: 247-252.

Aslan, M., Orhan, D.D., Orhan, N, Sezik, E. and Ysilada, E. 2007. *In vitro* antidiabetic and antioxidant potential of *Helichrysum plicatum* ssp. *plicatum* capitulums in streptozotocin-induced diabetic rats. J. Ethnopharmacol. **109**: 54-59.

Aslan, M., Orhan, D.D., Orhan, N, Sezik, E. and Ysilada, E. 2007. A study of antidiabetic and antioxidant effects of *Helichrysum graveolens* capitulums in streptozotocin-induced diabetic rats. J. Med. Food **10(2)**: 396-400.

Aslan, M., Orhan, N, Orhan, D.D. and Ergun, F. 2010. Hypoglycemic activity and antioxidant potential of some medicinal plants traditionally used in Turkey for diabetes. J. Ethnopharmacol. **128(2)**: 384-389.

Asmah, R., Vijay, K. and Huzaimah, A.S. 2000. Determination of total antioxidant activity in three types of vegetables shoots. In: Towards Bridging science and Herbal industry. Kaula Lumpur: FRIM.

Asolini, F.C., Tedesco, A.M., Carpes, S.T., Feraz, C. and de Alencar, S.M. 2006. Antioxidant and antibacterial activities of phenolic compound from extracts of plants used as tea. Brazilian J. Food Tech. **9(3)**: 209-215.

Assimopoulou, A.N., Bosckou, D. and Papageorgiou, V.P. 2004. Antioxidant activities of alkannin, shikonin and *Alkannia tinctoria* root extracts in oil substrates. Food Chem. **87(3)**: 433-438.

Assimopoulou, A.N., Sinakous, Z. and Papageorgiou, V.P. 2005. Radical scavenging activity of *Crocus sativus* L. extract and its bioactive constituents. Phytother. Res. **19(11)**: 997-1000.

Assimopoulou, A.N., Zlatanos, S.N. and Papageorgiou, V.P. 2005. Antioxidant activity of natural resins and bioactive triterpenes in oil substrates. Food Chem. **92(4)**: 721-727.

Astley, S.B. 2003. Dietary antioxidants – past, present and future? Trends in Food Sci. Tech. **14**: 93-98.

Atanassova, D., Kefalas, P. and Psillakis, E. 2005. Measuring the antioxidant activity of olive oil mill waste water using chemiluminescence. Environ. Int. **31(2)**: 275-280.

Atawodi, S.E. 2005. Antioxidant potential of African medicinal plants. African J. Biotechnol. **4 (2)**: 128-133.

Athanasas, K., Magiatis, P., Fokialakis, N., Skaltsounis, A.L., Pratsinis, H. and Klestas, D. 2004. Hyperjovinols A and B: Two new phloroglucinol derivatives from *Hypericum jovis* with antioxidant activity in cell cultures. J. Nat. Prod. **67(6)**: 973-977.

Athukorala, Y., Kim, K.N., and Jeon, Y.J. 2006. Antiproliferative and antioxidant properties of an enzymatic hydrolysate from brown alga, *Ecklonia cava*. Food Chem Toxicol. **44(7)**: 1065-1074.

Atoui, A.K., Mansouri, A., Boscou, G. and Kefalas, P. 2005. Tea and herbal infusions: their antioxidant activity and phenolic profile. Food Chem. **89**: 27-36.

Attaguile, G., Russo, A., Campisi, A., Savoca, F., Acquaviva, R., Ragusa, N. and Vanella, A. 2000. Antioxidant activity and protective effect on DNA cleavage of extracts from *Cistus incanus* L. and *Cistus monspeliensis* L. Cell Biol. Toxicol. **16(2)**: 83-90.

Auddy, B., Ferreira, M., Blasina, F., Lafon, L., Arredondo, F., Dajas, F., Tripathi, P.C., Seal, T. and Mukherjee B. 2003. Screening of antioxidant activity of three Indian medicinal plants, traditionally used for the management of neurodegenerative diseases. J. Ethnopharmacol. **84(2-3)**: 131-138.

Avci, G., Kupeli, E., Eryavuz, A., Yesilada, E. and Kucukkurt, I. 2006. Antihypercholesterolaemic and antioxidant activity assessment of some plants used as remedy in Turkish folk medicine. J. Ethnopharmacol. **107**: 418-423.

Avelessi, F., Alitonou, G., Sohounhlou, D.K., Menut, C. and Bessiere, J.M. 2005. Aromatic plants of tropical west Africa, part XIV. Chemical and biological investigation of *Lippia multiflora* Mold. essential oil from Benin. J. Essential Oil Res. **17(4)**: 405-407.

Awaad, A.S., El-Sayed, N.H., Maitland, D.J. and Mabry, T.J. 2006. Phenolic antioxidants from *Casimiroa edulis* leaves. Pharm. Biol. **44(4)**: 258-262.

Awah, F.M., Uzoegwu, P.N., Oyugi, J.O., Rutherford, J, Ifeonu, P., Yao, X-J., Fowke, K.R. and Eze, M.O. 2010. Free radical scavenging activity and immunomodulatory effect of *Stachytarpheta angustifolia* leaf extract. Food Chem. **119(4)**: 1409-1416.

Awaika, J.M., Rooney, L.W., Wu, X., Prior,.L. and Cisneros-Zavallos, L.2003. Screening methods to measure antioxidant activity of sorghum (*Sorghum bicolor*) and sorghum products. J. Agric. Food Chem. **51(23)**: 6657-6662.

Ay, M., Bahadori, F., Ozturk, M., Kolak, U. and G.Topcu. 2007. Antioxidant activity of *Erica arborea*. Fitoterapia **78**: 571-573.

Aydin, A., Sener, B., Cakici, I., Turan, N.N. and Erdemoglu, N. 2006. Antioxidant activities of some Lamiaceae plant extracts. Phytother. Res. **20**: 91-93.

Ayoola, G.A., Coker, H. A. B., Adesegun, S. A., Adepoju-Bello, A. A., Obaweya K., Ezennia, E.C. and Atangbayila, T.O. 2008. Phytochemical Screening and antioxidant activities of some selected medicinal plants used for malaria therapy in Southwestern Nigeria. Tropical J. Pharmaceutical Res. **7 (3)**: 1019-1024.

Ayoola, G.A., Folawewo, A.D., Adesegun, S.A., Abioro, O.O., Bello, A.A. and Coker, H.A.B. 2008. Phytochemical and antioxidant screening of some plants of apocynaceae from South West Nigeria. Afr. J. Plant Sci. **2(10)**: 124-128.

Ayoola, G.A., Ipav, S.S., Sofidiya, M.O., Adepoju-Bello, A.A., Coker, H.A.B. and Odugbemi, T.O. 2008. Phytochemical screening and free radical scavenging activities of the fruits and leaves of *Allanbackia floribunda* Oliv. (Guttiferae). Int.J. Health Res. **1(2)**: 87-93.

Ayoola, G.A., Sofidiya, T., Odukoya, O. and Coker, H.A.B. 2006. Phytochemical screening and free radical scavenging activity of some Nigerian medicinal plants. J. Pharm. Sci. and Pharm. Pract. **8**: 133-136.

Azad, F.F., Akbari, N.A., Zakheri, A., Andalib, S., Dijzaji, N.M., Bagheri, A.G., Haghjo, . A.G., Fakhrjo, A. and Garjani, A. 2009. Hypolipidemic and antioxidant effects of *Securigera securidaca* L. seed in high fat fed rats. Pharm. Sci. **15(4)**:

Azadmehr, A., Afshari, A., Baradarem, B., Haziaghaee, R., Rezazadeh, S. and Monsef-Esfahani, H. 2009. Suppression of nitric oxide production in Activated murine peritoneal macrophages *in vitro* and *ex vivo* by *Scrophularia striata* ethanolic extract. J. Ethnopharmacol. **124**: 166-169.

Azaizeh, H., Ljubuncic, P., Portnaya, I., Said, O., Cogan, U. and Bomzon, A. 2005. Fertilization-induced changes in growth parameters and antioxidant activity of medicinal plants used in traditional Arab medicine. Annals of Oncology. Evidence Based Complement Alternat. Med. **2**: 549-556.

Azuma, L., Nakayama, M., Koshioka, K., Ippoushi, Y., Yamaguchi, K., Kohata, Y., Yamaguchi, Y, Ito, H. and Higashio, H. 1994. Phenolic antioxidants from the leaves of *Corchorus olitorius* L. J. Agric. Food Chem. **47**: 396-399.

Babre, N.P., Debnath, S., Manjunath, Y.S., Rarameshwar, P., Wankhede, S.V. and Hariprakash, K. 2010. Antioxidant potential of hydroalcoholic extract of *Barringtonia acutangula* Linn. roots on streptozotocin-induced diabetic rats. Int. J. Pharmacy Pharmaceut. Sci. **2(4)**: 201-203.

Babu, B.H., Jayaram, H.N., Mary, N.K., Shylesh, B.S. and Jose, P. 2002. *In vitro* free radical scavenging, cytotoxic and antitumor activity of Tambulin – a flavone from *Zanthoxylum alatum*. Amala Res. Bull. **22**: 90-98.

Babu, B.H., Shylesh, B.S. and Padikkala, J. 2001. Antioxidant and hepatoprotective effect of *Acanthus ilicifolius* Fitoterapia, **72(3)**: 272-277.

Babu, R.K., Vinay, K., Sameena S.K., Prasad, S.V., Swapna, S. and Rao, A.C. 2009. Antihyperglycemic and antioxidant effects of *Talinum portulacifolium* leaf extracts in streptozotocin dibetic rats: A dose dependent study. Phcog. Mag. **5(19)**: 1-10.

Backhouse, N., Rosales, L., Apablaza, C., Goity, L., Erazo, S., Negrete, R., Theodoluz, C., Rodriguez, J. and Delporte, C. 2008. Analgesic, anti-inflammatory and antioxidant properties of *Buddleja globosa*, Buddlejaceae. J. Ethnopharmacol. **116(2)**: 263-269.

Badami, S., Gupta, M.K. and Suresh, B. 2003. Antioxidant activity of the ethanolic extract of *Striga orobanchoides*. J. Ethnopharmacol. **85**: 227-230.

Badami, S., Moorkoth, S., Rai, S.R., Elango, K. and Suresh, B. 2003. Antioxidant activity of *Caesalpinia sappan* heartwood. Biol Pharm. Bull. **26**: 1534-1537.

Badami, S, Moorkoth, S., Rai, S. R., Kannan, E. and Bhojraj, S. 2003. Antioxidant activity of *Caesalpinia sappan* heartwood. Biol. Pharm. Bull. **26(11)**: 1534-1537.

Badami, S., Prakash, O., Dongre, S.H. and B.Suresh. 2005. *In vitro* antioxidant properties of *Solanum pseudocapsicum* leaf extracts. Indian J. Pharmacol. **37**: 251-252.

Badami, S., Rai, S.R. and B. Suresh. 2005. Antioxidant activity of *Aporosa lindleyana* root. J. Ethnopharmacol. **101**: 180-184.

Badami, S., Vaijanathappa, J. and S.Bhojraj. 2008. *In vitro* antioxidant activity of *Argyreia cymosa* extracts. Fitoterapia **79(4)**: 287-289.

Bade, J.D., Jaju, A.H., Ahėr, A.N. and Pal, S.C. 2009. *In vitro* antioxidant and free radical scavenging activity of ethyl acetate extract of *Delonix regia* Rafin. Flowers. Asian J. Chem. **21(2)**: 1323-1329.

Badei, A., El-Akel, A., Faheid, S and Mahmoud, B. 2002. Application of some spices in flavoring and preservation of cookies, antioxidant properties of cardamom, cinnamon and clove. Deutsche Leben smittel-Rundschau **98**: 176-183.

Badrinathan, S., Suneeva, S.C., Shiju, T.M., Girish Kumar, C.P. and Pragasam, V. 2011. Exploration of a novel hydroxyl scavenger from *Sargassum myriocystum*. J. Med. Plants Res. **5(10)**: 1997-2005.

Bae, S-H. and Suh, H-J. 2007. Antioxdant activities of five different mulberry cultivars in Korea. LWT-Food Sci. Technol. **40**: 955-962.

Bafna, A. and Mishra, S. 2009. Antioxidant and immunomodulatory activity of the alkaloidal fraction of *Cissampelos pereira* Linn. Sci. Pharm. **78**: 21-31. doi: 10.3797/scipharm.0904-16.

Bafna, P.A. and Balaram, R. 2004. Anti-ulcer and antioxidant activity of Normacid, a herbomineral formulation. Indian J. Exp. Biol. **42(7)**: 674-680.

Bafna, P.A. and Balaram, R. 2005a. Anti-ulcer and antioxidant activity of pepticare, a herbomineral formulation. Phytomedicine **12(4)**: 264-270.

Bafna, P.A. and Balaram, R. 2005b. Antioxidant activity of DHC-1, an herbal formulation, in experimentally-induced cardiac and renal damage. Phytother. Res. **19(3)**: 216-221.

Bagchi, D., Kuszynski, C., Balmoori, J., Bagchi, M. and Stohs, S.J. 1998. Hydrogen peroxide-induced modulation of intracellular oxidized states in cultured macrophage J774A.1 and neuroactive PC-12 cells and protection by a novel grape seed proanthocyanidin extract. Phytother. Res. **12(8)**: 568-571.

Baggett, s., Protiva, P., Mazzola, S.P., Yang, H., Ressler, E.T., Basile, M.J., Weinstein, I.B. and Kennally, E.J. 2005. Bioactive benzophenones from *Garcinia xanthochymus* fruits. J. Nat. Prod. **68(3)**: 354-360.

Baggio, C.H., De Martinin Otofuji, G., de Souza, W.M., de Moraes Santos, C.A., Torres, L.M.B., Rieck, L., de Andrada Marques, M.C. and Nesia-Vela, S. 2005.

Gastroprotective mechanisms of indole alkaloids from *Himatanthus lancifolius*. Plant Medica **71(8)**: 733-738.

Bagri, P., Ali, Md., Aeri, V., Bhowmik, M and Sultana, S. 2009. Antidiabetic effect of *Punica granatum* flowers: Effect on hyperlipidemia, pancreatic cells lipid peroxidation and antioxidant enzymes in experimental diabetes. Food Chem. Toxicol. **47(1)**: 50-54.

Bagul, M.S., Ravishankara, M.N., Padh, H. and Rajani, M. 2003. Phytochemical evaluation and free radical scavenging properties of rhizome of *Bergenia ciliata* (Haw.) Stenb. Ferma ligulata Yoo. J. Nat. Remedies **3**: 83-89.

Bahorun, T., Neergheen, V.S. and Aruoma, O.L. 2005. Phytochemical constituents of *Cassia fistula*. Afr. J. Biotech. **4(13)**: 1530-1540.

Bai, X., Qiu, A., Guan, J. and Shi, Z. 2007. Antioxidant and protective effect of an oleanolic acid-enriched extract of *A.deliciosa* root on carbon tetrachloride induced rat liver injury. Asia Pacific J. Clin. Nutr. **16** suppl.: 169-173.

Bajpai, M., Pande, A., Tewari, S.K. and Prakash, D. 2005, Phenolic contents and antioxidant activity of some food and medicinal plants. Int. J. Food Sci. Nutr. **56 (4)**: 287-291.

Bajpai, V.K., Al-Reza, S.M., Choi, U.K., Lee, J.H. and Kang, S.C. 2009c. Chemical composition, antibacterial and antioxidant activities of leaf essential oil and extracts of *Metasequoia glyptostrobioides* Miki ex Hu. Food Chem. Toxicol. **47(8)**: 1876-1883.

Bajpai, V.K., Yoon, J.I. and Kang, S.C. 2009a. Antioxidant and antidermatophytic activities of essential oil and extracts of *Metasequoia glyptostroboides* Miki ex Hu. Food Chem. Toxicol. **47(6)**: 1355-1361.

Bajpai, V.K., Yoon, J.I. and Kang, S.C. 2009b. Antioxidant and antidermatophytic activities of essential oil and extracts of *Magnolia liliflora* Desr. Food Chem. Toxicol. **47(10)**: 2606-2612.

Bakasso, S., Lamien-Meda, A., Lamien, C.E., Kiendrebeogo, M., Milligo, J., Ouedraogo, A.G. and Nacoulma, O.G. 2008. Polyphenol contents and antioxidant activities of five *Indigofera* species (Fabaceae) from Burkina Faso. Pak. J. Biol. Sci. **11(11)**: 1429-1435.

Bakirel, T., Bakirel, U., Beles, O.U., Ulgen, S.G. and Yardibi, H. 2008. *In vivo* assessment of antidiabetic and antioxidant activities of rosemary (*Rosmarinus officinalis*) in alloxan-diabetic rabbits. J. Ethnopharmacol. **116(1)**: 64-73.

Balakrishnan, N., Panda, A.B., Raj, N.R., Shrivastava, A. and Prathani, R. 2009. The evaluation of nitric oxide scavenging activity of *Acalypha indica* Linn. root. Asian J. Res. Chem. **2(2)**: 148-150.

Balasubashini, M.S., Rukkumani, R, Viswanathan, P. and Venugopal, M. 2004. Ferulic acid alleviates lipid peroxidation in diabetic rats. Phytotherapy Res. **18**: 310-314.

Balasubramaniam, A., Manivannan, R. Paramaguru, R., Mazumder, P.M. and Vijayakumar, M. 2012. Evaluation of anti-inflammatory and antioxidant activities of stem bark of *Toddalia asiatica* (L.) Lam. using different experimental models. Pharmacologia, **3**: 144-149. DOI: 10.5567/pharmacologia.2012.144.149

Balasubramanian, T. and Chatterjee, T.K. 2010. Hepatoprotective and antioxidant effects of *Stereospermum suaveolens* on Carbon tetrachloride-induced hepatic damage in rats. J. Complement. Integr. Med. **7(1)**:

Balasundram, N., Ai, T.Y., Sambanthamurthi, R., Sundram, K. and Samman, S. 2005. Antioxidant properties of palm fruit extracts. Asia Pacific J. Clin. Nutr. **14(4)**: 319-324.

Balasundram, N., Sundram, K. and Sammar, S. 2006. Phenolic compounds in plants and agri-industrial by-products: Antioxidant activity, occurrence and potential uses. Food Chemistry **68**: 191-203.

Balazs, A., Apati, P., Szoke, E., Kery, A. and Blazovics, A. 2003. Comparison of free radical scavenging activity and phenoloid content of *Epilobium parviflorum* Schreb. Acta horticulture **597**: 185-190.

Balestrin, L., Dias, J.F.G., Miguel, O.G., Dall'Stella, D.S.G. and Miguel, M.D. 2008. Contribution to the phytochemical study of *Dorstenia multiformis* Miquel (Moraceae) with approach in antioxidant activity. Rev. Bras. Farmacogn. **18(2)**: 230-235.

Ballero, M., Foddis, C., Sanna, C., Scartezzini, P., Poli, F., Petitto, V., Serafini, M., Stanzioni, A., Bianco, A., Serilli, A.M., Spina, L., Longoni, R. and Kasture, S. 2010. Pharmacological activitie of *Ephedra nebrodensis* Tineo. Nat. Prod. Res. **24(12)**: 1115-1124.

Balu, M. and Pannelselvan. 2004. Age associated macromolecular damages in central nervous system of aged rats: Role of grape seed extract. Chemistry Biology interface: Synergistic New Frontiers, New Delhi, India p25-23, November 21-26, 2004.

Bandoniene, D. and Murkovic, M. 2002. The detection of radical scavenging compounds in crude extract of borage (*Borago officinalis* L.) by using an on-line HPLC-DPPH method. J. Biochem. Biophys. Methods **53(1-3)**: 45-49.

Bandoniene, D. and Murkovic, M. 2002. On-line HPLC-DAD-DPPH screening method for evaluation of radical scavenging phenols extracted from apples (*Malus domestica* L.). J. Agric. Food Chem. **50**: 2482-2487.

Bandoniene, D., Murkovic, M., Pfannhauser, W., Venskutonis, P.R. and Gruzdiene, D. 2002. Detection and activity evaluation of radical scavenging compounds by using DPPH free radical and on-line HPLC-DPPH methods. Eur. Food Res. Technol. **214**: 143-147.

Bandoniene, D., Pukalskas, A., Venskutonis, P.R. and Gruzdiene, D. 2000. Preliminary screening of antioxidant activity of some plant extracts in rapeseed oil. Food Res. Int. **33**: 785-791.

Bandoniene, D., Venskutonis, P.R., Gruzdiene, D. and Murkovic, M. 2002. Antioxidant activity of Sage (*Salvia officinalis* Linn.), Savroy (*Satureja hortensis* Linn.) and Borage (*Borago officinalis* Linn.) extracts in rapeseed oil. Europ. J. Lipid Sci. Technol. **104**: 286-292.

Bandyopadhyay, S. K., Pakrashi, S.C. and Pakrashi, A. 2000. The role of antioxidant activity of *Phyllanthus emblica* fruits on prevention from indomethacin induced gastric ulcer. J. Ethnopharmacol. **70(2)**: 171–176.

Bansal, P., Sannd, R., Srikanth, N. and Lavekar, G.S. 2007. Antioxidant activity of cooled neutraceutical rasayana products in the first clinical trial at Antarctica. J. Trop. Med. Plants **8(2)** 178-183.

Banerjee, A., Dasgupta, N. and De, B. 2005. *In vitro* study of antioxidant activity of *Syzygium cumini* fruit. Food Chem. **90**: 727-733.

Bannach, R., Valenzuela, A., Cassels, B.K., Nunez,-Verquera, L.J. and Speisky, H. 1996. Cytoprotective and antioxidant effects of boldine on tert-butyl hydroperoxide-induced damage to isolated hepatocytes. Cell Bio. Toxicol. **12(2)**: 89-100.

Banskota, A.H., Tezuka, Y., Nguyen, N.T, Awale, S., Nobukawa, T. and Kadota, S. 2003. DPPH radical scavenging and nitric oxide inhibitory activities of the constituents from the wood of *Taxus yunnanensis*. Planta Med. **69(6)**: 500-505.

Bao, H.N., Ochiai, Y, and Ohshima, T. 2010. Antioxidative activities of hydrophilic extracts prepared from the fruiting body of spent culture medium of *Flammulina velutipes*. Bioresourcee Tech. **101(15)**: 6248-6255. doi: 10.1016/ j.bioretech.2010.03.026

Bao, H.N., Ushio, H. and Ohshima, T. 2008. Antioxidative activity and antidiscoloration efficacy of ergothioneine in mushroom (*Flammulina velutipes*) extract added to beef and fish meats. J. Agric. Food Chem. **56(21)**: 10032-10040.

Bao, J., Cai, Y., Sun, M., Wang, G., Corke, H. 2005. Anthocyanins, flavonols and free radical scavenging activity of Chinese Bayberry (*Myrica rubra*) extracts and their color properties and stability. J. Agric. Food Chem. **53**: 2317-2332.

Bao, X., Wang, Z., Fang, J. and Li, X. 2002. Structural features of an immunostimulating and antioxidant acidic polysaccharide from the seeds of *Cuscuta chinensis*. Planta Med. **68(3)**: 237-243.

Baratta, M.T., Dorman, H.J.D., Deans, S.G., Biondi, D.M. and Ruberto, G. 1998. Chemical composition, antimicrobial and antioxidative activity of laurel, sage, rosemary, oregano and coriander essential oils. J. Essential Oil Res. **10**: 618-627.

Baratta, M.T., Dorman, H.J.D., Deans, S.G., Figueriredo, A.C., Barroso, J.G. and Ruberto, G. 1998a. Antimicrobial and antioxidant properties of some commercial oils. Flav. Fragr.J. **13**: 235-244.

Barbakadze, V., Kemertelidze, E., Targamadze, I., Mulkijayan, K., Shashkov, A.S. and Usov, A.I. 2005. Poly[3-(3,4-dihydroxyphenyl)glyceric acid], a new biologically active polymer from *Symphytum asperum* Lepech and *S. caucasicum* Bieb. (Boraginaceae). Molecules **10(9)**: 1135-1144.

Barbakadze, V., Van den Berg, A.J.J., Beukelman, C.J., Kemmink, J. and Van Ufford, H.C.Q.2009. Poly3-(3,4-dihydroxyphenyl)glyceric acid from *Symphytum officinale* roots and its biological activity. Chem. Nat. Compounds **45(1)**: 6-10.

Bariliari, J., Cervellati,R., Costa, S., Guerra, M.C., Speroni, E., Utan, A. and Iori, R. 2006. Antioxidant and choleretic properties of *Raphanus sativus* L. sprout (Kaiware Daikon) extract. *J. Agric. Food Chem.* **54 (26)**: 9773-9778.

Barla, A., Ozturk, M., Kultur, S., Oksuz, S. 2007. Screening of antioxidant activity of three *Euphorbia* species from Turkey. Fitoterapia **78**: 423-425.

Barlow, S.M. 1990. Toxicological aspects of antioxidants used as food additives. In: Food antioxidants. BJF Hudson (ed.). Elsevier London pp. 253-307.

Barra, A., Coroneo, V., Dessi, S., Cabras, P. and Angioni, A. 2007. Characterization of the volatile constituents in the essential oil of *Pistacia lentiscus* L. from different origins and its antifungal and antioxidant activity. J. Agric. Food Cehm. **55(17)**: 7093-7098.

Barra, A., Coroneo, V., Dessi, S., Cabras, P. and Angioni, A. 2010. Chemical variability, antifungal and antioxidant activity of *Eucalyptus camaldulensis* essential oil from Sardina. Nat. Prod. Comm. **5(2)**: 329-335.

Barrajon-Catalan, E., Fernandez-Arroyo, S. Saura, D., Guillen, E., Fernande-Gutierrez, A., Segura-Carretero, A. and Micol, V. 2010. Cistaceae aqueous extracts containing ellagitannins show antioxidant and antimicrobial capacity, and cytotoxic activity against human cancer cells. Food and Chemical Toxicol. **48(8-9)**: 2272-2282. doi: 10.1016/j.fct.2010.05.060.

Barrero, A.F., Herrador, M.M., Artega, P., Rosas-Romero, A. and Artega, J.F. 2000. Antioxidant activity of diterpenes and polyphenols from *Opryosporus heptanthus*. J. Agric. Food Chem. **54(7)**: 2537-2542.

Barros, L., Baptista, P. and Correia, J.S., Morais, D.M., Ferreira, I.C.F.R. 2007. Effects of conservation treatment and cooking on the chemical composition and antioxidant activity of Portuguese wild edible mushrooms. J. Agric. Food Chem. **55(12)**: 4781-4788. doi: 10.1016/j.foodchem.2006.07.038

Barth, S.S., Inselmann, G., Engcmann, R. and Heidemann, H.T.1991. Influence of *Ginkgo biloba* on cyclosporine A induced lipid peroxidation in human liver microsomes in comparison vitamin glutathione and N-acetylcysteine. Biochem. Pharmacol. **41**: 1521-1526.

Barthomeuf, C.M., Debiton, E., Barbakadze, V.V. and Kemertelidze, E.P. 2001. Evaluation of the dietic and therapeutic potential of high molecular weight hydroxycinnamate-derived polymer from *Symphytum asperum* Lepech. Regarding its antioxidant, antilipoperoxdant, anti-inflammatory and cytotoxic properties. J. Agric. Food Chem. **49(8)**: 3942-3946.

Bashir, S. and Gilani, A.H. 2009. Antiurolithic effect of *Bergenia ligulata* rhizome: an explanation of the underlying mechanisms. J. Ethnopharmacol. **122(1)**: 106-116.

Basile, A., Ferrara, L., Pezzo, M.D., Mele, G., Sorbo, S., Bassi, P. and Montesano, D. 2005. Antibacterial and antioxidant activities of ethanol extract from *Paullinia cupana* Mart. J. Ethnopharmacol. **102(1-2)**: 50-52.

Basile, A., Sorbo, S., Spadaro, V., Bruno, M., Maggio, A., Faraone, N. and Rosselli, S. 2009. Antimicrobial and antioxidant activities of coumarins from the roots of *Ferulago campestris* (Apiaceae). Molecules **14**: 939-952. doi: 10.3390/molecules14030939

Baskar, K., Malini, M.M., Varalakshmi, P., Balakrishna, K. and Bhima Rao, R. 1996. Effect of lupeol isolated from *Crataeva nurvala* stem bark against free radical-induced toxicity in experimental urolithiasis. Fitoterapia **67**: 121-125.

Baskar, R., Lavanya, R., Mayilavizhi, S. and Rajasekharan, P. 2008. Free radical scavenging activity of antitumor polysaccharide fractions isolated from *Ganoderma lucidum* (Fr.) P.Karst. Nat. Prod. Rad. **7**: 320-325.

Baskar, R., Rajeswari, V. and Satish Kumar, T. 2007. *In vitro* antioxidant studies in leaves of *Annona* species. Indian J. Biotech. **45**: 480-485.

Basma, A.A., Zakaria, Z., Latha, L.Y. and Sasidharaan, S. 2011. Antioxidant activity and phytochemical screening of the methanol extracts of *Euphorbia hirta* L. Asian Pacific J. Trop. Med. **4(5)**: 386-390.

Basniwal, P.K., Suthur, M., Rathore, G.S., Gupta, R., Kumar, V., Parekh, A. and Jain D. 2009. *In vitro* antioxidant activity of hot aqueous extract of *Helicteres isora* Linn. fruits. Natural Product Radiance **8(5)**: 483-487.

Basu, S. and Hazra, B. 2005. Evaluation of radical scavenging activity of *Rubia cordifolia* Linn. International Conference on Modern Trends in Plant sciences with special reference to the role of Biodiversity in Conservation, Amaravathi p. 166.

Basu, S. and Hazra, B. 2006. Evaluation of nitric scavenging activity, *in vitro* and *ex vivo*, of selected medicinal plants traditionally used in inflammatory diseases. Phytother. Res. **20**: 896-900

Basu, T.K., Temple, N.J. and Garg, M.L. (eds.). 1999. Antioxidants in human health and disease, CABI Publishing, New York.

Báthori, M., Zupko, I., Hunyadi, A., Gácsné-Baitz, E., Dinya, Z. and Forgó, P. 2004. Monitoring the antioxidant activity of extracts originated from various *Serratula* species and isolation of flavonoids from *Serratula coronata*. Fitoterapia **75(2)**: 162-167.

Bavarma, J. and Ambaga, M. 2010. Antioxidant activity of medicinal herb *Zygophyllum potalini* Maxim. on carbon tetrachloride induced liver injury in cats. www.scribd.com/./Antioxidant-Activity-of-Medicinal-Herb-Zygophyllum-Potaninii-Maxim-

Bayir, Y., Odabasoglu, F., Cakir, A., Aslan, A., Suleyman, H., Halici, M. and Kazaz, C. 2006. The inhibition of gastric mucosal lesion, oxidative stress and neutrophil-infiltration in rats by the lichen constituent diffractaic acid. Phytomedicine **13(8)**: 584-590.

Beara, I.V., Lesjak, M.M., Jovin, E.D., Balog, K.J., Anackov, G.T., Orcic, D.Z. and Mimica-Ducic, N.M. 2009. Plantain (*Plantago* L.) species as novel source of Flavonoid antioxidants. J. Agric. Food Chem. **57(19)**: 9268-9273.

Beauchamp, C. and Fridovich. I. 1971. Superoxide dismutase: improved assays and an assay applicable to acrylamide gel. Anal.Biochem. **44**: 276–87.

Beckman, K.B. and Ames, B.N. 1998. The free radical theory of Aging Matures. Physiol. Rev. **78**: 547-581.

Began, M., Varshavsky, L., Gottlieb, H.E. and Grossman, S. 2001. The antioxidant activity of aqueous Spinach extract: Chemical identification of active fractions. Phytochemistry **58**: 143-152.

Behbahani, M., Ali, A.M., Muse, R. and Mohd, N.B. 2007. Anti-oxidant and anti-inflammatory activities of leaves of *Barringtonia racemosa*. J. Med. Plant Res. 1: 95-102.

Behera, B.C., Verma, N., Sonone, A. and Makhija, U. 2005. Antioxidant and antibacterial activities of lichen *Usnea ghattensis in vitro*. Biotechnol Lett. **27(14)**: 991-995.

Behera, G.M., Satishkumar, B.N., Malay, B. and Panigrahi, G. 2010. Antihyperglycemic and antioxidant activity of *Glinus oppositifolius* (L.) Aug. DC. Pharmacology online **3**: 915-936.

Beknal., A.K., Korwar, P.G., Halkai, M.A., Kulkarni, U., Patil, B.S. and Soodam, SA.R. 2010. Phytochemical investigation and antioxidant activity study of *Drynaria quercifolia* Linn. rhizome. Int. J. Curr. Pharm. Res. **2(4)**:

Belinky, P.A., Aviram, M., Mahmood, S., and Vaya, J. 1998a. Structural aspects of the inhibitory effect of glabridin on LDL oxidation. Free Radical Biol. Med. **24**: 1419.

Belinky, P.A., Aviram, M., Fuhrman, B., Rosenblat, M. and Vaya, J. 1998b. The antioxidative effects of the isoflavan glabridin on endogenous constituents of LDL during its oxidation. Atherosclerosis **137(1)**: 49-61.

Benabadji, S.H., Wen, R., Zheng, J-B, Dong, X-C. and Yuan, S. G. 2004. Antioxidant activity of dindolylmethane derivatives. Acta Pharmacol. Sin. **25**: 666-671.

Bendaoud, H., Romdhane, M., Souchard, J.P., Cazaux, S., Bouajila, J. 2010. Chemical composition and anticancer and antioxidant activities of *Schinus molle* L. and *Schinus terebinthifolius* Raddi berries essential oils. J. Food Sci. **75**: 466-472. DOI: 10.1111/j.1750-3841.2010.01711.

Bendini, A., Cerretani, l., Pizzalanate, L., Toschi, T.G., Guzzo, F., Ceoldo, S., Marconi, A.M., Andreetta, F. and Levi, M. 2006. Phenol content related to antioxidant and antimicrobial activities of *Passiflora* spp. extracts. Europian Food Res. Tech. **223**: 102-109.

Benedí, J., Arroyo, R., Romero, C., Martín-Aragón, S. and Villar, A.M. 2004. Antioxidant properties and protective effects of a standardized extract of *Hypericum perforatum* on hydrogen peroxide-induced oxidative damage in PC12 cells. Life Sci. **75(10)**: 1263-1276.

Benkeblia, N. 2005. Free-radical scavenging capacity and antioxidant properties of some selected onions (*Allium cepa* L.) and garlic (*Allium sativum* L.) extracts. Braz. Arch. Biol. Technol. 48(5): doi: 10.1590/s1516-89132005000600011.

Benzie, I.F.F. and Szeto, Y.T. 1999. Total antioxidant capacity of teas by the Ferric Reducing Antioxidant Power assay. J. Agric. Food Chem. **47**: 633-636.

Benvenuti, S., Paellati, F., Melegari, M., Bertelli, D. 2004. Polyphenols, anthocyanins, ascorbic acid, and radical scavenging activity of *Rubus, Ribes* and *Aronia*. J. Food Sci. **69**: 164-169.

Benzel, L.V., Gargola, M.S., Ladnaya, L.Y. and Baran, E.I. 1988. Comparison of the flavonoid contents of thistles (*Carlina*) from different growth localities. Farmatsevitchni Zhurnal (Kiev.) **5**: 69-70.

Benzie, I.F.F. and Strain, J.J. 1999. Ferric reducing antioxidant power assay: Direct measure of total antioxidant activity of biological fluids and modified version for simultaneous measurement of total antioxidant power and ascorbic acid concentration. Methods in Enzymology **299**: 15-27.

Benzie, I.F.F. and Szeto, Y.T. 1999. Total antioxidant capacity of teas by the ferric reducing/antioxidant power assay. J. Agric. Food Chem. **47**: 633-636.

Berenguer, B., Sanchez, L.M., Quflez, A., Lopez-Barreiro, M., de Haro, O., Galvez, J. and Martin, M.J. 2006. Protective and antioxidant effects of *Rhizophora mangle* L. against NSAID-induced gastric ulcers. J. Ethnopharmacol. **103(2)**: 194-200.

Berk, S., Tepe, B. and Arslan, S. 2011. Screening of the Antioxidant, antimicrobial and DNA damage protection potentials of the aqueous extract of *Inula oculus-christi*. African J. Pharmacy and Pharmacol. **5(14)**: 1695-1702.

Berk, S., Tepe, B., Arslan, S. and Sarikurkcu. 2011. Screening of the antioxidant, antimicrobial and DNA damage protection potentials of the aqueous extract of *Asplenium ceterach* DC. African J. Biotech. **10**: 8902-8908.

Berker, K.I., Guclu, K., Tor, I., Apak, R. 2007. Comparative evaluation of Fe(III) reducing power based antioxidant capacity assays in the presence of phenanthroline, batho-phenanthroline, tripyridyltrazine (FRAP) and Ferriccyanide reagents. Talanta **72**: 1157-1165.

Bermudez-Soto, M.J. and Tomas-Barberan, F.A. 2004. Evaluation of commercial red fruit juice concentrates as ingredients for antioxidant functional. Eur. Food Res. Technol. **219**: 133-141.

Beta, T., Nam, S., Dexter, J.E. and Sapirstein. 2005. Phenolic content and antioxidant activity of pearled wheat and roller-milled wheat fractions. Cereal Chem. **82(4)**: 390-393.

Bezakova, L., Misik, V., Malekova, L., Svajdlenka, E. and Kostalova, D. 1996. Lipoxygenase inhibition and antioxidant properties of bisbenzylisoquinoline alkaloids isolated from *Mahonia aquifolium*. Pharmazie **51(10)**: 758-761.

Bhandari, U., Kanojia, R. and Pillai, K.K. 2005. Effect of ethanolic extract of *Zingiber officinale* on dyslipidemia in diabetic rats. J. Ethnopharmacol **97(2)**: 227-230.

Bhatia, A.L. and Jain, M. 2003. *Amaranthus paniculatus* Linn. improves learning after radiation stress. J. Ethnopharmacol. **16**: 1-13.

Bhattacharya, A., Chatterjee, A., Ghosal, S. and Bhattacharya, S.K. 1999. Antioxidant activity of active tannoid principles of *Emblica officinalis* (amla). Indian J. Exp Biol. **37**: 676-680.

Bhattacharya, A., Ghosal, S. and Bhattacharya, S.K. 2000. Antioxidant activity of tannoid principles of *Emblica officinalis* (amla) in chronic stress induced changes in rat brain. Indian J. Exp. Biol. **38(9)**: 877-880.

Bhattacharya, A., Ghosal, S. and Bhattacharya, S.K. 2001. Antioxidant effect of *Withania somnifera* glycowithanolides in chronic footshock stress-induced perturbations of oxidative free radical scavenging enzymes and lipid peroxidation in rat frontal cortex and striatum. J. Ethnopharmacol. **74(1)**: 1-6.

Bhattacharya, S., Haldar, P.K. and Md. K.Zaman. 2010. Anti-inflammatory and *in vitro* antioxidant property of *Zanthoxylum nitidum* root. Current Trends in Biotechnology and Pharmacy **4(3)**: 774-783.

Bhattacharya, S. K., Satyan K. S. and Ghosal, S. 1997. Antioxidant activity of glycowithanolides from *Withania somnifera*. Indian J. Exp. Biol. **35**: 236-240.

Bhatti, H.N., Ul-Arifen, W. and Jabbar, A. 2010. Total phenolics and antioxidant activity of Horsemint *Mentha longifolia*. Asian J. Chem. **22(6)**: 4629-4636.

Bhatti, H.N., Zafar, F. and Jamal, M.A. 2010. Evaluation of phenolic contents and antioxidant potential of methanolic extracts of green cardamom *Elettaria cardamomum*. Asian J. Chem. **22(6)**: 4787-4794.

Bhaumik, U.K. Kumar, A.D., Selvam, V.T., Saha, P., Gupta, M., and Mazumdar, U.K. 2008. Antioxidant and free radical scavenging property of methanol extract of *Blumea lanceolaria* leaf in different *in vitro* models. Pharmacology on line 2: 74-89.

Bheema Rao, N., Bhabani, S.J. and Prabhasankar, P. 2009. Influence of natural antioxidants and their carry-through property in biscuit processing. J. Sci. Food Agric. **89(2)**: 288-298.

Bhuiyan, M.A.R., Hoque, M.Z. and Hossain, S.J. 2009. Free radical scavenging activities of *Zizyphus mauritiana*. World J. Agric. Sci. **5(3)**: 318-322.

Biapa, P-C.N., Agbor, G.A., Oben, J.E. and Ngogang, J.Y. 2007. Phytochemical studies and antioxidant properties of four medicinal plants used in Cameroon. Afr. J. Trad. CAM **4 (4)**: 495-500.

Biglari, F., AlKarkhi, A.F.M. and Easa, A.M. 2009. Cluster analysis of antioxidant compounds in dates (*Phoenix dactylifera*): Effect of long-term cold storage. Food Chem. **112 (4)**: 998-1001.

Biglari, F., Alkarkhi, A.F.M. and Easa, A.M. 2008. Antioxidant activity and phenolic content of various date palm (*Phoenix dactylifera*) fruits from Iran. Food Chem. **107 (4)**: 1636-1641.

Bimla Singh, Kale, R.K. and Rao, A.R. 2004. Modulation of antioxidant potential of liver of mice by kernel oil of cashew nut (*Anacardium occidentale*) and its lack of tumor promoting ability in DMBA induced skin papilomagenesis. Indian J. Exp. Bio. **42(4)**: 373-377.

Biondi, D.M., Rocco, C. and Ruberto, G. 2003. New dihydrostilbene derivatives from the leaves of *Glycyrrhiza glabra* and evaluation of their antioxidant activity. J. Nat. Prod. **66(4)**: 477-480.

Bisson, J-F., Daubie, S., Hidalgo, S., Guillemet, D. and Linars, E. 2009. Diuretic and antioxidant effects of Cacti-Nea, a dehydrated water extract from prickly pear fruit, in rats. Phytother. Res. **24(4)**: 587-594.

Biswas, M., Bhattacharya, S., Ghosh, A.K. and Haldar, P.K. 2010. Evaluation of *in vitro* antioxidant and free radical scavenging effects of *Terminalia arjuna* leaf. Pharmacology online **3**: 392-400.

Biswas, R., Mandal., S.K., Dutta, S., Bhattacharya, S.S., Boujedaini, N. and Khuda-Bhukhsh, A.R.K. 2011. Thujone-rich fraction of *Thuja occidentalis* demonstrates major anti-cancer potentials: evidences from *in vitro* studies on A375 cells. Evidence Based Complement Alternat. Med. Doi: 10.1093/ecamneq042.

Biswas, S., S., Dasgupta, N. and De, B. 2010. Antioxidant activity of some Indian medicinal plants used for the treatment of diabetes mellitus. Pharmacology online **3**:

Bitis, L., Kultur, S., Melikoglu, G., Ozsoy, N. and Can, A. 2010. Flavonoids and antioxidant activity of *Rosa agrestis* leaves. Nat. Prod. Res. 24(6): 580-589. DOI: 10.1080/14786410903075507.

Bixby, M., Spieler, L., Menini, T. and Gugliucci, A. 2005. *Ilex paraguariensis* extracts are potent inhibitors of nitrosative stress: a comparative study with green tea and wines using a protein nitration model and mammalian cell cytotoxicity. Life Sci. **77 (3)**: 345-358.

Bizimenyera, E.S., Aderogba, M.A., Eloff, J.N. and Swan, G.E. 2006. Potential of neuroprotective antioxidant-based therapeutics from *Peltophorum africanum.* Afr. J. Trad. Compl. Alternat. Med. **4**: 99-106.

Blazovics, A., Feher, J., Feher, E., Kery, A. and Petri, G.1993. Liver protecting and lipid lowering effects of *Sempervivum tectorum* extract in the rats. Phytother. Res. **7(1)**: 98-100.

Blois, M.S. 1958. Antioxidant determination by the use of stable free radical. Nature **26**: 1199-1200.

Blois, M.S. 2001. Antioxdant activity of grape seed extracts on peroxidation models *in vitro*. J. Agric. Food Chem. **55**: 1018.

Bocco, A., Cuvelier, M.E., Richard, H. and Berset, C. 1998. Antioxidant activity and phenolic composition of citrus peel and seed extracts. J. Agric. Food chem. **46**: 2123-2129.

Bolshakova, I.V., Lozovskaya, E.L. and I.I.Sapezhinskii. 1999. Antioxidant properties of a number of plant extracts. Biofizika **43**: 186-188.

Bonina, F., Puglia, C., Ventura, D., Aquino, R., Tortora, S., Sacchi, A., Saija, A., Tomaino, A., Pellegrino, M.L. and de Caprariis, P. 2002. *In vitro* antioxidant and *in vivo* photoprotective effects of a lyophilized extract of *Capparis spinosa* L buds. J. Cosmet. Sci. **53 (6)**: 321-335.

Bonnely, S., Peyrat-Maillard, M.N., Rondini, L., Masy, D. and Berset, C. 2000. Antioxidant activity of malt rootlet extracts. J Agric Food Chem. **48 (7)**: 2785-2792.

Bonoli, M., Verardo, V., Marconi, E. and Caboni, M.F. 2004. Antioxidant phenols in barley (*Hordeum vulgare* L.) flour: comparative spectrophotometric study among extraction methods of free and bound phenolic compounds. J Agric. Food Chem. **52 (16)**: 5195-5200.

Bora, K.S. and Sharma, A. 2011. Evaluation of antioxidant and cerebroprotective effect of *Medicago sativa* Linn. against Ischemia an reperfusion insult. Evidence Based Compl Alt. Med. Doi: 10.1093/ecam/neq019

Borah, A., Yadav, R.N.S. and Unni, B.G. 2011. *In vitro* antioxidant and free radical scavenging activity of *Alternanthera sessilis*. Int. J. Pharm. Sci. Res. **2(6)**: 1502-1506.

Borchardt, J.R., Donald, L.W., Craig, C., Sheaffer, K.L., Kauppi, Fulcher, R.G., Nancy, J.E., David, D.B. and Russel, F.B. 2008. Antioxidant and antimicrobial activity of seed from plants of the Missisipi river basin. J. Med. Plants Res. **2(4)**: 81-93.

Borek, C. 2001. Antioxdant health effects of aged garlic extract. J. Nutr. **131 (3s)**: 1010S-5S.

Borges, G., Degeneva, A., Mullen, W. and Crozier, A. 2010. Identification of flavonoid and phenolic antioxidants in black currants, blueberries, raspberries, red currants and craberries. J. Agric. Food Chem. **58(7)**: 3901-3909.

Borneo, R., Aguirre, A. and Cantero, J.J 2008. *In vitro* antioxidant activity of species collected in Cordoba (Argentinaa). Electronic J. Environ. Agric. Food Chem. **7(8)**: 3168-3170.

Borneo, R., Leon, A.E., Aguirre, A., Ribotta, P. and Cantero, J.J. 2009. Antioxidant capacity of medicinal plants from province of Cordoba (Argentina) and their *in vitro* testing in a model food system. Food Chem. **112(3)**: 664-670.

Bors, W., Saran, M. and Elstner, E.F.1992. Screening for plant anti-oxidants. In: Linskens, H.F. and Jackson, J.F. (eds.) Modern Methods of plant analysis-Plant Toxin analysis-New Series, Vol. 13. Springer, Berlin.pp 277-295.

Borse, B.B., Kumar, H.V. and Rao, L.J. 2007. Radical scavenging conserves from unused fresh green tea leaves. J. Agric. Food Chem. **55(5)**: 1750-1754.

Bouayed, J., Piri, K., Rammal, H., Dicko, A., Desor, F., Younos, C., and Soulimani, R. 2007. Comparative evaluation of the antioxidant potential of some Iranian medicinal plants. Food Chem. **104(1)**: 364-368. doi: 10.1016 j.foodchem. 2006.11.069

Bouaziz, M., Jemai, H., Khabou, W. and Sayadi, S. 2010. Oil content, phenolic profiling and antioxidant potential of Tunisian olive drupes. J. Sci. Food Agric. **90(10)**: 1750-1758.

Bouchet, N., Laurence, B. and Fauconneau, B. 1998. Radical scavenging activity and antioxidant properties of tannins from *Guiera senegalensis*. Phytother. Res. **12(2)**: 159-162.

Bouhlel, I., Mansour, H.B., Limem, I., Sghaier, M.B., Mahmoud, A. *et al.*, 2007. Screening of antimutagenicity via antioxidant activity in different extracts from the leaves of *Acacia salicina* from the center of Tunisia. Env. Toxicol. Pharmacol. **23**: 56-63.

Bouhlel, I., Valenti *et al.*, 2008. Antimutagenic and antioxidant activities of *Acacia salicina* extracts (ASE) and modulation of cell gene expression by H_2O_2 and ASE treatment. Toxicol. In Vitro **22(5)**: 1264-1272.

Boveris, A.D., Galatro, A. and Puntarulo, S. 2000. Effect of nitric oxide and plant antioxidants on microsomal content of lipid radicals. Biol. Res. **33**: 159-165.

Boynes, J.W. 1991. Role of oxidative stress in the development of complication in diabetes. Diabetes **40**: 405-411.

Bozan, B., Kosar, M., Tunalier, Z., Ozturk, N. and Baser, K.H.C. 2003. Antioxidant and free radical scavenging activities of *Rhus coriaria* and *Cinnamomum cassia* extracts. Acta Alimentaria **32**: 53-61.

Bozan, B., Ozturk, N., Kosar, M. Tunalier, Z. and Baser, K.H.C. 2002. Antioxidant and free radical scavenging activities of eight *Salvia* species. Chem. Nat. Compd. **38(2)**: 198-200.

Bozin, B. Neda Mimica-Dukic, N. Isidora Samjlik and Emilija Jovin. 2007. Antimicrobial and antioxidant properties of rosemary and sage (*Rosmarinus officinalis* L. and *Salvia officinalis* L., Lamiaceae) essential oils. J. Agric. Food Chem. **55(19)**: 7879-7885.

Bozin, B., Mimica-Dukic, N., Simin, N. and Anackov, G. 2006. Characterization of the volatile oil composition of essential oils of some Lamiaceae species and the antimicrobial and antioxidant activities of the entire oils. J. Agric. Food Chem. **54**: 1822-1828.

B R, Y V, J A, M G and V R. 2008. Protective effect of *Phyllanthus polyphyllus* on acetaminophen induced hepatotoxicity in rats. Pak. J. Pharm. Sci. **21(1)**: 57-62.

Braca, A., Fico, G., Morelli, I., De Simone, F., Tomè, F. and De Tommasi, N. 2003. Antioxidant and free radical scavenging activity of flavonol glycosides from different *Aconitum* species. J. Ethnopharmacol. **86**: 63–67.

Braca, A., Sortino, C., Politi, M., Morelli, I. and Mendez, J. 2002. Antioxidant activity of flavonoids from *Licania licaniaeflora*. J. Ethnopharmacol. **79**: 379-381.

Braca A., Tommasi N. D., Bari, L.D., Piza C., Potil, K. and Morelli, I. 2001. Antioxidant principles from *Bauhinia tarapotensis*. J. Nat. Prod., **64**: 892: 895.

Bracht, A.K., Alvarez, M., and Bracht, A. 1985. Effects of *Stevia rebaudiana* natural products on rat liver mitochondria. Biochem. Pharmacol. **34(6)**: 873-882.

Bradford, R. and Allen, H. 1997. Oxidology. Chula Vista CA. R.W. Bradford Foundation pp. 64-65.

Braga, M.E., Leal, P.F., Carvalho, J.E. and Meireles, M. A. 2003. Comparison of yield, composition, and antioxidant activity of turmeric (*Curcuma longa* L.) extracts obtained using various techniques. J. Agric. Food Chem. **51(22)**: 6604-6611.

Braga, P.C., Dal Sasso, M., Culici, M., Spallino, A., Falchi, M., Bertelli, A., Morelli, R. and Lo Scalzo, R. 2009. Antioxidant activity of *Calendula officinalis* extract: inhibitory effects of chemiluminescence of human neutrophil bursts and electon paramagnetic resonance spectroscopy. Pharmacology **83(6)**: 348-355.

Braham, H., Mighri, Z., Jannet, H.B., Matthew, S. and Abreq, P.M. 2005. Antioxidant phenolic glycosides from *Moricandia arvensis*. J. Nat. Prod. **68(4)**: 517-522.

Brand-Williams, W., Cuvelier, M.E. and Berset, C. 1995. Use of free radical method to evaluate antioxidant activity. Lebensmittel Wissenschaft und Technologie **28**: 25-30.

Brantner, A.H., Pfeifhofer, H.W., Ercegovac, O., Males, Z. and Plazibat, M. 2005. Essential oil composition and antioxidant activity of *Thymus bracteosus* Vis. ex Benth. Flavour and Fragrance Journ. **20(6)**: 596-600.

Bravo, L., Goya, L. and Lecumberri, E. 2007. LC/MS characterization of phenolic constituents of mate (*Ilex paraguariensis* St. Hil.) and its antioxidant activity. Food Research International. **40 (3)**: 393-405.

Brem, B., Seger, C., Pacher, T., Hartl, M., Hadacek, F., Hofer, O., Vajrodaya, S. and Greger, H. 2004. Antioxidant dehydrotocopherols as a new chemical character of *Stemona* species. Phytochemistry **65(19)**: 2719-2729. doi: 10.1016/j.phytochem.2004.08.023

Breska, III, A.P., Takeoka, G.R., Hidalgo, M.B., Vilches, A., Vasse, J. and Ramming, D.W. 2010. Antioxidant activity and phenolic content of 16 raisin grape (*Vitis vinifera* L.) cultivars and selections. Food Chem. **121(3)**: 740-745.

Brettonnet, A., Hewavitarana, DeJong, S. and Lanari, M.C. 2010. Phenolic acids composition and antioxidant activity of canola extracts in cooked beef, chicken and pork. Food Chem. **121(4)**: 927-933.

Briante R.M., Patumi, S. Terenziani, E., Bismuto,F., Febbraio and Nucci, R. 2002. *Olea europaea* leaf extract and derivaties: Antioxidant properties. J. Agric. Food Chem., **50**: 4934-4940.

Bridi, R., Crossetti, F.P., Steffen, V.M., Henriques, A.T. 2001.The antioxidant activity of standardized extract of *Ginkgo biloba* (EGb 761) in rats. Phytother Res. **15(5)**: 449-451.

Brown, J.E. and Rice-Evans, C.A. 1998. Luteolin-rich artichoke extract protects low density lipoprotein from oxidation *in vitro*. Free Rad. Res. **29**: 247-255.

Bruni, R., Rossi, D., Muzzoli, M., Romagnoli, C., Paganetto, G., Besco, E., Choquecillo, Peralta, K., Lora, W.S. and Sacchetti, G. 2006. Antimutagenic, antioxidant and antimicrobial properties of *Maytenus krukovii* bark. Fitoterapia **77**: 538-545.

Bucur, L., Negreanu-Pirjol, T., Giurginca, M. and Istudor, V. 2008. Some new *Elaeagnus angustifolia* L. extracts and the pharmaceutical products antioxidant activities determined by the chemilumiscence method. Revue Roumaine de Chimie **53(10)**: 961-964.

Budinevi, M., Vrbaski, Z., Turkulovxyc, J. and Dimic, E. 1995. Antioxidant activity of *Oenothera biennis* L. Lipid-Fett. European J. Lipid Sci. Tech. **97(7-8)**: 277-280. doi: 10.1002/lipi.19950970709

Budryn, G. and Nebesny, E. 2005. Structure and antioxidant properties of coffee bean polyphenol. Bromat. Chem. Toksykol. **38(3)**: 203-209.

Builders, M.I., Wannang, N.N., Ajoku, G.A., Builders, P.F., Orisadipe, A. and Aguiyi, J.C. 2011. Evaluation of the antimalarial potential of *Vernonia ambigua* Kotschy and Peyr (Asteraceae). Int. J. Pharmacol. **7(2)**: 238-247. doi: 10.3923/ijp.2011.238-247.

Burdette, J.E., Chen, S.N., Lu, Z.Z. *et al.*, 2002. Black cohosh (*Cimcifuga racemosa* L.) protects against menadione-induced DNA damage through scavenging of reactive oxygen species: Bioassay directed isolation and characterization of active principles. J. Agric. Food Chem. **50**: 7022-7028.

Buricova, L. and Reblova, Z. 2008. Czech medicinal plants as possible sources of antioxidants. Czech J,. Food Sci. **26**: 132-138.

Burits, M., Asres, K., Bucar, F. 2001. The antioxidant activity of the essential oils of *Artemisia afra, Artemisia abyssinica* and *Juniperus procera*. Phytother Res. **15(2)**: 103-108.

Burits, M. and Bucar F. 2000. Antioxidant activity of *Nigella sativa* essential oil. Phytother. Res. **14(5)**: 323–328.

Burk, D.R., Cichacz, Z.A. and Daskalova, S.M. 2010. Aqueous extract of *Achillea millefolium* L. (Asteraceae) inflorescences suppresses lipopolysaccharide-induced inflammatory responses in RAW 264.7 murine macrophages. J. Med. Plants Res. **4(3)**: 225-234.

Burns, J., Gardner, P.T., Matthews, D., Duthie, G.G., Lean, M.E. and Crozier, A. 2001. Extraction of phenolics and changes in antioxidant activity of red wines during vinification. J. Agric. Food Chem. **49 (12)**: 5797-5808.

Burton, G.W. and Ingold, K.U. 1981. Antioxdation of biological molecules.1. The antioxidant activity of vitamin E and related chain-breaking phenolic antioxidants *in vitro*. J. Amer. Chem. Soc. **103**: 6472-6477.

Bushman, S.B., Phillips, B., Isbell, T., Ou, B., Crane, J.M. and Knapp, S.J. 2004. Chemical composition of craneberry (*Rubus* spp.) seeds and oils and their antioxidant potential. J. Agric. Food Chem. **52**: 7982-7987.

Bushra, S., Farooq A., Roman, P. 2007. Antioxidant activity of phenolic components present in barks of *Azadirachta indica, Terminalia arjuna, Acacia nilotica*, and *Eugenia jambolana* Lam. trees. Food chemistry. **104 (3)** : 1106-1114.

Busserolles, J., Gueux, E., BalasiDska, B., Piriou, Y., Rock, E., Rayssiguier, Y. and Mazur, A. 2006. *In vivo* antioxidant activity of procyanidin-rich extracts from grape seed and pine (*Pinus maritima*) bark in rats. Int. J. Vitam. Nutr. Res. **76(1)**: 22-27.

Butera, D., Tesoriere, L., Di Gaudio, F., Bongiorno, A., Allegra, M., Pintaudi, A., Kohen, R. and Livrea, M. 2002. Antioxidant Activities of Sicilian Prickly Pear (*Opuntia ficus indica*) fruit extracts and reducing properties of its betalains: betanin and indicaxanthin. J. Agric. Food Chem. **50**: 6895 -6901.

Butnaru, M., Iacobovici, A, Haulica, I. and Hefco, V. 1997. *Ginkgo biloba* extract (EGb 761) as an antioxidant factor exogenous origin. Rev. Med. Chir. Soc. Med. Nat. Iasi. **101(3-4)**: 192-196. (article in Romanian).

Butryee, C., Sungpuag, P. and Chitchumroonchokchai, C. 2009. Effect of processing on the flavonoid content and antioxidant capacity of *Citrus hystrix* leaf. Int. J. Food Sci. Nutr. **60(2)**: 162-174. doi: 10.1080/09637480903018816.

Buyukbalci, A. and El, S.N. 2008. Determination of *in vitro* antidiabetic effects, antioxidant activities and phenol contents of some herbal teas. Plant Foods Hum. Nutr. **63**: 27-33.

Cacace, J.E., and Mazza, G. 2002. Extraction of anthocyanins and other phenolics from black currants with sulfured water. J. Agric. Food Chem. **50 (21)**: 5939-5946.

Cai, L.Y., Zhou, J.J., Li, *et al.*, 2011. Study on the antioxidation of polysaccharide *in vitro* from *Alternanthera philoxeroides* (Mart.) Griseb. Food Sci. (Chinese).

Cai, Y.J., Dai, J.Q., Fang, J.G., Ma, L.P., Hou, L.F., Yang, L. and Liu, Z.L. 2002. Antioxidative and free radical scavenging effects of ecdysteroids from *Serratula strangulata*. Can. J. Physiol. Pharmacol. **80(12)**: 1187-1194.

Cai, Y.Z., Luo, Q., Sun, M. and Corke, H. 2004a. Antioxidant activity and phenolic compounds of 112 traditional Chinese medicinal plants associated with anticancer. Life Sciences 74: 2157-2184.

Cai, Y.Z., Sun, M. and Corke, H. 2003. Antioxidant activity of betalins from plants in the Amaranthaceae. J. Agric. Food Chem. **51**: 2288-2294.

Cai, Y.Z., Sun, M., Xing, J. and Corke, H. 2004b. Antioxidant phenolic constituents in roots of *Rheum officinale* and *Rubia cordifolia*: Structure-radical scavenging activity relationships. J. Agric. Food Chem. **52(26)**: 7884-7890.

Cai, Y.Z., Sun, M., Xing, J., Luo, Q. and Corke, H. 2006. Structure-radical scavenging activity relationships of phenolic compounds from traditional Chinese medicinal plants. Life Sci. **78**: 2872-2888.

Cai, Y.Z., Xing, J., Sun, M., Zhan, Z.Q. and Corke, H. 2005. Phenolic antioxidants (hydrolysable tannins, flavonols and anthocyanins) identified by LC-ESI-MS and MALDI-QIT-TOF MS from *Rosa chinensis* flowers. J. Agric. Food Chem. **53**: 9940-9948.

Cai, Z.F., Chen, Z.G., Guo, P., Zhang, S.M., Lian, L.X., Luo, L. and Hu, W.M. 1993. Scavenging effects of ginger on superoxide anion and hydroxyl radical. Chung Kuo, Chung Yao Tsa Chila **18**: 750-764.

Caker, J., Paric, D., Haveric, S., Maksimovic, M. and Bajrovic, K. 2011. Antioxidant and antiproliferative activities of *Helleborus odorus* Waldst. and Kit., *H. multifidus* Vis. and *H. hercegovinus* Martinis. Nat. Prod. Res. (epub. ahead of print.).

Cakir, A., Ahmet Yildisim, M.A., Duru, M.E., Harmander, M. and Kazaz, C. 2003. Isolation and characterization of antioxidant phenolic compounds from the aerial parts of *Hypericum hyssopifolium* L. by activity guided fractionation. J. Ethnopharmacol. **87**: 73-83.

Calder, A.I., Wright, B.J., Hurst, W.J. and van Breemen, R.B. 2009. Screening Antioxidants using LC-MS: Case study with cocoa. J. Agric. Food Chem. **57(13)**: 5693-5699.

Caldwell C. R. 2003. Alkylperoxyl radical scavenging activity of red leaf lettuce (*Lactuca sativa* L.) phenolics. J. Agric. Food Chem. **51 (16)**: 4589-4595.

Calliste, C.A., Kozlowski, D., Duroux, J.L., Champavier, Y., Chulia, A.J. and Trouillas, P. 2010. A new antioxidant from wild nutment. Food Cehm. **118(3)**: 489-496.

Campos, D., Noratto, G., Chirinos, R., Arbizu, C., Roca, W. and Cisneros, L. 2006. Antioxidant capacity and secondary metabolites in four species of Andean tuber crops: native potato (*Solanum* sp.), mashua (*Tropaeolum tuberosum* Ruiz and Pavon), Oca (*Oxalis tuberosa* Molina) and Ulluco (*Ullucus tuberosus* Caldas). J. Sci. Food Agric. **86**: 1481-1488. DOI: 10.1002/jsfa.2529

Campos, R., Garrido, A., Guerra, R., and Valenzuela, A. 1988. Silybin dihemisuccinate protects against glutathione depletion and lipid peroxidation induced by acetaminophen on rat liver. Plant Medica **55**: 417-419.

Canadanovic-Brunet, J.M., Djilas, S.M., Cetkovic, G.S. and Tumbas, V.T. 2005. Free radical scavenging activity of wormwood (*Artemisia absinthium* L.) extracts. J. Sci. Food Agric. **85**: 265-272.

Canadanovic-Brunet, J.M., Djilas, S.M., Cetkovic, G.S., Tumbas, V.T., Mandic, A.I. and Canadovic, V.M. 2006. Antioxidant activities of different *Teucrium montanum* L. extracts. Int. J. Food Sci. Technol. **41**: 667-673.

Can-Ake, R., Erosa-Rejon, G., May-Pat, F. and Pena-Rodriguez, L.M. 2004. Bioactive terpenoids from roots and leaves of *Jatropha gaumeri*. Rev. Soc. Quim. Mex. **48**: 11-14.

Candan, F. and Sökmen, A. 2004. Effects of *Rhus coriaria* L (Anacardiaceae) on lipid peroxidation and free radical scavenging activity. Phytother. Res. **18 (1)**: 84-86.

Candan, F. Unlu, M. Tepe, B. Daferera, D. Polissiou, M., Sokmen, A. and Akpulat, H.A. 2003. Antioxidant and antimicrobial activity of the essential oil and methanol extracts of *Achillea millefolium* subsp. *millefolium* Afan. J. Ethnopharmacol. **87 (2)**: 215-220.

Cao, G., Booth, S.L., Sadowski, J.A. and Prior, R.L. 1998. Increases in human plasma antioxidant capacity after consumption of controlled diets high in fruit and vegetables. Am. J. Clin. Nutr. **68**: 1081-1087.

Cao, G. and Prior, R.L. 1999. Measurement of oxygen radical absorbance capacity in biological samples. Methods in Enzymology. **299**: 50–62.

Cao, G., Shukitt-Hale, B., Bickford, P.C., Joseph, J.A., McEwen, J and Prior, R.L. 1999. Hyperoxia-induced changes in antioxidant capacity and the effect of dietary antioxidants. J. Appl. Physiol. **86**: 1817-1822.

Cao, G., Sofic, E and Prior, R.L. 1996. Antioxdant capacity of tea and common vegetables. J. Agric. Food chem. **44**: 3426-3431.

Cao, G., Sofic, E and Prior, R.L. 1997. Antioxdant prooxidant behaviour of flavonoids: structure-activity relationships. Free Radical Res. **36**: 575-581.

Cao, L., Si, J.Y., Liu, Y., Sun, H., Jin, W., Li, Z., Zhao, X.H. and Pan, R.L. 2009. Essential oil composition, antimicrobial and antioxidant properties of *Mosla chinensis* Maxim. Food Chem. **115(3)**: 801-805.

Cao, Z.F. 1996. Scavenging effect of tetrandrine of active oxygen radicals. Planta Med. **62**: 413-414. doi: 10.1055/s-2006-957928

Capecka, E., Mareczeck, A. and Leja, M. 2005. Antioxidant activity of fresh and dry herbs of some Lamiaceae species. Food Chem. **93**: 223-226.

Carbonneau, M.A., Leger, C.L., Descomps, B., Michel, F. and Monnier, L. 1998, Improvement in the antioxidant status of plasma and low-density lipoprotein in subjects receiving a red wine phenolics mixture. J. Am. Oil Chem. Sic. **75**: 235-240.

Cardador-Martinez, A., Loarca-Pina, G. and Ooman, B.D. 2002. Antioxidant activity in common beans (*Phaseolus vulgaris* L.). J. Agric. Food Chem. **50(24)**: 6975-6980.

Cardoso, C.L., Castro-Gamboa, I., Silva, D.H.S., Furlan, M., de A. Epifanio, R., da C. Pinto, A., de Rezende, C.M., Lima, A., and da Silva Bolzani, V. 2004. Indole glucoalkaloids from *Chimarrhis turbinata* and their evaluation as antioxidant agents and acetylcholinesterase inhibitors. J. Nat. Prod. **67(11)**: 1882-1885.

Cardoso, C.L., da Silva Bolzani, V., Silva, D.H.S., Ishii, H., Berova, N. and Nakanishi, K. 2006. The absolute configuration of 1-(3′4′-dihydroxycinnamoyl)cyclopentane-2,3-diol from the Amazonian tree *Chimarrhis turbinata*. J. Nat. Prod. **69(7)**: 1046-1050.

Carini, M., Aldini, G., Furlanetto, S., Stefani, R., and Facino, R.M. 2001. LC coupled to ion-trap MS for the rapid screening and detection of polyphenol antioxidants from *Helichrysum stoechas*. J. Pharm. Biomedic. Anal. **24**: 517-526.

Carini, M., Aldini, G.,Orioli, M., Faciono, R.M. 2002. Antioxidant and photoprotective activity of a lipophilic extract containing neolignans from *Krameria triandra* roots. Planta medica. **68(3)**: 193-197.

Carlsen, M.H., Halvorsen, B.L., Holte, K., Bohn, S.K., Umezono, Y., Sanada,., Willey, C., Senoo, H., Umezono, Y., Sanada, C., Barikmo, I., Berhel, N., Willett, W.C.,

Phillips, K.M., Jacobs, D.R. and Blomhoff, R. 2010. The total antioxidant content of more than 3100 foods, Beverages, spices, herbs and supplements used worldwide. Nutrition Journ. 9: 3 doi 0.1186/1475- ...1-9-3.

Carrillo, J.D. and Tena, M.T. 2006. Determination of volatile compounds in antioxidant rosemary extracts by multiple headspace solid-phase microextraction and gas chromatography. Flavour and Fragrance Journal **21(4)**: 626-633.

Casanova, E., Garcia-Mina, J.M. and Calvo, M.I. 2008. Antioxidant and antifungal activity of *Verbena officinalis* L. leaves. Plant Foods Human Nutr. **63**: 39-45.

Castrejon, A.D.R.Eicholz, I., Rohn, S., Kroh, L.W. and Huyskens-Keil, S. 2008. Phenolic profile and antioxidant activity of highbush blueberry (*Vaccinium corymbosum* L.) during fruit maturation and ripening. Food Chem. **109(3)**: 564-572.

Castro-Gamboa, I. and Castro, O. 2004. Iridoids from the aerial parts of *Verbena littoralis* (Verbenaceae). Phytochemistry **65(16)**: 2369-2372.

Cavin, A., Dyatmynko, W. and Hostettman, K. 1999. Screening of Indonesian plants for antifungal and free radical scavenging activities. Pharmaceutical Biol, **37(4)**: 260-268.

Cavin, A., Hostettman, K., Dyatmynko, W. and Potterat, O. 1998. Antioxidant and lipophilic constituents of *Tinospora crispa*. Planta Med. **64**: 393-396.

Cekic, C. and Ozgen, M. 2010. Comparison of antioxidant capacity and phytochemical properties of wild and cultivated red raspberries (*Rubus idaeus* L.). J. Food Composition Anal. (in press).

Celik, I. and Tuluce, Y. 2007. Elevation protective role of *Camellia sinensis* and *Urtica dioica* infusion against trichloroacetic acid-exposed in rats. Phytother. Res. **21(11)**: 1039-1044.

Cerón, M.C., García-Malea, M.C., Rivas, J., Acien, F.G., Fernandez, J.M., Del Río, E., Guerrero M.G. and Molina, E. 2007. Antioxidant activity of *Haematococcus pluvialis* cells grown in continuous culture as a function of their carotenoid and fatty acid content. Appl Microbiol Biotechnol. **74(5)**: 1112-1119.

Cervato, C., Carabelli, M., Gervasio, S., Cittera, A., Cazzola, R. and Cestaro, B. 2000. Antioxidant properties of Oregano (*Origanum vulgare*) leaf extracts. J. Food Chem. 24: 453-465.

Cervellati, R., Innocenti, G., Dall'Acqua, S., Costa, S. and Sartini, E. 2004. Polyphenols from *Polygala* spp. And their antioxidant activity. Chem. Biodivers. **1(3)**: 415-425.

Cetinus, S.A., Goze, I., Sarac, B. and Vural, N. 2006. Scavenging effect and antispasmodic activity of the essential oil of *Cyclotrichium niveum*. Fitoterapia **78**: 129-133.

Cetojevic-Simmon, D.D., Canandavic-Brunet, J.M., Bogdanovic, G.M., Djilas, S.M., Cetkovic, G.S., Tumbas, V.T. and Stojiljkovic, B.T. 2010. Antioxidative and antiproliferative activities of different horsetail (*Equisetumm arvense* L.) extracts. J. Med. Food **13(2)**: 452-459.

Chacha, M., Bojase, M.G. and Majinda, R.R.T. 2005. Antimicrobial and radical scavenging flavonoids from the stem wood of *Erythrina latissima*. Phytochemistry **66(1)**: 99-104.

Chae, S., Kim, J.S., Kang, K.A., Bu, H.D., Lee, Y., Hyun, J.W. and Kang, S.S. 2004. Antioxidant activity of Jionoside D from *Clerodendron trichotomum*. Biol. Pharm. Bull. **27**: 1504-1508.

Chae, S., Kim, J.S., Kang, K.A., Bu, H.D., Lee, Y., Seo, Y.R., Hyun, J.W. and Kang, S.S. 2005. Antioxidant activity of isoacteoside from *Clerodendron trichtomum*. J. Toxicol. Environ Health A. **68(5)**: 389-400.

Chaeda, V.S., Braicu, C. and Socaciu, C. 2010. Antioxidant, prooxidant activity of polyphenolic grape seed extract. Food Chem. **121(1)**: 132-139.

Chaieb, N., Gonzalez, J.L., Lopez-Mesas, M., Bouslama, M. and Valiente, M. 2011. Polyphenols content and antixodidant capacity of thirteen faba bean (*Vicia faba* L.) genotypes cultivated in Tunicia. Food Res. Int. **44(4)**: 970-977.

Chakraborty, D. and Shah, B. 2011. Antimicrobial, antioxidative and anti-hemolytic activity of *Piper betel* leaf extracts Int. J. Pharm. Pharm. Sci. **3(Suppl. 3)**: 192-199.

Chakraborty, G.S. 2009. Free radical scavenging activity of *Aesculus indica* leaves. Int. J. of PharmTech Res. **1(3)**: 524-526.

Chakraborty, G.S. and Ghorpade, P.M. 2010. Free radical scavenging activity of *Abutilon indicum* (Linn.) Sweet stem extracts. Int. J. ChemTech Res. **2(1)**: 526-531.

Chakraborty, M. and Mitra, A. 2008. The antioxidant and antimicrobial properties of the methanolic extract from *Cocos nucifera* mesocarp. Food Chem. **107 (3)**: 994-999.

Chaminda, T., Munasinghe, J., Seneviratne, C.K, Thabrew, M.I. and Abeysekere, A.M. 2001. Antiradical and antiperoxidative effects of some plant extracts used by Sri Lankan traditional practitioners for cardioprotection. Phytother. Res. **15**: 519-523.

Chamundeswari, D., Vasantha, J., Gopalakrishnan, S. and Sukumar, E. 2003 Free radical scavenging activity of the alcoholic extract of *Trewia polycarpa* roots in arthritic rats J. Ethnopharmacol. **88(1)**: 51-56

Chan, E.W.C. and Lim, Y.Y. 2006. Antioxidant activity of *Thunbergia laurifolia* tea. J. Trop. For. Sci. **18(2)**: 130-136.

Chan, E.W.C., Lim, Y.Y., Wong, K.K., Tan, S.P.Lianto, F.S. and Yong, M.Y. 2009. Effects of different drying methods on the antioxidant properties of leaves and tea of ginger species. Food Chem. **113(1)**: 166-172.

Chan, J.Y.W., Lam, F.C., Leung, P.C., Che, C.T. and Fung, K.P. 2009. Antihyperglycemic and antioxidant effects of a herbal formulation Radix Astragali, Radix Codonopsis and Cortex Lycii in a mouse model of type 2 diabetes mellitus. Phytother. Res. **23(5)**: 658-665.

Chanasut, U. and Rattanpanone, N. 2008. Screening methods to measure antioxidant activities of Thai egg plants. Acta Hort. (ISHS). **765**: 291-296.

Chanda, S., Dave, R. and Kaneria, M. 2011. *In vitro* Antioxidant Property of Some Indian Medicinal Plants. Res. J. Med. Plant **5**: 169-179. DOI: 10.3923/rjmp.2011.169.179.

Chander, R., Khanna, A.K. and Pratap, R. 2000. Antioxidant activity of guggulsterone, the active principle of guggulipid from *Commiphora mukul*. J. Med. and Aromatic Plant Sciences, **24**: 371-375.

Chander, R., Singh, K., Khanna, A.K., Kaul, S.M., Puri, A., Saxena, R., Bhatia, G., Rizvi, F and Rastogi, A.K. 2004. Antidyslipidemic and antioxidant activities of different fractions of *Terminalia arjuna* stem bark. Indian J. Clin. Biochem. **19(2)**: 141-148.

Chandra, A., Mahdi, A.A., Singh, R.K., Elhence, R. and Ahmad, S. 2003. Effect of herbal hypoglycemic agents on antioxidative potential and metal ion status STZ induced diabetic rats. National Symposium on Emerging Trends in Indian Medicinal Plants, Lucknow, p 57.

Chandra, S. and Sastry, M.S. 1990. Chemical constituents of *Aerva lanata*. Fitoterapia **61**: 188.

Chandrappa, C.P., Shilpashree, C.B., Karthik, M.R., Govindappa, M. and Sadananda, T.S. 2011. Antibacterial and antioxidant activities of *Adiantum pedatum* L. J. Phytology **3(1)**: 26-32.

Chandrasekhar, V.M., Muchandi, A.A., Sudi, S.V. and Ganapaty, S. 2010. Hepatoprotective activity of *Stereospermum suaveolens* against CCl_4-induced liver damage in albino rats. Pharmaceutical Biol. **48(5)**: 524-528.

Chang, C.H., Chyau, C.C., Hsieh, C.L., Wu, Y.Y., Ker, Y.B., Tsen, H.Y. and Peng, R.Y. 2008. Relevance of phenolic diterpene constituents to antioxidant activity of leaves of rosemary. Nat Prod Res. **22 (1)**: 76-90.

Chang Ching-Hüi and Liu Yung-Chuan. 2008. Evaluation of antioxidative performance of tomato extracts obtained by different methods. J. Sci. Food Agric. **88(4)**: 612-618.

Chang, H.C., Huang, G.J., Agarwal, D.C., Kuo, C.L., Wu, C.R. and Tsay, H.S. 2007. Antoxidant activities and polyphenol contents of six folk medicinal ferns used as "Gusuibu". Bot. Studies **48(4)**: 397-406.

Chang, H.Y., Ho, Y.L., Sheu, M.J., Lin, Y.H., Tseng, M.C., Wu, S.H., Huang, G.J. and Chang, Y.S. 2007. Antioxidant and free radical scavenging activities of *Phellinus merrillii* extracts. Botanical Studies **48(4)**: 407-417.

Chang, L.W., Yen, W.J., Huang, S.C. and Duh. P.D. 2002. Antioxidant activities of sesame seed coat. Food Chem. **78**: 347-354.

Chang Su-Feng, Hsieh Chiu-Lan and Yen Gowchin. 2008. The protective effect of *Opuntia dillenii* Haw. fruit against low-density lipoprotein peroxidation and its active compounds. Food Chem. **106(2)**: 569-575.

Chang, S.J., Lee, Y.C., Liu, S.-Y. and Chang, T.-W. 2004. Chinese yam (*Dioscorea alata* cv. Tainung No.2) feeding exhibited antioxidative effects in Hypercysteinemia rats. J. Agric. Food Chem **52(6)**: 1720-

Chang, S.T., Wu, J.H., Wang, S.Y. Kang, P.L., Yang, N.S. and Shyur, L.F. 2001. Antioxidant activity of extracts from *Acacia confusa* bark and heartwood. J. Agr. Food Chem. **49**: 3420-3424.

Chang, W.C., Kim, S.C., Hwang, S.S., Choi, B.K., Ahn, H.J., Lee, M.Y., Park, S.H. and Kim, S.K. 2002. Antioxidant activity and free radical scavenging capacity between Korean medicinal plants and flavonoids by assay-guided comparison. Plant Sci. **163**: 1161-1168.

Channabasavaraj, K.P., Badami, S. and Bhojraj, S. 2008. Hepatoprotective and antioxidant activity of methanol extract of *Ficus glomerata*. Nat. Med. (Tokyo) **62(3)**: 379-383.

Chantaro, P., Devahastin, S. and Chiewchan, N. 2008. Production of antioxidant high dietary fiber powder from carrot peels. LWT - Food Science and Technology, **41 (10)**: 1987-1994.

Chanwitheesuk, A., Teerawutgulrag, A. and Rakariyatham, N. 2005. Screening of antioxidant activity and antioxidant compounds of some edible plants of Thailand. Food Chem. **92**: 493-496.

Chao, G.R. 2001. Antioxidant properties and polysaccharide composition analysis of ear mushrooms. Master's thesis, National Chung-Hsing University, Taichung, Taiwan.

Chao, W.W., Kuo, Y.H., Li, W.C. and Lin, B.F. 2009. The production of nitric oxide and prostaglandin E2 in peritoneal macrophages inhibited by *Andrographis paniculata, Angelica sinensis* and *Morus alba* ethyl acetate fractions. J. Ethnopharmacol. **121 (1)**: 68-75.

Charde, R.M., Charde, M.S., Fulzele, S.V., Satturwar, P.M., Kasture, A.V. and Joshi, S.B. 2009. Evaluation of ethanolic extract of *Moringa oleifera* for wound healing, anti-inflammatory and antioxidant activities on rats. Hamdard Medicus **52(2)**: 59-67.

Chatterjee, M., Sarkar, K. and Parames, C. 2006. Sil herbal (*Phyllanthus niruri*) protein isolate protects liver from nimesulide induced oxidative stress. Pathophysiology **13**: 95-102.

Chatterjee, S., Niaz, Z., Gautam, S., Adhikari, S., Variyar, P.S. and Sharma, A. 2007. Antioxidant activity of some phenolic constituents from green pepper (*Piper nigrum* L.) and fresh nutmeg mace. Food Chemistry. **101 (2)**: 515-523.

Chatterjee, S.S., Jaggy, H. and Koch, E. 1996. Antioxidative and human neutrophil elastase inhibitory activity of a *Crataegus* extract (WS-1442) standardized on oligomeric procyanidins. Phytomedicine **3**(Suppl.): 61.

Chattopadhyay, N., Ghosh, T, Sinha, S., Chattopadhyay, K., Karmakar, P. and Ray, B. 2010. Polysaccharides from *Turbinaria conoides*: Structural features and antioxidant capacity. Food Chem. **118(3)**: 823-829.

Chattopadhyay, R.R. 1998. Possible biochemical mode of anti-inflammatory action of *Azadirachta indica* A. Juss. in rats. Indian J. Exp. Biol. **36**: 418.

Chaudhary, G.D., Kamboj, P., I. Singh and Kalia, A.N. 2010. Herbs as livers savers. Indian J. Nat. Prod. Resources **1(4)**: 397-408.

Chauhan, A.S., Negi, P.S. and Ramteke, R.S. 2007. Antioxidant and antibacterial activities of aqueous extract of Seabuckthorn (*Hippophae rhamnoides*) seeds. Fitoterapia **78**: 590-592.

Chawla, R., Arora, R., Kumar, R. Sharma, A., Prasad, J., Singh, S., Sagar, R., Chaudhary, P., Shukla, S., Kaur, G., Sharma, R.K., Puri, S.C., Dhar, K.L., Handa, G., Gupta, V.K. and Qazi, G.N. 2005. Antioxidant activity of fractionated extracts of rhizomes of high altitude *Podophyllum hexandrum*: Role in radiation protection. Mol. Cell Biochem. **273**: 193-208.

Chedea, V.S., Braicu, C. and Socaciu, C. 2010. Antioxidant/prooxidant activity of polyphenolic grape seed extract. Food Chem. **121**: 132-139.

Cheel, J., Theoduloz, C., Rodríguez, J. Caligari, P.D.S. and Schmeda-Hirshmann, G. 2007. Free radical scavenging activity and phenolic content in achenes and thalamus fiom *Fragaria chiloensis* ssp. *chiloensis*, *F. vesca* and *F. ananassa* cv. Chandler. Food Chem. **102**: 36-44.

Cheel, J., Theoduloz, C., Rodríguez, J. and Schmeda-Hirschmann, G. 2005. Free radical scavengers and antioxidants from Lemongrass (*Cymbopogon citratus* (DC.) Stapf). J. Agric. Food Chem. **53 (7)**: 2511-2517.

Chen, B-Q., Cui, X-Y., Zhao, X., Zhang, Y-H., Piao, H-S., Kim, J-H., Lee, B-C., Pyo, H-B. and Yun, Y-P. 2006. Antioxidative and acute anti-inflammatory effects of *Torrey grandis*. Fitoterapia **77**: 262-267.

Chen, C., Pearson, A.M. and Gray, J.I. 1992. Effects of synthetic antioxidants (BHA, BHT and PG) on the mutagenicity of IQ-like compounds. Food Chem. **43**: 177-183.

Chen, C.C., Huang, Y.L., Yeh, P.Y. and Ou, J.C. 2003. Cyclized geranyl stilbenes from the rhizomes of *Helminthostachys zeylanica*. Planta Medica **69(10)**: 964-967.

Chen, C-H., Chan, H-C, Chu, Y-T., Ho, H-Y., Chen, P-Y., Lee, T-H. and Lee, C-K. 2009. Antioxidant activity of some plant extracts towards Xanthine oxidase, lipoxygenase and tyrosinase. Molecules **14**: 2947-2958.

Chen, C.W. and Ho, C.T. 1995. Antioxidant properties of polyphenols extracted from green and black tea, J. Food lipids **2**: 35-46.

Chen, C.Y., Kuo, P.L., Chen, Y.H., Huang, J.C., Ho, M.L., Lin, R.J., Chang, J.S. and Wang, H.M. 2009. Tyrosinase inhibition, free radical scavenging, antimicroorganism and anticancer proliferation activities of *Sapindus mukorossi* extracts. J. Taiwan Inst. Chem. Eng. Doi: 10.1016/jjtice.2009.08.005.

Chen, F.A., Wu, A.B., Shieh, P., Kuo, D.H. and Hsieh, C.Y. 2006. Evaluation of the antioxidant activity of *Ruellia tuberosa*. Food Chem. **94**: 14-18.

Chen, H.-Y., Lin, Y.-C. and Hsieh, C.-L. 2007. Evaluation of antioxidant activity of aqueous extract of some selected nutraceutical herbs. Fod Chem. **104(4)**: 1418-1424. doi: 10.1016/j.foodchem.2007.02.004.

Chen, J.H. and Ho, C.T. 1997. Antioxidant activities of caffeic acid and its related hydroxycinnamic acid compounds. J. Agric. Food Chem.**45**: 2374-2378.

Chen J.J., Chang, Y.L. Teng, C.M., and Chen, I.S. 2001. Vasorelaxing and antioxidant constituents from *Hernandia nymphaefolia*. Planta Medica **67**: 593-598.

Chen, K., Plumb, G. W., Bennett R. N. and Bao, Y. 2005. Antioxidant activities of extracts from five anti–viral medicinal plants. J. Ethnopharmacol. **96**: 201–205.

Chen, P.Y., Tu, Y.X., Wu, C.T., Jong, T.T. and Chang, C.M. 2004. Continuous hot pressurized solvent extraction of 1,1-diphenyl-2-picrylhydrazyl free radical scavenging compounds from Taiwan yams (*Dioscorea alata*). J. Agric. Food Chem. **52 (7)**: 1945-1949.

Chen, Q.Y., Shi, H. and Ho, C.T.1992. Effect of rosemary extracts and major constituents of lipid peroxidation and soybean lipoxygenase activity. J. Amer. Oil Chem. Soc. **69(10)**: 999-1002.

Chen, S.J., Chung, J.G., Chung, Y.C. and Chou, S.T. 2008. *In vitro* antioxidant and antiproliferative activity of the stem extracts from *Graptopetalum paraguayense*. Amercan J. Chin. Med. **36 (2)**: 369-383.

Chen, Y., Huang, B., He, J., Han, L., Zhan, Y. and Wang, Y. 2011. *In vitro* and *in vivo* antioxidant effects of the ethanolic extract of *Swertia chirayita*. J. Ethnopharmacol. **136(2)**: 309-315.

Chen, Y., Wang, M., Rosen, R.T. and Ho, C.T. 1999. 2,2-Diphenyl-1-picrylhydrazyl radical scavenging active components from *Polygonum multiflorum* Thunb. J. Agric. Food Chem. **47(6)**: 2226-2228.

Chen, Y., Xie, M. and Gong, X. 2006. DPPH radical scavenging effect of *Ganoderma atrum*. Nat. Prod. Res. Dev. 2006-06.

Chen, Y.-H., Chang, F.R.,Lin, Y.J., Hsieh, P.W., Wu, M.J. and Wu, Y.C. 2008. Identification of antioxidants from rhizome of *Davallia solida* Food Chem. **107(2)**: 684-691.

Cheng, Z.J., Kuo, S.C., Chan, S.C., K., F.N. and Teng, C.M. 1998. Antioxidant properties of butein isolated from *Dalbergia odorifera*. Biochimia et Biophysica Acta **139(2-3)**: 291-299.

Cheong, K.O., Nam, K.S., Park, J.H., Kadota, S. and Moon, J.O. 1998. Development of the SOD mimics from the natural product by a novel biosystem-antiinflammatory effect of *Morus alba*. Korean J. Pharmacogn. **29(1)**: 1-7.

Cheung, L.M., Peter, C.K. and Vincent, E.C. 2003. Antioxidant activity and total phenolics of edible mushroom extracts. Food Chem. **81**: 249-255.

Chhajed, M., Jain, A., Gupta, S. and Radke, R. 2009. Current status and future prospects of free radicals and antioxidant in human health. Int. J. Pharmacol. Biol. Sci. **3(3)**: 143-160.

Chiang, P.-Y. and Ciou, J.Y. 2009. Effect of pulverization on the antioxidant activity of water caltrop (*Trapa taiwanensis* Nakai) pericarps. LWT – Food Sci. Tech. **43(2)**: 361-365. doi: 10.1016/j.lwt.2009.08.016.

Chiang, Y.M., Kuo, Y.H., Kuo, Y.H. and Oota, S. and Fukuyama, Y. 2003. Xanthones and bezophenones from the stems of *Garcinia multiflora*. J. Nat. Prod. **66(8)**: 1070-1073.

Chidambara Murthy, K.N., Jayaprakash, G.K. and Singh, R.P. 2002. Studies on antioxidant activity of Pomegranate (*Punica granatum*) peel extract using *in vivo* models. J. Agric. Food Chem. **50**: 4781-4795.

Chidambara Murthy, K.N., Singh, R.P. and Jayaprakasha, G.K. 2002. Antioxidant activities of grape (*Vitis vinifera*) pomace extracts. J. Agric. Food Chem. **50(21)**: 5909-5914.

Chin, Y.W., Jung, H.A., Liu, Y., Su, B.N., Castoro, J.A., Keller, W.J., Pereira, M.A., Douglas Kinghorn, A. 2007. Anti-oxidant constituents of the roots and stolons of licorice (*Glycyrrhiza glabra*). J. Agric. Food Chem. **55(12)**: 4691-4697.

Chinnici, F., Bendini, A., Gaiani, A. and Riponi, C. 2004. Radical scavenging activities of peels and pulps from cv. Golden delicious apples as related to their phenolic composition. J. Agric. Food Chem. **52(15)**: 4684-4689.

Chipault, J.R., Mizuno, G.R., Hawkins, J.M. and Lundberg, W.O. 1952. The antioxidant properties of natural spices. Food Res. **17**: 46-55.

Chipault, J., Mizumo, G. and Lundberg, W. 1956. The antoxidant properties of spice foods. Food Technol. **10**: 209-211.

Chirinos, R., Campos, D., Costa, N., Arbizu, C., Pedreschi, R. and Larondelle, Y. 2008. Phenolic profiles of Andean mashua (*Tropaeolum tuberosum* Ruiz et Pavon) tubers: Identification by HPLC-DAD and evaluation of their antioxidant activity. Food Chem. **106(3)**: 1285-1298. doi: 10.1016/j.foodchem.2007.07.024.

Chitra, V. and S. Leelamma, S. 1999. *Coriandrum sativum* changes the levels of lipid proxides and activity of antioxidant enzymes in experimental animals. Indian J. Biochm. Biophys. **36**: 59-61.

Chiu, C.Y., Li, C.Y., Chiu, C.C., Niwa, M., Kitanaka, S., Damu, A.G., Lee, E.J. and Wu, T.S. 2005. Constituents of leaves of *Phellodendron japonicum* Maxim. and their antioxidant activity. Chem. Pharm. Bull. **53(9)**: 1118-1121.

Chizzola, R., Michitsch, H. and Franz, C. 2008. Antioxidative properties of *Thymus vulgaris* leaves: comparison of different extracts and essential oil chemotypes. J. Agric. Food Chem. **56(16)**: 6897-6904.

Cho, E.J., Yokozawa, T., Rhyu, D.Y., Kim, S.C., Shibhara, N. and Park, J.C. 2003a. Study on the inhibitory effects of Korean medicinal plants and their main components on the 1,1-diphenyl 2-picrylhydrazyl radical. Phytomedicine 10(6-7): 544-551. doi: 10.1078/094471133322331520. PMID: 13678241.

Cho, E.J., Yokozawa, T., Rhyu, D.Y., Kim, S.C., Shibhara, N. and Park, J.C. 2003b. The inhibitory effects of 12 medicinal plants and their compounds on lipid peroxidation. Am. J. Chinese Med. **31(6)**: 907-917. doi: 10.1142/S0192415x03001648. PMID: 14992543.

Cho, J., Kim, H.M., Ryu, J.H., Jeong, Y.S., Lee, Y.S. and Jin, C. 2005. Neuroprotective and antioxidant effects of the ethyl acetate fraction prepared from *Tussilago farfara* L. Biol. Pharm. Bull. **28(3)**: 455-460.

Cho, S.M., Lee, H.Y., Yoo, S.M., Park, H.J., Yeum, K.J., Park, K.M. and Chun, H.K. 2008. Effects of Korean antioxidant plant extracts on immune activation. The FASEB Journal **22**: 732.

Cho, S.H., Cho, J.Y., Kang, S.E., Hong, Y.K., Ahn, D.H. 2008. Antioxidant activity of mojabanchromanol, a novel chromene, isolated from brown alga *Sargassum siliquastrum*. J Environ Biol. **29(4)**: 479-84.

Cho, Y.-J., Chun, S.-S., Yoon, S.-J. and Kim, J.-H. 2005. Antioxidant activity of herb plants extracts and antimicrobial activity against *Helicobacter pylori*. IFT annual meeting, July15-20, New Orleans, Louisiana.

Chobot, V., Kubicová, L., Nabbout, S., Jahodár, L. and Hadacek F. 2008. Evaluation of antioxidant activity of some common mosse. Z Naturforsch [C]. **63(7-8)**: 476-82.

Chobot, V., Kubicová, L., Nabbout, S., Jahodár, L. and Vytlcilova, J. 2006. Antioxidant and free radical scavenging activities of five moss species. Fitoterapia **77**: 598-600.

Choi, C.W., Jung, H.A., Kang, S.S. and Choi, J.S. 2007. Antioxidant constituents and a new triterpenoid glycoside from Flos Lonicerae. Arch. Pharm. Res. **30(1)**: 1-7.

Choi, E.M., Ding, V.N., Huu Tung, Park, S.-H. 2007. Antioxidant activity of Gomchi (*Ligularia fischeri*) leaves. Food Sci. Biotech. **16(5)**: 710-714.

Choi, E.M. and Hwang, J.K. 2004. Antiinflammatory, analgesic and antioxidant activities of the fruit of *Foeniculum vulgare*. Fitoterapia **75**: 557-565.

Choi, E.M. and Hwang, J.K. 2005. Effect of some medicinal plants on plasma antioxidant system and lipid levels in rats. Phytother. Res. **19(5)**: 382-386.

Choi, E.M. and Hwang, J.K. 2005. Screening of Indonesian medicinal plants for inhibitory activity on nitric oxide production of RAW 264.7 cells and antioxidant activity. Fitoterapia **76(2)**: 194-203.

Choi, J., Yoon, B., Han, Y.N., Lee, K.T., Ha, J., Jung, H.J. and Park, H.J. 2003. Antirheumatoid arthritis effect of *Rhus verniciflua* and of the acute component, sulfuretin. Planta Medica **69(10)**: 899-904.

Choi, J.S., Jung, M.J., Park, H.J., Chung, H.Y. and Kang, S.S. 2002. Further isolation of peroxynitrite and 1,1-diphenyl-2-picrylhydrazyl radical scavenging isorhamnetin 7–O-glucoside from the leaves of *Brassica juncea* L. Arch. Pharm. Res. **25(5)**: 625-627.

Choi, J.S., Yoon, T.J., Kang, K.R., Lee, K.H., Kim, W.H., Suh, Y.H., Song, J. and Jung, M.H. 2006. Glycoprotein isolated from *Acanthopanax senticosus* protects against hepatotoxicity induced by acute and chronic alcohol treatment. Biol. Pharm. Bull. **29(2)**: 306-314.

Choi, S.Y., Yang, K.M., Jeon, S.D., Kim, J.H., Khil, L.Y., Chang, T.S. and Moon, C.K. 1997. Brazilin modulates immune function mainly augmenting T cell activity in halothane administered mice. Planta Medica **63**: 405-408.

Choi, U.-K., Lee, O-H., Yim, J.H., Cho, C.-W., Rhee, Y.K., Lim, S. and Kim, Y-C. 2010. Hypolipidemic and antioxidant effects of Dandelion (*Taraxacum officinale*) root and leaf on Cholesterol-fed rabbits. Int. J. Mol. Sci. **11(1)**: 67-78.

Chomnawang, M.T., Surassmo, S., Nukoolkaran, V.S. and Gritsanapan, W. 2007. Effect of *Garcinia mangostana* on inflammation caused by *Propionibacterium acnes*. Fitoterapia **78(6)**: 401-408.

Chou, H.J., Kuo, J.L. and Lin, E.S. 2009. Comparative antioxidant properties of water extracts from different parts of Beef steak plant (*Perilla frutescens*). J. Food Drug. Ana. **17**: 489-496.

Chou, S.U., Heo, B.G., Park,Y.S., Kim, D.K. and Gorinstein, S. 2009. Total phenolics level, antioxidant activities and cytotoxicity of young sprouts of some traditional Korean salad plants. Plant Food Hum. Nutr. **64(1)**: 25-31.

Chou, S.C., Su, C.R., Ku, Y.C. and Wu, T.S. 2009. The constituents and their bioactivities of *Houttuynia cordata*. Chem. Pharm. Bull. **57(11)**: 1227-1230.

Chou, Y.-S., Ho, Y.-L., Ding, C.-W. and Chang, Y.-S. 2009. New antioxidant phenylethanol glycosides from *Torenia concolor*. J. Asian Nat. Prod. Res. **11(1-2)**: 109-114.

Choudhary, D. and Kale, R.K. 2002. Antioxidant and non toxic properties of *Piper betle* leaf extract *in vitro* and *in vivo* studies. Phytother. Res. **16**: 461-466.

Choudhary, G.P. 2006. Free radical scavenging activity of seeds of *Trigonella foenum-graecum* Linn. J. Nat. Remedies **6(1)**: 99-102.

Choudhary, M.I., Afsan begum, Abbaskhan, A., Ajaz, A., Shafique-ur-Rehman and Atta-ur-Rahman. 2005. Phenyl polypropanoids from *Lindelofia stylosa*. Chem. Pharm. Bull. **53(11)**: 1469-1471.

Choudhury, M.K., Venkatramana, S. and Upadhyay, L. 2011. Antioxidant and hypoglycemic property of *Streblus asper* in streptozotocin induced diabetic rats. J. Pharm. Res. **4(7)**:

Choudhury, R.P., Kumar, A. and Garg, A.N. 2006. Analysis of Indian mint (*Mentha spicata*) for essential oil, trace and toxic elements and its antioxidant behaviour. J. Pharm. Biomed. Anal. **41**: 825-832.

Choudhury, S. 2010. Studies on the phytochemistry and bioactivity of the Indian medicinal plant *Pajanelia longifolia*. 33rd Conference of Indian Botanical Society and International symposium on the new horizons of Botany, Shivaji University, Kolhapur, Abstract, p. 9.

Chougale, A.D., Bhosale, P.M., Jadhav, U. and Padul Manohar, V. 2011. Antibacterial and antioxidant activity of plant latex. J. Pharmacy Red. **4(2)**:

Christian, K.R., Nair, M.G. and Jackson, J.C. 2006. Antioxidant and cycloxygenase inhibitory activities of sorrel (*Hibiscus sabdariffa*). J. Food Comp. Anal. **19(8)**: 778-785.

Chu, Y. 2000. Flavonoid content of several vegetables and their antioxidant activity. J. Sci. Food and Agricult. **80**: 561-566.

Chu, Y.F., Sun, J., Wu, X. and Liu, R.H. 2002. Anioxidant and antiproliferative activities of common vegetables. J. Agric. Food Chem. **50(23)**: 6910-6919.

Chu, Y.H., Chang, C.L. and Hsu, H.F.2000. Flavonoid content of several vegetables and their antioxidant activity. J. Sci. Food Agric. **80**: 561-566.

Chua, M., Tung, Y. and Chang, S. (in press). Antioxdant activities of ethanolic extracts from twigs of *Cinnamomum osmophloeum*. Bioresource Technology.

Chuah Ali Mey, Lee Yachi, Yamaguchi Tomoko, Takamura Hitoshi,Yin Li-Jun and Matoba Teruyoshi. 2008. Effect of cooking on the antioxidant properties of pepper. Food Chem **111(1)**: 20-28.

Chumark, P., Khunawat, P., Sanvarinda, Y., Phornchirasilp, S., Morales, N.P., Phivthong-ngam, L., Ratanachamnong, P., Srisawat, S. and Pongrapeeporn, K. 2008. The *in vitro* and *ex vivo* antioxidant properties hypolipidaemic and antiatherosclerotic activities of water extract of *Moringa oleifera* Lam. leaves. J. Ethnopharmacol. **116 (3)**: 439-446.

Chun, O.K., Kim., D.O., Moon, H.Y., Kang, H.G. and Lee, C.Y. 2003. Contribution of individual polyphenolics to total antixodant capacity of plums. J. Agric. Food Chem. **51(25)**: 7240-7245.

Chun, O.K., Kim., D.O., Smith, N., Schroeder, D., Han, J.T. and Lee, C.Y. 2005. Daily consumption of phenolics and total antioxidant capacity from fruit and vegetables in the American diet. J. Sci. Food. Agric. **85**: 1715-1724.

Chung, C.K. and Jung, M.E. 2003. Ethanol fraction of *Aralia elata* Seeman enhances antioxidant activity and lowers serum lipids in rats when administered with beenzo(a)pyrene. Biol. Pharm. Bull. **26(10)**: 1502-1504.

Chung, S.K., Kim, Y.C., Akaya, Y., Terashima, K. and Niwa, M. 2004. Novel glycoside, 7-O-methyl mearnsitrin, from *Sageretia theezans* and its antioxidant effect. J. Agric. Food Chem. **52(15)**: 4664-4668.

Chung, S. K., Osawa, T. and Kawakishi, S. 1997. Hydroxyl radical scavenging effects of spices and scavengers from brown mustard (*Brassica nigra*). Biosci. Biotech. Biochem. **69**: 118–123.

Chung, Y.-C., Chen, S.-J., Hsu, C.-K., Chang, C.-T. and Chou, S.-T. 2005. Studies on the antioxidative activity of *Graptopetalum paruguayense* E. Walther. Food Chem. **91(3)**: 419-424. doi: 10.106/j.foodchem.2004.06.022

Chung, Y.S., Choi, Y.H., Lee, S.J., Choi, S., Lee, J.H., Kim, H. and Hong, E.K. 2005. Water extract of *Aralia elata* prevents cataractogenesis *in vitro* and *in vivo*. J. Ethnopharmacol. **101(1-3)**: 49-54.

Cigremis, Y., Ulukanli, Z., Ilcim, A. and Akgoz, M. 2010. *In vitro* antioxidant and antimicrobial assays of acetone extracts of *Nepeta meyeri* Bentham. Eur. Rev. Med. Pharmacol. Sci. **14(8)**: 661-668.

Cioffi, G., D'Auria, M., Braca, A., Mendez, J., Castillo, A., Morelli, I., De Simone, F. and De Tommasi, N. 2002. Antioxidant and free-radical scavenging activity of constituents of the leaves of *Tachigalia paniculata*. J. Nat. Prod. **65(11)**: 1526-1529.

Cioffi, G., Escobar, L.M., Braca, A. and Tommasi, N. 2003. Antioxidant chalcone glycosides and flavones from *Maclura* (*Chlorophora*) *tinctoria*. J. Nat. Prod. **66(8)**: 1061-1064.

Cioroi, M. and Dumitriu, D. 2009. Studies on total polyphenols content and antioxidant activity of aqueous extracts from selected Lamiaceae species. The Annals of the University of Dunerea de Jos of Galati Fascicle VI – Food Technol. **34(1)**: 42-46.

Citolu, G.S., Coban, T., Sever, B. and Ican, M. 2004. Antioxidant properties of *Ballota* species growing in Turkey. J. Ethnopharmacol. **92(2-3)**: 275-280.

Claiborne. A. 1985. Catalase activity. In: R.A. Greenwald. (ed.). CRC Handbook of Methods for Oxygen Radical Research. Boca Raton: CRC Press; 283–84.

Clark, J.R., Howard, L., and Talcott, S. 2002. Antioxidant activity of blackberry genotypes. Acta Hort. **585**: 475-480.

Co, M., Fagerlund, A., Engman, L., Sunnerheim, K., Sjoberg, P.J.R. and Turner, C. 2011. Extraction of antioxidants from spruce (*Picea abies*) bark using ecofriendly solvents. Phytochem. Anal. Doi: 10.1002/pca.1316

Collier, A., Wilson, R., Bradley, H., Thomson, J.A. and Small, M. 1990. Free radical activity in type 2 diabetes. Diabetes **7**: 27-30.

Compaore, M., Lamien-Meda, A., Mogosan, C., Lamien, C.E., Kiendrebeogo, M., Vostinaru, O., Vlase, L., Ionescu, C. and Nacoulma, O.G. 2011. Antioxidant, diuretic activities and polyphenol content of *Stereospermum kunthianum* Cham. (Bignoniaceae). Nat. Prod. Res. (epub ahead of print.)

Conforti, F., Loizzo, M.R., Marrelli, M., Monichini, F., Statti, G.A., Uzunov, D. and Menchini, F. 2010. Quantitative determination of Amaryllidaceae alkaloids from *Galanthus reginae-olgae* subsp. *vernalis* and *in vitro* activities relavant for neurodegenerative diseases. Pharm. Biol. **48(1)**: 2-9

Conforti, F., Loizzo, M.R., Statti, A. and Menichini, F. 2005. Comparative radical scavenging and antidiabetic activities of methanolic extract and fractions from *Achillea ligusticha* All. Biol. Pharm. Bull. **28(9)**: 1791-1794.

Conforti, F., Statti, G.A., Loizzo, M.R., Sacchetti, G., Poli, F. and Menichini, F. 2005. *In vitro* antioxidant effect and inhibition of alpha-amylase from two varieties of *Amaranthus caudatus* seeds. Biol. Pharm. Bull. **28(6)**: 1098-1102.

Conforti, F., Statti, G.A., Tundis, R., Loìzzo, M.R., Bonesi, M., Menichini, F., and Houghton, P.J. 2004. Antioxidant and cytotoxic activities of *Retama raetum* subsp. *gussonei*. Phytother. Res. **18(7)**: 585-587.

Conforti, F., Statti, G.A., Tundis, R., Loizzo, M.R. and Menichini, F. 2007. *In vitro* activities of *Citrus medica* L. cv. Diamante (Diamonte citron) relevant to treatment of diabetes and Alzheimer's disease. Phytother. Res. **21(5)**: 427-433.

Conforti, F., Statti, G.A., Tundis, R., Menichini, F. and Houghton, P. 2002. Antioxidant activity of methanolic extracts of *Hypericum triquetrifolium* Turra aerial part. Fitoterapia. **73(6)**: 479-83.

Connor, A.M., Finn, C.E. and Alspach, P.A. 2005a. Genotypic and environmental variation in antioxidant activity and total phenolic content among blackberry and hybridberry cultivars. J. Amer. Soc. Hort. Sci. **130**: 527-533.

Connoer, A.M., Finn, C.E., McGhie, T.K. and Alspach, P.A. 2005b. Genetic and environmental variation in anthocyanins and their relationship to antioxidant activity in blackberry and hybridberry cultivars. J. Amer. Soc. Hort. Sci. **130**: 680-687.

Connor, A.M., Luby, J.J. and Tong, C.B.S. 2002a. Variation and heritability estimates for antioxidant activity, total phenolic content and anthocyanin content in blueberry progenies. J. Amer. Soc. Hort. Sci. **127**: 82-88.

Connor, A.M., Luby, J.J. and Tong, C.B.S. 2002b. Variability in antioxidant activity in blueberry and correlations among different antioxidant activity assays. J. Amer. Soc. Hort. Sci. **127**: 238-244.

Connor, A.M., Luby, J.J., Tong, C.B.S., Finn, C.E., and Hancock, J.F. 2002. Genotypic and environmental variation in antioxidant activity, total phenolic content and anthocyanin content among blueberry cultivars. J.Amer. Soc. Hort. Sci. **127**: 89-97.

Connor, A.M., Stephens, M.J., Hall,.K. and Alspach, J.F. 2005a. Variation and heritabilities of antioxidant activity and total phenolic content estimated from a red raspberry factorial experiment. J. Amer. Soc. Hort. Sci. **130**: 403-411.

Connor, A.M., McGhie, T.K., Stephens, M.J., Hall, H.K., Alspach, P.A. 2005b. Variation and heritability estimates of anthocyanins and their relationship to antioxidant activity in a red raspberry factorial mating design. J. Amer. Soc. Hort. Sci. **130**: 534-542.

Constantino, L., Albasino, A., Rastelli, G. and Benvenuti, S. 1992. Activity of polyphenolic crude extracts as scavengers of superoxide radicals and inhibitors of xanthine oxidase. Planta Med. **58**: 342-344.

Cook, J.A., Van der Jagt, D.J., Dasgupta, A., Mounkaila, G, Glew, R.S., Blackwell, W. and Glew, R.H. 1998. Use of Trolox assay to estimate the antioxidant content of seventeen edible wild plants of Niger. Life Sci. **63(2)**: 105-110.

Cordova, C.A., Siquera, I.R. and Netto, C.A. 2002. Protective properties of butanolic extract of the *Calendula officinalis* L. (marigold) against lipid peroxidation of rat liver microsomes and action as free radical scavenger. Redox Report **7**: 95-102.

Correia, H., A. Gonzalez–Paramas, M. T. Amaral, C. Santos–Buelga and M. T. Batista. 2005. Characterisation of polyphenols by HPLC–PAD–ESI/MS and antioxidant activity in *Equisetum telmateia*. Phytochemical Analysis **16**: 380–387.

Corsino, J., Silva, D.H.S., Zanoni, M.V.B., Bolzani, V.da S., Franca, S.C., Pereira, A.M.S and Furlan, M. 2003. Antioxidant flavan-3-ols and flavonol glycosides from *Maytenus aquifolium*. Phytother. Res. **17(8)**: 913-916.

Coruh, N., Celep, A.G.S. and Ozgokce, F. 2007. Antioxidant properties of *Prangos ferulacea* (L.) Lindl., *Chaerophyllum macropodum* Boiss. and *Heracleum persicum*

Desf. from Apiaceae family used as food in Eastern Anatolia and their inhibitory effects on glutathione-S-transferase. Food Chem. **100**: 1237-1242.

Coruh, N., Gomez, A.A., Ercisli, S. and Bilen, S. 2007. Total phenolics, mineral elements, antioxidant and antibacterial activities of some edible wild plants in Turkey. Asian J. Chem. **19(7)**: 5755-5762.

Coruh, N., Sagdicoglu Celap, A.G., Ozgocke, F. and Iscan, M. 2007. Antioxidant capacities of *Gundelia tournefortii* L. extracts and inhibition on glutathione-S-transferase activity. Food Chem. **100**: 1249-1253.

Costa, D.A., Chaves, M.H., Silva, W.C.S. and Cost, C.L.S. 2010.Chemical constituents, total phenolics and antioxidant activity of *Sterculia striata* St. Hil. et Naudin. Acta Amaz. **40(10)**: 207-212. doi.org/10.1590/S0044-59672010000100026.

Costa, D.A.F., Sanagiotto, D., Rosa, R.M., Henriques, J.A.P., Ferraz, AB. and Richter, M.F. 2009. *In vivo* and *in vitro* antioxidant activities of *Alternanthera brasiliana* leaves extracts. Resumos do 55 Congresso Brasileiro de Genetica.

Costa, L.S., Fidelis, G.P., Telles, C.B.S., Dantas-Santos, N., Camaro, R.B.G., Codeiro, S.L., Costa, M.S.S.P., Almeida-Lima, J., Melo-Silveira, R.F., Oliveira, R.M., Albuquerque, I.R.L., Andrade, G.P.V. and Rocha, H.A.O. 2011. Antioxidant and antiproliferative activities of Heterofucans from the seaweed *Sargassum filipendula*. Mar. Drug **9**: 952-966. doi: 10.3390/md9060952.

Costantino, L., Rastelli, G., Rossi, T., Bertoldi, M. and Ibasini, A. 1994. Composition, superoxide radicals scavenging and antiperoxidant activity of some edible fruits. Fitoterapia **65**: 44-47.

Couladis, M., Tzakou, O., Verykokidou, E. and Harvala, C. 2003. Screening of some Greek Aromatic plants for antioxidant activity. Phytother. Res. **17**: 194-196.

Cox, S.D., Jayasinghe, K.C., Markham, J.L. 2005. Antioxidant activity in Australian native sarsaparilla (*Smilax glyciphylla*). J. Ethnopharmacol. **101 (1)**: 162-168.

Cozma, L.S. 2004. The role of antioxidant therapy in cardiovascular disease. Current opinion in Lipidology 15: 369-371.

Croft, K.D. 1999. Antioxidant effects of plant phenolic compounds. In: Basu, T.K., Temple, N.J. and Garg, M.L. (eds.). 1999. Antioxidants in human health and disease, CABI Publishing, New York. Pp. 109-112.

Cross, C.E., Halliwell, B., Borish, E.T., Pryor, W.A., Ames, B.N., Saul, R.L., McCord, J.M. and Harman, D. 1987. Oxygen radicals and human disease. Ann. International Med. **107**: 526-545.

Crozier, A. Burns, J., Aziz, A.A., Stewart, A.J., Rabiasz, H.S., Jenkins, G.I., Edwards, C.A. and Lean Mej. 2000. Antioxidant flavonols from fruits, vegetables and beverages: measurement and bioavailability. Biol. Res. **33**: 79-88.

Cuendet, M. Hostettmann, K. and Potterat. O. 1997. Iridoid glucosides with free radical scavenging properties from *Fagraea blumei*. Hel. Chim. Act. **80**: 1144–52.

Cuendet, M., Potterat, O., Salvi, A., Testa, B. and Hostettmann, K. 2000. A stilbene and dihydrochalcone with radical scavenging activities from *Loiseleuria procumbens*. Phytochemistry **54**: 871-874.

Cui Hengxiang, Liu Qin, Tao Yongzhen, Zhang Hongfeng, Zhang Linaand Ding Kan. 2008. Structure and chain confirmation of a (1!6)-α-d-glucan from the root of *Pueraria lobata* (Willd.) Ohwi and the antioxidant activity of its sulfated derivative. Carbohydr. Polym. **74(4)**: 771-778.

Cui, X.Y., Kim, J.H., Zhao, X., Chen, B.Q., Lee, B.C., Pyo, H.B., Yun, Y.P., Zhang, Y.H. 2006. Antioxdative and acute anti-inflammatory effects of *Campsis grandiflora* flower. J. Ethnopharmacol. **103(2)**: 223-228.

Cui, Y., Xie, H. and Wang, J. 2005. Potential biomedical properties of *Pinus massoniana* bark extract. Phytother. Res. **19(1)**: 34-38.

Currie, A., Langford, G., MsGhie, T., Apiolaza, L.A.,Snelling, C., Braithewaite, B. and Vather, R. 2006. Inheritence of antioxidants in a New Zealand Blackcurrant (*Ribes nigrum* L.) population. In: Meyer, C.F. (ed.) Breeding for Success: Diversity in Action, Proceedings of the 13th Australian Plant Breeding Conference, Christchurch, New Zealand, pp. 218-225.

Cuvelier, M.E., Berset, C. and Richard, H. 1994. Antioxidant constituents of sage (*Salvia officinalis*). J. Agric. Food Chem. **42**: 665-669.

Cuvelier, M.E., Richard, H and Berset, C. 1996. Antioxidative activity and phenolic compounds of pilot plant and commercial extracts of sage and rosemary. J. Am. Oil Chem. Soc. **73**: 645- 652.

Czinner, E., Hagymasi, K., Blazovics, A., Kery, A., Szoke, E. and Lemberkovics, E. 2001. The *in vitro* effect of *Helichrysi flos* on microsomal lipid peroxidation. J. Ethnopharmacol. **77**: 31-35.

D'Abrosca, B., Fiorentino, A., Scognamiglio, M., Pacifico, S., Piccolella, S. and Monaco, P. 2010. Structural characterization and radical scavenging activity of monomeric and dimeric cinnamyl glucose esters from *Petrorhagia velutina* leaves. Phytochem. Letters **3(1)**: 38-44. doi: 10.1016/j.phytol. 2009.11.001.

D'Abrosca, B. Pacifico, S. Cefarelli, G. Mastellone, C. Fiorentino, A. 2007. Limoncella' apple, an Italian apple cultivar: Phenolic and flavonoid contents and antioxidant activity Food Chem. **104 (4)**: 1333-1337.

Dai, F., Chen, W-F., Zhou, B., Yang, L. and Liu, Z-L. 2009. Antioxidative effects of curcumin and its analogues against the free-radical-induced peroxidation of linoleic acid in micelles. Phytother. Res. **23(9)**: 1220-1228.

Dai, S.J., Wu, Y, Wang, Y.H. He, W.Y., Chen, R.Y. and Yu, D.Q. 2004. New Diels-Alder type adducts from *Morus macroura* and their antioxidant activities. Chemical Pharmaceutical Bull. **52(10)**: 1190-1193.

Dalal, P.M., Duragkar, N.J., Katolkar, P.P., Bhodele, S.B. and Bhusari, K.P. 2010. *In vitro* antioxidant activity of methanolic extract of *Leea asiatica*. 2nd Intern. Seminar on Medicinal plants and Herbal Products, S.V.Univ., Tirupati, Abstract pp. 110.

Dalby-Brown, L., Barsett, H., Landbo, A.K., Meyer, A.S. and Mølgaard, P. 2005. Synergistic antioxidative effects of alkamides, caffeic acid derivatives, and polysaccharide fractions from *Echinacea purpurea* on *in vitro* oxidation of human low-density lipoproteins. J. Agric. Food Chem. **53(24)**: 9413-9423.

Dall'Acqua, S. and Innocenti, G. 2004. Antioxidant compounds from *Chaerophyllum hirsutum* extracts. Fitoterapia **75**: 592-595.

Dal Monte, R., Dal Tosa, R., Minghetti, A., Crespi, P. and Nicoletta, P.G. 2008. Extracts from *Ajuga reptans* cell lines comprising phenylpropanoids with antioxidant activity. European patent application EP20050425585, European Patent Office Bulletin 39, publication date 24-09-2008.

Damasceno, D.C., Volpato, G.T., Calderon, I.de M.P., Aguilar, R. and Rudge, M.V.C. 2004. Effect of *Bauhinia forficata* extract in diabetic pregnant rats: maternal repercussions. Phytomedicine **11(2-3)**: 196-201.

Dandlen, S.A., Lima, A.S., Mendes, M.D., Miguel, M.G., Faleiro, M.L., Sousa, M.J., Pedro, L.G., Barroso, J.G. and Figueiredo, A.C. 2010. Antioxidant activity of six thyme species essential oils. Flavour Fragr. J. **25**: 150-155.

Dang, M., Takacsova, M., Nguyen and Kristianova, K. 2001. Antioxidant activity of essential oils from various spices. Nahrung **45**: 64-66.

Daniel, J.F., Alves, K.Z., Jacques, D., Souza, P.V., de Carvalho, M.G., Freire, R.B., Ferreira, D.T. and Freire, M.F.I. 2005. Free radical scavenging activity of *Pfaffia glomerata* (Spreng.) Petterson (Amaranthaceae). www.journalonweb.com/ijp/Articles/IJP_482-04R1.doc

Daniel, R.S., Mathew, B.C. and Devi, K.S. 1998. Antioxidant effect of two flavonoids from the bark of *Ficus benghalensis* Linn. in hyperlipidemic rats. Indian J. Exp. Biol. **36**: 902-906.

Dapkevicius, A., van Beek, T.A., Lelyveld, G.P., van Veldhuizen, A., de Groot, A.E. and Linessen, J.P.H. 2002. Isolation and structure elucidation of radical scavengers from *Thymus vulgaris* leaves. J. Nat. Prod. **65**: 892-896.

Dapkevicius, A., Venskutonis, R., Van Beek, T.A. and Linssen, P.H. 1998. Antioxidant activity of extracts obtained by different isolation procedures from some aromatic herbs grown in Lithuania. J. Sci. Food and Agric. **77**: 140-146.

Dar, A., Faizi, S, Naqvi, S., Roome, T., Rehman, S.Z-U, Ali, M., Firdous, S. Moin, S.T. 2005. Analgesic and antioxidant activity of mangiferin and its derivatives: the structure activity relationship. Biol. Pharm. Bull. **28**: 596-600.

Das, S.K., Mukherjee, S. and Vasudevan, D.M. 2008. Medicinal properties of milk thistle with special reference to silymarin – An overview. Nat. Prod. Radiance **7(2)**: 182-192.

Dasgupta, N. and De, B. 2004. Antioxidant activity of Piper betle L. leaf extract *in vitro* Food Chem. **88 (2)**: 219-224.

Dasgupta, T., Rao, A.R. and Yadava, P.K. 2003. Modulatory effect of Henna leaf (*Lawsonia inermis*) on drug metabolizing phase I and phase II enzymes, antioxidant enzymes, lipid peroxidation and chemically induced skin and forestomach papillomagnesis in mice. Molecular and Cellular Biochem. **245**: 11-22.

Dasgupta, T., Rao, A.R. and Yadava, P.K. 2004. Chemonmodulatory efficacy of basil leaf (*Ocimum basilicum*) on drug metabolizing and antioxidant enzymes and on carcinogen-induced skin and forestomach papillomagenesis. Phytomedicine **11(2-3)**: 139-151.

Dash, S., Nath, L.K., Bhise, S. and Bhuyani, N. 2005. Antioxidant and antimicrobial activities of *Heracleum nepalense* D. Don root. Trop. J. Pharm. Res. **4(1)**: 341-347.

Da Silva, A.P., Rocha, R., Silva, C.M.L., Mira, L., Duarte, M.F. and Florencio, M.H. 2000. Antioxidants in medicinal plant extracts. A research study of the antioxidant capacity of *Crategus, Hamamelis* and *Hydrastis*. Phytother. Res. **14(8)**: 612-616.

Da Silva, J.M.R., Darmon, N., Fernandez, Y. and Mitjavila, S. 1991. Oxygen free radical scavenger capacity in aqueous models of different procyanidins from grape seeds. J. Agric. Food Chem. **39**: 1549-1552.

Da Silva, T.B.C., Souza, V.K., Da Silva, A.P., Lyra Lemos, R.P. and Conserva, L.M. 2010. Determination of the phenolic content and antioxidant potential of crude extracts and isolated compounds from leaves of *Cordia multispicata* and *Tournefortia bicolor*. Pharm. Biol. **48(1)**: 63-69.

Dastmalchi, K., Dorman, H.J.D., Laakso, I. and Hiltunen, R. 2007. Chemical composition and antioxidant activity of Moldavian balm (*Dracocephalum moldavica* L.) extracts. Food Sci. Tech. **40**: 1655-1663.

Dastmalchi, K., Dorman, H.J.D., Oinonen, P.P., Darwis, Y., Laakso, I. and Hiltunen, R. 2008. Chemical composition and *in vitro* antioxidative activity of a lemon balm (*Melissa officinalis* L.) extract. LWT **41**: 1-400.

Datt, S.C., Mathur, R.K. and Baruah, J.N. 1988. Minor alkaloids of *Coscinium fenestratum* root. Indian Drugs **25**: 350.

Dauksas, E., Venskutonis, P.R., Povilaityte,V. and Sivik, B. 2001. Rapid screening of antioxidant activity of sage (*Salvia officinalis* L.) extracts obtained by supercritical carbon dioxide at different extraction conditions. Nahrung. **45(5)**: 338-341.

Davino, S.C., Barros, S., Barros, S.B.M., Silva, D.H.S. and M.Yoshida. 1998. Antioxidant activity of *Iryanthera sagotiana* leaves. Fitotherapia **69(2)**: 185-186.

Davod, J.M., Barreiros, A.L.B.S. and David, J.P. 2004. Antioxidant phenypropanoid esters of triterpenes from *Dioclea lasiophylla*. Pharmaceutical Biol. **42(1)**: 36-38.

Davod, Y., Latif G., Rezaei, M.B., Massoud, T., Astaneh, S.A. and Iraj, R. 2006. Biochemical activities of Iranian *Mentha piperita* L. and *Myrtus communis* L. essential oils Phytochemistry **67(12)**: 1249-1255.

Deb, L., Dubey, S.K., Jain, A.K. and Pandian, G.S. 2009. Preventive effect of *Thuja occidentalis* Linn. on gastric ulcer – a novel role of free radical scavenger. J. Nat. Remedies **9(2)**: 152-158.

de-Campos, L.M., Leimann, F.V., Pedrosa, R.C. and Ferreira, S.R. 2008. Free radical scavenging of grape pomace extracts from Cabernet sauvingnon (*Vitis vinifera*). Bioresour Technol. **99(17)**: 8413-8420.

Deepak, M. and Handa, S.S. 2000. Antiinflammatory activity and chemical composition of extracts of *Verbena officinalis*. Phytother. Res, **14(6)**: 463-465.

Dehmlow, C., Murawski, N. and de Groot, H. 1996. Scavenging of reactive oxygen species and inhibition of arachidonic acid metabolism by silibinin in human cells. Life Sci. **58(18)**: 1591-1600.

Deighton, N., Brennan, R., Finn, Ch. And Davies, H.V. 2000. Antioxidant properties of domesticated and wild *Rubus* species. J. Sci. Food Agric. **80**: 1307-1313.

Deighton, N., Stewart, D., Davies, H.V., Gardner, P.T., Duthie, G.G., Mullen, W. and Crozier, A. 2002. Soft fruit as source of dietary antioxidants. Acta Hort. **585**: 459-465.

Delanty, N. and Dichter, M.A. 2000. Antioxidant therapy in neurologic disease. Arch. Neurol. **57(9)**: 1265-1270.

De las Heras, B., Slowing, K., Benedi, J., Carreto, E., Ortega, T., Toledo, C., Bermejo, P., Iglesias, I., Abad, M.J., Gomez-Serranillos, P., Liso, P.A., Villar, A. and Chiriboga, X. 1998. Antiinflammatory and antioxidant activity of plants used in traditional medicine in Ecuador. J. Ethnopharmacol. **61**: 161-166.

Delazar, A., Biglari, F., Esnaashari, S., Nazemiyeh, H., Talebpour, A.H., Nahar, L. and Sarker, S.D. 2006. GC-MS analysis of the essential oils, and the isolation of phenylpropanoid derivatives from the aerial parts of *Pimpinella aurea*. Phytochem. **67(19)**: 2176-2181.

Delazar, A., Talisch, B., Nazemiyeh, H., Rezazadeh, H., Nahar, L. and Sarker, S.D. 2006. Chrozophorin, a new acylated flavone glucoside from *Chrozophora tinctoria* (Euphorbiaceae). Brazilian J. Pharmacogn. **16(3)**: 286-290.

Deliorman-orhan, D., Senol, S., Kartal, M. and Orhan, I. 2009. Assessment of antiradical potential of *Calluma vulgaris* (L.) Hull and its major flavonoid. J. Sci. Food Agric. **89(5)**: 809-814.

Deliorman-orhan, D.S.S., Kartal, M. and Orhan, I. 2009. Assessment of antiradical potential of *Calluna vulgaris* (L.) Hull and its major flavonoid. J. Sci Food Agric. **89(5)**: 809-814.

Deliorman-orhan, D., Yesilada, E., Ergun, F., Tsuchiya, K. and Takaishi, Y. 2002. The antioxidant activity of the extracts and fractions from *Carduus nutans* L. J. Faculty of Pharmacy Gaza University **19(1)**: 61-66.

DellaGreca, M., Cutillo, F., D'Abrosca, B., Fiorentino, A., Pacifico, S. and Zarrelli, A. 2009. Antioxidant and radical scavenging properties of *Malva sylvestris*. Nat. Prod. Commun. **4(7)**: 893-896.

Del Valle, J.M., Godoy, C., Asencio, M. and Aguilera, J.M. 2004. Recovery of antioxidants from boldo (*Peumus boldus* M.) by conventional and supercritical CO_2 extraction. Food Res. Intern. **37(7)**: 695-702. doi: 10.1016/j.foodres.2003.03.001.

De Marino, S., Gala, F., Borbone, N., Zollo, F., Vitalini, S., Visioli, F. and Iorizzi, M. 2007. Phenolic glycosides from *Foeniculum vulgare* fruit and evaluation of antioxidative activity. Phytochem. **68(13)**: 1805-1812.

De Marino, S., Iorizzi, M. and Zollo, F., 2008. Antioxidant activity and biological properties of phytochemicals in vegetables and spices (*Capsicum, Laurus, Foeniculum*). Electronic J. Environ. Agric. Food Chem. **7(10)**: 3174-3177.

De Melo, J.G., Araujo, T.A.S., Castro, V.T.N.A., Cabral, D L.V., Rodrigues, M.D., Nascimento, S.C., Amorim, E.L.C. and Albuquerque, U.P. 2010. Antiproliferative activity, antioxidant capacity and tannin content in plants of semi-arid Northeastern Brazil. Molecules **15**: 8534-8542. doi: 10.3390/molecules15128534.

Demiray, S., Pintado, M.E. and Castro, P.M.L. 2009. Evaluation of phenolic profiles and antioxidant activities of Turkish medicinal plants: *Tilia argentea, Crataegi folium* leaves and *Polygonum bistorta* roots. World Academy of Science, Engineering and Technology **54**: 312-317.

Demirci, B., Kosar, M., Demirci, F., Dinc, M and Baser, K.H.C. 2007. Antimicrobial and antioxidant activities of the essential oil of *Chaerophyllum libanoticum* Boiss. et Kotschy. Food chem. **105**: 1512-1517.

Demo, A., Petrakis, C., Kefalas, P. and Boskou, D. 1999. Nutrient antioxidants in some herbs and Mediterranean plant leaves. Food Res. Intern. **31(5)**: 351-354.

Deny, S., Hasnah, M.S., Farediah, A., Rasadah, M.A., Norio, A. and Mariko, K. 2007. Antioxidant and cytotoxic flavonoids from the flowers of *Melastoma malabathricum* L. Food Chem. **103**: 710-716.

Deore S.L., Khadabadi S.S., Bhagure L., and Ghorpade, D.S. 2008. *In vitro* antimicrobial and antioxidant studies on *Enicostemma axillare* (Lam.) Raynal leaves. Natural Product Radiance **7(5)**: 409-412.

Derbel, S., Bouaziz, M., Dhouib, A., Sayadi, S. and Chaieb, M. 2010. Chemical composition and biological potential of seed oil and leaf extracts of *Henophyton desertii* Coss. et Durieu. Compte Rendus Chimie **13(4)**: 473-480.

Desmarchelier, C., Coussio, J. and Ciccia, G. 1998. Antioxidant and free radical scavenging effects in extracts of the medicinal herb *Achyrocline satureioides* (Lam.) DC. ("marcela"). Braz. J. Med. Biol. Res. **31(9)**: 1163-1170.

Desmarchelier, C., Mongelli, E., Conssio, J., and Ciccia G. 1997. Evaluation of the *in vitro* antioxidant activity in extract of *Uncaria tomentosa* (Willd.) DC. Phytother. Res. **11**: 254-256.

Desmarchelier, C., Novoa Bermudez, M.J., Coussio, J., Ciccia, G. and Boveris, A. 1997. Antioxidant and prooxidant activities in aqueous extracts of Argentine plants. Pharmaceutical Biol. **35 (2)**: 116-120.

Desmarchelier, C., Pacciaroni, A.V., Abate-Daga, D., Coussio, J., Gil, R.R. and Silva, G.L. 2005. Antioxidant and free radical scavenging activities of *Misodendrum punctulatum*, myzodendrone and structurally related phenols. Phytother. Res. **19(12)**: 1043-1047.

Desmarchelier, C., Romao, R.L., Coussio, J., Ciccia, G. 1999. Antioxidant and free radical scavenging activities in extracts from medicinal trees used in the 'Caatinga' region in northeastern Brazil. J. Ethnopharmacol. **67(1)**: 69-77.

De Sousa, E., Zanatta, L., Seifriz., I., Creczynski-Pasa, T.B., Pizzaloti, M.G., Szpoganicz, B. and Barreto Silva, F.R.M. 2004. Hypoglycemic effect and antioxidant potential of kaempferol-3,7-O(α)-dirhamnoside from *Bauhinia forficata* leaves. J. Nat. Prod. **67(5)**: 829-832.

De Sousa L.C., Luzdiogenes J.P., Periera, B.A., Souza R.A., Faria, O.G., Nato M.A. Alves R.S. De queiroz, M.G., Desousa C.F., Viana G.S., 2004. Antidibetic activity of *Bauhinia forficata* extracts in alloxan diabetic rats. Biol. Pharm. Bull. **27**: 125-127.

De Souza, A.C., Alviano, D.S., Blank, A.F., Alves, A.B., Alviano, C.S. and Gattass, C.R. 2004. *Melissa officinalis* L. essential oil: antitumoral and antioxidant activities. J. Pharm. Pharmacol. **56**: 667-681.

De Souza, P.A., Rodrigues, C., Santiago, A.P.S.A., de Lucas, N.C., Leitao, G.G. and Filho, A.G. 2011. Antioxidant activity of natural compounds from *Stachytarpheta cayennensis* (Rich.) Vahlm verbenaceae, by scavenger of mitochondrial reactive oxygen species. Rev. Bras. Farmacogn. **21(3)**: (e pub.)

Dessi, M.A., Deiana, M., Rosa, A., Piredda, M., Cottiglia, F., Bonsignore, L., Deidda, D., Pompei, R., Corongiu, F.P. 2001. Antioxidant activity of extracts from plants in Sardinia. Phytother. Res. **15(6)**: 511-518.

Devasagayam, T.P.A., Kamat, J.P., Mohan, H. and Kesavan, P.C. 1996. Caffeine as an antioxidant: Inhibition of lipid peroxidation induced by reactive oxygen species. Biochimica et Biophysica Acta (BBA) – Biomembranes **1282(1)**: 63-70.

Devasagayam, T.P.A. and Sainis, K.B. 2002. Immune system and antioxidants, especially those derived from Indian medicinal plants. Indian J. Exp. Biol. **40**: 639-655.

Devasagayam, T.P.A., Tilak, J.C. and Adhikari, S. 200. Antioxidant and radioprotective effect of an Indian medicinal plant. BARC Newsletter p. 17.

Devasagayam, T.P.A., Tilak, J.C., Boloor, K.K. *et al.*, 2004. Review: Free radical and antioxidants in human health. Curr. Stat. Fut. Pros. JAPI **53**: 794-804.

Devi, K.P., Suganthy, N., Kesika, P. and Pandian, S.K. 2008. Bioprotective properties of seaweeds: *in vitro* evaluation of antioxidant activity and antimicrobial activity against food borne bacteria in relation to polyphenolic content. BMC Complement Altern Med. **10**: 8: 38.

Devi, P.U. and Ganasoundari, A. 1999. Modulation and antioxidant enzymes by *Ocimum sanctum* and role in protection against radiation injury. Indian J. Exp. Biol. **37**: 262-268.

Devi, S. L., Kannappan, S. and Anuradha, C.V. 2007. Evaluation of *in vitro* antioxidant activity of Indian bay leaf, *Cinnamomum tamala* (Buch. -Ham.) T. Nees et Eberm using rat brain synaptosomes as model system. Indian J. Exp. Biol. **45(9)**: 778-784.

Devienne, K.F., Calgaro Helena, A.F., Dorta, D.J., Prado, I.M.R., Raddi, M.S.G., Vilegas, W., Uyemura, S.A., Santos, A.C. and Curti, C. 2007. Antioxidant activity of isocoumarins isolated from *Paepalanthus bromelioides* on mitochondria. Phytochemistry **68(7)**: 1075-1080.

Dhalwal, K., Deshpande, Y.S. and Purohit, A.P. 2007. Evaluation of *in vitro* antioxidant activity of *Sida rhombifolia* L. ssp. *retusa* L. J. Med. Food **10(4)**: 683-688.

Dhalwal, K., Deshpande, Y.S., Purohit, A.P. and Kadam, S.S. 2005. Evaluation of antioxidant activity of *Sida cordifolia*. J. Pharm. Biol. **43**: 754-761.

Dhanasekaran, M., Tharakan, B. and Manyam, B.V. 2007. Antiparkinson drug – *Mucuna pruriens* shows antioxidant and metal chelating activity. Phytother. Res. **22(1)**: 6-11.

Dharmishtha, A.M., Mishra, S.H. and Falguni, P.G. 2009. Antioxidant studies of methanolic extract of *Randia dumetorum* Lam. Pharmacology online **1**: 22-34.

Dharmishta, M., Srihari, M. and Falguni, G. 2010. Free radical scavenging activity of *Randia dumetorum* Lamk. Fruits. Int. Res. J. Pharm. **1(1)**: 122.

Dheeba, B., Sampath Kumar, P., Sathiya Priya, R.R. and Kannan, M. 2010. Phytochemical studies and evaluation of antioxidant potential of various extracts of *Aegle marmelos* bark. Pharmacology online **3**: 831-839.

Dhenge, R.M. and Itankar, P.R. 2008. Anti-inflammatory and antioxidant potential of ethanolic extract of *Indigofera trifoliata*. Indian J. Nat. Prod. **24(1)**: 3-7.

Dhobal, M.P., Sharma, M.C., Rao, S., Agrawwal, N. and Mangal, V. 2008. Evaluation of antioxidant activities of selected Indian medicinal plants.

Dhuley, J.N. 1998. Effect of ashwagandha on lipid peroxidation in stress induced animals. J. Ethnopharmacol. **60**: 173.

Dhuley, J.N. 1999. Antioxidant effect of Cinnamon (*Cinnamomum verum*) bark and greater Cardamom (*Amomum subulatum*) seeds in rats fed high fat diet. Indian J. Exp. Biol. **37**: 238-242.

Diallo, D, Marston, A., Terreaux, C., Toure, Y., Paulsen, B.S. and Hostettmann, K. 2001. Screening of Malian medicinal plants for antifungal, larvicidal, mollucidal, antioxidant and radical scavenging activities. Phytother. Res. **15(5)**: 401-406.

Diamond, B.J., Shiflett, S.C., Feiwel, N., Matheis, R.J., Noskin, O., Richards, J.A. and Schoenberger, N.E. 2000. *Ginkgo biloba* extract: mechanisms and clinical indications. Arch. Phys. Med. Rehabil. **81(5)**: 668-678.

Dianzani, M.U., Muzio, G., Biocca, M.E. and Canuto, R.A. 1991. Lipid peroxidation in fatty liver induced by caffeine in rats. Int. J. Tissue React. **13(2)**: 79-85.

Dicko, M.H., Gruppen, H., Traore, A.S., van Berkel, W J. and Voragen, A.G. 2005. Evaluation of the effect of germination of phenolic compounds and antioxidant activities in Sorghum varieties. J. Agric. Food Chem. **53**: 2581-2588.

Dickson, R.A., Houghton, P.J. and Hylands, P.J. 2007. Antibacterial and antioxidant cassane diterpenoids from *Caesalpinia benthamiana*. Phytochem. **68(10)**: 1436-1441.

Dickson, R.A., Houghton, P.J., Hylands, P.J. and Gibbons, S. 2006. Antimicrobial, resistance-modifying effects, antioxidant and free radical scavenging activities of *Mezoneuron benthamianum* Baill., *Securinega virosa* Roxb. and Willd. and *Microglossa pyrifolia* Lam. Phytother. Res. **20(1)**: 41-45.

Didi Jauhari Purwadiwarsa, Anas Subarnas, Cucu Hadiansyah and Supriyatna. 2000. The investigation on anti-mutagenic and antioxidant activity of *Schima wallichii* (DC.) Korth. leaves. Cermin Dunia Kodokt, **127**: 18-21.

Dimberg, L.H., Theander, O. and Lignert, H. 1993. Avenanthramides – a group of phenolic antioxidants in oats. Cereal Chem. **70**: 637-641.

Ding, Z.T., Fang, Y.S., Tai, Z.G., Yang, M.H., Xu, Y.Q., Li, F. and Cao, Q.E. 2008. Phenolic content and radical scavenging capacity of 31 species of ferns. Fitoterapia **79**: 581-583.

Dini, I., Gian Carlo Tenore, and Antonio Dini. 2006. New polyphenol derivative in *Ipomoea batatas* tubers and its antioxidant activity. J. Agric. Food Chem. **54 (23)**: 8733-8737.

Dinis, T.C.P., Madeir, V.M.C. and Almeida, M.L.M. 1994. Action of phenolic derivatives (acetoaminophen, salycilate, and 5-aminosalycilate) as inhibitors of membrane lipid peroxidation and as peroxyl radical scavengers. Arch.of Biochem. Biophys. **315**: 161.

Diouf, P.N., Stevanovic, T. and Cloutier, A. 2009. Antioxidant properties and polyphneol contents of trembling aspen bark. Wood Sci. Tech. **43(5-6)**: 457-470.

Divya, D., Sivakumar, V. and Priya, R.S. 2009. Anti-iron toxicity and antioxidant effect of *Camellia sinensis*. Asian J. Environ. Sci. **4(1)**: 58-61.

Divya, T. S., Latha, P.G., Ush, K., Anuja, G.I., Suja, S.R., Shyamal, S., Shine, V.J., S.Sini, Sikha, P. and Rajasekharan, S. 2009. Anti-inflammatory, analgesic and anti-lipid peroxidative properties of *Wattakaka volubilis* (Linn.f.) Stapf. Natural Product Radiance **8(2)**: 137-141.

Divyasree, P. and Nair, C.K.K. 2008. Antioxidant and radioprotecting property of *Coscinium fenestratum* Colebr. extract in mice exposed to whole body gamma irradiation. Amala Res. Bull. **28**: 62-68.

Dixit, P.P., Devasagayam, T.P.A. and Ghaskadbi, S. 2007. Formulated antidiabetic preparation Syndrex® has a strong antioxidant activity. Eur. J. Pharmacol. **58**: 216-225.

Dixit, P., Ghskadbi, S., Mohan, H. and Devasagayam, T.P.A. 2005. Antioxidant properties of germinated fenugreek seeds. Phytother. Res. **19(11)**: 977-983.

Dixit, S. and Ali, H. 2010. Antioxidant potential of some medicinal plants of Central India. J. Cancer Ther. **1**: 87-90.

Djeridane, B., Yousfi, M., Nadjemi, B., Boutassouna, D., Stocker, P. and Vidal, N. 2006. Antioxidant activity of some Algerian medicinal plants extracts containing phenolic compounds. Food Chem. **97(4)**: 654-660.

Djeridane, B., Yousfi, M., Vidal, N., Nadjemi, N., Lesgards, J.F. and Stocker, P. 2007. Screening of some Algerian medicinal plants for the phenolic compounds and their antioxidant activity. European Food Research and Technology **224**: 801-809.

Djilas, S.M., Markov, S.L., Cvetkovic, D.D., Canadanovic-Brunet, J.M., Cetkovic, G.S. and Tumbas, V.T. 2006. Antimicrobial and free radical scavenging activities of *Teucrium montanum*. Fitoterapia **77**: 401-403.

Dlamini, N.R., Rooney, L.W., Waniska, R.D. and Taylor, J.R.N. 2007. Effect of enzymatic hydrolysis on antioxidant activity and potential health benefits of traditional African sorghum-based food. www.sik.se/traditional grains/review/./Dlamini.pdf

D'Mello, P.M., Jadhav, M.A. and Jolly, C.I. 2000. Free radical scavenging activity of *Syzygium cumini* and *Ficus benghalensis* plants used in Ayurveda for Diabetes Mellitus. Indian Drugs. **37**: 518.

Dong, E., Yokazawa, T., Kashiwagi, H., Hattori, M., Watanabe, H. and Oura, H. 1996. The role of total Ginseng saponin and its components in antiperoxidation *in vitro*. Nat.Med. **50(2)**: 128-134.

Donovan, J.L. Meyer, A.S. and Waterhouse, A.L. 1998. Phenolic composition and antioxidant activity of Prunes and Prune juice (*Prunus domestica*). J. Agric. Food Chem. **46(4)**: 1247-1252.

Dordevic, S., Petrovic, S., Dobric, S., Milenkovic, M., Vucicevic, D., Zizic, S. and Kukic, J. 2007. Antimicrobial, anti-inflammatory, anti-ulcer and antioxidant activities of *Carlina acanthifolia* root essential oil. J. Ethnopharmacol. **109**: 458-463.

Dorman, H.J.D., Bachmayer, O., Kosar, M. and Hiltunen, R. 2004 Antioxidant properties of aqueous extracts from selected Lamiaceae species grown in Turkey. J. Agric. Food Chem. **52**: 762-770.

Dorman, H.J.D. and Deans S.G. 2004. Chemical composition antimicrobial and *in vitro* antioxidant properties of *Morinda citrodora* var. *citriodora*, *Myristica fragans*, *Origanum ovulgare* ssp. *hirtum*, *Pelargonium* sp. and *Thymus zygis* oils. Esstential oils. J. Res. **16**: 145-150.

Dorman, H.J.D., Deans S.G. and Noble, R.C. 1995. Evaluation of *in vitro* of plant essential oils as natural antioxidants. J. Essential Oil Res. **7**: 645-651.

Dorman, H.J.D., Hiltunen, R.P. and Tikkanen, M.J 2003. Characterization of antioxidant properties of de-odourised aqueous extracts from selected Lamiaceae herbs. Food Chem. **83**: 255-262.

Dorman, H.J.D., Kosar, M., Kahlos, K., Holm, Y. and Hiltunen, R. 2003. Antioxidant properties and composition of aqueous extracts from *Mentha* species, hybrids, varieties and cultivars. J. Agric. Food Chem. **51**: 4563-4569.

Dorman, H.J.D., Shikov, A.N., Pozharitskaya, O.N. and Hiltunen, R. 2011. Antioxidant and prooxidant evaluation of a *Potentilla alba* L. rhizome extract. Chem. Biodivers. Doi: 10.1002/cbdv.201100043.

dos Santos, J. G., Jr., F. H. M. do Monte, M. M. Blanco, V. M. do N. B. Lanziotti, F. D. Maia and L. K. de A. Leal. 2005. Cognitive enhancement in aged rats after chronic administration of *Equisetum arvense* L. with demonstrated antioxidant properties *in vitro*. Pharmacology Biochemistry and Behavior **81**: 593–600.

Doyama, J.T., Rodrigues, H.G., Novelli, E.L.B., Cereda, E. and Vilegas, W. 2005. Chemical investigation and effects of the tea of *Passiflora alata* on biochemical parameters in rats. J. Ethnopharmacol. **96(3)**: 371-374.

Drodge, W. 2002. Oxidative stress and aging. Adv. Exp. Med. Biol. **543**: 191-200.

Droesti, *I.E.* 2000. Antioxidant polyphenols in tea, cocoa and wine. Nutrition **16**: 692-694.

Du, Q., Zheng, J., Xu, Y. 2008. Composition of anthocyanins in mulberry and their antioxidant activity. J. Food Compos. Anal. **21 (5)**: 390-395.

Duan, L., Tao, H.W., Hao, X.J., Gu, Q.Q. and Zhu, W.M. 2009. Cytotoxic and antioxidative phenolic compounds from the traditional Chinese medicinal plant *Myristica fragrans*. Planta Medica **75(11)**: 1241-1245.

Duan, X., Wu, G. and Jiang, Y. 2007. Evaluation of the antioxidant properties of Litchi fruit phenolics in relation to pericarp browning prevention. Molecules **12**: 759-771.

Duan, X.-J., Zhang, W.-W., Li, X.-M. and Wang, B.-G. 2004. Evaluation of antioxidant property of extract and fractions obtained from red alga, *Polysiphonia ulceolata*. Food Chem. **95(1)**: 37-43.

Duan, X.J., Li, X.M., Wang, B.G. 2007. Highly brominated mono- and bis-phenols from the marine red alga *Symphyocladia latiuscula* with radical-scavenging activity. J Nat Prod. **70(7)**: 1210-3.

Duan, X.J., Zhang, W-W., Li, X.M. and Wang, B-G. 2006. Evaluation of antioxidant property of extract and fractions obtained from a red alga, *Polysiphonia urceolata*. Food Chem. **95**: 37-43.

Duarte-Almeida, J.M., Negri, G., Salatino, A., de Carvalh, J.E. and Lajolo, F.M.2007. Antiproliferative and antioxidant activities of a tricin acylate glycoside from sugarcane (*Saccharum officinarum*) juice. Phytochem. **68(8)**: 1165-1171.

Dubey, S.K. and Batra, A. 2009a. Antioxidant activities of *Thuja occidentalis* Linn. Asian J. Pharm. Clin. Res. **2(1)**: 73- 76.

Dubey, S.K. and Batra, A. 2009b. Role of phenolics in anti-atherosclerotic property of *Thuja occidentalis* Linn. Ethnobotanical leaflets **13**: 791-800.

Ducrey, B., Wolfender, J.L., Marston, A. and Hostettman, K. 1995. Analysis of flavonol glycosides of thirteen *Epilobium* species (Onagraceae) by LC-UV and thermospray LC-MS. Phytochemistry **38**: 129-137.

Dudonne, S., Pouporad, P., Coutieere, P., Woillez, M., Richard, T., Merillon, M. and Vitrac, X. 2011. Phenolic composition and antioxidant properties of poplar bud (*Populus nigra*) extract, Individual antioxidant contribution of phenolics and transcriptional effect on skin aging. J. Agric. Food Chem. **59(9)**: 4527-4536. doi: 10.1021/jf104791.

Dudonne, S., Vitrac, X., Coutiere, P., Woillez, M. and Merillon. J.M. 2009. Comparative study of Antioxidant properties and total phenolic content of 30 plant extracts of Industrial interest using DPPH, ABTS, FRAP, SOD and ORAC assays. J. Agric. Food Chem. **57(5)**: 1768-1774.

Duenas, M., Hernandez, T. and Estrella, I. 2006. Assessment of *in vitro* antioxidant capacity of the seed coat and the cotyledon of legumes in relation to their phenolic contents. Food Chem. **98**: 95-103.

Dufall, K.G., Ngadjui, B.T. Simeon, K.F., Abegaz, B.M. and Croft, K.D. 2003. Antioxidant activity of prenylated flavonoids from the West African medicinal plant *Dorstenia mannii*. J. Ethnopharmacol. **87 (1)**: 67-72.

Dufour, D., Pichettee, A., Mshvildadze, V., Bradette-Hebert, M., Lavoie, S., Longtin, A. Leprise, C. and Lequaet, J. 2007. Antioxidant, anti-inflammatory and anticancer activities of methanolic extracts from *Ledum groenlandicum* Retzius. J. Ethnopharmocol. **111**: 22-28.

Duh, P.D. 1998. Antioxidant activity of Budrock (*Arctium lappa* Linn.) its scavenging effect on free radical and active oxygen. J. Am. Oil Chem. Soc. **75**: 455-461.

Duh, P.D., Tu, Y.Y. and Yen, G.C. 1999. Antioxidant activity of water extract of Harng Jyur (*Chrysanthemum morifolium* Ramat). Lebensmittel-Wissenschaft und – Technologie. 32 **(5)**: 269-277.

Duncan, C.E., Gorbet, D.W., and Talcott, S.T. 2006. Phytochemical content and antioxidant capacity of water soluble isolates from peanut (*Arachis hypogaea* L.). Food Res. Int. **39(8)**: 898-904. doi: 10.1016/j.foodres. 2006.05.009

Dung, N.T., Kim, J.M. and Kang, S.C. 2008. Chemical composition, antimicrobial and antioxidant activities of the essential oil and the ethanol extract of *Cleistocalyx operculatus* (Roxb.) Merr. and Perry buds. Food Chem. Toxicol **46(12)**: 3632-3639.

Durmaz, G. and Alpaslan, M. 2005. Antioxidant properties of roasted apricot (*Prunus armeniaca* L.) kernel. Food Chem. **100(3)**: 1177-1181. doi: 10.16/j.foodchem.2005.10. 1067.

Duve, K.J. and White, P.J. 1991. Extraction and identification of antioxidants in oats. J. Am. Oil Chem. Soc. **68**: 365-370.

Dvorska, M., Zemlicka, M., Muselik, J., Karafiatova, J. and Suchy, V. 2007. Antioxidant activity of *Catalpa bignonioides*. Fitoterapia **78(6)**: 437-439.

Dykes, L., Rooney, L.W., Waniska, R.D. and Rooney, W.L. 2005. Phenolic compounds and antioxidant activity of Sorghum grains of varying genotypes. J. Agric. Food Chem. **53**: 6813-6818.

Eastwood, M.A. 1999. Interaction of dietary antioxidants *in vivo*: how fruit and vegetables prevent disease? Q. J. Med. **92**: 527-530.

Eberhardt, M.V., Lee, C.Y. and Liu, R.H. 2000. Antioxidant activity of fresh apples. Nature **405**: 903-904.

Ebrahimabadi, A.H., Ebrahimabadi, E.H., Djafari-Bidgoli, Z., Kashi, F.J., Mazoochi, A. and Batooli, H. 2010. Composition and antioxidant and antimicrobial activity of the essential oil and extracts of *Stachys inflata* Benth. from Iran. Food Chem. **119(2)**: 452-458.

Ebrahimabadi, A.H., Mazoochi, A., Kashi, F.J., Djafari-Bidgoli, Z. and Batooli, H. 2010. Essential oil composition and antioxidant and antimicrobial properties of the aerial parts of *Salvia eremophila* Boiss. from Iran. Food Chem. Toxicol. **48**: 1371-1376.

Ebrahimzadeh, M.A., Nabavi, S.F. and Nabavi, S. M. 2009a. Antioxidant activity of leaves and inflorescence of *Eryngium caucasicum* Trautv. at flowering stage. Pharmacogn. Res. **1(6)**: 435-439.

Ebrahimzadeh, M.A., Nabavi, S.F. and Nabavi, M. 2009b. Essential oil composition and antioxidant aactivity of *Pterocarya fraxinifolia*. Pak. J.Biol. Sci. **12(13)**: 957-963.

Ebrahimzadeh, M.A., Nabavi, S.F. and Nabavi, M. 2010. Antioxidant and antihemolytic activities of *Mentha longifolia*. Pharmacology online **2**: 464-471.

Ebrahimzadeh, M.A., Nabavi, S.M. and Nabavi, S.F. 2010. Biological activities of *Metha spicata* L. Pharmacology online **1**: 841-848.

Ebrahimzadeh, M.A., Nabavi, S.M., Nabavi, S.F., Bahramia, F. and Bekhradnia, A. 2010. Antioxidant and free radical scavenging activity of *H. officinalis* L. var. *angustifolius, V. udoarata, B. hurcana* and *C. speciosum*. Pak. J. Pharm. Sci. **23(1)**: 29-34.

Ebrahimzadeh, M.A., Nabavi, S.F., Nabavi, S. M. and Eslami, B. 2010a. Antioxidant and antihemolytic activities of *Mentha piperita*. Pharmacology online **1**: 744-752.

Ebrahimzadeh, M.A., Nabavi, S.M., Nabavi, S. F. and Eslami, B. 2010b. Biological activity of *Mentha aquatica* L. Pharmacology online **2**: 611-619.

Ebrahimzadeh, M.A., Nabavi, S.M., Nabavi, S. F. and Eslami, B. 2010. Antioxidant activity of the bulb and aerial parts of *Ornithogalum sintenisii* L. (Liliaceae) at flowering stage. Trop. J. Phar. Res. **9(2)**: 141-148.

Ebrahimzadeh, M.A., Pourmorad, F. and Bekhradnia, A.R. 2008. Iron chelating activity, phenol and flavonoid content of some medicinal plants from Iran. Afr. J. Biotechnol. **7(18)**: 3188-3192.

Ebrahimzadeh, M.A., Pourmorad, F. and Hafizi, S. 2008. Antioxdant activities of Iranian Silk. Turk. J. Biol. **32**: 43-49.

Economou, K.D., Oreopoulou, V. and Thomopoulos, C.D. 1991. Antioxidant activity of some plant extracts of the family Labiatae. JAOCS **68**: 235-244.

Eddy, E.P., Howard, P.C., McCoy, G.D. and Rosenkranz, H.S. 1987. Mutagenicity, unscheduled DNA sysnthesis, and metabolism of 1-nitropyrene in the human hepatoma cell line HepG2. *Canc. Res.* 47: 3163–3168.

Edziri, H., Mastouri, M., Cheraif, I. and Aouini, M. 2010. Chemical composition and antibacterial, antifungal and antioxidant activities of the flower oil of *Retaema raetum* (Forssk.) Webb. from Tunisia. Nat. Prod. Res. **24**: 789-796.

Efmorfopoilou, E. and Rodis, P.2005. Application of cyclodextrins to the extraction of antioxidant compounds from Olive oil. Chem. Nat. Compounds **41(1)**: 22-26.

Ehlenfeldt, M.K. and Prior, R.L. 2001. Oxygen radical absorbance capacity (ORAC) and phenolic and anthocyanin concentrations in fruit and leaf tissues of highbush blueberry. J. Agric. Food and Agri. **49**: 2222-2227.

Ekaprasada, M.T., Nurdin, H., Ibrahim, H. and Dachriyanus, D. 2009. Antioxidant activity of methyl gallate isolated from the leaves of *Toona sureni*. Indonesian J. Chem. **9(3)**:

Ekbote, M., Ramesh, C.K., Mahmood, R., Thippeswamy, B.S. and Veerapur, V. 2010. Hepatoprotective and antioxidant effect of *Azima tetracantha* Lam. leaves extracts against CCl_4-induced liver injury in rats. Indian J. Nat. Prod. Resources **1(4)**: 493-499.

El, S.N. and Karakaya, S. 2004. Radical scavenging and iron-chelating activities of some greens used in Traditional dishes in Mediterranean diet. Int. J. Food Sci. Nutr. **55**: 67-74.

El Babili, F., Bouajila, J., Fouraste, I., Valentin, A., Mauret, S., and Maulis, C. 2009. Chemical study, antimalarial and antioxidant activities and cytotoxicity to human breast cancer cells (MCF 7) of *Argania spinosa*. Phytomedicine **17(2)**: 157-160. doi: 10.1016/j.phymed.2009.05.014.

Eldahshan, O., Ayoub, N.A., Singab, A.N.B. and Azizi, M.M.A. 2008. Potential superoxide anion radical scavenging activity of Doum palm (*Hyphaene thebaica* L.) leaves extract. Rec. Nat. Prod. **2**: 83-93.

El Diwani, G., El Rafie, Sh. and Hawash, S. 2009. Antioxidant activity of extracts obtained from residues of nodes, leaves, stem and root of Egyptian *Jatropha curcas*. African J. Pharmacy Pharmacol **3(11)**: 521-530.

Eleazu, C.O., Amajor, J.U., Ikpeama, A.I. and Awa, E. 2011. Studies on the nutrient composition, antioxidant activities, functional properties and microbial load of the flours of 10 elite cassava (*Manihot esculenta*) varieties. Asian J. Clin. Nutr. **3**: 33-39. doi: 10.3923/ajc.2011.33.39.

El-Ghorab, A. 2003. Supercritical fluid extraction of the Egyptian rosemary (*Rosmarinus officinalis*) leaves and *Nigella sativa* L. seeds volatile oils and their antioxidant activities. J. Essential Oil Bearing plants **6(1)**: 67-77.

El-Ghorab, A.H., Mansour, A.F. and El-Massry, K.F. 2004. Effect of extraction methods on the chemical composition and antioxidant activity of Egyptian majoram (*Majorana hortensis* Moench). Flavour and Fragrance Journal **19(1)**: 54-61.

El-Ghorab, A., El-Massry, K.F., Anjum, F.M., Shahwarb, M.K. and Shibamoto, T. 2009. The chemical composition and antioxidant activity of essential oil of Pakistani *Eucalyptus camaldulensis* leaves. J. Essential Oil Bearing Plants **12(3)**: 262-272.

El-Ghorab, A., El-Massry, K.F. and Shibamoto, T. 2007. Chemical composition of the volatile extract and antioxidant activities of the volatile and nonvolatile extracts of Egyptian corn silk (*Zea mays* L.). J. Agric. Food Chem. **55 (22)**: 9124-9127.

El-Ghorab, A.H., El-Massry, K.F., Marx, F. and Fadel, H.M. 2003. Antioxidant activity of Egyptian *Eucalyptus camaldulensis* var. *brevirostris* leaf extracts. Nahrung. Feb. **47(1)**: 41-45.

El-Ghorab, A.H., El-Massry, K.F. and Shaaban, H.A. 2007. Effect of drying on the chemical composition of the Egyptian *Cupressus macrocarpa* (Hartw. ex Gordon) essential oils and their biological activities. J. Essential Oil Bearing plants **10(5)**: 399-411.

El-Ghorab, A.H., Nauman, M., Anjum, F.M., Hussain, S. and Nadeem, M. 2010. A comparative study on chemical composition and antioxidant activity of Ginger (*Zingiber officinale*) and Cumin (*Cuminum cyminum*). J. Agr. Food Chem. **58(14)**: 8231-8237.

El-Ghorab, A., Shibamoto, T. and Ozcan, M. 2007. Chemical composition and antioxidant activities of buds and leaves of capers (*Capparis ovata* Desf. var. *canescens*) cultivated in Turkey. J. Essential Oil Res. **19(1)**: 72-77.

El Hajaji, H., Lachkar, N., Alovi, K., Cherrah, Y., Farah, A., Ennalibili, A., El Bali, B. and Lachkar, M. 2011. Antioxidant activity, phytochemical screening and total phenolic content of extracts from three genders of carob tree barks growing in Morocco. Arabian J. Chem. **4(3)**: 321-324. doi: 10.1016/jarabjc.2010.06.053.

El-Hela, A. and Abdullah, A. 2010. Antioxidant and antimicrobial activities of methanol extracts of some *Verbena* species: *In vitro* evaluation of antioxidant and antimicrobial activity in relation to polyphenolic content. J. Appl. Sci. Res. **6(6)**: 683-689.

Ellnain-Wojtaszek, M., KruczyDski, Z., Kasprzak, J. 2003. Investigation of the free radical scavenging activity of *Ginkgo biloba* L. leaves. Fitoterapia. **74(1-2)**: 1-613.

Elmastas, M. and Gercekcioglu, R. 2006. Antioxidant activity of some Soft fruit species. II. National Soft Fruits Symposium. 14-16 September, 2006, Tokat pp. 295-298.

Elmastas, M., Gucin, I.,Beydemir, S., Kufrevioglu, O.I. and Aboul-Enein, H.Y. 2006. A study of the *in vitro* antioxidant activity of juniper (*Juniperus communis* L.) fruit extracts. Analytical Letter **39(1)**: 47-65.

Elmastas, M., Gucin, I., Ozturk, L. and Gokce, I. 2005. Investigation of antioxidant properties of spearmint (*Mentha spicata* L.). Asian J. Chem. **17**: 137-148.

Elmastas, M., Turkekul, I., Oztürk, L., Gülçin, I., Isildak, O., Aboul-Enein, H.Y. 2006. Antioxidant activity of two wild edible mushrooms (*Morchella vulgaris* and *Morchella esculanta*) from North Turkey. Comb Chem High Throughput Screen. **9(6)**: 443-448.

El-Safi, N.-E., Kollman, A., Khlifi, S. and Ducrot, P-H. 2007. Antioxidative effect of compounds isolated from *Globularia alypum* L. structure activity relationship. LWT Food Sci. Technol. **40(7)**: 1246-1252. doi: 10.1016/j.lwt.2006.08.019.

El-Sayed, M.M., Abdel-Hameed, S.S., Ahmed, W.S. and El-Wakil, E.A. 2008. Non-phenolic antioxidant compounds from *Buddleja asiatica*. Z. Naturforsch. C. **63(7-8)**: 483-491.

El-Sayed, N.H., Ammar, N.M., Al-Okbi, S.Y., El-Kassem, L.T. and Mabry, T.J. 2006. Antioxidant activity and two new flavonoids from *Washingtonia filifera*. Nat. Prod. Res. **20(1)**: 57-61.

Elzaawely, A.A., Xuan, T.D., Koyama, H. and Tawata, S. 2007. Antioxidant activity and contents of essential oil and phenolic compounds in flowers and seeds of *Alpinia zerumbet* (Pers.) B.L.Burtt. et R.M.Sm. Food Chem. **104(4)**: 1648-1653.

Elzaawely, A.A., Xuan, T.D. and Tawata, S. 2005. Antioxidant and antibacterial activities of *Rumex japonicus* Houtt. aerial parts. Biol. Pharm. Bull. **28(12)**: 2225-2230.

Emami, S.A., Asgary, S., Naderi, G.A., Shams Ardekan, M.R., Kasher, T., Aslani, S., Airin, A. and Sahebkar, A. 2011. Antioxidant activities of *Juniperus foetidissima* essential oils against several oxidative systems. Rev. Bras. Farmacogn **21(4)**: doi: 10.1590/s0102-695x2011005000120.

Emami, S.A., Asili, J., Mohagheghi, Z. and Hassanzadeh, M.K. 2007. Antioxidant activity of leaves and fruits of Iranian conifers. Advance Access Publication, ecam; 4, 313-319 doi: 10. 1093/ecam/nem011.

Emami, S.A., Fakhrjafary, M., Tafagodi, M. and Hassanzadesh, M.K. 2010. Chemical composition and antioxidant activity of the essential oils of different parts of *Cupressus arizonica* Greene. J. Essential oil Res. **22(3)**: 193-197.

Eminagaoglu, O., Tepe, B., Yumrutas, O., Akpulat, H.A., Daferera, D., Polissious, M. and Sokmen, A. 2007. The *in vitro* antioxidative properties of the essential oils and methanol extracts of *Satureja spicigera* (K.Koch) Boiss. and *Satureja cuneifolia* Ten. Food Chem. **100(1)**: 339-343. doi: 10.1016/j.foodchem.2005.09.054.

Emmons, C.L. and Peterson, D.M. 1999. Antioxidant activity and phenolic contents of oats, groats and hulls. Cereal Chem. **76**: 902-906.

Ena, A., Nelli, M., Pintucci, C., Faraloni, C., Paperi, R., Torzillo, G. and Pushparaj, B. 2009. Antioxidant potential of leaf extract of *Balsamita major* Desf. J. Med. Arom. Plant Sci. **31(3)**: 228-233.

Endrini, S. 2010. Anticarcinogenic properties and antioxidant activity of Mahkota Dewa (*Phaleria macrocarpa* (Scheff.) Boerl.). 2nd Intern. Seminar on Medicinal plants and Herbal Products, S.V.Univ., Tirupati, Abstract pp. 6.

Endrini, S., Rahmat, A., Ismail, P., Taufiq-Yap, Y.H. 2007. Comparing of the cytotoxicity properties and mechanism of *Lawsonia inermis* and *Strobilanthes crispus* extract against several cancer cell lines. J. Med. Sci. **7(7)**: 1098-1102.

Ennajar, M., Bouajila, J., Lebrihi, A., Mathieu, F., Abderraba, M., Raies, A. and Romdhane, M. 2009. Chemical composition and antmicrobial and antioxidant activities of essential oils and various extracts of *Juniperus phoenicea* L. (Cupressaceae). J. Food Sci. **74(7)**: 364-371.

Ennajar, M., Bouajila, J., Lebrihi, A., Mathieu, F.,Savagnac, A., Abderraba, M., Raies, A. and Romdhane, M. 2010. The influence of organ, season and drying method on chemical composition and antioxidant and antimicrobial activities of *Juniperus phoenicea* L. essential oils. J. Sci. Food Agr. **90**: 462-470.

Erasto, P., Bojase-Moleta, G. and Majinda, R.T. 2004, Antimicrobial and antioxidant flavonoids from the root wood of *Bolusanthus speciosus*. Phytochem. **65**: 875-880.

Erasto, P., Mbwambo, Z.H., Nondo, R.S.O. and Lall, N. 2011. Antimycobacterial, antioxidant activity and toxicity of extracts from roots of *Rauvolfia vomitoria* and. *R caffra*. Spatula **1(2)**: 73-80. doi: 10.5455/spatula.20110514043359.

Ercisli, S. and Orhan, E. 2007. Chemical composition of white (*Morus alba*), red (*Morus rubra*) and black (*Morus nigra*) mulberry fruits. Food Chem. **103**: 1380-1384.

Ercisli, S., Orhan, E., Ozdemir, O. and Sengul, M. 2007. The genotypic effects on chemical composition and antioxidant activity of sea buckthorn (*Hippophae rhamnoides* L.) berries grown in Turkey. Sci. Hortic. **115**: 27-33.

Ercisli, S., Ozdemir, O., Sengul, M., Orhan, E. and Gungor, N. 2007. Phenolic and antioxidant diversity among fruit species grown in Turkey. Asian J. Chem. **19(7)**: 5751-5754.

Erdemoglu, N., Turan, N.N., Cakc, I., Sener, B. and Aydin, A. 2006. Antioxidant activities of some Lamiaceae plant extracts. Phytother. Res. **20**: 9-13.

Erkan, N., Ayranci, G. and Ayranci, E. 2008. Antioxidant activities of rosemary (*Rosmarinus officinalis* L.) extract, blackseed (*Nigella sativa* L.) essential oil. Food Chem. **110 (1)**: 76-82.

Escudero, N.L., Albarracin, G.J., Lopezz, R.V.L. and Gimenez, M.S. 2011. Antioxidant activity and phenolic content of flour and protein concentrate of *Amaranthus cruentus* seeds. J. Food Biochem. 35(4): 1327-1341. doi: 10.1111/j.1745-4514.2010.00454x.

Eshrat, H.M. and Hussain, A. 2002. Hypoglycemic, hypolipidemic and antioxidant properties of combination of curcumin from *Curcuma longa* Linn. and partially purified product from *Abroma angusta* Linn. in streptozotocin-induced diabetes. Indian J. Clin. Biochem. **17(2)**: 33-43.

Esmaeili, M.A. and Soboli, A. 2010. Antioxidant, free radical scavenging activities of *Salvia brachyantha* and its protective effect against oxidative cardiac cell injury. Food and Chemical Toxicol. **48(3)**: 846-853.

Esmaeili, M.A., Soboli, A. and Noushabadi, M.A. 2010. Antioxidant and protective properties of six *Tanacetum* species against hydrogen peroxide-induced oxidative stress in K562 cell line: A comparative study. Food Chem. **121(1)**: 148-155.

Esmaeili, M.A., Zohari, F. and Sadeghi, H. 2009. Antioxidant and protective effects of major flavonoids from *Teucrium polium* β-cell destruction in a model of streptoztocin-induced diabetes. Planta Medica **75(13)**: 1418-1420.

Espin, J.C., Soler-Rivas, C., Wichers, J. and Viguera-Garcia, C. 2000. Anthocyanin-based natural colorants: A new source of antiradical activity for food stuff. J. Agric. Food Chem. **48**: 1588-1592.

Es-Safi, N.E., Khlifi, S., Kerhoas, L., Kollmann, A., El Abbouyi, A. and Ducrot, P.H. 2005. Antioxidant constituents of aerial parts of *Globularia alypum* growing in Morocco. J. Nat. Prod. **68**: 1293-1296.

Estiasih, T. and Kurniawan, 2006. Antioxidant activity of Javanese Ginseng (*Talinum triangulare* Willd.) root extracts. Jurnal Teknologi Dan Industri Pangal **17(3)**:

Etter, S.C. 2005. *Rosmarinus officinalis* and an antioxidant. J. Herbs Spices Med. Plants **11(1-2)**: 121-159.

Evanjelene, V.K. and Natarajan, D. 2011. Evaluation of free radical scavenging activity and biological properties of *Spinacia oleracea* L. Int. J. Eng. Sci. Tech. **3(1)**: 25-30.

Evelyne, H.A., Doolaege, Katleen Raes, Karen Smet, Mirjana Andjelkovic, Christof Van Poucke, Stefan De Smet and Roland Verche. 2007. Characterization of two unknown compounds in methanol extracts of rosemary oil. J. Agric. Food Chem. **55**: 7283-7287.

Exarchou, V., Nenadis, N., Tsimidou, Gerothanassis, I.P., Troganis, A. and Boskou, D. 2002. Antioxidant activities and phenolic composition of extracts from Greek Oregano, Greek Sage and Summer savory. J. Agric. Food Chem. **50(19)**: 5294-5299.

Fadhel, Z.A. and Amran, S. 2002. Effects of black tea extract on carbon tetrachloride induced lipid peroxidation in liver, kidneys, and testes of rats. Phytother. Res. **16**: S28-S32.

Falah, S., Suzuki, T. and Katayama, T. 2008. Chemical constituents from *Swietenia macrophylla* bark and their antioxidant activity. Pak. J. Biol. Sci. **11(16)**: 2007-2012.

Falcao, D.Q., Costa, E.R., Alviano, D.S., Alviano, C.S., Kuster, R.M. and Menzes, F.S. 2006. Antioxidant and anti-microbial activity from *Calceolaria chelidonioides* Humb. Bonpl. et Kunth. Brazilian J. Pharmacogn. **16(1)**: 73-76.

Fallarero, A., Peltoketo, A., Loikkanen, J., Tammela, P., Vidal, A. and Vuorela, P.2006. Effects of the aqueous extract of *Bryothamnion triquetrum* on chemical hypoxia and aglycemia-induced damage in GTT1-7 mouse hypothalamic immortalized cells. Phytomedicine **13(4)**: 240-245.

Faller, A.L.K. and Fialho, E. 2009. The antioxidant capacity and polyphenol content of organic and conventional retail vegetables after domestic cooking. Food Res. Int. **42(1)**: 210-215.

Faller, A.L.K. and Fialho, E. 2010. Polyphenol content and antioxidant capacity in organic and conventional plant foods. J. Food Composition Anal. **23(6)**: 561-568.

Fan, Y.-J., He, X.J., Zhou, S.D., Luo, A.X., He, Tao and Chun, Z. 2009. Composition analysis and antioxidant activity of polysaccharide from *Dendrobium denneanum*. Int. J. Biol. Macromol. **45**: 169-173.

Fang, S.H., Koteswara Rao, Y. and Tzeng, Y.M. 2008. Antioxidant and inflammatory mediator's growth inhibitory effects of compounds isolated from *Phyllanthus urinaria*. J. Ethnopharmacol. **116(2)**: 333-340.

Farag, R.S., Badei, A.Z., Heweij, F.M. and El-Baroty, G.S.A. 1986. Antioxidant activity of some spices essential oils on linoleic acid oxidation in aqueous media. J. Amer. Oil Chemists Soc. **66**: 792-799.

Farhoosh, R. and Golmovahhed, G.A. 2007. Antioxidant activity of various extracts of old tea leaves and black tea wastes (*Camellia sinensis* L.). Food Chem. **100**: 231-236.

Farias-Silva, E., Cola, M., Calvo, T.R., Barbastefano, V., Ferreira, A.L., De Paula Michelatto, D., Alves de Almeida, A.C., Hiruma-Lima, C.A., Vilegas, W. and Souza Brito, A.R.M. 2007. Antioxidant activity of Indigo and its preventive effect against ethanol-induced DNA damage in rat gastric mucosa. Planta Medica **73(12)**: 124-1246.

Farombi, E.O. 2009. Biochemical and molecular mechanisms of hepatoprotection by kolaviron, a natural antioxidant biflavonoid from *Garcinia kola* seed. In: Farombi, E.O. (ed.). *Nutritional Antioxidants in Cancer and Degenerative Diseases*. Transworld Research; Kerala, India: . pp. 153–166.

Farombi, E.O., Akanni, O.O. and Emerole, G.O. 2002. Antioxidative and scavenging activities of kolaviron *in vitro*. Pharm. Biol. **40**: 107-116.

Farombi, E.O. and Fakoya, A. 2005. Free radical scavenging and antigenotoxic activities of natural phenolic compounds in dried flowers of *Hibiscus sabdariffa* L. Mol. Nutr. Food Res. **49(12)**: 1120-1128.

Farombi, E.O. and Ige, O.O. 2007. Hypolipidemic and antioxidant effects of ethanolic extract from dried calyx of *Hibiscus sabdariffa* in alloxan-induced diabetic rats. Fundam. Clin. Pharmacol. **21(6)**: 601-609.

Farombi, E.O. and Nwaokeafor, I.A. 2005. Anti-oxidant mechanisms of action of kolaviron: Studies on serum lipoprotein oxidation, metal chelation and oxidative microsomal membrane damage in rats. Clin. Exp. Pharmacol. Phys. **32**: 667–674.

Farombi, E.O., Ogundipe, O. and Moody, J.O. 2001. Antioxidant and anti-inflammatory activities of *Mallotus oppositifolius* in model systems. Afr. J. Med. Sci. **30(3)**: 213-215.

Farombi, E.O. and Owoeye, O. 2011. Antioxidative and chemopreventive properties of *Vernonia amygdalina* and *Garcinia* biflavonoid. Int. J. Environ. Res. Public Health **8(6)**: 2533-2555.

Farsam, H., Amanlou, M., Dehpour, A.R. and Jahaniani, F. 2000. Anti-inflammatory and analgesic activity of *Biebersteinia multifida* DC. root extract. J. Ethnopharmacol. **71**: 443-447.

Farswan, M., Mazumdar, P.M. and Parche, V. 2009. Modulatory effect of an isolated compound from *Syzygium cumini* seeds on biochemical parameters of diabetes in rats. Int. J. Green Pharmacy **3(2)**: 128-133.

Fauconneau, B., Waffo-Teguo, P., Huguet, F., Barrier, L., Decendit, A. and Merillon, J.M. 1997. Comparative study of radical scavenger and antioxidant properties of phenolic compounds from *Vitis vinifera* cell cultures using *in vitro* tests. Life Sci. **61**: 2103-2110.

Fazilatun, N., Nornisah, M. and Zhari, I. 2004. Superoxide radical scavenging properties of extracts and flavonoids isolated from the leaves of *Blumea balsamifera*. Pharmaceutical Biol. **42(6)**: 404-408.

Fecka, I., Gasiorowski, K. and Brokos, B. 1997. Phytochemical analysis and immunotyropic activity of polyphenolic fraction from the common storks bill hers (*Erodium cicutarium* L'Herit.). Herba Polonica **43(3)**: 214-221.

Fejes, S., Blazovics, A., Lemberkovics, E., Petri, G., Szoke, E. and Kery, A. 2000. Free radical scavenging and membrane protective effects of methanol extracts from *Anthriscus cerefolium* L. and *Petroselenium crispum* (Mill.) Nym. ex A W. Hill. Phytother. Res. **14**: 362-365.

Fejes, S., Blazovics, A., Lugasi, A., Lemberkovics, E., Petri, G. and kery, A. 2000. *In vitro* antioxidant activity of *Anthriscus cerefolium* (L.) Hoffman extracts. J. Ethnopharmacol. **69(3)**: 259-265.

Feliciano, R.P., Bravo, M.N., Pires, M.M., Serra, A.T., Duarte, C.M., Boas, L.V. and Bronze, M.R. 2008. Phenolic content and antioxidant activity of Moscatel Dessert wines from the Setubal region of Portugal. Food Anal. Meth. **2(2)**: 149-161. DOI: 10.1007/s12161-008-9059-7.

Feng, X., Zhang, L. and Zhu, H. 2009. Comparative anticancer and antioxidant activities of different ingredients of *Ginkgo biloba* extract (EGb 761). Planta med. **75(8)**: 792-796.

Fernandes, A.C., Cromarty, A.D., Albrecht, C. and Jansen Van Rensburg, C.E. 2004. The antioxidant potential of *Sutherlandia frutescens*. J. Ethnopharmacol. **95(1)**: 1-5.

Fernandes, F., Valentão, P., Sousa, C., Pereira, J. A., Seabra, R M., and Andrade, P. B. 2007. Chemical and antioxidative assessment of dietary turnip (*Brassica rapa* var. *rapa* L.). Food Chem. **105**: 1003–1010.

Fernandez, F. and Torres, M. 2006. Evaluation of *Pluchea carolinensis* extracts as antioxidants by the epinephrine oxidation method. Fitoterapia **77**: 221-226.

Ferreres, F., Sousa, C., Valentão, P., Seabra, R. M., Pereira, J. A., and Andrade, P. B. 2007. Tronchuda cabbage (*Brassica oleraceae* L. var. *costata* DC) seeds: Phytochemical characterization and antioxidant potential. Food Chem. **101**: 549–558.

Ferreres, F., Pereira, D. M., Valent, P., Andrade, P. B., Seabra, R. M. and Sottomayor, M. **2008.** New phenolic compounds and antioxidant potential of *Catharanthus roseus* J. Agric. Food Chem. **56 (21)**: 9967-9974.

Fijesh, P.V. and Padikkala, J. 2008. Anti-inflammatory and anti-ulcer activity of *Cyclea peltata* Lam. Amala Res. Bull. **28**: 46-52.

Filip, R., Lotito, S.B., Ferraro, G. and Raga, C.G. 2000. Antioxidant activity of *Ilex paraguariensis* and related species. Nutrition Res. **20**: 1437-1446.

Filipe, P.M., Fernandes, A.C., Silava, J.N., Freitas, J.P. and Manso, C.F. 1997. Effect of Silibinin on oxidative damage of blood constituents. CR Seases Sco. Biol. Fil. **191(5-6)**: 821-835.

Filipek, J. 1994. Antioxidative properties of *Alchemilla xanthochlora, Salvia officinalis* and *Solidago virgaurea* water extracts. Biologica (Bratislava) **49(3)**: 359-364.

Fintelmann, V. 1996. Therapeutic profile and mechanisms of action of artichoke leaf extract: Hypolipidemic, antioxidant, hepatoprotective and choleretic properties. Phytomedicine **3**(Suppl.): 50.

Firuzi, O., Asadollahi, M., Gholami, M. and Javidnia, K. 2010a. Composition and biological activities of essential oils from four *Heracleum* species. Food Chem. **122(1)**: 117-122.

Firuzi, O., Javidnia, K., Gholami, M., Soltani, M. and Miri, R. 2010b. Antioxidant activity and total phenolic content of 24 Iranian Lamiaceae species growing in Iran. Nat. Prod. Commun. **5(2)**: 261-264.

Fisher, V.L. 2005. Indigenous *Salvia* species – An investigation of the antimicrobial activity, antioxidant activity and chemical composition of leaf extracts. Masters dissertation, University of Witwaterstrand, Johannesburg, South Africa.

Fortalezas, S., Tavares, L., Pimpao, R., Tyagi, M., Pontes, V., Alves, P.M., McDougall, G., Stewart, D., Ferreira, R.B. and Santos, C.N. 2010. Antioxidant properties and neuroprotective capacity of Strawberry tree fruit (*Arbutus unedo*). Nutrients **2(2)**: 214–229.

Finkel, T. and Halbrook, N.J. 2000. Oxidants, oxidative stress and the biology of ageing. Nature **408**: 239-247.

Fook, Y.C. and Kheng, Y.S. 2009. Antioxidative and antibacterial activities of *Pangium edule* seed extracts. Int. J. Pharmacol. **5(5)**: 285-297.

Foti, P., Erba, D., Riso, P., Spadafranca, A., Criscouli, F. and Testolin, G. 2005. Comparison between daizein and genistein antioxidant activity in primary and cancer lymphocytes. Archives Biochem. Biophys. **433(2)**: 421-427.

Frankel., E.N., Kanner, J., German, J.B., Parks, E. and Kinsella, J.E. 1993. Inhibition of oxidation of human low-density lipoprotein by phenolic substances in red wine. Lancet **341**: 454-457.

Frankel., E.N., Huang, S., Aeschbach, R. and Prior, E. 1996. Antioxidant activity of rosemary extract and its constituents, carnosic acid, carnosol and rosmarinic acid anoil-in-water emulsion. J. Agric. Food Chem. **44**: 131-135.

Frankel, E.N. and Meyer, A.S. 2000. The problems of using one dimensional methods to evaluate multifunctional food and biological antioxiants. J. Sci. Food Agric. **80**: 1925-1941.

Frankel., E.N., Waterhouse, A.L. and Teissedre, P.L. 1995. Principal phenolic phytochemicals in selected California wines and their antioxidant activity in inhibiting oxidation of human low-density lipoproteins. J. Agric. Food Chem. **43**: 890-898.

Fraternale, D., Giamperi, L., Bucchini, A. and Ricci, D. 2007. Essential oil composition and antioxidant activity of aerial parts of *Grindelia robusta* from central Italy. Fitoterapia **78(6)**: 443-445.

Fu, L., Xu, B.T., Xu, X.R., Qin, X.S., Gan, R.Y. and Li, H.B. 2010. Antioxidant capacities and total phenolic contents of 56 wild fruits from South China. Molecules **15(12)**: 8602-8617.

Fu, M., He, Z., Zhao, Y., Yang, J. and Mao, L. 2009. Antioxidant properties and involved compounds of daylily flowers in relation to maturity. Food Chem. **114(4)**: 1192-1197.

Fu, W., Chen, J., Cai, Y., Lei, Y., Chen, L., Pei, L., Zhou, D., Liang, X. and Ruan, J. 2010. Antioxidant, free radical scavenging, anti-inflammatory and hepatoprotective potential of the extract from *Parathelypteris nipponica* (Franch. et Sav.) Ching. J. Ethnopharmacol. **130(3)**: 521-528.

Fuhrman, B., Lavy, A. and Aviram, M. 1995. Consumption of red wine with meals reduce the susceptibility of human plasma and low-density lipoprotein to lipid peroxidation. Am. J. Clin. Nutr. **61**: 549-554.

Fujiki, H. 1999. Two stages of cancer prevention with green tea. J. Res. Clin. Oncol. **125(11)**: 589-597.

Fujita, Y., Vehara, L., Morimoto, Y., Nakashima, M. and Hanato, T. 1998. Studies on inhibition mechanism of auto oxidation by tannins and flavonoids. II. Inhibition mechanism of caffetannins isolated from leaves of *Artemisia* species on lipoxigenase dependent lipid peroxidation. Yakuguku zashi **108**: 129-135.

Fukai, T., Satoh, K. Nomura, T. and Sakagami, H. 2003a. Preliminary evaluation of antinephritis and radical scavenging activities of glabridin from *Glycyrrhiza glabra*. **74(7-8)**:: 624-629.

Fukai, T., Satoh, K. Nomura, T. and Sakagami, H. 2003b. Antinephritis and radical scavenging activity of prenylflavonoids. Fitoterapia **74(7-8)**: 720-724.

Fuki, T., Satoh, K., Nomura, and Sakagami, H. 2003. Preliminary evaluation of antinephrities and radical scavenging activity of glabridin from *Glycyrrhiza glabra*. Fitoterapia **74**: 624-629.

Fukumoto, L.R., and Mazza, G. 2000. Assessing antioxidant and proxidant activities and phenolic compounds. J. Agric. Food Chem. **48**: 3597-3604.

Fursule, R.A. and Patil, S.D. 2010. Hepatoprotctivee and antioxidant activity of *Phaseolus trilobus* Ait. on bile duct ligation induced liver fibrosiss in rats. J. Ethnopharmacol. **129(3)**: 416-419

Furuta, S., Nishiba, Y. and Suda, I. 1997. Fluorometric assay for screening antioxidative activity of vegetables. J. Food Sci. **62(3)**: 526-528.

Furuta, S., Suda, I., Nishiba, Y. and Yamakawa, O. 1998. High tert-butylperoxyl radical scavenging activities of sweet potato cultivars with purple flesh. Food Sci. Technol. Int. Tokyo **4**: 33-35.

Fylaktakidou, K.C., Hadjipalou-Litina, D.J., Litinas, K.E., Nicolaides, D.N. 2004. Natural and synthetic coumarin derivatives with anti-inflammatory/antioxidant activities. Curr. Pharm. Des. **10(30)**: 3813-3833.

Gabrieli, C.N., Kefalas, P.G. and Kokklou, E.L. 2005. Antioxidant activity of flavonoids from *Sideritis raeseri*. J. Ethnopharmacol. **96(3)**: 423-428.

Gachkar L., Yadegari D., Rezaei M.B., Taghizadeh M., Astaneh S.A. and Rasooli I. 2007. Chemical and biological characteristics of *Cuminum cyminum* and *Rosmarinus officinalis* essential oils. Food Chem. **102 (3)**: 893-904.

Gagan, V.D., Pradhan, P., Sipahimalani, A.T. and Banerji, A. 1994. Cordifolisides A, B, C: norditerpene furan glycosides from *Tinospora cordifolia*. Phytochemistry **37**: 781.

Gajera, H.P., Golakiya, B.A. and Patel, S.V. 2005. Antioxidant properties of some therapeutically active medicinal plants – an overview. J. Med. Aromatic Plant Sci. **27 (1)**: 91-100.

Galvez, M., Martin-Cordero, C., Houghton, P.J. and Ayuso, M.J. 2005a. Antioxidant activity of methanol extracts obtained from *Plantago* species. J. Agric. Food Chem. **53 (6)**: 1927-1933.

Galvez, M., Martin-Cordero, C., Houghton, P.J. and Ayuso, M.J. 2005b. Antioxidant activity of *Plantago bellardi* All. Phytother. Res. **19(12)**: 1074-1076.

Gamal-Eldeen, A.M., Kawasthy, S.A., Ibrahim, L.F., Shabana, M.M. and Negoumy, S.I.El. 2004. Evaluation of antioxidant, anti-inflammatory and antinociceptive properties of aerial parts of *Vicia sativa* and its flavonoids. J. Nat. Remedies **4(1)**: 81-96.

Gamez, E.J.C., Luyengi, L., Lee, S., Zhu, L., Zhou, B., Fong, H., Pezzuto, J. and Kinghorn, A.D. 1998. Antioxidant flavonoid glycosides from *Daphniphyllum calycinum*. J. Nat. Prod. **61**: 706-708.

Gamiotea-Turro, D., Cuesta-Rubio, O., Prieto-Ganzalez, S., De Simone, F., Passi, S. and Rastrelli, L. 2004. Antioxidative constituents from the leaves of *Hypericum styphelioides*. J Nat. Prod. **67(5)**: 869-871.

Gan, R.-Y., Kuang, L., Xu, X.-R., Zhang, Y., Xia, E.-Q., Song, F.-L. and Li, H.-B. 2010. Screening of natural antioxidants from traditional Chinese medicinal plants associated with treatment of rheumatic disease. Molecules **15**: 5988-5997.

Gan, R.-Y., Xu, X.-R., Song, F.-L., Kuang, L. and Li, H.-B. 2010. Antioxidant activity and total phenolic content of medicinal plants associated with prevention and treatment of cardiovascular and cerebrovascular diseases. J. Med. Plants Res. **4(22)**: 2438-2444.

Ganesan, P., Kumar, C.S. and Bhaskar, N. 2008. Antioxidant properties of methanol extract and its solvent fractions obtained from selected Indian red seaweeds. Bioresour. Technol. **99(8)**: 2717-23.

Gao, D., Li, Q., Li, Y., Liu, Z., Zhao, H., Li, J and Han, Z. 2009. Antidiabetic and antioxidant effects of oleanolic acid from *Ligustrum lucidum* Ait. in alloxan-induced diabetic rats. Phytother. Res. **23(9)**: 1257-1262.

Gao, J., Sun, C., Yang, J., Shi, J., Du, Y., Zhang, Y., Li, J. and Wan, H. 2011. Evaluation of hepatoprotective and antioxidant activities of *Rubus parvifolius* L. J. Zheijing Univ. Sci. B. **12(2)**: 135-142.

Gao, J., Igarashi, K. and Nukina, M. 2000. Three new phenylethanoid glycosides from *Caryopteris incana* and their antioxidative activity. Chem. Pharm. Bull. (Tokyo) **48(7)**: 1075-1078.

Gao, J., Liu, B., Ning, Z., Zhao, R., Zhang, A. and Wu, Q. 2009. Characterization and antioxidant activity of Flavonoid-rich extracts from leaves of *Ampelopsis grosssodentata*. J. Food Biochem. **33(6)**: 808-820. doi: 10.1111/j.1745-4514.2009.00253x.

Gao, J.J., Igalashi, K., Nukina, M. 1999. Radical scavenging activity of phenylpropanoid glycosides in *Caryopteris incana*. Biosci. Biotechnol. Biochem. **63**: 983-988.

Gao, X., Bjork, L., Trajkovski, V. and Uggla, M. 2000. Evaluation of antioxidant activities of rosehip ethanol extracts in different test systems. J. Agri. Food Chem. **80**: 2021-2027.

Gao, X., Uggla, M. and Rumpunen, K. 2000. Antioxidant activity of dried and boiled rosehips. ISHS Acta Horticulturae 690: Intern. Rose hip conference.

Gao, Z., Huang, K., Yang, X. and Xu, H. 1999. Free radical scavenging and antioxidant activities of flavonoids extracted from the radix of *Scutellaria baicalensis* Georgi. Biochim. Biophys. Acta. **1472(3)**: 643-650.

Garcia, R., Kaid, N., Vignaud, C. and Nicolas, J. 2000. Purification and some properties of catalase from wheat germ (*Triticum aestivum* L.). J. Agric. Food Chem. **48(4)**: 1050-1057.

Garcia, F., Pivel, J. P., Guerrero, A., Brieva, A., Martinez-Alcazar, M. P., Caamano-somoza, M., Gonzalez, S. 2006. Phenolic components and antioxidant activity of Fernblock®, an aqueous extract of the aerial parts of the fern *Polypodium leucotomos*. Methods and findings in experimental and clinical, **28(3)**: 157-160

Gardner, P.T., Tamsin, T.C., White, A.C., McPhail, D. and Duthie, G.D. 2000. The relative contributions of vitamin C, carotenoids and phenolics to the antioxidant potential of fruit juices. Food Chem. **68**: 471-474.

Garzon, G.A. and Wrolstad, R.E. 2009. Major anthocyanins and antioxidant activity of Nasturtium flowers (*Tropaeolum majus*). Food Chem. **114(1)**: 44-49. doi: 10.1016/j.foodchem.2008.09.013

Garza-Juarez, A., Salazar-Cavazos Mde, L., Salazar-Aranda, R., Perez-Mesquer, J. and de Torres, N.W. 2011. Correlation between Chromatographic fingerprint and antioxidant activity of *Turnera diffusa* (Daniana). Planta Medica **77(9)**: 958-963.

Gauthaman, K., Banerjee, S.K., Dinda, A.K., Ghosh, C.C. and Maulik, S.K. 2005. *Terminalia arjuna* (Roxb.) protects rabbit heart against ischemic-reperfuon injury: Role of antioxidant enzymes and heat shock protein. J. Ethnopharmacol. **96(3)**: 403-409.

Gawlekdziki, U., Swieca, M., Suguier, D. and Cichocka, J. 2009. Seeds of *Arnica montana* and *Arnica chamissonis* as a potential source of natural antioxidants. Herba Polonica **55(1)**: 60-71.

Gayathri, G. and Suresh Jadhav. 2010. *In vitro* antioxidant and *in vivo* antihyperglycemic potential of *Mimusops elengi* L. in Alloxan-induced diabetes in mice. J. Complementary and Integrative medicine **7(1)**: Article 4. http: / /www. bepress.com/jcim/vol7/iss1/4

Gayathri, V., Asha, V.V. and Subramoniam. 2005. A preliminary studies on the immunomodulatory and antioxidant properties of *Selaginella* species (*Selaginella involvens, S. delicatula* and *S. wightii*). Indian J. Pharmacol. **37**: 381-385.

Gazzani, G., Papetti, A., Massolini, G. and Daglia, M. 1998. Anti and prooxidant activity of water soluble components of some common diet vegetables and the effect of thermal treatment. J. Agric. Food Chem. **46**: 4118-4122.

Ge, H.-M., Yang, W.-H., Zhang, J. and Tan, R.-X. 2009. Antioxidant oligostilbenoids from the stemwood of *Hopea hainanensis.* J. Agric. Food Chem. **57(13)**: 5756-5761.

Gebhardt, R. 1996. Hepatocellular actions of artichoke extracts: Stimulation of biliary secretion, inhibition of cholesterol biosynthesis and antioxidant properties. Phytomedicine **3**(Suppl.): 51.

Gebhardt, R. 1997. Antioxidant and protective properties of extracts from the leaves of the Artichoke (*Cynara scolymus*) against hyperperoxide-induced oxidative stress in cultured rat hepatocytes. Toxicol. Appl. Pharmacol. **144(2)**: 279-286.

Geetha, S., Sai Ram, M., Singh, V., Ilavazhagan, G., Sawhney, R.C. 2002. Antioxidant and immunomodulatory properties of Seabuckthorn (*Hippophae rhamnoides*) - an *in vitro* study. J. Ethnopharmacol. **79**: 373-378.

Geetha *et al.*, 2003. Free radical scavenging action of Tulsi. Nat. Prod. Sci. **9(4)**: 223-225.

Georgiev, M., Alipieva, K., Pashova, S., Denev, P., Angelova, M., Kerns, G. and Bley, T. 2010. Antioxidant activity of devil's claw cell biomass and its active constituents. Food Chem. **121(4)**: 967-972.

Gerhardt, U. and Schreoter, A. 1983. Rosmarinic acid – a naturally occurring antioxidant in spices. Fleischwirtschaft **63**: 1628-1630.

Germano, M.P., D'Angelo,Biasini, T., Sanogo, R., De Pasquale, R. and Catania, S. 2006. Evaluation of the antioxidant properties and bioavailability of free and bound phenolic acids from *Trichilia emetica* Vahl. J. Ethnopharmacol. **105(3)**: 368-373.

Germano, M.P., D'Angelo, V., Sanogo, R., Morabito, A., Pergolizzi, S and De Pasquale, R. 2001. Hepatoprotective activity of *Trichilia roka* on carbon tetrachloride-induced liver damage in rats. J. Pharm. Pharmacol. **53(11)**: 1569-1574.

Geromel, V. *et al.*, 2001. Superoxide-induced massive apoptosis in cultures skin fibroblasts harboring the neurogenic ataxia retinitis pigmentosa (NARP) mutation in the ATPase-6 gene of the mitochondrial DNA. Human Mol. Genet. **10**: 1221-1228.

Gerothanassis, Exarchous, V., Lagouri V. Troganis A. Tsimidou M and Boskoud. 1998. Methodology for identification of phenolic acids in complex phenolic mixtures by High-Resolution Two dimensional nuclear magnetic Resonance. Application to methanolic extracts of two Oregano species. J. Agric. Food Chem. **46(10)**: 4185-4190.

Gerschman, R., Gilbert, D.L., Nye, S.W., Dwyer, P. and Fenn. W.O. 1954. Oxygen poisoning and X-irradiation; a mechanism in common. Science **119**: 623-626.

Gezer, K., Duru, M.E., Kivrak, I., Turkoglu, A., Mercan, N., Turkoglu, H. and Gulcan, S. 2006. Free radical scavenging capacity and antimicrobial activity of wild edible mushroom from Turkey. African J. Biotech. **5**: 1924-1928.

Ghafar, M.F.A., Prasad, K.N., Weng, K.K. and Ismail, A. 2010. Flavonoid, hesperiidine, total phenolic contents and antioxidant activities from *Citrus* species. Afr. J. Biotech. **9(3)**: 326-330.

Ghanta, S., Banerjee, A., Poddar, A. and Chattopadhyay, S. 2007. Oxidative DNA damage preventive activity and antioxidant potential of *Stevia rebaudiana* (Bertoni) Bertoni a natural sweetener. J. Agric. Food Chem. **55(26)**: 10962-10967.

Ghazghazi, H., Miguel, M.G., Hasnaoui, B., Sebei, H., Ksontini, M., Figueiredo, A.C., Pedro, L.G. and Barroso, J.G. 2010. Phenols, essential oils, and carotenoids of *Rosa canina* from Tunisia and their antioxidant activities. Afr. J. Biotechnol. **9**: 2709-2710.

Ghiselli, A., Serafini, M., Natella, F. and Caccini, C. 2000. Total antioxidant capacity as a tool to assess redox status: critical view and experimental data. Free Radical Biological Medicine **29(1)**: 1106-1114.

Ghoghari, A. M.,. Bagul, M. S. Anandjiwala, S. Chauhan, M. G. and Rajani, M. 2006. Free radical scavenging activity of *Aspidium cicutarium* rhizome. J. of Natural remedies **6**: 131-134.

Gholivand, M.B., Rahimi-Nasrabadi, M., Batooli, H. and Ebrahimbadi, A.H. 2010. Chemical composition and antioxidant activities of the essential oil and methanol extracts of *Psammogeton canescens*. Food Chem. Toxicol. **48**: 24-28.

Ghosal, S., Tripathi, V.K. and Chauhan, S. 1996. Active constituents of *Emblica officinalis*: Part I-The chemistry and antioxidative effects of two new hyderolysable tannins, emblicanin A and B. Indian J. Chem. Section B. **35(9)**: 941-948.

Ghosh, A., Mandal, S., Banerji, A., Kar, M. and Banerji, J. 2009. A new chalcone from *Pongamia pinnata* and its antioxidant properties. Nat.Prod. Commun. **4(2)**: 209-210.

Ghosh, A. and Sil, P.C. 2007. Anti-oxidative effect of a protein from *Cajanus indicus* L. against Acetaminophen-induced Hepato-nephro toxicity. Biochem. Mol. Bio. **40(6)**: 1039-1049.

Ghosh S.B., De-Strayo, Suprasanna, P., Devasagyam T.P.A. and Bapat, V.A. 2008. Antioxidant activity in sugar cane juice and its protective role against radiation induced DNA damage. Food Chem. **106(3)**: 1154-1160.

Ghosh, T., Maity, T. M. and Bose, A. 2009. *In vitro* free radical scavenging activity of ethanolic extracts of leaves of *Cajanus cajan* (L.) Millsp. J. Nat. Remed. **9(2)**: 228-234.

Ghosh, T., Maity, T. M., Das, M., Bose, A. and Dash, D.K. 2007. In *vitro* antioxidant and hepatoprotective activity of ethanolic extract of *Bacopa monnieri* Linn. aerial parts. Iranian Journal of Pharmacology and Therapeutics. **6**: 77-85.

Giamperi, A.L., Bucchini, Cara, P., Fraternale, D., Ricci, D., Genovese, S., Curini, M. and Epifano, F. 2009. Composition and antioxidant activity of *Nepeta foliosa* essential oil from Sardinia (Italy). Chem. Nat. Comp. **45(4)**: 4554-556.

Giamperi, A.L., Bucchini, A., Fraternale, D., Genovese, S., Curini, M. and Ricci, D. 2010. Composition and antioxidant activity of *Inula crithmoides* essential oil grown in central Italy (Marche region). Nat. Prod. Comm. **5(2)**: 315-318.

Giamperi, L. Fraternale, D. Bucchini, A. Ricci, D. 2004. Antioxidant activity of *Citrus paradisi* seeds glyceric extract. Fitoterapia, **75 (2)**: 221-224.

Gil, M.I., Ferreres, F. and Tomas-Barberan, F.A. 1999. Effect of postharvest storage and processing on the antioxidant constituents (flavonoids and vitamin C) of fresh-cut spinach. J. Agric. Food Chem. **47**: 2213-2217.

Gil, M.I., Tomas-Barberan, F.A., Hess-Pierce, B., Holcroft, D.M. and Kader, A.A. 2000. Antioxidant activity of pomegranate juice and its relationship with phenolic composition and processing. J. Agric. Food Chem. **48**: 4581-4589.

Gil, M.I., Tomas-Barberan, F.A., Hess-Pierce, B. and Kader, A.A. 2002. Antioxidant capacities, phenolic compounds, carotenoids and vitamin C contents of nectarine, peach and plum cultivars from California. J. Agric. Food Chem. **50**: 4976-4982.

Gillespie, K.M. 2007. Rapid measurement of total antioxidant capacity in plants. www.nature protocols.com/2007/05/24/rapid_measurement_of_total_ant_php.

Gini, K.C., Agi Thomson, M. and Kurup, M.G. 2006. Antioxidant effect of *Nostoc sphaeroides* against carbon tetrachloride induced liver dysfunction in rats. Indian J. Bot. Res. **2**: 83.

Gini, K.C. and Kurup, M.G. 2008. Antioxidant activity of *Aulosira fertilissima* on CCl_4 induced hepatotoxicity in rats. Indian J. Exp. Biol. **46**: 52.

Giorgi, A., Bombelli, R., Luini, A., Speranza, G., Cosentino, M., Lecchini, S. and Socucci, M. 2009. Antioxidant and cytoprotective properties of infusions from leaves and inflorescences of *Achillea collina* Becker ex Rchb. Phytother. Res. **23(4)**: 540-545.

Giorgi, A., Madeo, M., Speranza, G., Cocucci, M. 2010. Influence of environmental factors on composition of phenolic antioxidants of *Achillea collina* Becker ex Rchb. Nat. Prod. Res. **24(16)**: 1546-1559.

Gioti, E.M., Fiamegos, Y.C., Skalkos, D.C. and Stalikas, C. 2009. Antioxidant activity and bioactive components of the aerial parts of *Hypericum perforatum* L. from Epirus, Greece. Food Chem. **117(3)**: 398-404. doi: 10.1016/j.food.chem.2009.04.016.

Giovanelli, G. and Buratti, S. 2008. Comparison of polyphenolic composition and antioxidant activity of wild Italian blueberries and some cultivated varieties. Food Chem. **112(4)**: 903-908.

Gitika, B., Sai Ram, M., Sharma, S.K., Ilavazhagan, G. and Banerjee, P.K. 2006. Quercetin protects C_6 glial cells from oxidative stress induced by tertiary-butylhydroperoxidase. Free Radical Res. **40**: 95-102.

Gkinis, G., bozin, B., Mimica-Ducic, N. and Tzakou, O. 2010. Antioxidant activity of *Nepeta nuda* L. ssp. *nuda* essential oil rich in nepetalactones from Greece. J. Med. Food **13(5)**: 1176-1181.

Godevac, D., Zdunic, G., Savikin, K., Vajs, V. and Menkovic, N. 2008. Antioxidant activity of nine Fabaceae species growing in Serbia and Montenegro. Fitoterapia **79(3)**: 185-187.

Godkar, P., Gordon, R.K., Ravindran, A. and Doctor, B.P. 2003. *Celastrus paniculatus* seed water soluble extracts protect cultured rat forebrain neuronal cells from hydrogen peroxide-induced oxidative injury. Fitoterapia **74(7-8)**: 658-659.

Goel, R.K., Sairam, K. and Rao, V. 2001. Role of gastric antioxidant and antihelicobactor pylori activities in antiulcerogenic activity of plantain banana (*Musa sapientum* var. *paradisiaca*). Indian J. Exp. Biol. **39**: 719-722.

Gombau, L., Garcia, F., Lahoz, A., Fabre, M., Roda-Navarro, P., Majano, P., Alonso-Lebrero, J.L., Pivel, J.P., Castell, J.V., Gomez-Lechon, M.J. and Gonzalez, S. 2006. *Polypodium leucotomus* extract: antioxidant activity disposition. Toxicol. *In vitro* **20(4)**: 464-471.

Gomes, R.C., Bonamin, F., Darin, D.D., Slito, L.N., Di Stasi, L.C., Dokkedal, A.L., Vilegar, W., Brito, A.R.M.S. and Hiruma-Lima, C.A. 2009. Antioxidative action of methanolic extract and butanolic fraction of *Vochysia tucanorum* Mart. in the gastroprotection. J. Ethnopharmacol. **121 (3)**: 466- 471.

Gomez Ruiz, J.A., Leake, D.S. and Ames, J.M. 2007. *In vitro* antioxidant activity of coffee compounds and their metabolites. J. Agric. Food Chem. **55(17)**: 6962-6969.

Goncalves,., Landbo, A-K., Let, M., Silva, A.P., Rosa, E. and Meyer, A.S. 2004. Storage affects the phenolic profiles and antioxidant activities of the cherry (*Prunus avium* L.) on human low-density lipoproteins. J. S. Food Agric. 84: 1013-1020. doi: 10.1002/jsfa.1752.

Goncalves, C., Dinis, T. and Batista, M.T. 2005. Antioxidant properties of proanthocyanidins of *Uncaria tomentosa* bark decoction: a mechanism for anti-inflammatory activity. Phytochemistry **66**: 89-98.

Gonzalez, S., Joshi, P. C. and Pathak, M. A. 1994. *Polypodium leucotomos* extract as an antioxidant agent in the therapy of skin disorders. J. Investigative Dermatol. **102(4)**: 651.

Gopinathan, N., Srinivasan, K.K. and Mthew, J.E. 2004. Free radical scavenging properties of ethanol extract of *Saccharum spontaneum*. Indian Drugs **41**: 633-635.

Gordon, M.H. and Weng, C. 1992. Antioxidant properties of extracts from tanshen (*Salvia miltiorrhiza* Bunge). Food Chem. **44(2)**: 119-122. doi: 10.1016/0108-8146(92)90322-S.

Gorinstein, S., Leontowicz, H., Leontowicz, M., Krzeminsk, R., Gralak, M., Martin-Belloso, O., Delgado-Licon, E., Haruenkit, R., Katrich, E., Park, Y.S., Jung, S.T. and Trakhtenberg, S. 2004. Fresh Israeli Jaffa blond (Shamouti) ornage and Israeli Jaffa red Star Ruby (Sunrise) grapefruit juices affect plasma lipid metabolism and antioxidant capacity in rats fed added cholesterol. J. Agric. Food Chem. **52(15)**: 4853-4859.

Gorinstein, S., Yamamoto, K., Katrich, E., Leontowicz, H., Lojek, A., Leontowicz, M., Ciz, M., Goshev, I., Shalev, U. and Trakhtenberg, S. 2003. Antioxidative properties of Jaffa sweeties and Grape fruit and their influence on lipid metabolism and plasma antioxidative potential in rats. Biosci. Biotechnol. Biochem. **67**: 907-910.

Gourine, N., Yousfi, M., Bombarda, I., Nadjemi, B. and Gaydou, E. 2010a. Seasonal variation of chemical composition and antioxidant activity of essential oil from *Pistacia atlantica* Desf. leaves. J. Am. Oil Chem. Soc. **87**: 157-166.

Gourine, N., Yousfi, M., Bombarda, I., Nadjemi, B., Stocker, P. and Gaydou, E.M. 2010b. Antioxidant activities and chemical composition of essential oil of *Pistacia atlantica* from Algeria. Ind. Crops Prod. **31(2)**: 203-208.

Gouthamchandra, K., Mahmood, R. and Manjunatha, H. 2010. Free radical scavenging, antioxidant enzymes and wound healing activities of leaves extracts from *Clerodendrum infortunatum* L. Environmental Toxicol. Pharmacol. **30(1)**: 11-18.

Govil, J.N., Singh, V.K. and Ahmad, K. 2004. Recent Progress in medicinal plants. Series. Volume 14. Biopharmaceutical Studium Press.

Govindachari, T.R., Sathe, S.S. and Viswanathan, N. 1966. Gentianine, an artifact in *Enicostemma littorale*. Indian J. Chem. **4**: 201-202.

Govindappa, M., Naga Sravya, S., Poojashri, M.N., Sadananda, T.S. and Chandrappa, C.P. 2011. Antimicrobial, antioxidant and *in vitro* anti-inflammatory activity of ethanol extract and active phytochemical screening of *Wedelia trilobota* (L.) Hitch. J. Pharmcogn. Phyther. **3(3)**: 43-51.

Govindappa, M., Sadananda, T.S., Channabasava, R., Jeevitha, M.K., Pooja, K.S. and Raghavendra, V.B. 2011. Antimicrobial, antioxidant activity and phytochemical screening of *Tecoma stans* (L.) Juss. ex Kunth. J. Phytol. **3**: 68-76.

Govindarajan, R., Agnihotri, A.K., Rawat, A.K.S., Khatoon, S. and S. Mehrotra. 2004. Pharmacognostic evaluation of an antioxidant plant – *Acorus calamus* Linn. Nat. Prod. Sci. **9(4)**: 264-269.

Govindarajan, R., Rastogi, S., Vijayakumar, M., Rawat, A.K.S., Shirwaikar, A., Mehrotra, S. *et al.*, 2003. Studies on the antioxidant activities of *Desmodium gangeticum*. Biol. Pharmaceutical Bull. **26**: 1424-1427.

Govindarajan, R., Sreevidya, N., Vijayakumar, M., Thakur, M., Dixit, V. K., Mehrotra, S. 2005. *In vitro* antioxidant activity of ethanolic extract of *Chlorophytum borivilianum*. Nat. Prod. Sci. **11(3)**: 165-169.

Govindarajan, R., Vijayakumar, M., Rao, Ch. V., Rawat, A.K.S., Shirwaikar, A., Mehrotra, S. and Pushpangadan, P. 2004. Antioxidant potential of *Anogeissus latifolia*. Biol. Pharmaceutical Bulletin. **27(8)**: 1266-1269.

Govindarajan, R., Vijayakumar, M., Rawat, A.K.S. and Mehrotra, S. 2003. Free radical scavenging potential of *Picrorhiza kurroa* Royle ex Benth. Indian J. Exp. Biol. **41**: 875-879.

Govindarajan, R., Vijayakumar, M., Shirwaikar, A., Rawat, A.K.S., Mehrotra, S. and Pushpangadan, P. 2005. Activity guided isolation of antioxidant tannoid principles from *Anogeissus latifolia*. Nat. Prod. Sci. **11(3)**: 174-178.

Govindarajan, R., Vijayakumar, M., Singh, M., Rao, C.V., Shirwaikar, A., Rawat, A.K.S., and Pushpangadan, P. 2006. Antiulcer and antimicrobial activity of *Anogeissus latifolia*. J. Ethnopharmacol. **106(1)**: 57-61.

Gowri Shankar, N.L., Manavalan, R., Venkatappaya, D. and David Raj, C. 2008. Hepatoprotective and antioxidant effects of *Commiphora berryi* (Arn.) Engl. Bark extract against CCl_4-induced oxidative damage in rats. Food Chem. Toxicol. **46(9)**: 3182-3185.

Gowri, S.S., Pavithra, S. and Vasantha, K. 2011. Free radical scavenging capacity and antioxidant activity of young leaves and barks of *Acacia nilotica* (L.) Del. Int. J. Pharm. Pharm. Sci. **3(1)**: 160-164.

Gowri, S.S. and Vasantha, K. 2010. Free radical scavenging and antioxidant activity of leaves from Agathi (*Sesbania grandiflora* (L.) Pers.). American-Eurasian J. Sci. Res. **5(2)**: 114-119.

Goyal, A.K., Middha, S.K. and Sen, A. 2010. Evaluation of the DPPH radical scavenging activity, total phenolics and antioxidant activities in Indian wild *Bambusa vulgaris* "vittata" mehtanolic leaf extract. J. Nat. Pharmaceuticals **1(1)**: 40-45.

Goza, I., Alim, A., Sihoglu-Tepe, A., Sokmen, M., Sevgi, K. and Tepe, B. 2009. Screening of the antioxidant activity of the essential oil and various extracts of *Origanum rotundifolium* Boiss. From Turkey. J. Med. Plants Res. **3**: 246-254.

Goze, I., Alim, A., Cetinus, S.A., Cetin, A., Durmus, N., Atas, A.T. and Vural, N. 2010. *In vitro* antimicrobial, antioxidant and antispasmodic activities and the composition of the essential oil of *Origanum acutidens* (Hand-Mazz.) Ietswaart. J. Med. Food **13**: 705-709.

Goze, I., Alim, A., Tepe, A.S., Sokmen, M., Sevgi, K. and Tepe, B. 2009. Screening of the antioxidant activity of essential oil and various extracts of *Origanum rotundifolium* Boiss. from Turkey. J. Med. Plant. Res. **3(4)**: 246-254.

Granito, M., Paolini, M. and Pérez, S. 2008. Polyphenols and antioxidant capacity of *Phaseolus vulgaris* stored under extreme conditions and processed. LWT - Food Sci. Technol. 41(6): 994-999.

Grassmann, J., Hippeli, S., Vollmann, R., Elstner, E.F. 2003. Antioxidative properties of the essential oil from *Pinus mugo*. J. Agric. Food Chem. **51(26)**: 7576-82.

Graven, E.H., Deans, S.G., Svobada, K.P., Mari, and Gundidza, M.G. 1992. Antibacterial and antioxidative properties of the volatilte (essential oil) of *Artemesia afra* Jacq. Flavor. and Fragrance. **7**: 121-13.

Green, R.J. and Sanders, T.H. 2004. Antioxidant activity of peanut plant parts. National meeting of Institute of Food Technologists. Food Exp. IFT Book of Abstracts 67B: 37.

Greenwald, P. 2002. Science, Medicine and the future of Cancer chemoprevention. Brit. Med. J. **324**: 714-718.

Gregorio, M., Rene, D., Gema, P., Gabino, G., Alberto, N.S J.,Sonia, L.O.2000. Evaluation of the *in vitro* antioxidant activity of *Mangifera indica* L. extract (Vimang). Phytother. Res. **14(6)**: 424-427.

Grosso, C., Figueiredo, A.C., Burillo, J., Mainar, A.M., Urieta, J.S., Barroso, J.G., Coelho, J.A. and Palavra, A.M.F. 2010. Composition and antioxidant activity of *Thymus vulgaris* volatiles: comparison between supercritical fluid extraction and hydrodistillation. J. Sep. Sci. **33**: 2211-2218.

Grosso, C., Vinholes, J., Silva, L.R., de Pinho, P.G., Goncalvees, R.F., Valentao, P. Jaeger, A.K. and Andrade, P.B. 2011. Chemical composition and biological screening of *Capsella bursa-pastoris*. Rev. Bras. Farmacogno. Doi: 10.1590/s0102-695x2011005000107.

Gu, L. and Weng, X. 2001. Antioxidant activity and components of *Salvia plebeia* R.Br. – a Chinese herb. Food Chem. **73**: 299-305.

Guan, S., Bao, Y.-M., Jiang, B. and An, L.-J. 2006. Protective effect of protocatechuic acid from *Alpinia oxyphylla* on hydrogen peroxide-induced oxidative PC12 ceell death. Eur. J. Pharmacol. **538(1-3)**: 73-79.

Guclu, K., Altun, M., Özyurek, M., Karademir, S.E. and Apak, R. 2006. Antioxidant capacity of fresh, sun- and sulfited-dried Malaya apricot (*Prunus armeniaca*) assayed by CUPRAC, ABTS, TEAC and Folin methods. Int. J. Food Sci.Tech. **41**: 76-85.

Gude, R., and Shukla, V. 2004. Theophylline, methylxanthine an antimetastitic agent: Its role through antioxidative pathway. Chemistry, Biology Interface: Synergistic Frontiers, New Delhi, India. OP – 7, November 21-26, 2004.

Guerra, L., Pereira, C., Andrade, P.B., Rodriques, A., Ferreres, F., De Pinho, P.G., Seabra, R.M. and Valentao, P. 2008. Targeted metabolite analysis and antioxidant potential of *Rumex induratus*. J. Agric. Food Chem. **56(17)**: 8184-8194.

Guerra, N.B., Melo, E.A. and Filho, J. 2005. Antioxidant compounds from coriander (*Coriandrum sativum* L.) ether extract. J. Food Composition and Analysis **18**: 193-199.

Guha, G., Rajkumar, V. and Ashok Kumar, R. 2010a. Polyphenolic constituents of methanolic and aqueous extracts of *Vitex negundo* render protection to Hep3B cells against oxidative cytotoxicity. Food Chem. Toxicol. **48(8-9)**: 2133-2138. doi: 10.1016/j.fct.2010.05.017.

Guha, G., Rajkumar, V., Ashok Kumar, R. and Mathew, L. 2009a. Antioxidant activity of *Lawsonia inermis* extracts inhibits Chromium (VI)-induced cellular and DNA toxicity. Evidence Based Complement alternat. Med. Doi: 10.1093/ecam/nep205.

Guha, G., Rajkumar, V., Ashok Kumar, R. and Mathew, L. 2011. Therapeutic potential of polar and non-polar extracts of *Cynanthillium cinerum in vitro*. Evid. Based Complement. Alternat. Med. doi: 10.1093/e cam/neP155.

Guha, G., Rajkumar, V., Mathew, L. and Ashok Kumar, R. 2010b. Antioxidant potential and DNA protection efficiency of Indian tribal medicinal plants. Turk. J. Biol. Doi: 10.3906/biy-0906-64

Guil-Guerrero, J.L. and Rebolloso-Fuentes, M.M. 2009. Nutrient composition and antioxidant activity of eight tomato (*Lycopersicon esculentum*) varieties. Journal of Food Composition and Analysis. **22(2)**: 123-129.

Gülçin, I. 2005. The antioxidant and radical scavenging activities of black pepper (*Piper nigrum*) seeds. Int. J. Food Sci. Nutr. **56 (7)**: 491-499.

Gülçin, I. 2006. Antioxidant activity of caffeic acid (3,4-dihydroxycinnamic acid). Toxicology **217(2-3)**: 213-220.

Gülçin, I., Alici, H.A. and Cesur. 2005. Determination of *in vitro* antioxidant and radical scavenging activities of propofol. Chem. Pharm. Bull. (Tokyo). **53(3)**: 281-285.

Gülçin, I., Berashvili, D. and Gepdiremen, A. 2005. Antiradical and antioxidant activity of total anthocyanins from *Perilla pankinensis* Decne. J. Ethnopharmacol. **101(1-3)**: 287-293.

Gülçin, I., Beydemir, S., Sat, I.G., Kufrevioglu, O.I. 2005. Evaluation of antioxidant activity of cornelian cherry (*Cornus mas* L.). Acta Alimentaria **34(2)**: 193-202.

Gülçin, I., Büyükokuroglu, M.E., Oktay, M. and Küfrevioglu, O.I. 2003. Antioxidant and analgesic activities of turpentine of *Pinus nigra* Arn. subsp. *pallsiana* (Lamb.) Holmboe. J Ethnopharmacol. **86(1)**: 51-58.

Gülçin, I., Elias, R., Gepdiremen, A., Chea, A., and Topal, F. 2010. Antioxidant activity of bisbenzylisoquinoline alkaloids from *Stephania rotunda*: Cepharanthine and fangchinoline. J. Enzyme Inhib. Med. Chem. **25(1)**: 44-53.

Gülçin, I., Elias, R., Gepdiremen, A., Taoubi, K. and Koksal, E. 2007. Antioxidant secoiridoids from fringe tree (*Chionanthus virginicus* L.). Wood Sci. Technol. **43(3-4)**: 195-212. doi: 10.1007/s00226-008-0234-1.

Gülçin, I., Elmastas, M. and Aboul-Enein, H.Y. 2007. Determination of antioxidant radical scavenging activity of Basil (*Ocimum basilicum* L. Family: Lamiaceae) assayed by different methodologies. Phytother. Res. **21(4)**: 354-361.

Gülçin, I., Kufrevioglu, O.I., Oktay, M. and Buyukuroglu, M.E. 2004. Antioxidant, antimicrobial, antiulcer and analgesic activities of nettle (*Urtica dioica* L.). J. Ethnopharmacol. **90(2-3)**: 205-215.

Gülçin, I., Mshvildadze, V., Gepdiremen, A. and Elias, R. 2004. Antioxidant activity of saponins isolated from Ivy: alpha-hederin, hederasaponin-C, hederacolchiside-E and hederacolchiside-F. Planta Medica **70(6)**: 561-563.

Gülçin, I., Mshvildadze, V., Gepdiremen, A. and Elias, R. 2006. Screening of antiradical and antioxidant activity of monodesmosides and crude extract from *Leontice smirnowii* tuber. Phytomedicine **13(5)**: 343-351.

Gülçin, I., Oktay, M., Kirecci, E. and Kufrevioglu, I.O. 2003. Screening of antioxidant and antimicrobial activities of anise (*Pimpinella anisum* L.) extracts. *Food Chem.* **83**: 371-382.

Gülçin, I., Oktay, M., KüfrevioOktay, M., KüfrevioOktay, M., Küfrevio. 2002. Determination of antioxidant activity of lichen *Cetraria islandica* (L) Ach. J. Ethnopharmacol. **79 (3)**: 325-329.

Gülçin, I., Sat, I.G., Beydemir S., Elmastas M. and Kufrevioglu O.I. 2004. Comparison of antioxidant activity of clove (*Eugenia caryophyllata* Thunb.) buds and lavender (*Lavandula stoechas* L.). Food Chem. **87 (3)**: 393-400.

Gülçin, I., Uguz, M.T., Oktay, M., Beydemir, S. and Kufrevioglu, O.I. 2003. Antioxidant and antimicrobial activities of *Teucrium polium* L. J. Food Tech. **1**: 9-17.

Gulluce, M., Aslan, A., Sokmen, M., Sahin, F., Adiguzel, A., Agar, G. and Sokmen, A. 2008. Screening the antioxidant and antimicrobial properties of the lichens *Parmelia saxatilis, Platismatia glauca, Ramalina pollinaria, Ramalina polymorpha* and *Umbilicaria nylanderiana*. Phytomedicine. **13(7)**: 515-21.

Gulluce, M., Sokmen, M., Daferena, D., Agar, G., Ozkhan, H. and Kartal, N. 2003. *In vitro* antibacterial, antifungal and antioxidant activities of the essential oil and methanol extracts of herbal parts and callus cultures of *Satureja hortensis* L. J. Agric. Food Chem. **51**: 3958-3965.

Gulumser, A., Dogan, N.M., Duru, M.E. and Kivrak, I. 2010. Phenolic profiles, antimicrobial and antioxidant activity of the various extracts of *Crocus* species in Anatolia. African J. Microbiol. Res. **4(11)**: 1154-1161.

Gunjegaonkar, S.M., Saraswathi, C.D., Hrishikesavan, H.J., Harish, M.S. and Nargund, L.V.G. 2010. Hepatoprotective and antioxidant activity of *Tephrosia purpurea* whole plant aqueous extract. Pharmacology online **2**: 568-574.

Guo, C., Yang, J., Wei, J., Li, Y., Xu, J. and Jiang, Y. 2003. Antioxidant activities of peel, pulp and seed fractions as determined by FRAP assay. Nutr. Res. 23: 1719-1726.

Guo, D.-J., Cheng, H.-L., Chan, S.-W. and Yu, H.-F. 2009. Antioxidative activities and the total phenolic contents of tonic Chinese Medicinal Herbs. Inflammopharmacology **16(5)**: 201-207. DOI: 10.1007/s10787-008-8016-9.

Guo, J. and Wang, M.H. 2007. Antioxidant and antidiabetic activities of *Ulmus davidiana* extracts. Food Sci. Biotech. **16(1)**: 55-61.

Guo, L., Cho, S.Y., Kang, S.S., Lee, S.H., Baek, H.Y. and Kim,Y.S. 2007. Orthogonal array design for optimizing extraction efficiency of active constituents from Jakyak-Gamcho Decoction, the complex formula of herbal medicines, *Paeoniae radix* and *Glycyrrhizae radix*. J. Ethnopharmacol. **113 (2)**: 306-311.

Guo, P., Xu, J., Xu, S. and Wang, K. 1997. Inhibition of hydrogen peroxide production in chondrocytes induced by fulvic acid by ginger volatile oil. China, J. Chinese Materia Medica **22**: 559-561.

Guo, X.-F., Yue, Y-D., Tang, F., Wang, J. and Yao, X. 2008. Detection of antioxidative capacity of Bamboo leaf extract by scavenging organic free radical DPPH. Spectroscopy and Spectral Analysis. 2008-07.

Guo, Z.L., Zhao, H.Y. and Zheng, X.H. 1995. An experimental study of the mechanism of *Andrographis paniculata* Nees (APN) in alleviating the Ca^{2+}-overloading in the process of myocardial ischemic reperfusion. J. Tongji Med. Univ. **15**: 205.

Gupta, D., Bharadwaj, R. and Gupta, R.K. 2011. *In vitro* antioxidant activity of extracts from the leaves of *Abies pindrow* Royle. Afr. J. Trad. Compl. Med. **8(4)**:

Gupta, M., Mazumder, U.K., Manikandan, L., Bhattacharya, S., Senthilkumar, G.P. and Suresh, R. 2005. Anti-ulcer activity of ethanol extract of *Terminalia pallida* Brandis in Swiss albino rats. J. Ethnopharmacol. **97(2)**: 405-408.

Gupta, M., Mazumder, U.K., Sambath Kumar, R., Sivakumar, T., Gomathi, and Rajeshwar, Y. 2005. Antioxidant defense system induced by a methanol extract of *Caesalpinia bonducella*. Pharm. Biol. **43(5)**: 411-419.

Gupta, M., Mazumdar, U.K., Sivakumar, T, Vamis, M.L.M.Karkis, S., Sambathkumar, R. and Manikandan, L. 2003. Antioxidant and anti-inflammatory activities of *Acalypha fruticosa*. Nig. J. Nat. Prod. Med. pp. 25-29.

Gupta, M., Mukhtar, H.M. and Ahmad, S. 2011. *In vitro* evaluation of antioxidant activity and total phenolic content of *Tylophora indica* (Burm.f) Merril. Int. J. Pharm. Sci. Res. **2(1)**: 121-126.

Gupta, R.K., Kesari, A.N., Diwakar, S., Tyagi, A., Tandon, V., Chandra, R. and Watal, G. 2008. *In vivo* evaluation of antioxidant and anti-lipidemic potential of *Annona squamosa* aqueous extract in Type 2 diabetic models. J. Ethnopharmacol. **118(1)**: 21-25.

Gupta, R.K., Tyagi, A., Watal, G. and Tandon, V., 2004. Antioxidant and hypolipidemic effect of aqueous leaves extract of *Annona squamosa* L. in normal and STZ induced diabetic rats. Chemistry, Biology Interface: Synergistic Frontiers, New Delhi, India. P 25– 3, November 21-26, 2004.

Gupta, S. and Prakash, J. 2009. Studies on Indian green leafy vegetables for their antioxidant activity. Plant Foods Human Nutr. **64**: 39-45.

Gupta, S.D. (ed.). 2010. Reactive Oxygen species and antioxidants in Higher Plants. Science Publishers, Enfield, U.S.A.

Gupta, V. and Sharma, M. 2011. *In vitro* antioxidant studies on methanolic extracts of *Pongamia pinnata* (L.) Pierre: A great versatile leguminous plant. J. Pharm. Res. **4(7)**:

Gupta, V.K. 2006. Plants as natural antioxidants. Nat. Prod. Radiance **5(4)**: 326-334.

Gurav, S., Deshkar, N., Gulkari, V., Duragkar, N. and Patil, A. 2000. Free radical scavenging activity of *Polygala chinensis* Linn. http: //www.unisa.it/download/1966_10305_2551855_24_Gurav.pdf

Guri, A., Kefalas, P., Roussis, V. 2006. Antioxidant potential of six pine species. Phytother Res. **20(4)**: 263-266.

Gursoy, N., Sarikurkcu, C., Cengiz, M. and Solak, M.H. 2009. Antioxidant activities, metal contents, total phenolics and flavonoids of seven *Morchella* species. Food Chem. Toxicol. **47(9)**: 2381-2388.

Gursoy, N., Sarikurkcu, C., Tepe, B. and Solak, M.H. 2010. Evaluation of the antioxidant activities of three edible mushrooms *Ramaria flava*, *Rhizopogon roseolus*, *Russula delica*. Food Sci. Biotech. **19**: 691-696.

Gursoy, N., Sihoglu Tepe, A. and Tepe, B. 2009. Determination of *in vitro* antioxidative and antimicrobial properties and total phenolic contents of *Ziziphora clinopodioides, Cyclotrichum niveum* and *Mentha longifolia* subsp. *typhoides* var. *typhoides*. J. Med. Food 12: 684-689.

Gursoy, N. and Tepe, B. 2009. Determination of antimicrobial and antioxidative properties and total phenolics of two "endemic" Lamiaceae species from Turkey: *Ballota rotundifolia* L. and *Teucrium chamaedrys* C. Koch. Plant Foods for Human Nutr. **64**: 135-140.

Gursoy, N., Tepe, B. and Sokmen, M. 2010. Evaluation of the chemical composition and antioxidant activity of the peel oil of *Citrus nobilis*. Int. J. Food Properties **3**: 983-991.

Guruvayoorappan, C., Afira, A.H. and Kuttan, G. 2006. Antioxidant potential of *Biophytum sensitivum* extract *in vitro* and *in vivo*. J. Basic Clin. Physiol. Pharmacol. **17(4)**: 255-267.

Gutierrez, R., Alvarado, J.L., Presno, M., Perez-Veyna, O., Serrano, C.J. and Yahuaca, P. 2010. Oxidative stress modulation by *Rosmarinus officinalis* in CCl_4-induced liver cirrhosis. Phytother. Res. **24**: 595-601.

Gutierrez, R.M.P., Navarro, Y.T.G. 2010. Antioxidant and hepatoprotective effects of the methanol extract of the leaves of *Satureja macrostoma*. Phcog. Mag. **6(22)**: 125-131.

Gutteridge, J.M.C. 1994. Free radicals and aging. Rev. Clin. Gerontol. **4**: 279-288.

Guveller, M.E., Brest, C. and Richard, H. 1994. Antioxidant constituents in sage (*Salvia officinalis*). J. Agric. Food Chem. **42**: 665-669.

Guveller, M.E., Richard, H. and Berst, C. 1996. Antioxidative activity and phenolic composition of pilot-plant and commercial extract of sage and rosemary. J. Amer. Chem. Soc. **73**: 645-652.

Guzman, S., Gato, A. and Calleja, J.M. 2001. Antiinflammatory, analgesic and free radical scavenging activities of the marine microalgae *Chlorella stigmatophora* and *Phaeodactylum tricornutum*. Phytother. Res. **15(3)**: 224-230.

Gyamfi, M.A., Yonamine, M. and Aniya, Y. 1999. Free-radical scavenging action of medicinal herbs from Ghana: *Thonningia sanguinea* on experimentally-induced liver injuries. Gen. Pharmacol. **32(6)**: 661-667.

Ha, B.J., Lee, S.H., Kim, H. and Lee, J.Y. 2006. The role of *Salicornia herbacea* in ovariectomy-induced oxidative stress. Biol. Pharm. Bull. **29(7)**: 1305-1309.

Ha, H.H., Park, S.Y., Ko, W.S. and Kim, Y. 2008. *Gleditsia sinensis* thorns inhibit the production of NO through NF-kappaB suppression in LPS stimulated macrophages. J. Ethnopharmacol. **118(3)**: 429-434.

Habbu, P.V., Mahadevan, K.M., Kulkarni, P.V., Daulatsingh, C., Veerapur, V.P. and Shastry, R.A. 2010. Adaptogenic and *in vitro* antioxidant activity of flavonoids and other fractions of *Argyreia speciosa* (Burm.f.) Boj. in acute and chronic stress paradigms in rodents. Indian J. Exp. Biol. **48(1)**: 53-60.

Habib, M.R., Sayeed, M.A., Rahman, M.M., Hasan, M.R. and Saha, A. 2010. *In vitro* evaluation of cytotoxic, antibacterial, antioxidant and phytochemical screening of petroleum ether extract of *Phyllanthus acidus*. Int. J. Pharm. Ci. Res. **2(4)**: 875-881.

Habibuddin, M., Daghriri, H.A., Humaira, T., AlQahtani, M.S. and Hefzi, A.A.H. 2008. Antidiabetic effect of alcoholic extract of *Caralluma sinaica* L. on streptozotocin-induced diabetic rats. J. Ethnopharmacol. **117 (2)**: 215-220.

Habila, J.D., Bello, I.A., Dzikwe, A.A., Ladan, Z. and Sabiu, M. 2011. Comparative evaluation of phytochemicals, antioxidant and antimicrobial activity of four medicinal plants native to Northern Nigeria. Austr. J. Basic Appl Sci. **5(5)**: 537-543.

Habila, J.D., Bello, I.A., Dzikwe, A.A., Musa, H. and Abubakar, N. 2010. Total phenolics and antioxidant activity of *Tridax procumbens* Linn. African J. Pharm. Pharmacol. **4(3)**: 123-126.

Habsah, M., Amran, M., Mackeen, M.M. Lajis, N.H. Kikuzaki, H., Nakatani, N., Rahman, A.A., Ghafar and Ali, A.M. 2000. Screening of Zingiberaceae extracts for antimicrobial and antioxidant activities. J. Ethnopharmacol. **72(3)**: 403-410.

Habsah, M., Lajis, N.H., Ali, A.M., Sukari, M.A., Hin, Y.Y., Kikuzaki, H. and Nakatani, N. 2003. The antioxidative components from *Alpinia nutans*. Pharmaceutical Biol. **41(1)**: 7-9.

Habtemariam, S. and Dagne, E. 2010. Comparative antioxidant and prooxidant and cytotoxic activity of sigmoidin and eriodictyol. Planta Medica **76(6)**: 589-594.

Hadafi, A., Ismail Alaoui, M., Chaouch, A., Benjilal, B. and Zrira, S. 1998. Antioxidant activity and phenolic acid content in rosemary (*Rosmarinus officinalis* L.) and myrtle (*Myrtus communis*). Riv. Ital. EPPOS pp. 325-340.

Hadaruga, D.I. and Hadaruga, N.G. 2009a. Antioxidant activity of *Chelidonium majus* L. extracts from the Banat county. J. Agroalimentary Processes and Technologies **15(3)**: 396-402.

Hadaruga, D.I. and Hadaruga, N.G. 2009b. Antioxidant activity of hepatoprotective silymarin and *Silybum marianum* L. extract. Chem. Bull. Politechnica. Univ. Timisora **54(68)**: 104-110.

Hagat, R., Kanupriya, Prasad, D. and Prasad, S. 2004. Cytoprotective and antioxidant activity of *Rhodiola imbricata* on U937 human macrophages. Chemistry Biology Interface: Synergistic New frontiers, New Delhii, India, P25-51, November 21-26, 2004.

Hajji, M., Masmoudi, O., Souissi, N., Triki, Y., Kammoun, S. and Nasri, M. 2010. Chemical composition, angiotensin 1-converting enzyme (ACE) inhibitory, antioxidant and antimicrobial activities of the essential oil from *Periploca laevigata*. Food Chem. **121(3)**: 724-731.

Hajra, S., Mehta, A., Pandey, P., Vyas, P. 2011. Antioxidant and antidiabetic potential of ethanolic extract of *Swietenia mahagoni* (Linn.) seeds. Int. J. Pharm. Res. Dev. **3(3)**: 180-186.

Hakiman, M. and Maziah, M. 2009. Non enzymatic and enzymatic antioxidant activities in aqueous extract of different *Ficus deltoidea* accessions. J. Med. Plants Res. 3(3): 120-131.

Häkkienen, S.H., Kärenlampi, S.O., Heinonen, I.M., Mykkänen, H.M. and Törrönen, A.R. 1999. Content of the flavonols quercetin, myricetin and kaempferol in 25 edible berries. J. Agr.Food Chem. **47**: 2274-2279.

Häkkinen, S.H., Mykkanen, H.M., Karenlampi, S.O. and Toronen, A.R. 1998. Phenolic profiles in Finnish berries. Cost 916 1st Workshop: Polyphenols in Food. April 1997. Office for Official Publication of the European Communities, Luxembourg.

Halder, N., Joshi, S., Nag, T.C., Tandon, R. and Gupta, S.K. 2009. *Ocimum sanctum* extracts attenuate hydrogen peroxide induced cytotoxic ultrastructural changes in human lens epithelial cells. Phytother. Res. **23(12)**: 1734-1737.

Halici, M., Odabasoglu, F., Suleyman, H., Cakir, A., Aslan, A. and Bayir, Y. 2005. Effect of water extract of *Usnea longissima* on antioxidant enzyme activity and mucosal damage caused by indomethacin in rats. Phytomedicine **77**: 398-405.

Halim, M.E. and Mishra, A. 2011. The effects of the aqueous extract of *Pterocarpus* heartwood and vitamin E supplementation in streptozotocin-induced diabetic rats. J. Med. Plants Res. **5(3)**: 398-409.

Halliwell, B. 1991. Reactive oxygen species in living systems: Source, biochemistry, and role in human disease. Am. J. Med. **91**: 14-22.

Halliwell, B. 1994. Free radicals, antioxidants and human disease: curiosity, cause or consequence? Lancet 344: 721-724.

Halliwell, B. 1996. Antioxidants in human health and disease. Annual Review of Nutrition **16**: 33-50.

Halliwell, B. 1997. Antioxidants: the basics – what they are and how to evaluate them. Advances in Pharmacol. **38**: 3-20.

Halliwell, B. and Guetteridge, J.M.C. 1999. Free radicals in Biology and Medicine. Oxford Univ. Press.

Halliwell, B., Gutteridge J.M.C. and Amoma, O.1987. The deoxy ribose method: A simple test tube assay for the determination of rate constant for reaction of hydroxyl radical. Anal. Biochem. **165**: 215-19.

Haloui, E., Marzouk, Z., Marzouk, B., Bouffira, I., Bouraoui, A. and Fenina, N. 2010. Pharmacological activities and chemical composition of the *Olea europaea* L. leaf essential oils from Tunisia. J. Food Agr. Environ. **8**: 204-208.

Halvorsen, B.L., Holte, K., Myhrstad, M.C.W., Barikmo, I., Hvattum, E., Remberg, S.F., Wold, A., Haffner, K., Bougerod, H., Anderson, L.F., Moskaug, J.O., Jacobs, D.R., Blomhoff, R. 2002. A systematic screening of total antioxidants in dietery plants J. Nutr. **132**: 461-471.

Hamad, I., Erol-Dayi, O., Pekmez, M., Onay-Ucar, E. and Arda, A. 2007. Free radical scavenging activity and protective effects of *Alchemilla vulgaris* L. J. Biotechnol. **131**: s40-s41.

Hamada, A., Yoshioka, S., Takima, D., Yokota, J., Cuit, T., Kusunose, M., Miyamura, M., Kyotani, S. and Nishioka, Y. 2004. The effect of *Eriobotrya japonica* seed extract on oxidative stress in adriamycin-induced nephropathy in rats. Biol. Pharamaceutical Bull. **27(12)**: 1961-1964.

Hamauzu, Y., Yasui, H., Inno, T., Kume, C. and Omanyuda, M. 2005. Phenolic profile, antioxidant property, and anti-influenza viral activity of Chinese quince (*Pseudocydonia sinensis* Schneid.) Quince (*Cydonia oblonga* Mill) and apple (*Malus domestica* Mill.) fruits. J. Agric. Food Chem. **53**: 928-934.

Hamerski, L., Bomm, M.D., Silva, D.H.S., Young, M.C.M., Furlan, M., Eberlin, M.N., Castro-Gamboa, I., Cavalheiro, A.J., da Silva Bolzani, V. 2005. Phenylpropanoid glucosides from leaves of *Coussarea hydrangeifolia* (Rubiaceae). Phytochem. **66(16)**: 1927-1932.

Hamzeh Amiri. 2007. Chemical Composition, antibacterial and antioxidant Activity of the essential oil of *Tanacetum polycephalum* Schutz. Bip. Intern. J. Bot. **3**: 321-324.DOI: 10.3923/ijb.2007.321.324

Han, H., Baik and Byung-Kee. 2008. Antioxidant activity and phenolic content of lentils (*Lens culinaris*), chickpeas (*Cicer arietinum* L.), peas (*Pisum sativum* L.) and soybeans (*Glycine max*), and their quantitative changes during processing. Int. J. Food Sci. and Technol. **43 (11)**: 1971-1978.

Han, J., Weng, X. and Kaishun Bi. 2008. Antioxidants from a Chinese medicinal herb – *Lithospermum erythrorhizon*. Food Chem. **106(1)**: 2-10.

Han, S.J., Ryu, S.N. and Kang, S.S. 2004. A new 2-arylbenzofuran with antioxidant activity from the black colored rice (*Oryza sativa* L.) bran. Chem. Pharmaceutical Bull. **52(11)**: 1365-1366.

Hanachi, P. 2009. Using HPLC to determination the composition and antioxidant activity of *Berberis vulgaris*. European J. Sci. Res. **29 (1)**: 47-54.

Hanen, F., Riadh, K., Samia, O., Sylvain, G., Christian, M. and Cgedly, A. 2009. Interspecific variability of antioxidant activities and phenolic composition in *Mesembryanthemum* genus. Food Chem. Toxicol. **47(9)**: 2308-2313.

Haque, R. 2003. Studies on modulatory role of antioxidants against immunotoxicity of anticancer drugs. Ph.D. thesis. Jamia Hamdard, New Delhi.

Haraguchi, H., Inoue, J., Tamura, Y and Mizutani, K. 2000. Inhibition of mitochondrial lipid peroxidation by bakuchiol, a meroterpene from *Psoralea corylifolia*. Planta Medica **66**: 569-571.

Haraguchi, H., Inoue, J., Tamura, Y and Mizutani, K. 2002. Antioxidative components of *Psoralea corylifolia* (Leguminosae). Phytother. Res. **16(6)**: 539-544.

Haraguchi, H., Ishikawa, H. and Kubo, I. 1997. Antioxidative action of diterpenoids from *Podocarpus nagi*. Planta Med. **63(3)**: 213-215.

Haraguchi, H., Ishikawa, H., Mizutani, K., Tamura, Y. and Kinoshita, T. 1998. Antioxidative and superoxide scavenging activities of retrochalcones in *Glycyrrhiza inflata*. Biorg. Med. Chem. **6(3)**: 339-347.

Haraguchi, H., Ishikawa, H., Sakai, S., Ying, B.P. and Kubo, I. 1996. Inhibition of lipid peroxidation by diterpenoid from *Podocarpus nagi*. Experientia **52(6)**: 573-576.

Haraguchi, H., Ishikawa, H., Sanchez, Y., Ogura, T., Kubo, Y. and Kubo, I. 1997. Antioxidative constituents in *Heterotheca inuloides*. Bioorg. Med. Chem. **5(5)**: 865-871.

Haraguchi, H., Saito, T., Ishikawa, Date, H., Kataoka, S., Yamura, Y. and Mizutani, K. 1996. Antiperoxidative components in *Thymus vulgaris*. Planta Med. **62**: 217-221.

Haraguchi, H., Saito, T., Ishikawa, H., Sanchez, Y., Ogura, T. and Kubo, I. 1996. Inhibition of lipid peroxidation by sesquiterpenoid in *Heterotheca inuloides*. J. Pharm. Pharmacol. **48(4)**: 441-443.

Haraguchi, H., Saito, T., Okamura, N. and Yagi, A. 1995. Inhibition of lipid peroxidation and superoxide generation by diterpenoids from *Rosmarinus officinalis*. Planta Med. **61(4)**: 333-336.

Harika, G.V., Patil, C.R. and Patil, M.R. 2007. Protective effect of *Kalanchoe pinnata* Pers. (Crassulaceae) on gentamicin-induced nephrotoxicity in rats. Indian J. Pharmacol. **39(4)**: 201-205.

Harish, R. and Shivanandappa, T. 2006. Antioxidant activity and hepatoprotective potential of *Phyllanthus niruri*. Food Chem. **95(2)**: 180-185.

Harman, D. 1956. Aging: a theory based on free radical and radiation chemistry. J. Gerontol. **11**: 298-300.

Harman, D. 1972. The Biologic clock: the mitochondria? J. Am. Geriatr. Soc. **20**: 145-147.

Harman, D. 1992. Free radical theory of aging. Mutat. Res. **275(3-6)**: 257-266.

Harman, D. 1993. Free radical involvement in aging. Pathophysiol. Therap. Implications. Drugs Aging **3(1)**: 60-80.

Harput, U.S., Genc, Y., Khan, N. and Saracoglu, I. 2011. Radical scavenging effects of different *Veronica* species. Rec. Nat. Prod. **5(2)**: 100-107.

Harzallah, H.J., Neffati, A., Skandrani, I., Maaloul, E., Chekir Ghedira, L. and Mahjoub, T. 2010. Antioxidant and antigenotoxic activities of *Globularia alypum* leaves extracts. J. Med. Plants Res. **4(19)**: 2048-2053.

Hasan, M.S., Ahmed, M.I., Mondal, S., Uddin, S.J., Masud, M.M., Sadhu, S.K. and Ishibashi, M. 2006. Antioxidant, antinociceptive activity and general toxicity study of *Dendrophthoe falcata* and isolation of quercitrin as the major component. Orient Pharmacy and Experminetal medicine **6**: 355-360.

Hasan, S.M.R., Hossain, M.M., Akter, R., Jamila, M., Mzaumder, M.E.H. and Rahman, S. 2009. DPPH free radical scavenging activity of some Bangladeshi medicinal plants. J. Med. Plants Res. **3(11)**: 875-879.

Hasan, S.M.R., Jamila, M., Majumder, M.M., Akter, R., Hossain, M.M., Mazumder, M.E.H., Alam, M.A., Jahangir, R., Rana, M.S. and Rahman, S. 2009. Analgesic and antioxidant activity of the hydromethanolic extract of *Mikania scandens* (L.) Willd. leaves. Amer. J. Pharmacol. Toxicol. **4(1)**: 1-7.

Haseena Banu, B., Prasad, K.V.S.R.G., Bharathi, K. and Veena. 2010. Evaluation of hyperlipidemic and antioxidant activities of *Inula racemosa* roots. 2nd Intern. Seminar on Medicinal plants and Herbal Products, S.V.Univ., Tirupati, Abstract pp. 103.

Hashem, F.A. 2008. Free radical scavenging activity of the flavonoids isolated from *Tecoma radicans*. J. Herbs Spices Med. Plants **13(2)**: 1-10.

Hashim, F.A. 2007. Phenolic compounds of *Erucaria microcarpa* Boiss. and their effect as scavengers for siglet oxygen. J. Herbs, Spices Med. Plants **12(3)**: 27-41.

Hashim, M.S., Lincy, S., Remya, V., Teena, M. and Anila, L. 2005. Effect of polyphenolic compounds from *Coriandrum sativum* on H_2O_2 induced oxidative stress in human lymphocytes. Food Chem. **92**: 653-660.

Hashimoto, T., Quang, D.N., Kuratsune, M. and Asakawa, Y. 2006. Hydnellins A and B, nitrogen containing terphenyls from the mushrooms *Hydnellum suaveolens* and *Hydnellum geogerirum*. Chem. Pharm. Bull. **54(6)**: 912-914.

Hassanpour, H., Yousef, H., Jafar, H. and Mohammad, A. 2011. Antioxidant capacity and phytochemical properties of cornelian cherry (*Cornus mas* L.) genotypes in Iran. Scientia Horticulturae. Doi: 10.1016/j.scienta.2011.04.017.

Hatano, T., Edamatsu, R. and Mori, A. 1989. Effect of interaction of tannins with co-existing substances, VI, Effects of tannins and related polyphenols on superoxide anion radical and DPPH radical. Chem. Pharm. Bull. **37**: 2016-2021.

Hatano, T., Miyatake, H., Natsume, M., Osakabe, N., Takizawa, T. and Ito, T. *et al.*, 2002. Proanthocyanidin glycosides and related polyphenols from cacao liquor and their antioxidant effects. Phytochemistry **59**: 749-758.

Hassig, A., Liang, W.X., Shwabl, K. and Stampfl, K. 1999. Flavonoids and tannins: plant based antioxidants with vitamin character. Med. Hypotheses. **52**: 471-481.

Hatano, T., Kagawa, H., Yasuhara, T. and Okuda, T. 1988. Two new flavonoids and other constituents in licorice root: their relative astringency and radical scavenging effects. Chem. Pharm. Bull. **36**: 1090-1097.

Hay, A.E., Aumond, M.C., Mallet, S., Dumonet, V., Litaudon, M., Rondeau, D. and Richomme, P. 2004. Antioxidant xanthones from *Garcinia vieillardii*. J. Nat. Prod. **67(4)**: 707-709.

Hayder, N., Abdelwahed, A., Kilani, S., Ammar, R.B., Mahmoud, A., Ghedira, K. and Chekir-hedira, L. 2004. Anti-genotoxic and free-radical scavenging activities of extracts from (Tunisian) *Myrtus communis*. Mut. Res. **564(1)**: 89-95.

Hayet, E., Maha, M., Samia, A., Ali, M.M., Souhir, B., Abderaout, K., Mighri, Z. and Mahjoub, A. 2009. Antibacterial, antioxidant and cytotoxic activities of extract of *Conyza canadensis* (L.) Cronquist growing in Tunisia. Med. Chem. Res. **18**: 447-454.

Haznagy-Radnai, E., Czigle, Sz., Zupko, I., Falkay, Gy. And Mathe, I. 2006. Comparison of antioxidant activity in enzyme-independent system of six *Stachys* species. Fitoterapia. **77**: 521-524.

Hazra, B., Biswas, S., and Mandal, N. 2008. Antioxidant and free radical scavenging activity of *Spondias pinnata*. BMC complimentary and Alternative Medicine 8: 63. doi: 10.1186/1472-6882-8-63.

Hazra, B., Sarkar, R., Biswas, S. and Mandal, N. 2010. The antioxidant, iron chelating and DNA protective properties of 70 per cent methanolic extract of 'Katha' (Heartwood extract of *Acacia catechu*). J. Complementary Integr. Med. 7(1). DOI 10.2202/1553-3840.1335.

Hazra, B., Sarkar, R., Biswas, S. and Mandal, N. 2010. Comparative study of the antioxidant and reactive oxygen species scavenging properties in the extracts of the fruits of *Terminalia chebula*, *Terminalia belerica* and *Emblica officinalis*. BMC Complement Altern. Med. **10**: 20.

Hazzit, M., Baaliouamer, A., Faleiro, M.L. and Miguel, M.G. 2006. Composition of the essential oils of *Thymus* and *Origanum* species from Algeria and their antioxidant and antimicrobial activities. J. Agric. Food Chem. **54(17)**: 6314-6321.

He, Y.-Q., Ma, Z.-Y., Zhang, J., Du, B.-Z. and Yao, B.-H. 2011. Antioxidant activity of the chemical constituents from the leaves of *Quercus macrocarpa*. Chem. Nat. Compounds **47(3)**: 472-473. doi: 10.1007/s10600-011-9969-2.

He, Z.D., Lau, K.M. and But, P.P.H. 2003. Antioxidative glycosides from the leaves of *Ligustrum robustum*. J. Nat. Prod. **66(6)**: 851-854.

Hegde, K. and Joshi, A.B. 2009. Hepatoprotective effect of *Carissa carandas* Linn. root extract against CCl_4 and paracetamol induced hepatic oxidative stress. Indian J. Exp. Biol. **47**: 660-667.

Heilman, J., Winkelmann, K. and Sticher, O. 2003. Studies on the antioxidative activity of phloroglucinol derivatives isolated from *Hypericum* species. Planta Medica **69(3)**: 202-206.

Heinen, C.G., Haenen, G.R., Oostveen, R.M., Stalpers, E.M. and Bast, A. 2002. Protective of flavonoids against lipid peroxidation: the structure activity relationship revisited. Free Radical Res. **36**: 575-581.

Heinonen, I.M., Lehtonen, P.J. and Hopia A.I. 1998a. Antioxidant activity of berry and fruit wines and liquors. J. Agric. Food Chem. **46**: 25-31.

Heinonen, I.M., Meyer, A.S. and Frankel, E.N. 1998b. Antioxidant activity of berry phenolics on human low-density lipoprotein and liposome oxidation. J. Agric. Food Chem. **46**: 4107-4112.

Helen, A., Krishnakumar, K., Vijayammal, P.L., Augusti, K.T. 2000. Antioxidant effect of onion oil (*Allium cepa* Linn.) on the damages induced by nicotine in rats as compared to alpha-tocopherol. Toxicol. Lett. **116(1-2)**: 61-68.

Helen, A, Rajasree, C.R., Krishnakumar, K, Augusti, K.T. and Vijayammal, P.L. 1999. Antioxidant role of oils isolated from garlic (*Allium sativum* Linn.) and onion

(*Allium cepa* Linn.) on nicotine-induced lipid peroxidation. Vet. Hum. Toxicol. **41**: 316.

Henderson, D.E., Slickman, A.M. and Hendersen, S.K. 1999. Quantitative HPLC determination of the antioxidant activity of Capsaicin on the formation of Lipid Hydroperoxides of Linoleic Acid: A comparative study against BHT and Melatonin. J. Agric. Food Chem. **47(7)**: 2563-2570.

Henkel, R., Fransman, W., Hipler, U.C., Wiegand, C., Schreber, G., Menkveld, R., Weitz, F. and Fisher, D. 2011. *Typha capensis* (Rohrb.) N.E.Br. (bulrush) extract scavenges free radicals, inhibits collagenase activity and affects human sperm motility and mitochondrial membrane potential *in vitro*: a pilot study. Andrologia doi: 10.1111/j.1439-0272.2011.01179.x.

Henriquez, C., Almonacid, S., Escobar, B., Chiffelle, I., Gomez, M. and Speisky, H. 2009. Antioxidant content and activity in different structures of five apple cultivars grown in Chile. Acta Hort. **841**: 275-280.

Henry, G.E., Campbell, M.S., Zelinsky, A.A., Liu, Y., Bowen-Forbes, C.S., Li, L., Nair, m.G., Rowley, D.C. and Seeram, N.P. 2009. Bioactive acylphloroglucinols from *Hypericum densiflourus*. Phytother. Res. **23(12)**: 1759-1762.

Heo, K.S., Lee, S.J. and Lim, K.T. 2004. Cytotoxic effect of glycoprotein isolated from *Solanum nigrum* L. through the inhibition of hydroxyl radical-induced DNA-binding activities of NF-Kappa B in HT-29 cells. Env. Toxicol. Pharmacol. **17(1)**: 45-54.

Heo, S.-J., Cha, S.-H., Lee, K.-W. and Jeon, Y.J. 2006. Antioxidant activity of Red Algae from Jeju island. Algae **21(1)**: 149-156.

Heo, S.J., Kim, J.P., Jung, W.K., Lee, N.H., Kang, H.S., Jun, E.M., Park, S.H., Kang, S.M., Lee, Y.J., Park, P.J. and Jeon, Y.J. 2008. Identification of chemical structure and free radical scavenging activity of diphlorethohydroxycarmalol isolated from a brown alga, *Ishige okamurae*. J Microbiol. Biotechnol. **18(4)**: 676-81.

Heo, S.J., Park, E.J., Chon, S.U., Lee, S.Y., Cho, J.Y. and Gorinstein, S. 2007. Antioxidant activity and cytotoxicity of methanol extracts from aerial parts of Korean salad plants. Biofactors **30(2)**: 79-89.

Heo, S.J., Park, E.J., Lee, K.W. and Jeon, Y.J. 2005. Antioxidant activities of enzymatic extracts from brown seaweeds. Bioresour Technol. **96(14)**: 1613-23.

Herodez, S.S., Hadolin, M. and Knez, Z. 2003. Solvent extraction study of antioxidants from balm (*Melissa officinalis* L.) leaves. Food Chem. **80**: 275-282.

Herrero, M., Ibáñez, E., Cifuentes, A. and Señoráns, J. 2004. Pressurized liquid extracts from *Spirulina platensis* microalga. Determination of their antioxidant activity and preliminary analysis by micellar electrokinetic chromatography. J. Chromatogr A. **1047 (2)**: 195-203.

Herrmann, K. 1961. The antioxidative activity of Labiatae drugs and their constituent Labiatae acid. Z. Lebensmittelunters-Forsch. **116**: 224-228.

Hertog, M.G.L., Feskens, E.J.M., Hollman, P.C.H., Katan, M.B. and Kromhout, D. 1993. Dietary antioxidant flavonoids and risk of coronary heart disease: the Zutphen elderly study. Lancet **342**: 1007-1011.

Hevesi Toth, B. 2009. Phytochemical and *in vitro* biological evaluation of potentially active compounds in *Epilobium* species. Ph. D. thesis, Semmelweis Univ. Budapest, Hungary.

Hevesi Toth, B., Blazics, B. and Kery, A. 2009a. Polyphenol composition and antioxidant capacity of *Epilobium* species. J. Pharm Biomed. Anal. **49(1)**: 26-31.

Hevesi Toth, B., Houghton, P.J., Habtemariam, S. and Kery, A. 2009b. Antioxidant and anti-inflammatory effect of *Epilobium parviflorum* Schreb. Phytother. Res. **23**: 719-724.

Hewamanna, R. *et al.*, 2004. Antioxidant flavonoids from *Vitex peduncularis.* J. Trop. Med. Plants **5**: 217-222.

Hibatallah, J., Carduner, C. and Poelman, M.C. 1999. *In vivo* and *in vitro* assessment of the free-radical-scavenger activity of Ginkgo flavone glycosides at high concentration. J. Pharm. Pharmacol. **51(12)**: 1435-1440.

Hierman, A., Reidlinger, M., Juan, H. and Sametz, W. 1991. Isolierung des antiphlogistischen Wirkprinzips von *Epilobium angustifolium*. Planta Med. **57**: 357-360.

Higashi-Okai, K., Kamimoto, K., Yoshioka, A. and Okai, Y. 2002. Potent suppressive activity of fresh and dried peels from Satsuma mandarin *Citrus unshiu* (Marcorv.) on hydroperoxide generation from oxidized linoleic acid. Phytother Res. **16 (8)**: 781-784.

Higashi-Okai, K., Kanbara, K., Amano, K., Hagiwara, A., Sugita, C., Matsumoto, N. and Okai, Y. 2004. Potent antioxidative and antigenotixic activity in aqueous extract of Japanese rice bran – association with peroxidase activity. Phytother. Res. **18(8)**: 628-633.

Higashi-Okai, K., Yamazaki, M., Nagamori, H. and Okai, Y. 2001. Identification and antioxidant activity of several pigments from the residual green tea (*Camellia sinensis*) after hot water extraction. J. UOEH. **23 (4)**: 335-344.

Hinneburg, I., H.J.D.Dorman and R.Hiltunen. 2005. Antioxidant activities of extracts from selected culinary herbs and spices. Food Chem. **97(1)**: 122-129.

Hiramatsu, M., Yoshikawa, T. and Inoue, M. (eds.). 1997. Free radicals and diseases. Plenum Press. New York.

Hirano, R., Sasamoto, W., Matsumoto, A., Itakura, H., Igarashi, O. and Kondo, K.2001. Antioxidant ability of various flavonoids against DPPH and LDL peroxidation. J. Nutr. Sci. Vitaminol. **47**: 357-362.

Hirasa, K. and Takemasa, M. 1998. Spice science and Technology. Marcel Dekker, New York.

Hirunpanich, V., Utaipat, A., Morales, N.P. Bunyapraphatsara, N., Sato, H., Herunsale, A. and Suthisisang, C. 2006. Hypocholesterolemic and antioxidant effects of aqueous extracts from the dried calyx of *Hibiscus sabdariffa* L. in hypercholesterolemic rats. J. Ethnopharmacol. **103(2)**: 252-260.

Hisamoto, M., Kikuzaki, H., Ohigashi, H. and Nakatani, N. 2003. Antioxidant compounds from the leaves of *Peucedanum japonicum* Thunb. J. Agric. Food Chem. **52**: 445-450.

Hitoshi, T., Kobayashi, Y., Takashi, G, Kasemura, K. and Nomura, M. 2005. DPPH Radical-scavenging effect of several phenylpropanoid compounds and their glycoside derivatives. Yokugaku Zasshi **125(4)**: 371-375.

Hnatgyszyn, O., Martino, V., Gerraro, G. and Cousio, J. 1996. *Phyllanthus sellowianus*: Bioassays in the search for biological activities. Phytomedicine **3**(Suppl.): 232.

Ho, C.-H., Noryati, I., Sulaiman, S.F. and Rosma, A. 2010. *In vitro* antibacterial and antioxidant activities of *Orthosiphon stamineus* Benth. extracts against food-borne bacteria. Food Chem. **122(4)**: 1168-1172.

Ho, C.T. 1992. Phenolic compounds in food: an overview. In: M.T.Huang, C.T.Ho and C.Y.Lee (eds.). Phenolic compounds in food and their effects on health II. Antioxidants and cancer prevention. Amer. Chemical Society, Washington, pp. 2-7.

Hohman, J., Zupko, I., Redei, D., Csanyi, M., Falkay, G., Mathe, I. and Janicsak, G. 1999. Protective effects of the aerial parts of *Salvia officinalis*, *Melissa officinalis* and *Lavandula angustifolia* and their constituents against enzyme-dependent and enzyme-independent lipid peroxidation. Planta Medica **65**: 576-578.

Hollman, P.C., Van Trijp, J.M, Mengelers, H.J., Van Trijp, J.M. and Katan, M.B. 1997. Bioavailability of the dietary antioxidant flavonol quercetin in man. Cancer Lett. **114**: 139-140.

Homhual, S., Zhang, H.J., Bunyapraphatsara, N., Kondratyuk, T.P., Santarsiero, B.D., Mesecar, A.D., Herunsalee, A., Chaukul, W., Pezzuto, J.M., Fong, H.H.S. 2006. Bruguiesulfurol, a new sulfur compound from *Bruguiera gymnorhiza*. Planta Medica **72(3)**: 255-260.

Hong, C., Mi, X., Piao, X. and Li, X. 2009. Studies on antioxidant activities from *Ligularia fischeri*. J. Central Univ. Nationalities (Nat. Sci.) 2009-01.

Hong, Y., Lin, S., Jiang, Y. and Ashraf, M. 2008. Variation in contents of total phenolics and flavonoids and antioxidant activities in the leaves of 11 *Eriobotrya* species. Plant Foods Hum. Nutr. **63**: 200-204.

Hort, M.A., Dalbo, S., Brighente, I.M., Pizzalaatti, M.G., Pedrosa, R.C. and Ribeiro-do-valle, R.M. 2008. Antioxidant and hepatoprotectiv effects of *Cyatheaa phalerata* Mart. (Cyatheaceae). Basic Clin. Pharmacol. Toxicol. **103(1)**: 17-24.

Hosein, H.K.M. and Zinab, D. 2007. Phenolic compounds and antioxidant activity of Henna leaves extract (*Lawsonia inermis*). World J. Dairy and Food Sci. **2(1)**: 38-41.

Hossain, M.M., Jahangir, R., Hasan, S.M.R., Akter, R., Ahmed, T., Islam, M.I. and Faruque, A. 2009. Antioxidant, analgesic and cytotoxic activity of *Michelia champaca* Linn. leaf. S.J.Pharm. Sci. **2(2)**: 1-7.

Hosseinimehr, S.J., Pourmorad, F., Shahabimajd, N., Shahrbandy, K. and Hosseinzadeh, R. 2007. *In vitro* antioxidant activity of *Polygonum hyrcanicum, Centaurea depressa, Sambucus ebulus, Mentha spicata* and *Phytolacca americana*. Pak. J. Biol. Sci. **10(4)**: 637-640.

Hou, W.C., Chen, Y.C., Chen, H.J., Lin, Y.H., Yang, L.L. and Lee, M.H. 2001. Antioxidant activities of trypsin inhibitor, a 33 KDa root storage protein of sweet potato (*Ipomoea batatas* (L.) Lam cv. Tainong 57). J. Agric. Food Chem. **49(6)**: 2978-2981.

Hou, W.C., Hsu, F.L. and Lee, M.H. 2002. Yam (*Dioscorea batatas*) tuber mucilage exhibited antioxidant activities *in vitro*. Planta Medica **68**: 1072-1076.

Hou, W.C., Lee, M.H., Chen, H.J., Liang, W.L., Han, C.H., Liu, Y.W. and Lin, Y.H. 2001. Antioxidant activities of dioscorin, the storage protein of yam (*Dioscorea batatas* Decne) tuber. J. Agric. Food Ce. **49**: 4956-4960.

Hou, W.C., Lin, R.D., Cheng, K.T., Tsung, H.Y., Cho, C.H., Chen, C.H., Hwang, S.Y. and Lee, M.H. 2003. Free radical-scavenging activity of Taiwanese native plants. Phytomedicine **10**: 170-175.

Hou, W.C., Lin, R.D., Lee, T.H., Huang, Y.H., Hsu, F.L. and Lee, M.H. 2005. The phenolic constituents and free radical scavenging activities of *Gynura formosana* Kitamura. J. Sci. Food Agric. **85**: 615-621.

Hou, W.C., Wu, W.C., Yang, C.Y., Chen, H.J., Liu, S.Y. and Lin, Y.H. 2004. Antioxidant activities of methanolic and hot-water extracts from leaves of three cultivars of Mai-Men-Dong (*Liriope spicata* L.) Bot. Bull. Acad. Sin. **45**: 285-290.

Houghton, P.J., Hylands, P.J., Mensah, A.Y., Hensel, A. and Deters, A.M. 2005. *In vitro* tests and ethnopharmacological investigations: wound healing as an example. J. Ethnopharmacol. **100(1-2)**: 100-107.

Howard, L.R., Clark, J.R. and Brownmiller, C. 2003. Antioxidant capacity and phenolic content in blueberries as affected by genotype and growing season. J. Sci. Food Agr. **83**: 1238-1247.

Howard, L. R., Talcott, S. T., Brenes, C. H. and Villalon, B. 2000. Changes in phytochemical and antioxidant activity of selected pepper cultivars (*Capsicum* species) as influenced by maturity. J. Agric. Food Chem. **48(5)**: 1713-1720.

Hras, A.R., Hadolin, M., Knez, Z. and Bauman, D. 2000. Comparison of antioxidative and synergistic effects of rosemary extract with alpha-tocopherol ascorbyl palmitate and citric acid in sunflower oil. Food Chem. **71**: 229-233.

Hseieh, C. L., Chiang S Y., Cheng, K. S. Lin, Y. H., Tang, N. Y., Lee, C. J., Pon, C. Z. and Hsieh,C.T 2001. Anticonvulsive and free radical scavenging activities of *Gastrodia elata* Bl. in kainic acid-treated rats. The American journal of Chinese medicine. **29(2)**: 331-41.

Hseieh, C.L. and Yen, G.C. 2000. Antioxidant actions of du-zhong (*Eucommia ulmoides* Oliv.) toward oxidative damage in biomolecules. Life Sci. **66**: 1387-1400.

Hseieh, M.-C., Shen, Y.-J., Kuo, Y.-H., and Huang, L.S. 2008. Antioxidative activity and active components of longan (*Dimocarpus longan* Lour.) flower extracts. J. Agric. Food Chem. **56(16)**: 7010-7016. doi: 10.1021/jf 801155j. PMID: 18662011.

Hseu, Y.C., Chang, W.H., Chen, C.S., Liao, J.W., Huang, C.J., Lu, F.J., Chia, Y.C., Hsu, K.H., Wu, J.J. and Yang, H.L. 2008. Antioxidant activities of *Toona sinensis* leaves extracts using different oxidant models. Food Chem. Toxicol. **46**: 105-114

Hseu, Y.C., Chen, S.C., Yech, Y.J., Wang, L. and Yang, H.L. 2008. Antioxidant activity of *Antrodia camphorata* on free radical induced endothelial damage. J. Ethnopharmacol. **118 (2)**: 237-245.

Hsiao, G., Shen, M.Y., Lin, K.H., Lan, M.H., Wu, L.Y., Chou, D.S., Lin, L.H., Su, C.H. and Sheu, J.R. 2003. Antioxidative and hepatoprotective effects of *Antrodia camphorata* extract. J. Agric. Food Chem. **51(11)**: 3302-3308.

Hsu, B.,Coupar, I.M. and Ng, K. 2006. Antioxidant activity of hot water extract from the fruit of the Doum palm, *Hyphaene thebaica*. Food Chem. **98(2)**: 317-328. doi: 10.1016/j.foodchem.2005.05.077.

Hsu, C.-L., Chen, W., Weng, Y.-M. and Tseng, C.-Y. 2003. Chemical composition, physical properties and antioxidant activity of yam flours as affected by different drying methods. Food Chem. **83(1)**: 85-92.

Hsu, C.-L., Hong, B.-H., Yu, Y.S. and Yen, G.C. 2010. Antioxidant and anti-inflammatory effects of *Orthosiphon aristatus* and its bioactive compounds. J. Agric. Food Chem. **58(4)**: 2150-2156.

Hsu, C.Y. 2006. Antioxidant activity of extract from *Polygonum aviculare* L. Biol. Res. **39**: 281-288.

Hsu, C.Y. 2008. Antioxidant activity of *Pyrrosia petiolosa*. Fitoterapia **79(1)**: 64-66.

Hsu, H.-Y., Chen, J.Y., Yang, J.J. and Lin, C.C. 1999. Evaluation of the antioxidant activities of *Erycibe obtusifolia*. Amer. J. Chinese Medicine **27(1)**: 117-122.

Hsu, T.Y., Sheu, S.C., Liaw, E.T., Wang, T.C. and Lin, C.C. 2005. Anti-oxidant activity and effect of *Pinus morrisonicola* Hay. on the survival of leukemia cell line U937. Phytomedicine. **12(9)**: 663-669.

Hsu, Y.M., Lai, C.H., Chang, C.Y., Fan, C.T., Chen, C.T. and Wu, C.H. 2008. Characterizing the lipid-lowering effects and antioxidant mechanisms of tomato paste. Biosci. Biotech. Biochem. **72(3)**: 677-685.

Htay, H.H., Kawai, M., MacNaughton, L.E., Katsuda, M. and Juneja, L.R. 2006. Tea in Myanmar, with special reference to pickled tea. Int. J. Tea Sci. **5(3-4)**: 11-18.

http: //www.naturalhealthandwellnesscenter.com/herbs.htm

Hu, C. and Kitts, D.-D. 2000. Studies on the antioxidant activity of *Echinacea* root extract. J. Agric. Food Chem. **48**: 1466-1472.

Hu, C. and Kitts, D.-D. 2005. Dandelion (*Taraxacum officinale*) flower extract suppresses both reactive oxygen species and nitric oxide and prevents lipid oxidation *in vitro*. Phytomedicine **12**: 588-197.

Hu, T. and Liu, D., Chen, Y., Wu, J. and Wang, S. 2010. Antioxidant activity of sulfated polysaccharide fractions extracted from *Undaria pinnatifida in vitro*. Int. J. Biol. Macromol. **46**: 193-198.

Hu, Q., Xu, J., Chen, S. and Yang, F. 2004. Antioxidant activity of extracts of black sesame seed (*Sesamum indicum* L.) by supercritical carbon dioxide extraction. J. Agric. Food Chem. **52 (4)**: 943-947.

Hu, X.P., Shin, J.W., Wang, J.H., Cho, J.H., Son, J.Y., Cho, C.K. and Son,.G. 2008. Antioxidative and hepatoprotective effect of CGX, an herbal medicine, against toxic injury in mice. J. Ethnopharmacol. **120 (1)**: 51-55.

Hu, W. and Wang, M.H. 2009. Antioxidant activity of *Aralia elata* seeds fractions. Hort. Environ. Biotech. **50(3)**: 253-257.

Hu, Y. Xu, J. and Hu, Q. 2003. Evaluation of antioxidant potential of *Aloe vera* (*Aloe barbadensis* Miller) extracts. J. Agric. Food Chem. **51**: 7788-7791. doi: 10.1021/jf034255i

Huang, B., Ban, X.Q., He, J.S., Tong, J., Tian, J. and Wang, Y.W. 2010. Hepatoprotective and antioxidant activity of the ethanol extracts of edible lotus (*Nelumbo nucifera* Gaertn.) leaves. Food Chem. **120(3)**: 873-878. doi: 10.1016/j.foodchem.2009.11.020.

Huang, B., Ban, X., He, J.S., Zeng, H., Zhang, P. and Wang, Y. 2010. Hepatoprotective and antioxidant effects of the methanolic extract from *Halenia elliptica*. J. Ethnopharmacol. **131(2)**: 276-281.

Huang, D., Ou, B., Hampsch-Woodill, M.F., Judith A. and R.L. Prior. 2002. High-throughput assay of oxygen radical absorbance capacity using a multichannel liquid handling system coupled with a microplate fluorescence reader in 96-well format. J. Agri. Food Chem. **50**: 4437–44.

Huang, D., Ou, B. and Prior, R.L. 2005. The chemistry behind antioxidant capacity assays. J. Agric. Food Chem. **53**: 1841-1856.

Huang, D.J., Chen, H.J., Hou, W.C., Lin, C.D. and Lin, Y.H. 2004. Active recombinant thioredoxin protein with antioxidant activities from sweet potato (*Ipomoea batatas* L. Lam. Tainong 57) storage roots. J. Agric. Food Chem. **52(15)**: 4720-4724.

Huang, D.J., Chen, H.J., Hou, W.C., Lin, C.D. and Lin, Y.H. 2006. Sweet Potato (*Ipomoea batatas* (L.) Lam. Tainong 57) storage root mucilage with antioxidant activities *in vitro*. Food Chem. **98**: 774-781.

Huang, D.J., Chen, H.J., Lin, C.D. and Lin, Y.H. 2005. Antioxidant and antiproliferative activities of water spinach (*Ipomoea aquatica* Forsk.) constituents. Bot. Bull. Acad. Sin. **46**: 99-106.

Huang, D.J., Lin, C.D., Chen, H.J. and Lin, Y.H. 2004. Antioxidant and antiproliferative activities of sweet potato (*Ipomoea batatas* (L.) Lam. 'Tainong 57') constituents. Bot. Bull. Acad. Sin. **45**: 179-186.

Huang, H.Y., Chang, C.K., Tso, T.K., Huang, J.J., Chang, W.W. and Tsai, Y.C. 2004a. Antioxidant activities of various fruits and vegetables produced in Taiwan. Int. J. Food Sci. Nutr. **55(5)**: 423-429.

Huang, H.Y., Huang, J.J., Tso, T.K., Tsai, Y.C. and Chang, C.K. 2004b. Antioxidant and angiotension-converting enzyme inhibition capacities of various parts of *Benincasa hispida* (wax gourd). Nahrung **48(3)**: 230-233.

Huang, M.-H., Huang, S.-S., Wang, B.-S., Wu, C.-H., Sheu, M.-J., Hou, W.-C., Lin, S.-S. and Huang, G.-J. 2011. Antioxidant and anti-inflammatory properties of *Cardiospermum halicacabum* and its reference compounds *ex vivo* and *in vivo*. J. Ethnopharmacol. **133(2)**: 743-750. doi: 10.1016/j.jep.2010.11.005.

Huang, M.T., Ho, C.T. and Lee, C.Y. (eds.). 1992. Phenolic compounds in food and their effects on health II. Antioxidants and cancer prevention. Amer. Chemical Society, Washington.

Huang, P. V., Maeda, T., Miyatake, K. and Morita, N. 2009. Total phenolic compounds and antioxidant capacity of wheat graded flours by polishing methods. Food Res. Int. **42(1)**: 185-190.

Huang, S.-S., Huang, G.-J., Ho, Y.-L., Lin, Y.-H., Hung, H.-J., Chang, T.-N., Chang, M.-J., Chen, J.-J. and Chang, Y.-S. 2008. Antioxidant and antiproliferative activities of the four *Hydrocotyle* species from Taiwan. Bot. Stud. **49**: 311-322.

Huang, T.-C., Hsu, H.-H., Fu, H-Y. and Ho, C.-T. 2004. Effect of lactic acid fermentation on quercetin composition and antioxidative properties of *Toona sinensis* leaves. In: Shahidi, F. and Naczk, M. (eds.) Phenolics in Food and Neutraceuticals. CRC Press, Boca Raton. Pp. 46-56.

Huang, T.M., Na, M.K., Min, B.S., Zhang, X.F., Lee, I.S., Ngoc, T.M., Thuong, P.T., Sok, D.E. and Bae, K.H. 2007. Protective effect of magnoflorine isolated from *Coptidis* Rhizoma on Cu^{2+} induced oxidation of human low density lipoprotein. Planta Medica **73(12)**: 1281-1284.

Huang, T.M., Na, M., Thoung, P.T., Su, N.D., Sok, D., Song, K.S., Seong, Y.H. and Bae, K. 2006. Antioxidant activity of caffeoyl quinic acid derivatives from the roots of *Dipsacus asper* Wall. J. Ethnopharmacol. **108(2)**: 188-192.

Huang, X., Liu, Y., Song, F., Liu, Z. and Liu, S. 2009. Studies on principal components and antioxidant activity of different Radix astragali samples using high-performance liquid chromatography/electrospray ionization multiple-stage tandem mass spectrometry. Talanta **78(3)**: 1090-1101.

Huang, X.Z., Wang, Y.H., Yu, S.S., Fu, G.M., Liu, Y., Fan, L.H. and Hu, Y.C. 2005. Iridioid glycosides and grayanane diterpenoids from the roots of *Craibiodendron henryi*. J. Nat. Prod. **68(11)**: 1646-1650.

Huang, Y.L., Tsai, W.J., Shen, C.C. and Chen, C.C.2005. Resveratrol derivatives from the roots of *Vitis thunbergii*. J. Nat. Prod. **68(2)**: 217-220.

Huang, Y. L., Yeh, P. Y., Shen, C. C. and Chen, C. C. 2003. Antioxidant flavonoids from the rhizomes of *Helminthostachys zeylanica*. Phytochem. **64**: 1277-1283.

Huang, W.Y., Cai, Y.Z., Xing, J., Corke, H. and Sun, M. 2007. A potential antioxidant resource: Endophytic Fungi from medicinal plants. Economic Bot. **61(1)**: 14-30.

Huda-Faujan, N., Noriham, A., Norrakiah, A.S. and Babji, A.S. 2007. Antioxidative activities of water extracts of some Malaysian herbs. ASEAN Food J. **14**: 61-68.

Huda-Faujan, N., Noriham, A., Norrakiah, A.S. and Babji, A.S. 2009. Antioxidant activity of plants methanolic extracts containing phenolic compounds. African J. Biotech. **8(3)**: 484-489.

Hudson, B.J.F. 1990. Food antioxidants. Elsevier Applied Science, London.

Hunter, D.C, Greenwood, J., Zhang, J. nd Skinner, M.A. 2011. Antioxidant and 'natural protective' properties of kiwifruit. Curr. Trop. Med. Chem. **11(14)**: 1811-1820.

Hurrell, R.F., Manju, R. and Cook, J.D. 1999. Inhibition of non-haem iron absorption in man by polyphenolic-containing beverages. British J. Nutr. **81**: 289-295.

Hussain, A.I., Anwar, F., Sherazi, S.T.H. and Przybylski, R. 2008. Chemical composition, antioxidant and antimicrobial activities of basil (*Ocimum basilicum*) essential oils depends on seasonal variations. Food chem. **108**: 986-995.

Hussain, H.E.M.A. 2004. Hypoglcemic poglycemic, hypolipidemic and antioxidant properties of combination of curcumin from *Curcuma longa* Linn., and partially purified product from *Abroma angusta* Linn. in streptozotocin induced diabetes mellitus. Chemistry Biology Interface: Synergistic New Frontiers, New Delhi, India P25-16, November 21-26, 2004.

Hussain, K., Ismail, Z., Sadikun, A. and Ibrahim, P. 2009. Antioxidant and anti-TB activities, phenolic and amide content of standardized extracts of *Piper sarmentosum* Roxb. Nat. Prod. Res. **23(3)**: 238-249.

Hussain, K., Ismail, Z., Sadikun, A. and Ibrahim, P. 2010. Standardization and *in vivo* antioxidant activity of ethanol extracts of fruit and leaf of *Piper sarmentosum.* Planta Medica **76(5)**: 418-425.

Hussain, M.A. 2011. Antidiabetic and antioxidant activity of *Jasione montana* extract in streptozotocin-induced diabetic rats. Saudi Pharmaceutical Journ. **16(3-4)**: 214-221.

Hussain, M.A. and Othman, S.M. 2010. Structure and antioxidant activity relationship and free radical scavenging capacity of hesperidin. J. Med. Chem. **1**: 1-10.

Hussain, S., Jamil, M., Ullah, F., Khan, A., Ullah, F., Arfan, M., Ahmad, S. and Khatoon, L. 2010. Antimicrobial and antioxidant activities of the plant *Heliotropium strigosum.* Afr. J. Biotech. **9(45)**: 7738-7743.

Hutadilok-Towatana, N., Chaiyamutti, P., Panthong, K., Mahabusarakam W. and Rukachaisirikul, V. 2006. Antioxidative and free radical scavenging activities of some plants used in Thai folk medicine. Pharm. Biol. **44**: 221-228.

Hwang, B.Y., Kam, H.S., Lee,.H., Hong, Y.S., Ro,J.S., Lee, K.S., Lee, J.J. 2001. Antioxidant benzoylaed flavan-3-ol glycoside from *Celastrus orbiculatus*. J. Nat. Prod. **64(1)**: 82-84.

Hwang, E.I., Kim, J.R., Jeong, T.S., Lee, S., Rho, M.C. and Kim, S.U. 2006. Phellinsin A from *Phellinus* sp. PL3 exhibits antioxidant activities. Planta Medica **72(6)**: 572-575.

Hwang, J-K, Choi, E-M. and Lee, J.H. 2005. Antioxidant activity of *Litsea cubeba.* Fitoterapia **76**: 684-686.

Hwang, J.M., Wang, C.J., Chou, F.P., Tseng, T.H., Hsieh, Y.S., Lin, W.L. and Chu, C.Y. 2002. Inhibitory effect of berberine on tert-butyl hydroperoxide-induced oxidative damage in rat liver. Arch. Toxicol. 76: 664-670. doi: 10.1007/s00204-002-0351-9.

Hwang, P.A., Wu, C.H., Gau, S.Y., Chien, S.Y. and Hwang, D.F. 2010. Antioxidant and immune stimulating activities of hot-water extract from seaweed *Sargassum hemiphyllum*. J. Mar. Sci. Technol. **18**: 41-46.

IARC 1996. Preamble to the IARC handbook of cancer chemoprevention. In: B.W.Stewart, McGregor and P. Keihues (eds.) Principles of Chemoprevention. IARC Scientific publication, No. 139, pp 1-12. Intl. Agency for Res. Cancer, Lyon, France.

Ibanes, E., Aranzazu Oca, Gonzalo de Murga, sara Lopez-Sebastian and Guillermo Reglero. 1999. Supercritical fluid action and fractionation of different preprocessed rosemary plants. J. Agric. Food Chem. **47**: 1400-1404.

Ibañez, E., Kubátová, A., Señoráns, F.J., Cavero, S., Reglero, G. and Hawthorne, S.B. 2003. Subcritical water extraction of antioxidant compounds from rosemary plants. J. Agric Food Chem. **51(2)**: 375-382.

Ichikawa, M., Ryu, K., Yoshida, J., Ide, N., Kodera, Y., Sasaoka, T. and Rosen, R.T. 2003. Identification of six phenylpropanoids from garlic skin as major antioxidants. J. Agric. Food Chem. **51(25)**: 7313-7317.

Ichikawa, M., Yoshida, J., Ide, N., Sasaoka, T., Yamaguchi, H. and Ono, K. 2006. Tetrahydro-beta-carboline derivatives in aged garlic extract show antioxidant properties. J. Nutr. **136 (3)**: 726-731.

Ide, N. and Itakura, Y. 1996. Scavenging effect of aged garlic extract and its constituents on active oxygen species. Phytother. Res. **10**: 40-41.

Ide, N., Ryu, K., Imai, J., Moriguchi, T., Nagae, S., Sumiyoshi, H., Matsuura, H. and Itakura, Y. 1996. Antioxidant and radical scavenging effects of aged garlic extract and its constituents. Phytomedicine **3**(Suppl.): 137.

Igile, G.O., Oleszek, W., Jurzysta, M., Burda, S., Fafunso, M. and Fasanmade, A.A. 1994. Flavonoids from *Vernonia amygdalina* and their antioxidant activities. J. Agric. Food Chem. **42(11)**: 2445-2448. doi: 10.1021/jf00047a015.

Ilavarasan, R., Mallika, M. and Venkataraman, S. 2005. Anti-inflammatory and antioxidant activities of *Cassia fistula* Linn. bark extracts Afr. J. Trad. CAM **2(1)**: 70-85.

Ilavarasan, R.M., Vasudavan, Anbazhagan, S. and Venkataraman, S. 2003. Antioxidant activity of *Thespesia populnea* bark extract against carbon tetrachloride induced liver injury in rats. J. Ethanpharmcol. **87**: 227-230.

Iliya, I., Ali, Z., Tanaka, T., Iinuma, M., Furusawa, M., Nakaya, K., Murata, J., Darnaedi, D., Matsuura, N. and Ubukata, M. 2003. Stilbene derivatives from *Gnetum gnemon* Linn. Phytochem. **62(4)**: 601-606. doi: 10.1016/s0031-9422(2)00570-2.

Ill Min, C., Song Joong, Y., Jung Tae, K., Jae Gyun, G., Jae Duck, S. and Huang Soo, S. 1995. Test of superoxide dismutase characteristics and antioxidant activity in *Perilla* leaves. Korean J. Crop Sci. **40**: 504-511.

Imai, J., Ide, N., Nagae, S., Moriguchi, T., Matsuura, H. and Itakura, Y. 1994. Antioxidant and radical scavenging effects of aged garlic extracts and its constituents. Planta Med. **60(5)**: 417-420.

Inatani, R., Nakatani, N., Fuwa, H. and Seto, H. 1982. Structure of a new antioxidative phenolic diterpene isolated from rosemary. Agric. Biol. Chem. **46**: 1661-1666.

Inselman, G., Blohmer, A., Kottany, W., Nellessen, U., Hanel, W. and Heidman, H.T. 1995. Modification of cisplatin-induced renal *p*-aminohippurateuptake alteration and lipid peroxidation by thiols, *Ginkgo biloba* extract, deferoxamine and torbafylline. Nephron **70(4)**: 425-429.

Institute of Medicine (US). 1998. Dietary reference intakes. Proposed definition and plan for review of dietary antioxidants and related compounds. Food and nutrition board. Washington DC: IOM. National Academy Press, Washington DC, USA.

Ioana, V.S., Prokish, J. and socaciu, J. 2007. Variation in antioxidant activity of fresh and dried leaves of *Viscum album* using automatic FRAP Assay. Bull. USAMV-CN **63**: 367-372.

Ionska-Rys, E.J., Zalewska-Korona, M. and Kalbarczyk, J. 2009. Antioxidant capacity, ascorbic acid and phenolics content in wild edible fruits. J. Fruit Ornamental Plant Res. **17(2)**: 115-120.

Iorizzi, M., Lanzotti, V., De Marino, S., Zollo, F., Blanco-Molina, M., Macho, A., Munoz, E. 2001. New glycosides from *Capsicum annuum* L. var. *acuminatum*. Isolation, structure determination, and biological activity. J. Agric. Food Chem. **49**: 2022-2029.

Ip, S.P., Tse, A.S.M., Poon, M.K.T., Ko, K.M. and Ma, C.Y. 1997. Antioxidant activities of *Polygonum multiflorum* Thunb. *in vivo* and *in vitro*. Phytother. Res. **11(1)**: 42-44.

Ip, S.P., Woo, K.Y., Lau, O.W., Che, C.T. 2004. Hepatoprotective effect of *Sabina przewalski* against menadione-induced toxicity. Phytother. Res. **18(4)**: 329-331.

Iqbal, S., Asghar, M.N., Khan, I.U. and Zia, I. 2010. Antioxidant potential profile of extracts from different parts of black mulberry. Asian J. Chem. **22(1)**: 353-364.

Iqbal, S., Bhanger, M.I., Akhtar, M., Anwar, F., Ahmed, K.R. and Anwer, T. 2006. Antioxidant properties of methanolic extracts from leaves of *Rhazya stricta*. J. Med Food. Summer. **9 (2)**: 270-275.

Isabelle, M., Lee, B.L., Lim, M.T., Koh, W.-P., Huang, D. and Ong, C.N. 2010. Antioxidant activity and profiles of common fruits in Singapore. Food Chem. **123**: 77-84.

Islam, K.M., Howlader, M.A., Kundu, G.C., Bulbul, I.J. and Ahsan, M.R. 2010. Free radical scavenging activity of chloroform and ethyl acetate extracts of leaves of *Piper betle* Linn. Libyan Agr. Res. Center J. Int. **1(6)**: 384-387.

Ismail, A., Marjan, Z.M. and Foong, C.W. 2004. Total antioxidant activity and phenolic contents in selected vegetables. Food Chem. **87**: 581-586.

Ismail, C., Atill, T. and Ismail, I. 2009. Hepatoprotective role and antioxidant capacity of pomegranate (*Punica granatum*) flowers infusion against trichloroacetic acid-exposed in rats. Food Chem. Technol. **47(1)**: 145-149.

Ismail, H.I., Chan, K.W., Mariod, A.A. and Ismail, M. 2010. Phenolic content and antioxidant activity of cantaloupe (*Cucumis melo*) methanolic extracts. Food Chem. **119(2)**: 643-647.

Ismail, M., Manickam, E., Danial, A.M., Rahmat, A. and Yahaya, A. 2000. Chemical composition and antioxidant activity of *Strobilanthes crispus* leaf extract J. Nutr. Biochem. **11 (11)**: 536-542.

Ismaili, H., Milella, L., Fkih-Tetouani, S., Ilidrissi, A., Camporese, A., Sosa, S., Altinier, G., Della Loggia, R. and Aquino, R. 2004. *In vivo* topical anti-inflammatory and *in vitro* antioxidant activities of two extracts of *Thymus satureioides* leaves. J. Ethnopharmacol. **91(1)**: 31-36.

Ivanova, D., Gerova, D., Chervenkov, T. and Yankova, T. 2005. Polyphenols and antioxidant capacity of Bulgarian medicinal plants. J. Ethnopharmacol. **96 (1-2)**: 145-150.

Ivanova, S.A., Nomura, K., Malfanov, I.L., Sklyar, I.V. and Ptitsyn, L.R. 2010. Isolation of a novel catechin from *Bergenia* rhizome that has pronounced lipase-inhibiting and antioxidative properties. Fitoterapia **82**: 212-218.

Iwai K. 2008. Antidiabetic and antioxidant effects of polyphenols in brown alga *Ecklonia stolonifera* in genetically diabetic KK-A(y) mice. Plant Foods Hum Nutr. **63(4)**: 163-169.

Iwai, K., Onodera, A. and Matsue, H. 2001. Antioxidant activity and inhibitory effect of Gamazumi (*Viburnum dilatatum* Thunb.) on oxidative stress in rats. Int. J. Food Sci. Nutr. **52(5)**: 443-451.

Iwai, K., Onodera, A. and Matsue, H. 2003. Mechanism of preventive action of *Viburnum dilatatum* Thunb. (gamazumi) crude extract on oxidative damage in rats subjected to stress. J. Sci. Food Agric. **83(15)**: 1593-1599. doi: 10.1002/jsfa.1580.

Iwalewa, E.O., Adewunmi, C.O., Omisore, N.O.A., Adebanji, O.A., Azike, C.K., Adigun, A.O., Adesina, O.A. and Olowoyo, O.G. 2005. Pro- and anti-oxidant effects and cytoprotective potentials of nine edible vegetables in South West Nigeria. J. Med. Foods **8**: 539-544.

Iwalokun, B.A., Efedede, B.U., Alabi-Sofunde, J.A., Oduala, T., Magbagbeola, O.A. and Akinwande, A.I. 2006. Hepatoprotective and antioxidant activities of *Vernonia amygdalina* on acetaminophen induced hepatic damage in mice. J. Med. Food **9**: 524-530.

Iwasaki, M., Ishikawa, C., Maesuura, Y., Ohhashi, E. and Harada, R. 1984. Antioxidants of all spices and clove. Sagami Joshi Daigaku Kiyo **48**: 1-6.

Iwashima, M., Mori, J., Ting, X., Matsunaga, T., Hayashi, K., Shinoda, D., Saito, H., Sankawa, U. and Hayashi, T. 2005. Antioxidant and antiviral activities of plastoquinones from the brown alga *Sargassum micracanthum*, and a new chromene derivative converted from the plastoquinones. Biol Pharm Bull. **28(2)**: 374-377.

Izawa, S. and Inoue, Y. A. 2004. Screening system for antioxidants using thioredoxin-deficient yeast: discovery of thermostable antioxidant activity from *Agaricus blazei* Murill. Appl Microbiol. Biotechnol. **64(4)**: 537-42.

Jabrane, A., Jannet, H.B., Mastouri, M., Mighri, Z. and Casanova, J. 2010. Chemical composition and *in vitro* evaluation of antioxidant and antibacterial activities of the root of *Ridolfia segetum* (L.) Moris from Tunisia. Nat. Prod. Res. **24**: 491-499.

Jacob, R. 1995. The integrated antioxidant system. Nutr. Res. **15**: 755-766.

Jadhao, K.D., Badwe, D.V., Wadekar, M.P. and Iqbal, M. 2009. Evaluation of antioxidants from *Tinospora cordifolia*. Biosci. Biotech. Res. Asia **6(2)**: 803-806.

Jadhav, H.R. and Bhutani, K.K. 2002. Antioxidant properties of Indian medicinal plants. Phytother. Res. **16(8)**: 771-773.

Jaehrig, S.C., Rohn, S., Kroh, L.W., Fleischer, L.G. and Kurz, T. 2007. *In vitro* potential antioxidant activity of (1→3),(1→6)-beta-D-glucan and protein fractions from *Saccharomyces cerevisiae* cell walls. J. Agric. Food Chem. **55(12)**: 4710-4716.

Jagetia, G.C. and Baliga, M.S. 2004. The evaluation of nitric oxide scavenging activity of certain Indian medicinal plants *in vitro*. A preliminary study. J. Med. Food **7**: 343-348.

Jagetia, G.C., Baliga, M.S., Venkatesh, P., Ullor, J.N. 2003. Influence of ginger rhizome (*Zingiber officinale* Rosc.) on survival, glutathione and lipid peroxidation in mice after whole-body exposure to gamma radiation. Radiat. Res. **160**: 584-592.

Jagetia, G.C., Rao, S.K., Baliga, M.S. and Babu, K.S. 2004. The evaluation of nitric oxide scavenging activity of certain herbal formulations *in vitro*: a preliminary study. Phytother. Res. **18(7)**: 561-565.

Jagetia, G.C. and Venkatesha, V.A.K. 2005. Enhancement of radiation effect by *Aphanamixis polystachya* in mice transplanted with ehrlich ascites carcinoma. Biol. Pharm. Bull. **28(1)**: 69-77.

Jagtap, U.B., Panaskar, S.N. and V.A.apat. 2010. A preliminary evaluation of phytochemicals and antioxidant activity of jackfruit pulp. Intern. Symp. on the New Horizons of Botany, Shivaji University, Kolhapur, Abstract. Pp. 276.

Jain, A., Jain, S., Deb, L., Jain, A. and Dutt, K.R. 2007. Antioxidant activity of green herb *Tabernaemontana divaricata* leaves. Int. J. Green Pharm. **1(2)**: 46-48.

Jain, A., Soni, M., Deb, L., Jain, A., Rout, S.P., Gupta, V.B., Krishna, K.L. 2008. Antioxidant and hepatoprotective activity of ethanolic and aqueous extracts of *Momordica dioica* Roxb. leaves. J Ethnopharmacol. **115(1)**: 61-66.

Jain, P. and Agarwal, R.K. 2009. *In vitro* antioxidant activity of some Indian medicinal plants. Hamdard Medicus **52(20)**: 79-85.

Jain. P.G. 2009. Antioxidant activity of ethyl acetate extract of *Cassia auriculata* leaves. J. Pharmacy Res. **2(9)**:

Jain, S., Dixit, V.K., Malaviya, N. and Ambawatia, V. 2009. Antioxidant an hepatoprotective activity of ethanolic and aqueous extracts of *Amorphophallus campanulatus* Roxb. tubers. Acta Pol. Pharm. **66(4)**: 423-428.

Jain, S., Gupta, A., Malviya, N. and Suhur, H. 2009. Comparative antioxidant potential screening of polyherbal formulations. Adv. Pharmacol. Toxicol. **10(1)**: 101-110.

Jain, S.C., Jain, R. and Singh, R. 2009. Antioxidant activity of mimenosides isolated from *in vitro* regenerated plants of *Mimosa hamata* Willd. Indian J. Plant Physiol. **14(2)**: 124-129.

Jain,S.C., Singh, R. and Jain, R. 2008. Antimicrobial and Antioxidant potentials of *Verbesina encelioides* (Cav.) Benth. et Hook. Fil ex Gray. Research Journal of Medicinal Plant, **2**: 61-65.

Jain, V.C., Patel, N.M., Shah, D.P., Patel, P.K. and Joshi, B.H. 2010a. Antioxidant and antimicrobial activities of *Bryophyllum calycinum* Salisb. leaf. Pharmacology online **1**: 393-405.

Jain, V.C., Patel, N.M., Shah, D.P., Patel, P.K. and Joshi, B.H. 2010b. Antioxidant and antimicrobial activities of *Terminalia crenulata* Roth bark. Pharmacology online **2**: 204-217.

Jainu, M. and Shyamala Devi. 2004. Antioxidant effect of methanolic extract of *Solanum nigrum* berries on aspirin induced gastric mucosal injury. Indian J. Clinical Biochem. **19(1)**: 57-61.

Jainu, M., Vijay Mohan, K. and Shyamala Devi, C.S. 2006. Protective effect of *Cissus quadrangularis* on neutrophil mediated tissue injury induced by aspirin in rats. J. Ethnopharmacol. **104(3)**: 302-305.

Jaishree, V. and Badami, S. 2009. Antiedematogenic and free radical scavenging activity of Swertiamarin isolated from *Enicostemma axillare*. Pharmacol. **75(1)**: 12-17.

Jaishree, V. and Badami, S. 2010. Antioxidant and hepatoprotective effect of swertiamarin from *Enicostemma axillare* against D-galactosamine induced acute liver damage in rats. J. Ethnopharmacol. **130(1)**: 103-106.

Jaishree, V., Badami, S., Krishnamurthy, P.T. 2010. Antioxidant and hepatoprotective effect of the ethyl acetate extract of *Enicostemma axillare* (Lam.) Raynal against CCl_4-induced liver injury in rats. Indian J. Exp. Biol. **48(9)**: 896-904.

Jaishree, V., Badami, S. and Suresh, B. 2008. *In vitro* antioxidant activity of *Enicostemma axillare*. J. Health Sci. **54**: 524.

Janakkat, S. and Al-Thanibat, O. 2008. Antilipidoperoxidative effect of three edible plants extracts: *Viscum album, Arum dioscoridis* and *Eminium spiculatum*. J. Food Quality **31**: 1-12. DOI: 10.111/j.1745-4557.2007.00180.x

Jang, H.-D., Chang, K.-S., Chang, T.-C. and Hsu, C.-L. 2010. Antioxidant potentials of buntan pumelo (*Citrus grandis* Osbeck) and its ethanolic and acerified fermentation products. Food Chem. **118(3)**: 554-558.

Jang, H.S., Kim, K.R., Choi, S.W., Woo, M.H. and Choi, J.H. 2007. Antioxidant and antithrombus activities of enzyme treated *Salicornia herbacea* extracts. Ann. Nutr. Metab. **51(2)**: 119-125.

Jang, H.W., Ka, M.-H. and Lee, K.-G. 2008. Antioxidant activity and characterization of volatile extracts of *Capsicum annuum* L. and *Allium* spp. Flavour and Fragrance J. **23(3)**: 178-184. doi: 10.1002/ffj. 1872.

Jang, I.C. Park, J.H., Park, E., Park, H.R. and Lee, S.C. 2008. Antioxidative and antigenotoxic activity of extracts from Cosmos (*Cosmos bipinnatus*) flowers. Plant Foods Hum. Nutr. **63**: 205-210.

Janicsak, G., Hohmann, J., Zupko, I., Forgo, P., Redei, D., Falkay, G. and Mathe, I. 2003. Diterpenes from the aerial parts of *Salvia candelabrum* and their protective effects against lipid peroxidation. Planta medica **69(12)**: 1156-1159.

Jariyawat, S., Kigpituck, P., Suksen, K., Chuncharunee, A., Chaovanalikit, A.and Piyachaturawat, P. 2009. Protection against cisplatin-induced nephrotoxicity in mice by *Curcuma comosa* Roxb. ethanol extract. J. Nat. Medicines 63($): 430-436.

Jatwa, R. and Kar, A. 2009. Amelioration of metformin-induced hypothyroidism by *Withania somnifera* and *Bauhinia purpurea* extracts in Type 2 diabetic mice. Phytother. Res. **23(8)**: 1140-1145.

Javanmardi, J., Stushnoff, C., Locke, E. and Vivanco, J.M. 2003. Antioxidant activity and total phenolic content of Iranian *Ocimum* accessions. Food Chem. **83**: 547-550.

Javidnia, K., Gholami, M., Firuzi, O., Oser, N.H. and Kandemir, A. 2010. Antimicrobial and antioxidant activity and chemical composition of essential oil of *Tanacetum macrophllum*. J. Essential Oil Res. **22(2)**: 186-188.

Jayajothi, E., Elavarasu, T., Hamsaveni, M. and Sridhar, S.K. 2004. Antioxidant activity and total phenolic content of Triphala churna. Nat. Prod.Sci. **10(1)**: 16-19.

Jayakumar, T., Ilayaraja, M., Sridhar, M.P., Sivalokanathan, S. and Bharath Prasad, T.R. 2009. Protective effect of *Adhatoda vasica* on D-galactosamine induced liver damage in rats with reference to lipid peroxidation and antioxidant status. J. Nat. Remedies **9(1)**: 91-98.

Jayakumar, T., Ramesh, E. and Geraldine, P. 2006. Antioxidant activity of the oyster mushroom, *Pleurotus ostreatus*, on CCl_4-induced liver injury in rats. Food Chem. Toxicol. **44(12)**: 1989-1996.

Jayakumar, T., Sakthivel, M., Thomas, P.A. and Geraldine, P. 2008. *Pleurotus ostreatus*, an oyster mushroom, decreases the oxidative stress induced by carbon tetrachloride in rat kidneys, heart and brain. Chem. Biol. Interact. **176(2-3)**: 108-120.

Jayakumar, T., Thomas, P.A. and Geraldine, P. 2007. Protective effect of ant extract of the oyster mushroom, *Pleurotus ostreatus*, on antioxidants of major organs of aged rats. Exp. Gerontol **42(3)**: 183-191.

Jayamalar, P. and Suhaila, M. 1998. Antioxidative activities of Malaysian plants. Malays. Appl. Biol. **27**: 56-58.

Jayaprakasha, G.K., Girennavar, B. and Patil, B.S. 2008. Radical scavenging activity of Rio red grapefruits and sour orange fruit extracts in different *in vitro* model system. Bioresour. Technol. **36**: 703-725.

Jayaprakasha, G.K. and Jaganmohan Rao, L. 2000. Phenolic constituents from lichen

Parmotrema stuppeum (Nyl.) Hale and their antioxidant activity. Z. Naturforsch. **55c**: 1018-1022.

Jayaprakasha,.G.K., Jena, B.S., Negi, P.S. and Devi, K. 2002. Evaluation of antioxidant activities and antimutagenicity of turmeric oil: a byproduct from curcumin production. A. Naturforsch., C. Biosci. **57c**: 828-835.

Jayaprakasha, G.K., Rao, L.J. and Sakariah, K.K. 2004. Antioxidant activity of Flavidin in different *in vitro* model systems. Bioorganic Med. Chem. **12**: 5141-5146.

Jayaprakash, G.K., Selvi, T. and Sakariah, K.K. 2003. Antibacterial and antioxidant activities of grape (*Vitis vinifera*) seed extract. Food Res. Int. **36**: 117-122.

Jayaprakasha, G.K., Singh, R.P. and Sakariah, K.K. 2001. Antioxidant activity of grape seed extracts on peroxidation models *in vitro*. Food. Chem. **73**: 285-290.

Jayashree, G. K., Muraleedhara, G., Sudarslal, S. and Jacob, V.B. 2003. Anti-oxidant activity of *Centella asiatica* on lymphoma-bearing mice. Fitoterapia **74(5)**: 431-434.

Jayasinghe, C., Gotoh, N., Aoki, T. and Wada, S. 2003. Phenolics composition and antioxidant activity of sweet basil (*Ocimum basilicum* L.). J. Agric. Food Chem. **51**: 4442-4449.

Jayasinghe, L., Balasooriya, B.A., Padmini, W.C., Hara, N. and Fujimoto, Y. 2004a. Geranyl chalcone derivatives with antifungal and radical scavenging properties from the leaves of *Artocarpus nobilis*. Phytochem. **65(9)**: 1287-1290.

Jayasinghe, L., Puvanendran, S., Hara, N. and Fujimoto, Y. 2004b. Stilbene derivatives with antifungal and radical scavenging properties from the stem bark of *Artocarpus nobilis*. Nat. Prod. Res. **18(6)**: 571-574.

Jayathilakan, K., Sharma, G.K., Radhakrishna, K. and Bawa, A.S. 2007. Antioxidant potential of synthetic and natural antioxidants and its effect on warmed-over-flavour in different species of meat. Food Chem. **105(3)**: 908-916.

Je, J.-Y., Park, P.-J., Kim, E.-K., Park, J.-S., Yoon, H.D., Kim, K.-R. and Ahn, C.B. 2009. Antioxidant activity of enzymatic extracts from the brown seaweed *Undaria pinnatifida* by electron spin resonance spectroscopy. LWT-Food Sci. Tech. **42(4)**: 874-878.

Jeetendra, N. and Manish, B. 2011. Correlation of antioxidant activity with phenolic content and isolation of antioxidant compound from *Lygodium flexuosum* (L.) Sw. extracts. Int. J. Pharmacy and Pharmaceutical Sci. **3(2)**:

Jeller A.H., Silva D.H.S., Liao, L.M., De bolzani V. and Furlan M. 2004. Antioxidant phenolic and quinonemethide triterpernes from *Cheiloclinum cognatum*. Phytochem. **65**: 1977-1982.

Jensen, S.R., Gotfredsen, C.H., Harput, U.S. and Saracoglu, I. 2010. Chlorinated iridoid glucosides from *Veronica longifolia* and their antioxidant activity. J. Nat. Prod. **73(9)**: 1593-1596.

Jeong, C-H., Choi, G.N., Kim, J.H., Kwak, J.H., Kim, D.O., Kim, Y.J. and Heo, H.J. 2010. Antioxidant activities from the aerial parts of *Platycodon grandiflorum*. Food Chem. **118(2)**: 278-282.

Jeong, C.H., Heong, H.R., Choi, G.N., Kim, D,O, Lee, U and Heo, H.J. 2011. Neuroprotective and antioxidant effects of caffeic acid isolated from *Erigeron annuus* leaf. Chin. Med. **6**: 25.

Jeong, G.S., Lee, D.S., Li, B., Byun, E., Kwon, D.Y., Park, H. and Kim, T.C. 2010. Protective effect of sauchinone by upregulating heme oxygenase-1, via the P38 MAPK and Nrf2/ARE pathways in HepG2 cells. Planta Medica **76(1)**: 41-46.

Jeong, I.Y. Lee, H.J., Park, Y.D., Jin, C.H., Choi, D.S. and Byun, M.W. 2009. Effects of gamma irradiation on total polyphenols, radical scavenging activities and decolourization of *Nelumbo nucifera* extracts. Radiation Phys. and Chem. **78(7-8)**: 575-577.

Jeong, J.B., De Lumen, B.O. and Jeong, H.J. 2010. Lunasin peptide purified from *Solanum nigrum* L. protects DNA from oxidative damage by suppressing the generation of hydroxyl radical via blocking fenton reaction. Cancer Lett. **293(1)**: 58-64.

Jeong, J.B., Ju, S.Y., Park, J.H., Lee, J.R., Yun, K.W., Kwon, S.T., Lim, J.H., Chung, G.Y. and Jeong, H.J. 2009a. Antioxidant activity in essential oils of *Cnidium officinale* Makino and *Ligusticum chuanxiong* Hort and their inhibitory effects on DNA damage and apoptosis induced by ultraviolet B in mammalian cell. Cancer Epidemiol. **33(1)**: 41-46.

Jeong, J.B., Park, J.H., Lee, H.K., Ju, S.Y., Hong, S.C., Lee, J.R., Chung, G.Y., Lim, J.H. and Jeong, H.J. 2009b. Protective effect of the extracts from *Cnidium officinale* against oxidative damage induced by hydrogen peroxide via antioxidant effect. Food Chem. Toxicol. **47(3)**: 525-529.

Jerez, M., Selga, A, Sineiro, J., Torres, J.L. and Nune, M.J. 2007. A comparison between bark extracts from *Pinus pinaster* and *Pinus radiata*: Antioxidant activity and procyanidin composition. Food Chem. **100**: 439-444.

Jeyakumar, S., Nalaini, N. and Venugopal, M. 1999. Antioxidant activity of ginger in rats fed a high fat diet. Med. Sci. Res. 27: 341-344.

Jia, H.L., Ji, Q.L., Xiang, S.L., Zhang, P.H., Zhu, G.L. and Wang, X.H. 2010. Chemical composition and antioxidant, antimicrobial activities of the essential oils of *Thymus marschallinus* Will. and *Thymus proximus* Serg. J. Food sci. **75**: E59-E65.

Jiang, B., Yang,., Nuntanakorn, P., Balick, M.J., Kronenberg, F. and Kennelly, E.J. 2005. The value of plant collections in ethnopharmacology: a case study of an 85 year old black cohosh (*Actaea racemosa* L.) sample. J. Ethnopharmacol. **96(3)**: 521-528.

Jiang, H., Zhan, W.Q., Liu, X., Jiang, S.X. 2008. Antioxidant activities of extracts and flavonoid compounds from *Oxytropis falcata* Bunge. Nat. Prod. Res. **22(18)**: 1650-1656.

Jiang, P.-P. *et al.*, 2010. Extraction of antioxidant active components of *Crotalaria mucronata* Desv. and optimization of activity determination system. J. Anhui Agric. Sci. DOI: CNKI: SUN: AHNY.0.2010-28-052.

Jiang, R.W., Lau, K.M., Lam,.M., Yam, W.S., Leung, L.K., Choi, K.L., aye, M.M.Y., Mak, T.C.W., Woo, K.S. and Fung, K.P. 2005. A comparative study on aqueous root extracts of *Pueraria thomsonii* and *Pueraria lobata* by antioxidant assay and HPLC fingerprint analysis. J. Ethnopharmacol. **96(1-2)**: 133-138.

Jiang, S.-H., Wang, C.-L., Chen, Z.Q., Chen, M.-H., Wang, Y.R., Liu, C.-J., Zhou, Q.-L. and Li, Z.J. 2009. Antioxidant properties of the extract and subfractions from old leaves of *Toona sinensis* Roem. (Meliaceae). J. Food Biochem. 33(3): 425-441. doi: 10.1111/j.1745-4514.2009.00226.x

Jiang, Y.L., Piao, H.S. and Li, G. 2008. Study on antioxidant activity of constituents from mulberry leaf. Zhong Yao Cai **31(4)**: 519-522. (in Chinese).

Jimenez, I., Garrido, A., Bannach, R., Gotteland, M. and Speisky, H. 2000. Protective effects of bolding against free radical-induced erythrocyte lysis. Phytother. Res. **14(5)**: 339-343.

Jimenez-Escrig, A., Dragsted, L.O., Daneshvar, B., Pulido, R. and Saura-Calixto, F. 2003. *In vitro* antioxidant activities of edible artichoke (*Cynara scolymus* L.) and effect on biomarkers of antioxidants in rats. J. Agric. Food Chem. **51(18)**: 5540-5545.

Jimoh, F.O., Adedapo, A.A. and Afolyan, A.J. 2011. Comparison of the nutritive value, antioxidant and antibacterial activities of *Sonchus asper* and *Sonchus oleraceus*. Rec. Nat. Prod. **5(1)**: 29-42.

Jimoh, F.O., Sofidiya, M.O. and Afolayan, A.J. 2007. Antioxidant properties of the methanol extracts from the leaves of *Paullinia pinnata*. J. Ned. Food **10(4)**: 707-711.

Jin, W.Y., Thuong, P.T., Su, N.D., Min, B.S., Son, K.H., Chang, H.W., Kim, H.P., Kang, S.S., Sik, D.E. and Bae, K. 2007. Antioxidant activity of cleomiscosins A and C isolated from *Acer okamotonum*. Arch. Pharm. Res. **30(3)**: 275-281.

Jin, Y.S., Heo, S.I., Lee, M.J., Rhee, H.I and Wang, M.H. 2005. Free radical scavenging and hepatoprotective actions of *Quercus aliena* acorn extract against CCl_4-induced liver. Free Radic. Res. **39(12)**: 1351-1358.

Jin, Y.S., KLee, M.J., Han, W., Heo, S.I., Sohn, S.I. and Wang, M.H. 2006. Antioxidant effects and hepatoprotective activity of 2,5-dihydroxy-4,3′-di(beta-d-glucopyranosyloxy)-trans-stilbene from *Morus bombycis* Koidzumi roots on CCl_4-induced liver damage. Free Radic. Res. **40(9)**: 986-992.

Jiratanam, T. and Liu, R.H. 2004. Antioxidant activity of processed table beets (*Beta vulgaris* var. *conditiva*) and green beans (*Phaseolus vulgaris* L.). J. Agric. Food Chem. **52(9)**: 2659-2670.

Jitendra, Y.N. 2009. *In vitro* antioxidant activity of *Achyranthes aspera* L. J. Pharm. Res. **2**: 1402-1403.

Jitoe, A., Masuda, T., Tengah, I.G.P., Suprapta, D.N., Gara, I.W., and Nakatani, N. 1992. Antioxidant activity of tropical ginger extracts and analysis of the contained curcuminoids. J. Agric. Food Chem. 40: 1337-1340.

Jodynis-Liebert, J., Matlawska, I., Bylka, W. and Murias, M. 2005. Protective effect of *Aquilegia vulgaris* L. on APAP-induced oxidative stress in rats. J. Ethnopharmacol. **97(2)**: 351-358.

Joharapurkar, A.A., Zambad, S.P., Wanjari, M.M. and Umathe, S.N. 2003. *In vitro* evaluation of antioxidant activity of alcoholic extract of *Rubia cordifolia* Linn. and its influence on ethanol-induced immunosuppression. Indian J. Pharmacol. **35**: 232-236.

Johnson, M.K., Alexander, K.E., Lindquist, N. and Loo, G. 1999. A phenolic antioxidant from the fresh water orchid, *Habenaria repens*. Comp. Biochem. Physiol. Pharmacol.Toxicol. Endocrinol **122(2)**: 211-214.

Jollow, D.J., Mitchell, J.R., Zampaglione, N. and Gillete, J.R. 1974. Bromobenzene induced liver necrosis: protective role of glutathione and evidence for 3,4 bromobenzeneoxide as the hepatotoxic intermediate. Pharmacol. **11**: 151–169.

Jordan, M.J., Sanchez-Gomez, P., Jimenez, J.F., Quilez, M. and Sotomayor, J.A. 2010. Chemical composition and antiradical activity of the essential oil from *Satureja intricata, S. obovata* and their hybrid *Satureja* x *delpozoi*. Nat. Prod. Commun. **5(4)**: 629-634.

Jose, J.K. and Kuttan, R. 1995. Antioxidant activity of *Emblica officinalis*. J. Clin. Biochem. Nutr. **19**: 63-70.

Jose, N., Ajith, T.A. and Janardhanan, K.K. 2002. Antioxidant, anti-inflammatory and antitumor activities of *Pleurotus florida*. Curr. Sci. **79**: 941-943.

Jose, N. and Janardhan, K.K. 2000. Antioxidant and antitumor activity of *Pleurotus florida*. Curr. Sci **78(7)**: 941-943.

Joseph., J.A., Shukitt-Hale, B., Denisova, N.A., Bielinski, D., Martin, A., McEwen, J.J. and Bickford, P.C. 1999. Reversal of age related declines in neuronal signal transduction, cognitive and motor behavioral deficits with blue berry, spinach or strawberry dietary supplementations. J. Neurosci. **19**: 8114-8121.

Joseph, S., Sabulal, B., George, V., Smina, T.P. and Janardhanan, K.K. 2009. Antioxidative and anti-inflammatory activities of the chloroform extract of *Ganoderma lucidum* found in South India. Sci. Pharm. **77**: 111-121.

Joshi, J.P., Kamat, J., Mohan, H., Chintalwar, G. and Chattopadhyay, S. 2003. Antioxidant effects of *Convolvulus pluricaulis* in rat brain mitochondria against oxidative damage induced by gamma radiation and photosensitization. BARC Newslatter **249**: 183-187.

Joshi, R., Kamat, J.P. and Mukherjee, T. 2007. Free radical scavenging reactions and antioxidant activity of embelin: biochemical and pure radiolytic studies. Chem. Biol. Interact. **167(2)**: 125-134.

Joshi, R., Sharma, A. and Lal, B. 2010. Analysis of antioxidant activity in extracts of *Calotropis procera* (Ait.) R.Br. J. Appl. Biosci. **17**: 899-903.

Joshi, S.C., Verma, A.R., Mathela, C.S. 2010. Antioxidant and antibacterial activities of the leaf essential oils of Himalayan Lauraceae species. J. Chem. Toxicol. **48**: 37-40.

Joubert, E., Richards, E.S., Merwe, J.D., De Beer, D., Manley, M. and Gelderblom, W.C. 2008. Effect of species variation and processing on phenolic composition and *in vitro* antioxidant activity of aqueous extracts of *Cyclopia* spp. (Honeybush tea). J. Agric. Food Chem. **56(3)**: 954-963.

Joubert, E., Winterton, P., Britz, T.J., Gelderblom, W.C. 2005. Antioxidant and pro-oxidant activities of aqueous extracts and crude polyphenolic fractions of rooibos (*Aspalathus linearis*). J. Agric. Food Chem. **53(26)**: 10260-10267.

Jovanovic, S.V., Steenken, S., Simic, M.G. and Hara, Y. 1993. Antioxidant properties of flavonoids: reduction potentials and electron transfer reactions of flavonoids radicals. In: Rice-Evans and Packer, L. (eds.). Flavonoids in Health and disease. Marcel Dekker Inc. New York, pp 139-161.

Joy, K.L., Kuttan, G. and Kuttan, R. 1999. Antioxidant and anticarcinogenic activity of Lycovinan indigenous herbal preparation. Indian J. Exp. Biol. **37**: 1177-1181.

Joyeux, M., Moitier, F. and Fleurentin, J. 1995. Screening of antiradical, antilipidoperoxidant and hepatoprotective effects of nine plant extracts used in Carribean folk medicine. Phytother. Res. **9**: 228-230.

Ju, E.M., Lee, S.E., Hwang, H.J. and Kim, J.H. 2004. Antioxidant and anticancer activity of extract from *Betula platyphylla* var. *japonica*. Life Sci. **74(8)**: 1013-1026.

Ju, H.K., Chung, H.W., Hong, S.-S, Park, J.H., Lee, J. and Kwon, S.W. 2010. Effect of steam treatment on soluble phenolic content and antioxidant activity of the Chaga mushroom (*Ionotus obliquus*). Food Chem. **119(2)**: 619-625.

Juan-Badaturuqe, M., Habtemariam, S., Jackson, C. and Thomas, M.J. 2009. Antioxidant principles of *Tnacetum vulgare* L. aerial parts. Nat. Prod. Commun. **4(11)**: 1561-1564.

Juice Beauty Antioxdant serum – 2 oz – AM/PH (http: //www.skin-one.com/juice-beauty-antioxdant-serum.html).

Juliani, H.R. and Simon, J.E. 2002. Antioxidant activity of basil. In: Janick, J. and Whipkey, A. (eds.). Trends in new crops and new uses. ASHS Press, Alexandria, Va.pp. 575-579.

Jun, M., Tohru, U., Jianzhang, L. and Takeshi, F. 2004. Identification and evaluation of antioxidant activities of bamboo extracts. For Stud. China **6**: 1-5.

Jun, W.J., Seong, H.S., Chun, H., Lim, E.J., Kim, K.I. and Cho, H.Y. 2007. Determination of antioxdative potentials of *Acanthopanax sessiliflorus* (Rupr. et Maxim.) Seem. In differentiated HL-60 cells. Food Chem. **105(4)**: 1557-1563.

Jung, C.H., Jun, C.Y., Lee, S., Park, C.H., Cho, K. and Ko, S.G. 2006. *Rhus verniciflua* Stokes extract: Radical scavenging activities and protective effects on H_2O_2-induced cytotoxicity in macrophage RAW264.7 cell lines. Biol. Pharm. Bll. **29(8)**: 1603-1607.

Jung, H.A., Jung, M.J., Kim, J.Y., Chung, H.Y. and Choi, J.S. 2003. Inhibitory activity of flavonoids from *Prunus davidiana* and other flavonoids on total ROS and hydroxyl radical generation. Archives of Pharmaceutical Res. **26(10)**: 809-815.

Jung, H.A., Kim, A.R., Chung, H. and Choi, J.S. 2002. *In vitro* antioxidant activity of some selected *Prunus* species in Korea. Arch Pharm. Res. **25(6)**: 865-872.

Jung, H.A., Kim, J.E., Chung, H.Y. and Choi, J.S. 2003. Antioxidant principles of *Nelumbo nucifera* stamens. Arch. Pharm. Res. **26(4)**: 279-285.

Jung, H.A., Lee, H.J., Kim, Y.A., Park, K.E., Ahn, J.W., B.J.Lee, Moon, S.G. and Seol, Y.W. 2004. Antioxidant activity of *Artemisia capillaris* Thunberg. Food Science and Biotechnology **13**: 328-331.

Jung, H.A., Woo, J.J., Jung, M.J., Hwang, G.S. and Choi, J.S. 2009. Kaempferol glycosides with antioxidant activity from *Brassica juncea*. Arch. Pharm. Res. **32(10)**: 1379-1384.

Jung, H.W., Ka, M.-H. and Lee, K.-G. 2008. Antioxidant activity and characterization of volatile extracts of *Capsicum annuum* L. and *Allium* spp. Flavour Fragrance J. **23(3)**: 178-184. doi: 10.1002/ffj.1872.

Jung, K.A., Song, T.C., Han, D., Kim, I.H., Kim, Y.E. and Lee, C.H. 2005. Cardiovascular protective properties of kiwifruit extracts *in vitro*. Biol. Pharm. Bull. **28(9)**: 1782-1785.

Jung, M., Chung, H., Choi, J., Jung, M.J., Chung, H.Y. and Choi, J.S. 2001. Antioxidant activity of roasted defatted perilla seed. Nat. Prod. Sci. **7**: 72-75.

Jung, M.J., Chung, H.Y., Choi, J.H., Choi, J.S. 2003. Antioxidant principles from the needles of red pine, *Pinus densiflora*. Phytother Res.**17(9)**: 1064-1068.

Jung, M.J., Chung, H.Y., Kang, S.S., Choi, J.H., Bae, K.S., Choi,.S. 2003. Antioxidant activity from the stem bark of *Albizia julibrissin*. Arch. Pharm. Res. **26(6)**: 458-462.

Jung, M.J., Heo, S.I. and Wang, M.-H. 2008. Free radical scavenging and total phenolic contents from methanolic extracts of *Ulmus davidiana*. Food Chem. **108(2)**: 482-487. doi: a0.1016/j.foodchem.2007.10.081.

Jung, M.J., Heo, S.I. and Wang, M.-H. 2010. HPLC analysis and antioxidant activity of *Ulmus davidiana* and some flavonoids. Food Chem. **120(1)**: 313-318.

Jung, M.S., Kang, K.A., Zhang, R., Chae, S., Yoo, B.S., Yang, Y.T., Lee, N.H., Park, J.W. and Hyum, J.W. 2006. Protective activity against ionizing radiation of antioxidative plants indigenous to Korea. Nat. Prod. Sci. **12(1)**: 1-7.

Jung, S.H., Lee, J.M., Lee, H.J., Kim, C.Y., Lee, E.H. and Um, B.H. 2007. Aldose reductase and advanced glycation end products inhibitory effect of *Phyllostachys nigra*. Biol. Pharm. Bull. **30(8)**: 1569-1572.

Jung, S.H., Lee, Y.S., Lim, S.S., Lee, S., Shin, K.H. and Kim, Y.S. 2004. Antioxidant activities of isoflavones from the rhizomes of *Belamcanda chinensis* on carbon tetrachloride-induced hepatic injury in rats. Archives of Pharmacal. Res. **27(2)**: 184-188.

Jung, S.J., Kim, D.H., Hong, Y.H., Lee, J.H., Song, H.N., Rho, Y.D., and Baek, N.I. 2007. Flavonoids from the flower of *Rhododendron yedoense* var. *poukhanense* and their antioxidant activities. Arch. Pharm. Res. **30 (2)**: 146-150.

Juntachote, T. and Berghofer, E. 2005. Antioxidative properties and stability of ethanolic extracts of Holy basil and Galangal. Food Chem. **92(2)**: 193-202. doi: 10.1016/j.foodchem.2004.04.044.

Juntachote, T., Berghofer, E., Siebenhandl, S. and Bauer, F. 2006. The antioxidative properties of Holy basil and Galangal in cooked ground pork. Meat Sci. **72(3)**: 446-456.

Juteau, F., Masotti, V., Bessiere, J.M., Dherbomez, M. and Viano, J. 2002. Antibacterial and antioxidant activities of *Artemisia annua* essential oil. Fitoterapia **73**: 532-535.

Jyothi, T.M., Shankaraiah, M.M., Prabhu, K., Lakshminarasu, S., Śrinivas, G.M. and Ramachandra, S. 2008. Hepatoprotective and antioxidant activity of *Euphorbia tirucalli.* Iranian J. Pharmacol. Therapeutics **7**: 25-30.

Ka, M.H., Choi, E.H., Chun, H.S. and Lee, K.G. 2005. Antioxidant activity of volatile extracts isolated from *Angelica tenuissimae* roots, peppermint leaves, pine needles and sweet flag leaves. J. Agric. Food Chem. **53(10)**: 4124-4129

Kadam U. S., Ghosh S. B., De Strayo, Suprasanna, P., Devasagayam, T. P. A., Bapat, V. A. 2008. Antioxidant activity in sugarcane juice and its protective role against radiation induced DNA damage. Food Chem. **106(3)**: 1154-1160.

Kadifkova-Panovska, T., Kulevanova, S. and Stefova, M. 2005. *In vitro* antioxidant activity of some *Teucrium* species (Lamiaceae). Acta Pharm. **55**: 207-214.

Kähkonen, M.P., Hopia, A.I., Heinonen, M. 2001. Berry phenolics and their antioxidant activity. J. Agric. Food Chem. **49**: 4076-4083.

Kähkonen, M.P., Hopia, A.I. and Heinonen, M. 2001. Berry phenolics and their antioxidant activity. J. Agr. Food Chem. **49**: 4076-4082.

Kähkonen, M.P., Hopia, A.I., Vuorela, H.J., Rauha, J.P., Pihlaja, K., Kujala, T.S. and Heininen, M. 1999. Antioxidant activity of plant extracts containing phenolic compounds. J. Agric. Food Chem. **47**: 3954-3962.

Kaleem, M., Asif, M., Ahmed, Q. U. and Bano, B. 2006. Antidiabetic and antioxidant activity of *Annona squamosa* extract in streptozotocin induced diabetic rats. J. Singapore Med. **47(8)**: 670-675.

Kaleem, M., Sheema, Sarmad, H. and Bano, B. 2005. Protective effects of *Piper nigrum* and *Vinca rosea* in alloxan induced diabetic rats. Indian J. Physicl. Pharmacol. **49(1)**: 65-71.

Kalia, K., Sharma, K., Singh, H.P., and Singh, B. 2008. Effects of extraction methods on phenolic contents and antioxidant activity in aerial parts of *Potentilla atrosanguinea* Lodd. and quantification of its phenolic constituents by RP-HPLC. J. Agric. Food Chem. **56 (21)**: 10129-34.

Kalim, M.D., Bhattacharya, D., Banerji, A. and Chattopadhyay, S. 2010. Oxidative DNA damage preventive activity and antioxidant potential of plants used in Unani system of medicine. BMC Complementary and Alternative Medicine. **10**: 77.

Kalkar, S. and Sambare, S. 2005. Phytochemical analysis of *Aegle marmelos* (L) Corr. In relation to its antioxidant activity. International Conference on Modern Trends in Plant Sciences with special reference to the role of Biodiversity in conservation, Amaravathi, p 8-907.

Kallimanis, G. and Philiianos, S. 1980. Plant Med. Phytother. **14**: 233.

Kalliora, A.C. and Andrikopoulous, K. 2005. Effect of *Alkanna albugam* root on LDL oxidation. A comparative study with species of the Lamiaceae family. Phytother. Res. **19(12)**: 1077-1079.

Kalt, W., Forney, C.F., Martin, A. and Prior, R.L. 1999. Antioxidant capacity, vitamin C, phenolics and anthocyanins after fresh storage of small fruits. J. Agric. Food chem. **47**: 4638-4644.

Kalt, W., McDonald, J.E. and Donner, H. 2000. Anthocyanins, phenolics and antioxidant capacity of processed lowbush blueberry products. J. Food Sci. **65**: 390-393.

Kamalakkannan, N., Mainzen, S. and Prince, P. 2003. Effect of *Aegle marmelos* Correa (Bael) fruit extract on tissue antioxidants in streptozotocin diabetic rats. Indian J. Exp. Biol **44**: 1285-1288.

Kamat, J.P., Boloor, K.K., Devasagayam, T.P.A. and Venkatachalam, S.R. 2000. Antioxidant properties of *Asparagus racemosus* against damage induced by gamma-radiation in rat liver mitochondria. J. Ethnopharmacol. **71**: 425-435.

Kamath, B.S., Srikanta, B.M., Dharmesh, S.M, Sarada, R. and Ravishankar, G.A. 2008. Ulcer preventive and antioxidative properties of astaxanthin from *Haematococcus pluvialis*. Eur. J. Pharmacol. **590(1-3)**: 87-95.

Kamath, V. and Rajini, P.S. 2007. The efficacy of cashew nut (*Anacardium occidentale* L.) skin extract as a free radical scavenger. Food Chem. **103**: 428-433.

Kamatou, G.P.P. 2006. Indigenous *Salvia* species: An investigation of their pharmacological activities and phytochemistry. Ph.D. thesis, Faculty of Health Sciences, University of Whitewaterstrand, South Africa.

Kamatou, G.P.P, Viljoen, A.M. and Steenkamp, P. 2010. Antioxidant, anti-inflammatory activities and HPLC analysis of South African *Salvia* species. Food Chem. **119(2)**: 684-688.

Kamble, S.M., Kamalakar, P.L. and Vaidya, S. 1998. Assessment of antioxidant potentials of *Coccinia indica* in human diabetes. Indian Med. Gaz. pp. 221-223.

Kambouche, N., Merah, B., Bellahouel, S., Bouayed, J., Dicko, A., Derdour, A., Younos, C. and Soulimani, R. 2008. Chemical composition and antioxidant potential of *Ruta montana* L. essential oil from Algeria. J. Medicinal Food. **11(3)**: 593-595. doi: 10.1089/jmf.2007.0515.

Kamden Waffo, A.F., Mulholland, D., Wansi, J.D., Mbaze, L.M., Powo, R., Mpondo, T.N., Fomum, Z.T., Konig, W. and Nkengfack, A.E. 2006. Afzelixanthones A and B, two new prenylated xanthones from *Garcinia afzelii* Engl. (Guttiferae). Chem. Pharm. Bull. **54(4)**: 448-451.

Kamiloglu, O., Ercisli, S., Sengul, M., Toplu, C. and Serce, S. 2009. Total phenolics and antioxidant activity of jujube (*Ziziphus jujube* Mill.) genotypes selected from Turkey. Afr. J. Biotechnol. **8(2)**: 303-307.

Kamkaen, N. and Wilkinson, J.M. 2009. The antioxidant activity of *Clitoria ternatea* flower petal extracts and eye gel. Phytother. Res. **23(11)**: 1624-1625.

Kamkar, A., Javan, A.J., Asadi, F. and Kamalinejad, M. 2010. Antioxidative effect of Iranian *Mentha pulegium* extracts and essential oil in suflower oil. Food and Chemical Toxicol. **48(7)**: 1796-1800.

Kamkar, A., Monfared, M., Javan, A.J., Asadi, F. and Aknodzadeh, A. 2010. Antioxidative effects of liquid and organic extracts from Iranina nettle (*Urtica dioica* L.) As. J. Food Ag-Ind. **3(5)**: 491-497.

Kanatt, S.R., Chander, R. and Sharma, A. 2007. Antioxidant potential of mint (*Mentha spicata* L.) in radiation processed lamb meat. Food chem. **100**: 451-458.

Kandaswami, C. and Middleton, E. Jr. 1994. Free radical scavenging and antioxidant activity of plant flavonoids. Adv. Exp. Med. Biol. **366**: 351-376.

Kandaswami, C. and Middleton, E. Jr. 1997. Flavonoids as antioxidants. In F.Shahid (ed.). Natural antioxidants, Chemistry, Health effects and application. AOCS Press pp. 182-203.

Kang, H.W., Yu, K.W., Jun, W.J., Chang, I.S., Han, S.B., Kim, H.Y.and Cghao, H.Y. 2002. Isolation and characterization of alkyl peroxyl radical scavenging compound from leaves of *Laurus nobilis*. Biol. Pharm. Bull. **25**: 102-108.

Kang, H.M., Kim, J.R., Jeong, T,.S., Choi, S.G., Ryu, Y.H., Oh, G.T., Baek, N.I. and Kwon, B.M. 2006. Cyclic diarylheptanoids inhibit cell-mediated low-density lipoprotein oxidation. Nat. Prod. Res. **20(2)**: 139-143.

Kang, H.J., Chawla, S.P., Jo, C., Kwon, J.H., and Byun, M.W. 2006. Studies on the development of functional powder from citrus peel. Bioresour Technol. **97 (4)**: 614-620.

Kang, H.M. and Saltveit. 2002. Antioxidant enzymes and DPPH-radical scavenging activity in chilled and heatshocked rice (*Oryza sativa* L.) seedling radicles. J. Agric. Food Chem. **50**: 513-518.

Kang, J., Chen, R.Y. and Yu, D.Q. 2006. Five new diels-alder type adducts from the stem and root bark of *Morus mongolica*. Planta Medica **72(1)**: 52-59.

Kang, K.A., Bu, H.D., Park, D.S., Go, G.M., Jee, Y., Shin, T. and Hyun, J.W. 2005. Antioxidant activity of ethanol extract of *Callophyllis japonica*. Phytother. Res. **19(6)**: 506-510.

Kang, K.K.A., Bu, H.D., Park, D.S., Go, G.M., Jee, Y., Shin, T. and Hyun, J.W. 2005. Antioxidant activity of ethanol extract of *Callophyllis japonica*. Phytother. Res. **19(6)**: 506-510.

Kang, K.S., Yamabe, N., Kim, H.Y. and Yokozawa, T. 2007. Effect of sun ginseng methanol extract on lipopolysaccharide induced liver injury in rats. Phytomedicine **14(12)**: 840-845.

Kang, K.W., Oh, S.J., Ryu, S.Y., Song, G.Y., Kim, B.-H., Kang, J.S. and Kim, S.K. 2010. Evaluation of the total oxyradical scavenging capacity of catechins isolated from green tea. Food Chem. **121(4)**: 1089-1094.

Kang, T.H., Chao, H., Oh, D.H, Sohan and Kim, Y.C. 2005. Flavonol glycosides with free radical scavenging activity of *Saurus chinensis.* Fitoterapia **76**: 115-117.

Kang, W.-Y., Li, C.-F. and Liu, Y.-X. 2009. Antioxidant phenolic compounds and flavonoids of *Mitragyna rotundifolia* (Roxb.) Kuntze *in vitro*. Med. Chem. Res. **19(9)**: 1222-1232. doi: 10.1007/s00044-009-9265-x.

Kansci, G., Dongo, E. and Genot, C. 2003. 2,2-Diphenyl-1-picrylhydrazyl (DPPH) test demonstrates antiradical activity of *Dorstenia psilurus* and *Dorstenia ciliata* plant extracts. Nahrung **47(6)**: 434-437.

Kanter, M., Coskun, O. and Budancamanak, M. 2005. Hepatoprotective effects of *Nigella sativa* L. and *Urtica dioica* L. on lipid peroxidation, antioxidant enzyme systems and liver enzymes in carbon tetrachloride-treated rats. World J. Gastroenterol. **11(42)**: 6684-6688.

Kanter, M., Meral, I., Dede, S., Cemek, M., Ozbek, H., Ugyan, I. and Gunduz, H. 2003. Effects of *Nigella sativa* L. and *Urtica dioica* L. on lipid peroxidation, antioxidant enzyme systems and some liver enzymes in CCl_4-treated rats. J. Veterinary Med. A **50**: 264-268.

Kao, M.-W.S. 2006. A comparative study of antioxidant and physicochemical properties of blackberry and kiwifruit. M.Sc. Thesis, Auburn Univerrsity.

Kapasakalidis, P.G., Rastall, R.A. and Gordon, M.H. 2006. Extraction of polyphenols from processed black currant (*Ribes nigrum* L.) residues. J. Agric. Food Chem. **54 (11)**: 4016-4021.

Kapoor, I.P.S., Singh, B., Singh, G., Isidorov, V. and Szczepaniak, L. 2008. Chemistry, antifungal and antioxidant activities of Cardamom (*Amomum subulatum*) essential oil and oleoresins. Int. J. Essent. Oil Therap. **2(1)**: 29-40.

Kapoor, I.P.S., Singh, B., Singh, G., Isidorov, V. and Szczepaniak, L. 2009. Chemistry, antimicrobial and antioxidant potentials of *Cinnamomum tamala* Nees et Eberm. (Tejpat) essential oil and oleoresins. Nat. Product Radiance **8(2)**: 106-116.

Kapur, V., Pillai, K.K., Hussain, S.Z. and Balani, D.K. 1994. Hepatoprotective activity of jigrine on liver damage caused by alcohol, carbon tetrachloride and paracetamol in rats. Indian J. Pharmacol. **26**: 35-40.

Karagaezyan, K.G., Vartayam, G.S. and Panosyan, A.G. 1981. Effect of *Bryonia alba* root extract on lipid peroxidation in the liver of rats with alloxan diabetes. Bull. Exp. Biol. Med. **92**: 1044-1046.

Karagozler, A.A., Erdag, B, Emek, C. and Uygun, D.A. 2008. Antioxidant activity and proline content of leaf extracts from *Dorystoechas hastata*. Food Chem. **111(2)**: 400-407.

Karami, M., Salehi, H. and Nghshvar, F. 2008. Study of histopathology and antioxidant activity of methanolic extract of *Feijoa sellowiana* against dosage induced by MDMA in mouse liver. Pharmacology on line **3**: 315-321.

Kardasova, A. and Machova, E. 2006. Antioxidant activity of medicinal plant polysaccharides. Fitoterapia **77**: 367-373.

Karlidag, H., Ercisli, S., Sengul, M. and Tosun, M. 2009. Physico-chemical diversity in fruits of wild-growing sweet cherries (*Prunus avium* L.). Biotechnol. and Biotechnol. Eq. 23: 1325-1329.

Karou, D., Dicko, M.H., Simpore, J. and Traore, A.S. 2005. Antioxidant and antibacterial activities of polyphenols from ethnomedicinal plants of Burkina Faso. African J. Biotech. **4**: 823-828.

Kartal, N., Sokmen, M., Tepe, B., Daferera, D., Polissou, M. and Sokmen, A. 2007. Investigation of the antioxidant properties of *Ferula orientalis* L. with suitable extraction procedure. Food Chem. **100**: 584-589. doi: 10.1016/j.foodchem.2005.09.084.

Karthikeyan, J. and Rani, P. 2003. Enzymatic and non-enzymatic antioxidants in selected *Piper* species. Indian J. Exp. Biol. **41**: 135-140.

Karuppusmy,S., Muthuraja, G. and Rajasekaran, K.M. 2011. Antioxidant activity of selected lesser known edible fruits from Western Ghats of India. Indian J. Nat. Prod. Res. **2(2)**: 174-178.

Karyagina, T., Gukasova, E. and Bairamashvili, D. 2011. Antioxidant activity of extracts from *Thalictrum minus* suspension culture. Russian J. Plant Physiol. **58(4)**: 715-720. doi: 10.1134/S1021443711040078.

Kasar Rahul, P., Laddha, K.S. and Chaudhary, J. 2006. Isolation of marsupin from *Pterocarpus marsupium* and the scavenging activity. J. Trop. Med. Plants **7(2)**: 257-260.

Kataki, M.S. 2010. Antibacteerial activity, *in vitro* antioxidant activity and anthelmintic activity of ethanolic extracts of *Ananas comosus* L. tender leaves. Pharmacology online **2**: 308-319.

Kataki, M.S., Ahmad, M.Z., Awasthi, D., Tomar, B., Mehra, P., Yadav, R.S. and Rajak, P. 2012. *In vitro* Antioxidant profile of *Wedelia calendulacea* leaves. Pharmacologia, **3**: 75-83. DOI: 10.5567/pharmacologia.2012.75.83

Katalinic, V., Milos, M., Kulisic, T and Jukic, M. 2006. Screening of 70 medicinal plant extracts for antioxidant capacity and total phenols. Food Chem. **94**: 550-557.

Katalinic, V., Mozina, S.S., Skroza, D., Generalic, I., Abramovic. H., Milos, M., Ljubenkov, I., Piskernik, S., Pezo, I., Terpinc, P. and Boban, M. 2010. Polyphenolic profile, antioxidant properties and antimicrobial activity of grape skin extracts of 14 *Vitis vinifera* varieties grown in Dalmatia (Croatia). Food Chem. **119(2)**: 715-723.

Katerere, D.R. and Eloff, J.N. 2004. Variation in the chemical composition, antibacterial and anti-oxidant activity of fresh and dried *Acacia* leaf extracts. South Afr. J. Bot. **70**: 303-305.

Katerere, D.R, and Eloff, J.N. 2005. Antibacterial and antioxidant activity of *Sutherlandia frutescens* (Fabaceae), a reputed anti-HIV/AIDS phytomedicine. Phytother. Res. **19 (9)**: 779-81.

Katerere, D.R, and Eloff, J.N. 2008. Antibacterial and anti-oxidant activity of *Hypoxis hemerocallidea* (Hypoxidaceae): can leaves be substituted for corms as a conservation strategy. South African J. Bot. **74(4)**: 613-616.

Kath, R.K., Gupta, R.K. 2006. Antioxidant activity of hydroalcoholic leaf extract of *Ocimum sanctum* in animal models of peptic ulcer. Indian J. Physiol. Pharm. **50(4)**: 391-396.

Kathiriya, A., Das, K., Kumar, E.P. and Mathai, K.B. 2010. Evaluation of antitumor and antioxidant aactivity of *Oxalis corniculata* Linn. against Ehrlich Ascites carcinoma on mice. Iranian J. Cancer Prev. **4**: 157-165.

Katsube, T., Imawaka, N., Kawano, Y., Yamazaki, Y., Shiwaku, K., Yamane, Y. 2006. Antioxidant flavonol glycosides in mulberry (*Morus alba* L.) leaves isolated based on LDL antioxidant activity. Food Chem. **97 (1)**: 25-31.

Katsude, T., Tabata, H., Ohta, Y., Yamasak, Y., Anuurad, E., Shiwaku, K. and Yamane, Y. 2004. Screening for antioxidant activity in edible plant products: Comparison of low-density lipoprotein oxidation assay, DPPH radical scavenging assay, and foli-ciocalteu assay. J. Agric. Food Chem. **52(8)**: 2391-2396.

Kaur, C. and Kapoor, H.C. 2001. Antioxidants in fruits and vegetables – the millenium's health. Int. J. Food Sci. and Tech. **36**: 703-72.

Kaur, C. and Kapoor, H.C. 2002. Antioxidant activity and total phenolic content of some Asian vegetables. Int. J. Food Sci. Tech. **37(2)**: 153-162.

Kaur, G., Alam, M.S., Jabbar, Z., Javed, K. and Athar, M. 2006. Evaluation of antioxidant activity of *Cassia siamea* flowers. J. Ethnopharmacol. **108 (3)**: 340-348.

Kaur, G. Athar, M. and Alam, M.S. 2008. *Quercus infectoria* galls possess antioxidant activity and abrogates oxidative stress-induced functional alterations in murine macrophages. Chem. Biol. Interactions **171(3)**: 272-282.

Kaur, P and Arora, S. 2011. Superoxide anion radical scavenging activity of *Cassia siamea* and *Cassia javanica*. Med. Chem. Res. **20(1)**: 9-15. doi: 10.1007/s00044-009-9274-9.

Kaur, P., Singh, B., Kumar S. and Kaur, S. 2008. *In vitro* evaluation of free radical scavenging activity of *Rubia cordifolia* L. J. Chinese Clinical Medicine. **3(5)**: 278-284.

Kaur, R., Arora, S. and Singh, B. 2008. Antioxidant activity of the phenol rich fractions of leaves of *Chukrasia tabularis* A. Juss. Bioresource Technology. **99 (16)**: .7692-7698. doi: 10.1016/j.biortech.2008.01070.

Kausik, D., Khokra, S.L., Kausik, P., Sharma, C. and Aneja, K.R. 2010. Evaluation of antioxidant and antimicrobial activity of *Abutilon indicum*. Pharmacology online **1**: 102-108.

Kausik, P., Kaushik, D., Sukhbir, L.K., Chetan Sharma, Aneja, K.R., Shalia Khan, Ravinder and Singh, G. 2010. Evaluation of antioxidant and antimicrobial activity of *Madhuca indica*. Pharmacology online **2**: 1-8.

Kavalcioglu, N., Acik, L., Demirci, f., Demirci, B., Demir, H. and Basher, K.H. 2010. Biological activities of *Bellis perennis* volatiles and extracts. Natural Prod. Comm. **5(1)**: 147-150.

Kaviarasan, K., Kalaiarasi, P. and Pugalendi, V. 2008. Antioxidant efficacy of flavonoid-rich fraction from *Spermacoce hispida* in hyperlipidemic rats. J. Appl. Biomed. **6**: 165-176.

Kaviarasan, S., Naik, G. H., Gangabharathi, R., Anuradha, C. V., and Priyadarsini, K.I. 2007. *In vitro* studies on antiradical and antioxidant activities of fenugreek (*Trigonella foenum-graecum*) seeds. Food chem. **103(1)**: 31-37.

Kavimani, S., Keerthiseelan, V.B., Mounnisamy, V.M. and Vetrichelvan, T. 2004. Antioxidant activitiy of medicinal plants – a review. Antiseptic **101(6)** 227-228.

Kawamura, F. and Okada, M. 1992. Antioxidant effect of ginger on the peroxidation of lard in boiled water. Nippon Kasei Gakkaishi **43(1)**: 31-35.

Kaya, G.I., UnverSomer, N., Konyalioglu, S., Yalcin, H.T., Yavasoglu, N.V.K., Sarikaya, B. and Onur, M.A. 2010. Antioxidant and antibacterial activities of *Ranunculus marginatus* var. *trachycarpus* and *R. sprunerianus*. Turk. J. Biol. **34**: 139-146. doi: 10.3906/biy-0809-13

Kayano, S., Kikuzaki, H., Fukitsuka, N., Mitani, T. and Nakaatani, N. 20022. Antioxidant activity of prune (*Prunus domestica* L.) constituent and a new synergist. J. Agric. Food Chem. **50(13)**: 308-3712.

Kayano, S.I., Kikuzaki, H., Yamada, N.F., Aoki, A., Kasamatsu, K., Yamasaki, Y., Ikami, T., Suzuki, T., Mitani, T., Nakatani, N. 2004. Antioxidant properties of prunes (*Prunus domestica* L.) and their constituents. BioFactors. **21(1-4)**: 309-313.

Keawpradub, N., Dej-Dej-adisai, S. and Yuenyongsawad, S. 2005. Antioxidant and cytotoxic activities of Thai medicinal plants names Khaminkhurea: *Arcangelsia flava*, *Coscinium blumeanum* and *Fibraurea tinctoria*. Songklanakari J. Sci. Technol. **27** (Suppl.2): 455-467.

Kedage, W., Tilak, J.C., Dixit, G.B., Devasagayam, T.P. and Mhatre, M. 2007. A study of antioxidant properties of some varieties of grapes (*Vitis vinifera* L.) Crit. Rev. Food Sci. Nutr. **47(2)**: 175-185.

Kedzierska, M., Olas, B., Wachowicz, B., Stochmal, A., Oleszek, W., Jeziorski, A., Piekarski, J. and Glowacki, R. 2009. An extract from berries of *Aronia melanocarpa* modulates the generation of superoxide anion radicals in blood platelets from breast cancer patients. Planta Medica **75(13)**: 1405-1409.

Kelebek, H., Canbas, C. and Selli, S. 2008. Determination of phenolic composition and antioxidant capacity of blood orange juices obtained from cvs. Moro and Sanguinello (*Citrus sinensis* (L.) Osbeck) grown in Turkey. Food Chem. **107**: 1710-1716.

Kelen, M. and Tepe, B. 2007. Screening of antioxidative properties and total phenolic compounds of various extracts of three different seed of grape varieties (*Vitis vinifera* L.) from Turkish Flora. Pak. J. Biol. Sci. 10(3): 403-408. doi: 10.3923/pjbs. 2007.403.408.

Kelen, M. and Tepe, B. 2008. Chemical composition, antioxidant and antimicrobial properties of the essential oils of three *Salvia* species from Turkish flora. Bioresource Technol **99(10)**: 4096-4104.

Kelm, M.A, Nair, M.G, Strasburg, G.M. and DeWitt, D.L. 2000. Antioxidant and cyclooxygenase inhibitory phenolic compounds from *Ocimum sanctum* Linn. Phytomedicine.**7 (1)**: 7-13.

Kelmer, B.A., Alvarez, M. and Bracht, A. 1985. Effects of *Stevia rebaudiana* natural products on rat liver mitochondria. Biochem. Pharmacol. **34**: 873-882.

Kenjale, R.D., Sha, R.K. and Sathaye, S.S. 2007. Anti-stress and anti-oxidant effects of roots of *Chlorophytum borivilianum* Santapau et Fernandes. Indian J. Exp. Bio. **45(11)**: 974-979.

Kerem, Z., Regev-Shoshani, G., Flaishman, M.A. and Sivan, L. 2003. Resveratrol and two monomethylated stilbenes from Israeli *Rumex bucephalophorus* and their antioxidant potential. J. Nat. Prod. **66(9)**: 1270-1272.

Kery, A., Balazs, A., Kursinszki, L., Apati, P., Szoke, E., Blazovics, A., Hagymasi, K. and Nagy, E. 2008. Radical scavenger and antioxidant activities of selected medicinal plants. International conference on Medicinal and aromatic plants (part II).

Kery, A., Lugasi, A., Balazs, A., Feher, E., Pronai, L., Czinner, E., Hagymasi, K., Apati, P., Papp, I., Hevesi-Toth, B. and Blazovics, A. 2004. Free radical scavenger and lipid peroxidation inhibiting effects of medicinal plants used in phytotherapy. Acta Pharma Hung **74(3)**: 158-165.

Kerry, N.L. and Abbey, M. 1997. Red wine and fractionated phenolic compounds prepared from red wine inhibit low density lipoprotein oxidation *in vitro*. Atherosclerosis **135**: 93-102.

Khadeer Ahmed, M.B., Krishna, V. and Dandin, C.J. 2010. *In vitro* antioxidant and *in vivo* prophylactic effects of two gamma-lactones isolated from *Grewia tiliaefolia* against hepatotoxicity in carbon tetrachloride intoxicated rats. Eur. J. Pharmacol. **631(1-3)**: 42-52.

Khadri, A., Serralheiro, M.L.M., Nogueira, J.M.F., Neffati, M., Smiti, S. and Araujo, M.E.M. 2008. Antioxidant and antiacetylcholinesterase activities of essential oils from *Cymbopogon schoenanthus* (L.) Spreng. Food Chem. **109 (3)**: 630-637.

Khadri, A., Neffati, M., Smiti, S., Fale, P., Lino, A.R.L., Serralheiro, M.L.M. and Araujo, M.E.M. 2010. Antioxidant, antiacetylcholinesterase and antimicrobial activities of *Cymbopogon schoenanthus* (L.) Spreng. (lemon grass) from Tunisia. LWT – Food Sci. Technol. **43(2)**: 331-336.

Khairullina, V.R., Garifullina, G.G., Gerchikov, A.Ya., Ostroukhova, L.A. and Babkin, V.A. 2006. Quantitative antioxidant activity of the ethyl acetate extract of *Larix sibirica* bark and its individual components. Chem. Nat. Compounds **42(2)**: 160-163.

Khalaf, N.A., Shakhya, A.K., Al-Oathman, A., El-Agbar, Z. and Farah, H. 2008. Antioxidant activity of some common plants. Turk. J. Biol. **32**: 51-55.

Khalil, M.Y., Moustafa, A.A. and Naguib, N.Y. 2007. Growth, phenolic compounds and Antioxidant activity of some medicinal plants grown under organic farming condition. World Journal of Agricultural Sciences **3(4)**: 451-457.

Khalil, N.M., Petacci, F., Leite, R.R.S., Silva, E., de Souza, G.F., de Oliveira, R.E.L., Costa, F., Rochaa, V.A. and Mendonca Filho, V. 2006. Antioxidant activity of *Vochysia discolor* Mart. flower. Acta Farm. Bonaerense **25(4)**: 564-566.

Khan, B.A., Abraham, A. and Leelamma, S. 1997. Antioxidant effects of curry leaf, *Murraya koenigii* and mustard seeds, *Brassica juncea* in rats fed with high fat diet. Indian J. Exp. Biol. **35(2)**: 148-150.

Khan, M.A., Shahwar, D., Ahmad, N., Khan, Z.D. and Ajib, M. 2010. Chemical constituents of *Carissa opaca* extracts and their evaluation as antioxidants and preservatives in edible oils. Asian. Chem. **22(1)**: 379-388.

Khanavi, M., Hajimahmoodi, M., Cheranghi-Niroomand, M., Kargar, Z., Ajani, Y., Hadjiakhoondi, A. and Oveisi, M.R. 2009. Comparison of the antioxidant activity and total phenolic contents in some *Stachys* species. African J. Biotech **8(6)**: 1143-1147.

Khanum, H., Ramalakshmi, K., Srinivas, P. and Borse, B.B. 2011. Synergistic antioxidant action of Oregano, ajowan and Borage extracts. Food Nutr. Sci. 2(5): doi: 10.4236/fns.2011.25054.

Khare, C.P. 2007. Indian medicinal plants, an illustrated dictionary. Springer India Pvt. Ltd., New Delhi pp. 766.

Khairul, F.K. Nurul Haniza, M. and Zhari, I. 2005. Antioxidative properties of various extract of *Labisia pumila* (Kacip Fatimah). Curr. Trends Persp. **4**: 306-312.

Khlifi, S., El Hachimi, Y., Khalil, A., Es Safi, N. and El Abbouyi, A. 2005. *In vitro* antioxidant effect of *Globularia alypum* L. hydromethanolic extract. Indian J. Pharmacol. **37(4)**: 227-231.

Khomsug, P., Thonfjaroenbuangam, W., Pakdeenarong, N., Suttajit, M. and Chantratikul, P. 2010. Antioxidative activities and phenolic content of extracts from Okra (*Abelmoschus esculentus* L.). Res. J. Biol.Sci. **5(4)**: 310-3113.

Khopde, S. M., K. Indira Priyadarsini, K., Mohan, H., Gawandi, V.B., Satav, J.G., Yakhmi, J.V., Banavaliker, M. M., Biyani, M.K., and Pittal. J.P. 2001. Characterizing the antioxidant activity of amla (*Phyllanthus emblica*) extract. Curr. Sci. **81 (2)**: 185-190.

Kiendrebeogo, M., Dijoux-Franca, M.-G., Lamien, C.E., Meda, A., Wouessidjewe, D. and Nacoulma, O.G. 2005. Acute toxicity and antioxydent property of *Striga hermonthica* (Del.) Benth. (Scrophulariaceae). Afr. J. Biotech **4(9)**: 919-922.

Kiessoun, K., Souza, A., Meda, N.T.R., Coulibay, A.Y., Kiendrebeogo, M., Lamien-Meda, A., Lamidi, M., Millogo-Rasolodimby, J. and Nacoulma, O.G. 2010. Polyphenol contents, antioxidant and anti-inflammatory activities of six Malvaceae species traditionally used to treat Hepatitis B in Burkina Faso. Eur. J. Sci. Res. **44(4)**: 570-580.

Kikuzaki, H., Kawai, Y. and Nakatani, N. 2001. 1,1-Diphenyl-2-picrylhydrazyl radical-scavenging active compounds from greater cardamom (*Amomum subulatum* Roxb.) J. Nutr. Sci. Vitaminol (Tokyo) **47(2)**: 167-171.

Kikuzaki, H., Kawasaki, Y. and Nakatani, N. 1994. Structure of antioxidant compounds in ginger. In: Ho, C.T., Osawa, T., Huang, M.T., Rosen, R.T. (eds.). Food phytochemicals for cancer prevention II. Teas, spices and herbs. American chemical society, Washington DC, pp. 237-243.

Kikuzaki, H., Kayano, S.I., Fukutsuka, N., Aoki, A., Asamatsu, K., Yamasaki, Y., Mitani, T. and Nakatani, N. 2004. Abscisic acid related compounds and lignans in prunes (*Prunus domestica* L.) and their oxygen radical absorbance capacity (ORAC). J. Agric. Food Chem. **52(2)**: 344-349.

Kikuzaki, H. and Nakatani, N. 1988. Structure of a new antioxidative phenolic acid from oregano (*Origanum vulgare* L.). Agric. Biol. Chem. **53**: 519-524.

Kikuzaki, H. and Nakatani, N. 1993. Antioxidant effect of some ginger constituents. J. Food Sci. **58**: 1407-1410.

Kilani, S., Ammar, R.B., In'es, Bouhlel, Abdelwahed, A., Hayder, N., Mahmoud, A. *et al.*, 2005. Investigation of extracts from (Tunisian) *Cyperus rotundus* as antimutagens and radical scavengers. Environmental Toxicology and Pharmacology **20**: 478-484.

Kim, B.J., Kim, J.H., Kim, H.P. and Heo, M.Y. 1997. Biological screening of 100 plant extracts for cosmetic use (II): antioxidative activity and free radical scavenging activity. Int. J. Cosmetic Sci. **19**: 299-307.

Kim, C.Y., Lee, M.-Y. and Park, I. 2006. Antioxidative activities of fractions from *Sedum sarmentosum*. J. Food Sci. Nutr. **11(1)**: 6-9.

Kim, D., Jeong, S. and Lee, Ch. 2003. Antioxidant capacity of phenolic phytochemicals from various cultivars of plums. Food Chem. **81**: 321-326.

Kim, D.K. 2001. Antioxidative components from the aerial parts of *Lactuca scariola* L. Arch. Pharm. Res. **24(5)**: 427-430.

Kim, D.O., Chun, O.K., Kim, Y.J., Moon, H.Y. and Lee, C.Y. 2003. Quantification of polyphenolics and their antioxidant capacity in fresh plums. J. Agr.Food Chem. **51**: 6509-6515.

Kim, D.O. and Zakour, O.I.P. 2004. Jam processing effect on phenolics and antioxidant capacity in Anthocyanin-rich fruits: Cherry, Plum, and raspberry. J. Food Sci. **69(9)**: 395-400.

Kim, G.-N., Shin, J.-G. and Jang, H.-D. 2009. Antioxidant and antidiabetic activity of Dangyuja (*Citrus grandis* Osbeck) extract treated with *Aspergillus saitoi.* Food Chem. **117(1)**: 35-41.

Kim, H., Choi, H.-K., Moon, J.Y., Kim, Y.S., Mosadik, A. and Cho, S.K. 2011. Comparative antioxidant and antiproliferative activities of red and white pitayas and their correlation with flavonoid and polyphenol content. J. Food Sci. **76(1)**: 38-45.

Kim, H., Moon, J.Y., Kim, H., Lee, D.-S., Cho, M., Choi, H.-K. Kim, Y.S., Mosaddik, A. and Cho, S.K. 2010. Antioxidant and antiproliferative activities of mango (*Mangifera indica* L.) flesh and peel. Food chem. **121(2)**: 4229-436.

Kim, H., Yokozawa, T., Cho, E.J., Cheigh, H.S., Choi, J.S. and Chung, H.Y. 2003. *In vitro* and *in vivo* antioxidant effects of mustard leaf (*Brassica juncea*). Phytother. Res. **17(5)**: 465-471.

Kim, H.J., Chen, F., Wang, X., Chung, H.Y. and Jin, Z. 2005. Evaluation of antioxidant activity of vetiver (*Vetiveria zizanioides* L.) oil and identification of its antioxidant constituents. J. Agric. Food Chem. **53(20)**: 7691-7695.

Kim, H.J., Chen, F., Wu, C., Wang, X., Chung, H.Y. and Jin, Z. 2004. Evaluation of antioxidant activity of Australian tea tree (*Melaleuca alternifolia*) oil and its components. J. Agric. Food Chem. **52(10)**: 2849-2854.

Kim, H.J. and Jung, S.H. 2008. Growth characteristics, vitamin C content and antioxidative activity among local strains of *Sedum sarmentosum*. J. Bio-Environmental Control **17(2)**:: 110-115.

Kim, H.J., Kim, E.J., Seo, S.H., Shin, C.G., Jin, C. and Lee, Y.S. 2006. Vanillic acid glycoside and quinic acid derivatives from *Gardenia fructus*. J. Nat. Prod. **69(4)**: 600-603.

Kim, H.J. and Lee, Y.S. 2005. Identification of new dicaffeoylquinic acid from *Chrysanthemum morifolium* and their antioxidant activities. Planta Medica **71(9)**: 871-876.

Kim, H.J., Saleem, M., Seo, S.H., Jin, C. and Lee, Y.S. 2005. Two new antioxidant stilene dimers, parthenostilbenins A and B from *Parthenocissus tricuspidata*. Planta Medica **71(10)**: 973-976.

Kim, I.S., Lee,O.-H. and Kang, S.-N. 2011. The antioxidant activity and the bioactive compound content of *Stevia rebaudiana* water extracts. LWT – Food Sci. Tech. **44(5)**: 1328-1332.

Kim, J.E., Kim, A.R., Chung, H.Y., Han, S.Y., Kim, B.S. and Choi, J.S. 2003. *In vitro* peroxynitrite scavenging activity of diarylheptanoids from *Curcuma longa*. Phytother. Res. **17(5)**: 481-484.

Kim, J.H., Bae, C.H., Park, S.Y., Lee, S.J. and Kim, Y. 2010. *Uncaria rhynchophylla* inhibits the production of nitric oxide and interleukin-1β through blocking nuclear factor kB, Akt, and mitogen-activated protein kinase activation in macrophages. J. Med. Food. **13(5)**: 1133-1140.

Kim, J.H., Lee, H.J., Yeon, S.C., Choi, D.H., Lee, S.S., Kang, J.K., Chae, C.H., Paik, N.W., Lee, K.H. and Cho, M.H. 2004. Antioxidative effects of 7-hydroxy-3-methoxy-cadalene extracted from *Zelkova serrata* on 4-(-methylinitros amino)-1-(3-pyridyl)-1-butanone-induced oxidative stress in A/J mice. Phytother. Res. **18(5)**: 425-427.

Kim, J.H., Ryu, Y.B., Kang, N.S., Lee, B.W., Heo, J.S., Jeong, I.Y. and Park, K.H. 2006. Glycosidase inhibitory flavonoids from *Sophora flavescens*. Biol. Pharm. Bull. **29(2)**: 302-305.

Kim, J.I., Kim, H.H., Kim, S., Lee, K.T., Ham, I.H. and Whang, W.K. 2008. Antioxidative compounds from *Quercus salicina* Blume stem. Arch. Pharm. Res. **31(3)**: 274-278.

Kim, J.S., Kang, S.S., Choi, J.S., Lee, M.W. and Lee, T.S. 1998. Antioxidant components from *Aralia continentalis*. Korean J. Pharmacogn. **28(1)**: 13-17.

Kim, J.W., Lee, B.C., Lee, J.H., Nam, K.C. and Lee, S.C. 2008. Effect of electron-beam irradiation on the antioxidant activity of extracts from *Citrus unshiu* pomaces, Radiation Physics and Chemistry, **77 (1)**: 87-91.

Kim, J.-Y., Kim, S.-S., Oh, T.-H., Baik, J.S., Song, G., Lee, N.H. and Hyun, C.-G. 2009. Chemical composition, antioxidant, anti-elastase and anti-inflammatory activities of *Illicium anisatum* essential oil. Acta Pharm. **59**: 289-300.

Kim, K.-W., Suh, S.-J., Kim, J.-D., Kim, S.-S., Lee, I.-S., Kim, J.K., Chang, G.-T., Kim, D.-S. and Kim, C.-H. 2008. Effects on lipid peroxidation and antioxidative enzymes of *Euonymus alatus* in cultured rat hepatocytes. Basic and Clin. Pharmacol. Toxicol. **104(1)**: 60-70. doi: 10.1111/j.1742-7843.2007.00152.x.

Kim, M.J., Hyun, J.N., Kim, J.A., Park, J.C., Kim, M.Y., Kim, J.G., Lee, S.J., Chun, S.C. and Chung, I.M. 2007. Relationship between phenolic compounds, anthocyanins content and antioxidant activity in colored barley germplasm. J. Agric. Food Chem. **55(12)**: 4802-4809.

Kim, M.R., Lee, J.Y., Lee, H.H., Aryal, D.K., Kim, Y.G., Kim, S.K., Woo, E.R. and Kang, K.W. 2006. Antioxidative effects of quercetin-glycosides isolated from the flower buds of *Tussilago farfara* L. Food Chem. Toxicol. **44**: 1299-1307.

Kim, M.-Y., Iwai, K., Onodera, A. and Matsue, H. 2003. Identification and antiradical properties of anthocyanins in fruits of *Viburnum dilatatum* Thunb. J. Agric. Food Chem. 51(21): 6173-6177. doi: 10.1021/jf034647p.

Kim, M.Y., Seguin, P., Ahn, J.K., Kim, J.J., Chun, S.C., Kim, E.H., Seo, S.H., Kang, E.Y., Kim, S.L., Park, Y.J., Ro, H.M. and Chung, I.M. 2008. Phenolic compound concentration and antioxidant activities of edible and medicinal mushrooms from Korea. J. Agric. Food. Chem. **56(16)**: 7265-70.

Kim, N.H., Tung, S.K., Ra, J.C., Zhao, Y.Z., sohn, D.H. and Kim, Y.H. 2010. Antioxidative and hepatoprotective diarylheptanoids from the bark of *Alnus japonica*. Planta Medica **76(6)**: 626-629.

Kim, N.M., Kim, J., Chung, H.Y. and Choi, J.S. 2000. Isolation of luteolin 7-O-rutinoside and esculetin with potential antioxidant activity from the aerial parts of *Artemisia montana*. Arch. Pharm. Res. **23 (3)**: 237-239.

Kim, O.K., Murakami, A., Nakamura, Y., Ohigashi, H. 1998. Screening of edible Japanese plants for nitric oxide generation inhibitory activities in RAW 264.7 cells. Cancer Lett. **125**: 199-207.

Kim, S.E., Rhyu, D.Y., Yokozawa, Takako and Park, J.C. 2007. Antioxidant effect of *Alisma plantago-aquatica* var. *orientale* and its main component. Korean J. Pharmcol. **38(4)**: 372-375.

Kim, S.-H., Choi, D.-S., Athukorala, Y., Jeon, Y.J., Senevirathne, M. and Rha, C.K. 2007. Antioxidant activity of sulfated polysaccharides isolated from *Sargassum fulvellum*. J. Food Sci. Nutr. **12**: 65-75.

Kim, S.H., Hyun, S.H. and Choung, S.Y. 2008. Antioxidative effects of *Cinnamomum cassiae* and *Rhodiola rosea* extracts in liver of diabetic rats. BioFactors **26**: 209-219.

Kim, S.I., Sim, K.H. and Choi, H.-Y. 2010. A comparative study of antioxidant activity in some Korean medicinal plants used as food materials. Mol. Cell. Toxicol. **6(3)**: 279-285. doi: 10.1007/s13273-010-0038-x.

Kim, S.J., Jin, S. and Ishi, G. 2004. Isolation and structural elucidation of 4-(β-glucopyranosyldisulfanyl) butylglucosinolate from leaves of rocket salad (*Eruca sativa* L.). Bioscience, Biotechnol. Biochem. **68(12)**: 2444-2450.

Kim, S.T., Kim, J.D., Ahn, S.H., Ahn, G.S., Lee, Y.I. and Jeong, Y.S. 2004. Hepatoprotective and antioxidant effects of *Alnus japonica* extracts on acetaminophen-induced hepatotoxicity in rats. Phytother. Res. **18(12)**: 971-975.

Kim, S.Y., Gao, J.J., Lee, W.C., Ryu, K.S., Lee, K.R. and Kim, Y.C. 1999. Antioxidative flavonoids from the leaves of *Morus alba*. Arch. Pharm. Res. **22(1)**: 81-85.

Kim, S.Y., Jeong, S.M., Park, W.P., Nam, K.C., Ahn, D.U. and Lee, S.C. 2006. Effect of heating conditions of grape seeds on the antioxidant activity of grape seed extracts. Food Chem. **97(3)**: 472-479.

Kim, Y., Di Silvestro, R. and Clinton, S. 2004. Effects of lycopene-beadlet or tomato-powder feeding on carbon tetrachloride-induced hepatotoxicity in rats. Phytomedicine **11(2)**: 152-156.

Kim, Y.G. and Park, H.Y. 2004. The effects of pycnogenol on DNA damage *in vitro* and expression of superoxide dismutase and HP1 in *Escherichia coli* SOD and catalase deficient mutant cells. Phytother. Res. **18(11)**: 900-905.

Kimura, T., Suzuki, M., Takenaka, M., Yamagishi, K. and Shinmoto, H. 2004. L-O-caffeoylhomoserine from *Mateuccia struthiopteris*. Phytochem. **65(4)**: 423-426.

Kinghorn, A.D., Su, B.N., Jang, D.S., Chang, L.C., Gu, J.Q., Carcache-Blanco, E.J., Pawluss, A.D., Lee, S.K., Park, E.J., Cuendet, M., Gills, J.J., Bhat, H.S., Meta Greenwood, E., Song, L.L., Jang, M. and Pezzuto, J.M. 2004. Natural inhibitors of carcinogenesis. Planta Med. **70(8)**: 691-705.

Kinoshita, K., Togawa, T., Hiraishi, A., Nakajima, Y., Koyama, K., Narui, T., Wang, L. and Takahashi, K. 2010. Antioxidant activity of red pigments from the lichens *Lethariella sernanderi*, *L. cashmeriana* and *L. sinensis*. J. Nat. Medicines **64(1)**: 85-88.

Kinoshita, S., Inou, Y., Nakama, S., Ichiba, T. and Aniya, Y. 2007. Antioxidant and hepatoprotective actions of medicinal herb, *Terminalia catappa* L. from Okinawa Island and its tannin corilagin. Phytomedicine **14(11)**: 755-762.

Kirakosyan, A., Seymour, E.M., Noon, K.R., Llanes, D.E.U., Kaufman, P.B., Warber, S.L. and Bolling, S.F. 2010. Interactions of antioxidants isolated from tart cherry (*Prunus cerasus*) fruits. Food Chem. **122(1)**: 78-83.

Kirby, A.J. and R.J.Schmidt. 1997. The antioxidant activity of Chinese herbs for eczema and of placebo herbs. J. Ethnopharmacol. **56**: 103-103.

Kirca, A. and Arslan, E. 2008. Antioxidant capacity and total phenolic content of selected plants from Turkey. Int. J. Food Sci. Technol. **43(1)**: 2038-2046.DOI: 10.1111/j.1365-2621.2008.01818.x.

Kirmizibekmez, H., Bassarello, C., Piacente, S., Celep, E., Atay, I., Mercanoglu, G. and Yesilada, E. 2009. Phenolic compounds from *Hypericum calycinum* and their antioxidant activity. Nat. Prod. Commun. **4(4)**: 531-534.

Kiselova, Y., Ivanova, D., Chervenkov, T, Gerova, D., Galunska, B. and Yankova, T. 2006. Correlation between the *in vitro* antioxidant activity and polyphenol content of aqueous extracts from Bulgarian herbs. Phytother. Res. **20**: 961-965.

Kishore, G. S., Kumar, S.B., Ramachandran, M., Saravan, M. and Sridhar, S.K. 2001. Antioxidant and wound healing properties of *Glycyrrhiza glabra* root extracts. Indian Drugs. **38(7)**: 355-357.

Kittagawa, I., Chen W.Z., Hori, K., Harada, E., Yasuda, N., Yoshikawa, M. and Ren, J. 1994. Chemical studies of Chinese-licorice roots. I. Elucidation of five new flavonoid constituents from the roots of *Glycyrrhiza glabra* L. collected in Xinjiang. Chem. Pharm. Bull. Tokyo **42**:

Kitts, D.D., Wijewickreme, A.R. and Hu, C. 2010. Antioxidant properties of a North American ginseng extract. Mol. Cell. Biochem. **203(1-2)**: 1-10.

Kizil, S., Hasimi, N., Tolan, V., Kilinc, E., Karatas, H. 2010. Chemical composition, antimicrobial and antioxidant activities of Hyssop (*Hyssopus officinalis* L.) essential oil. Not. Bot. Hort. Agrobot. Cluj. **38(3)**: 99-103.

Klimczak, I., Malecka, M., Szlachta, M. and Gliszczynska-Swiglo, A. 2007. Effect of storage on the content of polyphenols, vitamin C and the antioxidant activity of orange juices. J. Food Compos. Anal. 20: 313-322.

Klopotek, Y., Otto, K. and Bohm, V. 2005. Processing strawberries to different products alters contents of vitamin C, total phenolics, total anthocyanins and antioxidant capacity. J. Agric. Food Chem. **53**: 5640-5646.

Knorr, R. and Hamburger, N. 2004. Quantitative analysis and anti-inflammatory and radical scavenging triterpenoid esters in evening primrose oil. J. Agric. Food Chem. **52(11)**: 3319-3324.

Ko, H.J., Song, A., Lai, M.N. and Ng, L.T. 2009. Antioxidant and antiradical activities of Wu Ling Shen in a cell free system. Am. J. Chinese Med. **37(4)**: 815-828.

Ko, J.H., Lee, S.J. and Lim, K.T. 2007. Hypolipidemic effect and antioxidant activity of glycoprotein isolated from *Ulmus davidiana* Nakai in Triton WR-1339-treated mouse. Cell Biochem Funct. **25(5)**: 495-500.

Ko, K.M., Yick, P.K., Poon, M.K.T., Chet, C.T., Ng, K.H. and Kong, Y.C. 2006. *Schisandra chinensis*-derived antioxidant activities in 'Sheng Mai San' a compound formulation, *in vivo* and *in vitro*. Phytother. Res. **9(3)** 203-206.

Ko, T.F., Weng, Y.M. and Chiou, R.Y. 2002. Squalene content and antioxidant activity of *Terminalia catappa* leaves and seeds. J. Agric. Food Chem. **50 (19)**: 5343-5348.

Koca, I. and Karadeniz, B. 2009. Antioxidant properties of blackberry and blueberry fruits grown in the Black sea region of Turkey. Sci. Hort. **121**: 447-450.

Koch, H.P. and Loffler, E. 1985. Influence of silymarin and some flavonoids on lipid peroxidation in human platelets. Methods and Findings in Experimental and Clinical Pharmacology **7(1)**: 13-18.

Koduru, S., Grierson, D.S., Aderogba, M.A., Eloff, J.N. and Afolayan, A.J. 2006. Antioxidant activity of *Solanum aculeastrum* (Solanaceae) berries. Int. J. Pharmacol. **2(2)**: 262-264.

Koelzer, J., Pareira, D.A., Dalmarco, J.B., Pizzolatti, M.G. and Frode, t.S. 2009. Evaluation of the anti-inflammatory efficacy of *Lotus corniculatus*. Food chem. **117(3)**: 444-450.

Koflar, M., Bozan, B., Temelli, F. and Baser, K.H.C. 2007. Antioxidant activity and phenolic composition of sumac (*Rhus coriaria* L.) extracts. Food Chem. **103**: 952-959.

Koganov, M. 2009. Parthenolide free bioactive ingredients from feverfew (*Tanacetum parthenium)* and process for their production. US Patent 7537,791.

Kokila, N.R., Chetankumar, N., Gangadhara, N.S., Harsha, R., Dinesh, R. and Thammanna Gowda, S.S. 2010. Antiglycation and antioxidant activity of polysaccharides isolated from fruit extract of pomegranate (*Punica granatum*). Pharmacology online **1**: 821-829.

Koksal, E. and Gulcin, I. 2008. Antioxidant activity of Cauliflower (*Brassica oleracea* L.) Turkish J. Agric. Forest **32**: 65-78.

Kolak, U., Ozturk, M., Ozgokee, F. and Ulubelen A. 2006. Norditerpene alkaloids from *Delphinium linearilobum* and antioxidant activity. Phytochem. **67(19)**: 2170-2175.

Kolayali, S., Kucuk, M., Duran, C., Candam, F. and Dincer, B. 2003. Chemical and antioxidant properties of *Laurocerasus officinalis* Roem. (cherry laurel) fruit grown in the black sea region. J. Agric. Food Chem. 51(25): 7489-7494.

Koleva, I.I., van Beek, T., Linssen, J.P.H., de Groot, A. and Evstatieva, L.N. 2002. Screening of plant extracts for antioxidant activity: a comparative study of three testing methods. Phytochem. Anal. **13**: 8-17.

Koleva, I.I., Linssen, J.P.H., van Beek, T.A., Evstatieva, L.N., Kortenska, V. and Kandjieva, N. 2003. Antioxidant activity screening of extracts from *Sideritis* species (Labiatae) grown in Bulgaria. J. Sci. Food Agric. **83**: 809-819.

Kolhir, V.K., Bykov, V.A., Baginska, A.I., Sokolov S.Y., Glazova, N.G., Leskova, T.E., Sakolovich, G.S., Tjukavkina, N.A., Kolesnik, Yu, A. and Rulanko, I.A. 1996. Antioxidant activity of a dihydroquercetin isolated from *Larix gmelinii* (Rupr.) Rupr. Wood. Phytother. **10(6)**: 478-482.

Konczak, I., Zabaras, D., Dunstan, M., Aguas, P. 2010. Antioxidant capacity and phenolic compounds in commercially grown native Australian herbs and spices. Food Chem. **122(1)**: 260-266.

Kondo, K., Matsumoto, A., Kurata, H., Tanahasi, H., Koda, H., Amachi, T. and Takura, H. 1994. Inhibition of oxidation of low density lipoprotein with red wine. Lancet **344**: 1152.

Konyalioglu, S. and Karamenderes, C. 2004. Screening of total flavonoid, phenol contents and antioxidant capacities of some *Achillea* L. species growing in Turkey. Acta Pharmaceutica Turcia **46**: 163-170.

Konyalioglu, S. and Karamenderes, C. 2005. The protective effects of *Achillea* L. species native to Turkey against H_2O_2-induced oxidative damage in human erythrocytes and leucocytes. J. Ethnopharmacol. **102(2)**: 221-227.

Kordali, S., Kotan, R., Mavi, A., Cakir, A., Ala, A. and Yildirin, A. 2005. Determination of the chemical composition and antioxidant activity of the essential oil of *Artemisia dracunculus* and of the antifungal and antibacterial activities of Turkish *Artemisia absinthium, Artemisia dracunculus, Artemisia santonicum* and *Artemisia spicigera* essential oils. J. Agric. Food Chem. **53**: 9552-9558.

Korotkova, E.I., Avramchik, O.A., Yusubov, M.S., Belousov, M.V. and Andreeva, T.I. 2003. Determination of antioxidant activity of plant extracts by means of cathode voltametry. Pharmaceutical Chem. Journ. **37(9)**: 511-512.

Korycinska,M., Karolina, C., Anna, J. and Arkadiusz, K. 2009. Antioxidant activity of rye bran alkylresorcinols and extracts from whole-grain cereal products. Food Chem. **116(4)**: 1013-1018.

Kosar M., Altinas, A., Kirimer, N., and Baser, K.H.C. 2003. Determination of free radical scavenging activity of *Lycium* extracts. Chem. Nat. Comp. **39**: 531-535.

Kosar M., Bozan B., Temelli F. and Baser K.H.C. 2007. Antioxidant activity and phenolic composition of sumac (*Rhus coriaria* L.) extracts. Food Chem. **103 (3)**: 952-959.

Kosar, M., Demirci, B., Demirci, F. and Baser, K.H. 2008. Effect of maturation on the composition and biological activity of the essential oil of a commercially important *Satureja* species from Turkey: *Satureja cuneifolia* Ten. (Lamiaceae). J. Agric. Food Chem. **56(6)**: 2260-2265.

Kosar, M., Dorman, H.J.D., Bachmayer, O., Baser, K.H.C. and Hiltunen, R. 2003. An improved on-line HPLC-DPPH method for the screening of free radical scavenging compounds in water extracts of lamiaceae plants. Chem. Nat. Comp. **39**: 161-166.

Kosar, M., Dorman, H.J.D., Baser, K.H.C. and Hiltunen, R. 2004. Screening of free radical scavenging compounds in water extracts of *Mentha* samples using a post-column derivation method. J. Agric. Food Chem. **52**: 5004-5010.

Kosar, M., Dorman, H.J.D. and Hiltunen, R. 2005. Effect of an acid treatment of the phytochemical and antioxidant characteristics of extracts from selected Lamiaceae species. Food Chem. **91**: 525-533.

Kosar, M., Goger, F. and Baser, K.H.C. 2008. *In vitro* antioxidant properties and phenolic composition of *Salvia virgata* Jacq. from Turkey. J. Agric. Food Chem. **56**: 2369-2374.

Kose, E.O., Deniz, I.G., Sarikurkcu, C., Aktas, O. and Yavuz, M. 2010. Chemical composition, antimicrobial and antioxidant activities of the essential oils of *Sideritis erythrantha* Boiss. et Heldr. (var. *erythrantha* and *cedretorum* P.H.Davis) endemic in Turkey. Food Chem. Toxicol. **48(10)**: 2960-2965. doi: 10.1016/j.fct.2010.07.033.

Kose, K. and Dogan, P. 1995. Lipoperoxidation induced by hydrogen peroxide in human erythrocyte membranes. 1. Protective effect of *Ginkgo biloba* extract (EGb 761). J. Int. Med. Res. **23(1)**: 108.

Kose, K., Dogan, P., Ascioglu, M. and Ascioglu, C. 1997. *In vitro* antioxidant effect of *Ginkgo biloba* (EGb 761) extract on lipid peroxidation induced by hydrogen peroxides in erythrocytes of Behcet's patients. J. Pharmacol. **75(3)**: 253-258.

Koseoglu, M., Isleten, F., Atay, A. and Kaplan, Y.C. 2009. Effects of acute and subacute garlic supplement administration on serum total antioxidant capacity and lipid parameters in healthy volunteers. Phytother. Res. **24(3)**: 374-378.

Kostova, I. 2001. *Fraxinus ornus* L. Fitoterapia **72(5)**: 471-480.

Kostyuk, V., Potapovich, A., Suhan, T., De Luca, C., Pressi, G., Dal Tosa, R. and Korkina, L. 2008. Plant polyphenols against UV-C-induced cellular death. Planta Medica **74(5)**: 509-514.

Kotamballi, N., Chidambara, M., Ravendra, P., Shingh and Jayaprakasha, G.K. 2002. Antioxidant activities of grape (*Vitis vinifera*) pomace extracts. Agric. Food Chem. **50**: 5909-5914.

Koti Reddy, T., Seshadri, P., Reddy, K.K.R., Jagetia, G.C. and Damodar Reddy, C. 2008. Effect of *Terminalia arjuna* extract on adriamycin-induced DNA damage. Phytother. Res. **22(9)**: 1188-1194.

Kouakou-Siransy, G., Sahpaz, S., Irie-Nguessan, G., Datte,.J., Kablan. J., Gressier, B and Bailleul, F. 2010. Oxygen species scavenger activities and phenolic contents of four West African plants. Food Chem. **118(2)**: 430-435.

Kranner, I., Beckett, R.P., Wornik, S., Zorn, M. and Pfeihofer, H.W. 2002. Revival of a resurrection plant correlates with its antioxidant status. Plant J. **31(1)**: 13-24.

Kranthi, G., Reddy, A.R.N., Kumar, P., Rajyalakshmi, G., Umate, P., Reddy, Y.N. and Sadanandam, A. 2009. Antioxidant and antidabetic activities from leaf extracts of *Streblus asper* Lour. Biotechnology, **3(4)**:

Kren, L., Presser, A., Pradhan, R., Bahr, B., Paper, D.H., Meyer, K.K. and Koopp, B. 2003. Sulfemodin 8-O-β-D-Glucoside, a new sulfated anthraquinone glycoside, and antioxidant phenolic compounds from *Rheum emodi*. J. Nat. Prod. **66(8)**: 1107-1109.

Krigstein, P. and Cederbaum, A.I. 1995. Boldine prevents human liver microsomal lipid peroxidation and inactivation of cytochrome P4502E1. Free Radic. Bio. Med. **18(3)**: 559-563.

Krishna, M.S. and Nair, A.J. 2010. Antibacterial, cytotoxic and antioxidant potential of different extracts from leaf, bark and wood of *Tectona grandis*. Int. J. Pharm. Sci. Drug Res. **2(2)**: 155-158.

Krishna, M.S. and Nair, A.J. 2011. Anthraqunones from leaves of *Tectona grandis*: A detailed study on its antibacterial activity and their biological properties. Int. J. Phytomedicine **3(1)**:

Krishnamoorthy, P., Bhuvaneswari, A., Vimal Rani, A. and Vaithinathan, S. 2003. Lipid peroxidation and antioxidant enzyme activities in rat testis after administration of extract of *Piper longum*. National symposium on Emerging Trends in Indian Medicinal Plantas, Lucknow p. 30.

Krishnamoorthy, P., Vaithinathan, S. and Bhuvaneswari, A. 2006. Protective effect of *Syzygium cumini* (Linn.) Skeels seed extract on lipid peroxidation in alloxan induced diabetic rats. Nat. Prod. Radiance **5(2)**: 103-107.

Kroyer, G.T. 2004. Red clover extract as antioxidant active and functional food ingredient. Innov. Food Sci. Emerg. Tech. **5**: 101-105.

Kruawan, K. and Kangsadalampai, K. 2006. Antioxidant activity, phenolic compound contents and antimutagenic activity of some water extracts of herbs. Thai J. Pharmaceut. Sci. **30**: 28-35.

Kshirsagar, R. and Upadhyay, S. 2009. Free radical scavenging activity screening of medicinal plants from Tripura, Northeast India. Nat. Product Radiance **8(2)**: 117-122.

Ku, C.S., Jang, J.P. and Mun, S.P. 2007. Exploitation of polyphenol-rich pine barks for potent antioxidant activity. J. Wood Sci. **53(6)**: 524-528.

Kubavat, J.B. and Asdaq, S.M.B. 2009. Role of *Sida cordifolia* L. leaves on biochemical and antioxidant profile during myocardinal injury. J. Ethnopharmacol. **124(1)**: 162-165.

Kubo, I., Chaudhuri, S.K., Kubo, Y., Sanchez, Y., Ogura, T., Saito, T., Ishikawa, H. and Haraguchi, H. 1996. Cytotoxic and antioxidative sesquiterpenoids from *Heterotheca inuloides*. Planta Med. **62(5)**: 427-430.

Kubola, J. and Siriamornpun, S. 2008. Phenolic contents and antioxidant activities of bitter gourd (*Momordica charantia* L.) leaf, stem and fruit fraction extracts *in vitro*. Food Chem. **110 (4)**: 881-890.

Kucukboyaci, N., Orhan, I., Sener, B., Nawaz, S.A. and Choudhary, M.I. 2009. Assessment of enzyme inhibitory and antioxidant activities of lignans from *Taxus baccata* L. Z. Naturforsch. **65**: 187-194.

Kucukboyaci, N. and Sener, B. 2010. Biological activities of lignans from *Taxus baccata* L. growing in Turkey. J. Med. Plants Res. **4(12)**: 1136-1140.

Kuglerova, M., Marsik, P., Vanek, T., Kokoska, L., Wanyana-Magayi, O., Van Damme, P. and Grade, J. 2008. Antioxidative activity of Ugandan medicinal plants. African J. Tradi. Complement. Altern. Med. Abstracts of the World Congress on Medicinal and Aromatic plants, Cape Town.

Kukic, J., Petrovic, S.and Niketic, M. 2006. Antioxidant activity of four endemic *Stachys* taxa. Biol. Pharm. Bull. **29(4)**: 725-729.

Kukongviriyapan, U., Luangaram, S. and Leekhaosoong, K. 2007. Antioxidant and vascular protective activities of *Cratoxylum formosum*, *Syzygium gratum* and *Limnophila aromatica*. Biol. Pharm. Bull. **30(4)**: 661-666.

Kukongviriyapan, V., Janyacharoen, T., Kukongviriyapan, U., Laupattarakasaem, P. and Kanokmedhakul, S. 2003. Hepatoprotective and antioxidant activities of *Tetracera loureiri*. Phytother. Res. **17(7)**: 717-721.

Kulisic-Bilusic, T., Blazevic, I., Dejanovic, B., Milos, M. and Pifat, G. 2010. Evaluation of the antioxidant activity of essential oils from Caper (*Capparis spinosa*) and sea fennel (*Crithmium maritimum*) by different methods. J. Food Biochem. **34**: 286-302.

Kulkarni, A., Suzuki, S. and Etoh, H. 2007. Antioxidant compounds from *Eucalyptus grandis* biomass by subcritical liquid water extraction. J. Wood Sci. **54(2)**: 153-157.

Kulkarni, M.P. and Juvekar, A.R. 2009. Effecct of *Alstonia scholaris* (Linn.) R.Br. on stress and cognition in mice. Indian J. Exp. Biol **47(1)**: 47-52.

Kulkarni, R.R., Virkar, A.D. and D'Mello, P. 2008. Antioxidant and anti-inflammatory activity of *Vitex negundo*. Indian J. Pharmaceutical Sci. **70 (6)**: 838-840.

Kulkarni, S.D., Tilak, J.C., Acharya, R., Rajurkar, N.S., Devasagayam, T.P.A. and Reddy, A.V. 2006. Evaluation of the antioxidant activity of wheatgrass (*Triticum aestivum* L.) as a function of growth under different conditions. Phytother. Res. **20 (3)**: 218-227.

Kumar, A. 2009. Antioxidant effect of *Adiantum capillus-veneris* Linn. on human lymphocyte: An *in vitro* study. J. Cell Tiss. Res. **9(2)**: 1899-1902.

Kumar, A., Chauhan, P.K., Bhardwaj, V.S., Kumar, R., Tyagi, A. 2011. *In vitro* antioxidant and phytochemical investigations of ethanolic extracts of *Viola serpens* and *Morus nigra*. J. Chem. Pharm. Res. **3(4)**: 166-171.

Kumar, D. 2011. Anti-inflammatory, analgesic and antioxidant activity of methanolic wood extract of *Pterocarpus santalinus* L. J. Pharmacol. Pharmaacother. **2(3)**: 200-202.

Kumar, D.L., Rao, K.N.V., Madhavi, B.B., Kumar, D.S. and Banji, D. 2011. Antioxidant activity of *Wrightia tinctoria* Roxb. and *Schrebera swietenioides* Roxb. bark extract. J. Pharm. Res. **4(2)**: 396-397.

Kumar, G., Sharmila Banu, G. and Murugesan, A.G. 2008. Effect of *Helicteres isora* bark extracts on heart antioxidant status and lipid peroxidation in streptozotocin diabetic rats. J. Appl. Biomed. **6**: 89-95.

Kumar, G., Sharmila Banu, G., Murugesan, A.G. and Rajasekhara Pandian, N. 2007. Antihyperglycemic and antiperoxidative effect of *Helicteres isora* L. bark extracts in streptozotocin-induced diabetic rats. J. Applied Biomed. **5**: 97-104.

Kumar, G.P. and Singh, S.B. 2011. Antibacterial and antioxidant activities of ethanol extracts of Trans Himalayan medicinal plants. Eur. J. Appl. Sci. **3(2)**: 53-57.

Kumar, K.B.H. and Kuttan, R. 2005. Chemoprotective activity of an extract of *Phyllanthus amarus* against cyclophosphamide induced toxicity in mice. Phytomedicine **12**: 494-500.

Kumar, M.H.V. and Gupta, Y.K. 2002. Antioxidant property of *Celastrus paniculatus* Willd. a possible mechanism in enhancing cognition. Phytomedicine **9**: 302-311.

Kumar, P.P. and Kuttan, G. 2009. *Vernonia cinerea* scavenges free radicals and regulates nitric oxide and proinflammatory cytokines profile in carragenan induced paw edema model. Immunopharmacol. Immunotoxicol. **31(1)**: 94-102.

Kumar, R.S., Sivakumar, T., Sunderam, R.S., Gupta, M., Mazumdar, U.K., Gomathi, P., Rajeshwar, Y., Saravanan, S., Kumar M.S., Murugesh, K. and Kumar K.A. 2005. Antioxidant and antimicrobial activities of *Bauhinia racemosa* L. stem bark. Brazilian J. Med. Biol. Res. **38**: 1015-1024.

Kumar, S., Kumar, D., Singh, N. and Vashist, B.D. 2007. *In vitro* free radical scavenging and antioxidant activity of *Moringa oleifera* pods. J. Herbal Medicine and Toxicol. **1 (2)**: 17-22.

Kumar, S., Malhotra, R. and Kumar, D. 2010. Antidiabetic and free radicals scavenging potential of *Euphorbia hirta* flower extract. Indian J. Pharm. Sci. **72(4)**: 533-537. doi: 10.4103/0250-474x.73921.

Kumar V., Khanna, A.K., Khan, M.M., Singh, R., Singh, S., Chander, R., Mahdi, F., Saxena, J.K., Saxena, S., Singh, V.K. and Singh, R.K. 2009. Hypoglycaemic, lipid lowering and antioxidant activities in root extract of *Anthocephalus indicus* in Alloxan induced diabetic rats. Indian J. Clinical Biochem. **24(1)**: 65-69.

Kumar, V.P., Sashidhara, S., Kumar, M.M. and Sridhara, B.Y. 2000. Effects of *Luffa echinata* on lipid peroxidation and free radical scavenging activity. J. Pharm. Pharmacol. **52**: 891.

Kumar, V.P., Sashidhara, S., Kumar, M.M. and Sridhara, B.Y. 2001. Cytoprotective role of *Solanum nigrum* against gentamycin-induced kidney cell (vero cells) damage *in vitro*. Fitoterapia. **72(5)**: 481-486.

Kumaran, A. and Karunakaran, R.J. 2006a. Antioxidant and free radical scavenging activity of an aqueous extract of *Coleus aromaticus*. Food Chem. **97 (1)**: 109-114.

Kumaran, A. and Karunakaran, R.J. 2006b. Antioxidant activities of the methanolic extract of *Cardiospermum halicacabum*. Pharm. Biol. **44(2)**: 146-151.

Kumaran, A. and Karunakaran, R.J. 2007a. Activity-guided isolation and identification of free-radical-scavenging components from an aqueous extract of *Coleus aromaticus*. Food Chem. **100(1)**: 356-361. doi: 10.1016/j.foodchem.22005.09.051

Kumaran, A and Karunakaran, R.J 2007. Antioxidant activity of *Cassia auriculata* flowers. Fitoterapia **78(1)**: 46-47.

Kumaran, A. and Karunakaran, R. J. 2007. *In vitro* antioxidant activities of methanol extracts of *Phyllanthus* species from India. LWT Food Sci. Tech. **40**: 344-352.

Kumarasamy, Y., Byres, M., Cox, P.J., Delazar, A., Jaspars, M., Nahar, L., Shoeb, M. and Sarker, S.D. 2004. Isolation, structure elucidation and biological activity of flavone 6-C-glycosides from *Alliaria petiolata*. Chem. Nat. Compounds **40(2)**: 122-128.

Kumaraswamy, M.V., Raghavendra, M.P. and Satish, S. 2010. Antioxidant and anti-inflammatory activity of isolated phytoconstituent from *Woodfordia fruticosa* Kurz. J. Pharm. Res. **3(7)**: 1492-1495.

Kumaraswamy, M.V. and Satish, S. 2008. Free radical scavenging activity and lipoxygenase inhibition of *Woodfordia fruticosa* Kurz. and *Betula utilis* Wall. Afr. J. Biotech. **7**: 2013-2016.

Kumaraswamy, Y., Cox, P.J., Jaspars, M., Nahar, L. and Sarker, S.D. 2002. Biological activity of *Glechoma hederacea*. Fitoterapia **73(7-8)**: 721-723.

Kumaraswamy, Y., Cox, P.J., Jaspars, M., Nahar, L. Wilcock, C.C. and Sarker, S.D. 2003. Biological activity of *Euonomus europaeus*. Fitoterapia **74(3)**: 305-307.

Kumaraswamy, Y., Cox, P.J., Jaspars, M., Rashid, M.A. and Sarker, S.D. 2003. Bioactive flavonoid glycosides from the seeds of *Rosa cania*. Pharmaceutical Biol. **41(4)**: 237-242.

Kumari, K. and Augusti, K T. 2002. Antidiabetic and antioxidant effects of S-methyl cysteine sulfoxide isolated from onions (*Allium cepa* Linn.) as compared to standard drugs in alloxan diabetic rats. Indian J. Exp. Biol. **40(9)**: 1005-1009.

Kumazawa, S., Taniguchi, M. Suzuki, Y., Shimura, M., Kwon, M. and Nakayama, T. 2002. Antioxidant activity of polyphenols in carob pods. J. Agric. Food Chem. **50(2)**: 373-377.

Kumpulainen, J.T. and Salonen, J.T. 1999. Natural antioxidants and carcinogens in Nutrition, Health and Disease. The Royal Society of Chemistry, UK, pp. 178-187.

Kunchandy, E. and Rao, M.N.A. 1990. Oxygen radical scavenging activity of curcuminoid. Int. J. Pharmacognosy **58**: 237-240.

Kundakovic, T., Dukic, N.M. and Kovacevic, N. 2005. Free radical scavenging activity of *Achillea alexandri-regis* extracts. Fitoterapia **76**: 574-576.

Kunduraci, B.S., Bayrak, A. and Kiralan, M. 2010. Effects of essential oil and extracts from Oregano (*Origanum onites* L.) leaves on the oxidative stability of refined sunflower oil. Asian J. Chem. **22(2)**: 1377-1386.

KunduSen, S., Saha, P., Bhattacharya, S., Bala, A., Mazumder, U.K., Gupta, M. and Haldar, P.K. 2010. Evaluation of *in vitro* antioxidant activity of *Citrus limetta* and *Citrus maxima* on reactive oxygen and nitrogen species. Pharmacology online **3**: 850-857.

Kunzemann, J. and Hermann, K. 1977. Isolation and identification of flavon (ol)-O-glycosides in caraway (*Carum carvi* L.), fennel (*Foeniculum vulgare* Mill.) anise (*Pimpinella anisum* L.) and coriander (*Coriandrum sativum* L.) and of flavon-C-glycosides in anise. I. Phenolics of spices. Zeitschrift fur Lebensmitteluntersuchung und forschung A **164**: 194-200.

Kuo, C.C., Chiang, W., Liu, G.P., Chen, Y.L., Chang, J.Y., Lee, C.K., Lo, J.M., Huang, S.L., Shin, M.C. and Kuo, Y.H. 2002. 2,2-diphenyl-1-picrylhydrazyl radical-scavenging active components from adlay *Coix lachryma-jobi* L. var. *ma-yuen* Stapf) hulls. J. Agric. Food Chem. **50(21)**: 5850-5855.

Kuo, J., Yeh, D. and Sun Pan, B. 1999. Rapid photometric assay evaluating antioxidative activity in edible plant material. J. Agric. Food Chem. **47**: 3206-3209.

Kuo, Y.H., Chien, S.C. and Kuo, C.C. 2002. Antioxidative 7-oxodehydropodocarpane-type trinorditerpenes from the bark of *Taiwania cryptomerioides*. Planta Med. **68(11)**: 1020-1023.

Kupeli, E. Goger, F., Kosar, M. and Baser, K.H.C. 2008. Phenolic composition and biological activity of *Salvia halophila* and *Salvia virgata* from Turkey. Food Chem. **108**: 942-949.

Kupeli, E., Kartal, M., Aslan, S. and Yesilada, E. 2006. Comparative evaluation of the anti-inflammatory and antinociceptive activity of Turkish *Eryngium* species. J. Ethnopharmacol. **107(1)**: 32-37.

Kuppusamy, U.R., Indran, M. and Balraj, N.R.S. 2002. Antioxidant effects of local fruits and vegetable extracts. J. Trop. Med. Plants **3(1)**: 47-54.

Kuriakose, G.C. and Kurup, M.G. 2010. Antioxidant and hepatoprotective activity of *Aphanizomenon flos-aquae* Linn. against paracetamol intoxication in rats. Indian J. Exp. Bio. **48(11)**: 1123-1130.

Kurian, G.A., Rajamani, T., Ramnarayan, P. and Padikkala, J. 2009. A comparative study on *in vitro* and *in vivo* antioxidant activities of aqueous extract of *Desmodium gangeticum* (Leguminosae) root. Intern. J. Green Pharmacy **3(4)**: 324-331.

Kuroyanagi, M., Shmomae, M., Nagashima, Y., Muto, N., Okuda, T., Kawahara, N., Nakane, T. and Sano, T. 2005. New diarylheptanoids from *Alnus japonica* and their antioxidative activity. Chem. Pharm. Bull. **53(12)**: 1519-1523.

Kusar, A., Zupancic, A., Sentjurc, M. and Baricevic, D. 2006. Free radical scavenging activities of yellow gentian (*Gentiana lutea* L.) measured by electron spin resonance. Hum. Exp. Toxicol. **25(10)**: 599-604.

Kusirisin, W., Jaikang, C., Chaiyasut, C. and Narongchai, P. 2009. Effect of polyphenolic compounds from *Solanum torvum* on plasma lipid peroxidation, superoxide anion and cytochrome P450 2E1 in human liver microsomes. Med. Chem. **5(6)**: 583-588.

Kusku-Kiraz, Z., Mehmetcik, G., Dogru-Abbasoglu, S. and Uysal, M. 2009. Artichoke leaf extract reduces oxidative stress and lipoprotein dyshomeostasis in rat fed on high cholesterol diet. Phytother. Res. **24(4)**: 565-570.

Kuti, J.O. and Kunuru, H.B. 2004. Antioxidant capacity and phenolic content in leaf extracts of tree spinach (*Cnidoscolus* spp.). J. Agric. Food Chem. **52(1)**: 117-121.

Kuwahara, H., Kanazawa, A., Wakamatu, D., Morimura, S., Kida, K., Akakike, T. and Hiroshi, M. 2004. Antioxidative and antimutagenic activities of 4-vinyl-2,6-dimethoxyphenol (canolol) isolated from Canola oil. J. Agric. Food Chem. **52(14)**: 4380-4387.

Kvasnicka, F., Biba, B., Sevcik., R., Voldrich, M. and Kratka, J. 2003. Analysis of the active components of silymarin. J. Chromatogr. **990**: 239.

Kwak, J.H., Kim, H.J., Lee, K.H., Kang, S.C. and Zee, O.P. 2009. Antioxidative iridoid glycosides and phenolic compounds from *Veronica peregrina*. Arch. Pharm. Res. **32(2)**: 207-213.

Kwon, H.S., Kim, M.-J., Jeon, H.J., Yang, M.S., Park, K.H., Jeong, T.S. and Lee, W.S. 2008. Low-density lipoprotein (LDL)-antioxidant lignans from *Myristica fragrans* seeds. Bioorganic and Med. Chem. Lett. 18(1): 194-198. doi: 10.1016/j.bmcl.22007.10.098.

Kwon, Y.S., Kim, E.Y., Kim, W.J., Kim, W.K. and Kim, C.M. 2002. Antioxidant constituents from *Setaria viridis*. Arch. Pharm. Res. **25(3)**: 300-305. doi: 10.1007/BF02976630.

Kweon, M.H., Hwang, H.J. and Sung, H.C. 2001. Identification and antioxidant activity of novel chlorogenic acid derivatives from bamboo (*Phyllostachys edulis*). J. Agric. Food Chem. **49 (10)**: 4646-55.

Kyselova, Y. and Ivanova, D. 2006. Correlation between the *in vitro* antioxidant activity and polyphenol content of Bulgarian herbs. Phytother. Res. 20: 961-965.

Labuckas D.O., Maestri D.M., Perello M., Martinez M.L. and Lamarque A.L. 2008. Phenolics from walnut (*Juglans regia* L.) kernels: Antioxidant activity and interactions with proteins. Food Chem. **107 (2)**: 607-612.

Lachance, P.A., Nakat, Z and Jeong, W.S. 2001. Antioxidants: an integrative approach. Nutrition **17**: 835-838.

Lachman, J., Hamouz, K., Orsak, M., Pivec, V. and Dvorak, P. 2008. The influence of flesh colour and growing locality on polyphenolic content and antioxidant activity in potatoes. Scientia Horticulturae, **117(2)**: 109-114.

Lacopini, P., Baldi, M., Storchi, P. and Sebastiani, L. 2008. Catechin, epicatechin, quercetin, rutin and reservatrol in red grape: Content, *in vitro* antioxidant activity. J. Food Compos. Anal. 21(8): 589-598.

Lado, C., Then, M., Varga, I., Szoke, E. and Szentmihaly, K. 2004. Antioxidant property of volatile oils determined by the Ferric Reducing Ability. Zeitschrift fur Naturforschung **59c**: 354-358.

Lagnika, L., Anago, E. and Sanni, A. 2011. Screening for antibacterial, antioxidant activity and toxicity of some medicinal plants used in Benin folkloric medicine. J. Med. Pl. Res. **5(5)**: 773-777.

Lagouri, V., Blekas, G., Tsimidou, M., Kokkini, S., and Boskou, D. 1993. Composition and antioxidant activity of essential oils from Oregano plants grown wild in Greece. Zeitschrift fur Lebensmitteluntersuchung und Forschung A. **197 (1)**: 20-23.

Lai, H.Y., Lim, Y.Y. and Tan, S.P. 2009. Antioxidative, tyrosinase inhibiting and antibacterial activities of leaf extracts from medicinal ferns. Biosci. Biotech. Biochem. **73(6)**: 1362-1366.

Lai, L.S., Chou, S.T. and Chao, W.W. 2001. Studies on the oxidative activities of Hsiantsao (*Mesona procumbens* Hems.) leaf gum. J. Agric. Food Chem. **49**: 963-968.

Lai, S.C., Ho, Y.L., Huang, S.C., Huang, T.H., Lai, Z.R., Wu, C.R., Lian, K.Y., Chang, Y.S. 2010. Antioxidant and antiproliferative activities of *Desmodium triflorum* (L.) DC. Am. J. Chin. Med. **38(2)**: 329-342.

Lako, J., Trenerry, V.C., Wahlqvist, M., Wattanapenpaiboon, N., Sotheeswaran, S. and Premier, R. 2006. Phytochemical flavonols, carotenoids and the antioxidant properties of a wide selection of Fijian fruit, vegetables and other readily available foods. Food Chem. **101**: 1727-1741.

Lakshmi, B., Tilak, J.C., Adhikari, S., Devasagayam, T.P.A. and Jananrdhanan, K.K. 2004. Evaluation of antioxidant activity of selected Indian mushrooms. Pharm. Biol. **42**: 179-185. doi: 10.1080/13880200490514023.

Lam, U.D.P., Hoang, D.N., Lee, H.-B., Kim, B., Lee, J.D., Shin, J.H. and Kim, E.-K. 2011. Depigmenting effect of *Sterculia lynchnophera* on B16F10 melanoma and C57BL/6 melan-a cella. Korean J. Chem. Engineering **28(4)**: 1074-1077. doi: 10.1007/s11814-010-0484-0.

Lamaison, J.L., Carnat, A. and Petitjean-Freyetet, C.1991. Main flavonoid contents of commercial samples of *Filipendula ulmaria* (L.) Maxim. Plant. Med. Phytother. **25**: 1-5.

Lamaison, J.L., Petitjean-Freyetet, C. and Carnat, A. 1991. Lamiaceae medicinals antioxidant properties, rosmarinic acid as potential source. Pharmaceutica Acta Helvetiae **66(7)**: 185-188.

Lamien-Meda A., Lamien C. E., Compaoré M. M., Meda R. N., Kiendrebeogo M., Zeba B., Millogo J. F. and Nacoulma O. G. 2008. Polyphenol content and antioxidant activity of fourteen wild edible fruits from Burkina Faso. Molecules. **13(3)**: 581-594.

Lana, M.M. and Tijskens, L.M.M. 2006. Effects of cutting and maturity on antioxidant activity of fresh-cut tomatoes. Food Chem. **97 (2)**: 203-211.

Landbo, A.K. and Meyer, A.S. 2001. Enzyme-assisted extraction of antioxidative phenols from black currant juice press residues (*Ribes nigrum*). J. Agric. Food Chem. **49 (7)**: 3169-77.

Larson, R.A. 1988. The antioxidants of higher plants. Phytochemi. **27(4)**: 969-978.

Larson, R.A. 1997. Naturally occurring antioxidants. Lewis publishers, Boca Raton.

Latte, K.P. and Kolodziej, H. 2004. Antioxidant properties of phenolic compounds from *Pelargonium reniforme*. J. Agric.Food Chem. **52(15)**: 4899-4902.

Laupattarksem, P., Houghton, P.J. and Hoult, J.R.S. 2004. Anti-inflammatory isoflavonoids from the stems of *Derris scandens*. Planta Medica **70**: 496-501.

Lavanya, G., Sivajyothi, R., Manjunath, M. and Parthasarathy, P.R. 2009. Fate of biomolecules during carbon tetrachloride induced oxidative stress and protective nature of *Ammannia baccifera* Linn.: A natural antioxidant. Int. J. Green pharmacy **3(4)**: 300-305.

Lavanya, Y., Swapna, P. and Thakur, S.R. 2010. *In vitro* antioxidant studies of *Tinospora cordifolia* stems. 2nd Intern. Seminar on Medicinal plants and Herbal Products, S.V.Univ., Tirupati, Abstract pp. 106.

Laxane, S.N., Swarnkar, S.K. and Setty, M.M. 2008. Antioxidant studies on the ethanolic extract of *Zornia gibbosa*. Pharmacology online **1**: 319-330

Lean, M.E., Noroozi, M., Kelly, I., Burns, J., Talward, D, Sattar, N. and Crozier, A. 1999. Dietary flavonols protect diabetic human lymphocytes against oxidative damage to DNA. Diabetes **48**: 176-181.

Leccese, A., Bartolini, S. and Viti, R. 2007. Total antioxidant capacity and phenolics content in apricot fruits. Int. J. Fruit Sci. **7(2)**: 3-16.

Le Claire, E., Schwaiger, S., Banaigs, B., Stuppner, H. and Gafner, F. 2005. Distribution of a new rosmarinic acid derivative in *Eryngium alpinum* L. and other Apiaceae. J. Agric. Food Chem. **53(11)**: 4367-4372.

Lee, B.B., Park, S.R., Han, C.S., Han, D.Y., Park, E.J., Park, H.R. and Lee, S.C. 2008. Antioxidant activity and inhibition activity against α-amylase and α-glucosidase of *Viola mandshurica* extracts. J. Kor. Soc. Food Sci. Nutr. **37(4)**: 405-409.

Lee, C-Y.J., Kim, M., Yoon, S-W. and Lee, C-H. 2003. Short-term control of capsacin on blood and oxidative stress of rats *in vivo*. Phytother. Res. **17(5)**: 454-458.

Lee, H.J., Hyun, E.A., Yoon, W.J., Kim, B.H., Rhee, M.H., Kang, H.K., Cho, J.Y. and Yoo, E.S. 2006. *In vitro* anti-inflammatory and anti-oxidative effects of *Cinnamomum camphora* extracts. J. Ethnopharmacol. **103(2)**: 208-216.

Lee, H.S., Won, N.H., Kim, K.H., Lee, H., Jun, W. and Lee, K.W. 2005. Antioxidative effects of aqueous extract of *Terminalia chebula in vivo* and *in vitro*. Biol. Pharm. Bull. **28(9)**: 1639-1644.

Lee, I. and Ahn, S.Y. 1985. The antioxidant activity of gingerol. Korean J. Food Sci. Tech. **17(2)**: 55-59.

Lee, I.K., Jung, J.Y., Seok, S.J., Kim, W.G. and Yun, B.S. 2006a,. Free radical scavengers from the medicinal mushroom *Inonoutus xeranticus* and their proposed biogenesis. Bioorg. Med. Chem. Lett. **16(2)**: 5621-5624.

Lee, I.K., Seok, S.J., Kim, W.K. and Yun, B.S. 2006b. Hispidin derivatives from the mushroom *Inonotus xeranticus* and their antioxidant activity. J. Nat. Prod. **69(2)**: 299-301.

Lee, I.K. and Yun, B.S. 2006. Hispidin analogs from the mushroom *Inonotus xeranticus* and their free radical scavenging activity. Bioorg. Med. Chem. Lett. **16(9)**: 2376-2379.

Lee, I.K. and Yun, B.S. 2007. Highly oxygenated and unsaturated metabolites providing a diversity of hispidin class antioxidants in the medicinal mushrooms *Inonotus* and *Phellinus*. Bioorg. Med. Chem. Lett. **16(10)**: 3309-3314.

Lee, J. and Finn, C.E. 2007. Anthocyanins and other polyphenols in American elderberry (*Sambucus candensis*) and European elderberry (*S.nigra*) cultivars. J. Sci. Food Agric. **87**: 2665-2675.

Lee, J., Finn, C.E. and Wrolstad, R.E. 2004. Comparison of anthocyanin pigment and other phenolic compounds of *Vaccinium membranaceum* and *Vaccinium ovatum* native to the Pacific Northwest of North America. J. Agric. Food Chem. **52 (23)**: 7039-44.

Lee, J., Koo, N. and Min, D.B. 2004. Reactive oxygen species, aging and antioxidative nutraceuticals.Comp. Rev. Food Sci. Food Saftey **3**: 21-33.

Lee, J.H., Jeon, W.J., Yoo, E.S., Kim, C.M. and Kwon, Y.S. 2005. The chemical constituents and their antioxidant activity of the stem of *Rhododendron mucronulatum*. Nat. Prod. Sci. **11(2)**: 97-102.

Lee, J.H., Park, J.C. and Choi, J.S., 1996. The antioxidant activity of *Ecklonia stolonifera*. Arch. Pharm.Res. **19**: 223-227.

Lee, J.H., Renita, M., Fioritto, R.J., Martin, S.K., Schwartz, S.J. and Vodovotz, Y. 2004. Isoflavone characterization and antioxidant activity of Ohio soybeans. J. Agric. Food Chem. **52(9)**: 2647-2651.

Lee, J.H. and Talcott, S.T. 2004. Fruit maturity and juice extraction influences ellagic acid derivatives and other antioxidant polyphenolics in muscadine grapes. J. Agric. Food Chem. **52(2)**: 361-366.

Lee, J.Y., Chang, E.J., Kim, H.J., Park, J.H. and Choi, S.W. 2002. Antioxdative flavonoids from leaves of *Carthamus tinctorius*. Archives of Pharmacol. Res. **25(3)**: 313-319.

Lee J.-Y., Yoon, J.-W. and Kim, C.-T. 2004. Antioxidant activity of phenylpropanoid esters isolated and identified from *Platycodon grandiflorum* A. DC. Phytochemistry. **65 (22)**: 3033-3039.

Lee, K.G., Mitchell, A.E. and Shibamoto, T. 2000. Determination of antioxidant properties of aroma extracts from various beans. J. Agric. Food Chem. **48**: 4817-4820.

Lee, K.-G. and Shibamoto, T. 2002. Determination of antioxidant potential of volatile extracts isolated from various herbs and spices. J. Agric. Food Chem. **50**: 4947-4952.

Lee, K.H., Kim, A.J. and Choi, E.M. 2009. Antioxidant and antiinflammatory activity of pine pollen extract *in vitro*. Phytother Res. **23(1)**: 41-48.

Lee, K.W., Kim, Y.J., Lee, H.J. and Lee, C.Y. 2003. Cocoa has more phenolic phytochemicals and a higher antioxidant capacity than has teas and red wine. J. Agric. Food Chem. **51**: 7292-7295.

Lee, M-H., Jiang, C-B., Juan, S-H., Lin, R-D. and Hou, W-C. 2006. Antioxidant and heme oxygenase-1(HO-1)-induced effects of selected Taiwanese plants. Fitoterapia **77**: 109-115.

Lee, M-H., Lin, R.-D., Lee-Y. S., Yang, L.L., Yen, K.Y. and Hou, W. C. 2001. Monoamine Oxidase B and free radical scavenging activities of natural flavonoids in *Melastoma candidum* D. Don. J. Agric. Food Chem. **49 (11)**: 5551-5555.

Lee, S., Son, D., Ryu, J., Lee, Y.S., Jung, S.H., Kang, J., Lee, S.Y., Kim, H.S. and Shin, K.H. 2004. Antioxidant activities of *Acanthopanax senticosus* stems and their lignan components. Arch. Pharm. Res. **27(1)**: 106-10.

Lee, S.-C., Jeong, S.-M., Kim, S.-Y., Park, H.R., Nam, K.C. and Ahn, D.U. 2006. Effect of far infrared radiation and heat treatment on the anticxidant activity of water extracts from peanut hulls. Food Chem. **94**: 489-493.

Lee, S.E., Hwang, H.J., Ha, J.S., Jeong,.S. and Kim, J.H. 2003a. Screening of medicinal plant extracts for antioxidant activity. Life Sci. **73(2)**: 167-179.

Lee, S.E., Ju, E.M. and Kim, J.H. 2001. Free radical scavenging and antioxidant enzyme fortifying activities of extracts from *Smilax china* root. Exp. Mol. Med. **33 (4)**: 263-8.

Lee, S.E., Ju, E.M. and Kim, J.H. 2002. Antioxidant activity of extracts from *Euryale ferox* seed. Exp. Mol. Med. **34(2)**: 100-106.

Lee, S.E., Kim, G.S., Han, H.S., Lee, E.S., Kim, Y.O., Lee, J.H., Song, N.S., Lee, S.W. and Kim, Y.C. 2010. Biological activity of organic solvent fractions from *Lythrum salicaria* L. (root). Korean J. Medicinal Crop Sci. **18(5)**: 323-328.

Lee, S.E., Shin, H.T., Hwang, H.J. and Kim, J.H. 2003b. Antioxidant activity of extracts from *Alpinia katsumadai* seed. Phytother. Res. **17(9)**: 1041-1047.

Lee, S.H. 1992. Antioxidant activity of browing reaction products isolated from storage-aged orange juice. J. Agric. Food Chem. **40(4)**: 550-552.

Lee, S.J., Song, E.J., Lee, S.Y., Kim, K.B.W.R., Kim, S.J., Ycon, S.Y., Lee, C.J. and Ahn, D.H. 2009. Antioxidant activity of leaf, stem and root extracts from *Otostachys japonicus* and their heat and pH stabilities. J. Kor. Soc. Food Sci. Nutr. **38(11)**: 1571-1579.

Lee, S.K. and Bae, Y.S. 2006. Antioxidant activity of the extractives from the needles of *Picea abies* Karsten. J. Korean For. Soc. **95(4)**: 429-434.

Lee, S.K., Mbwambo, Z.H., Chung, H.S., Luyengi, L., Games, E.J.C., Mehta, R.G., Kinghorn, A.D. and Pezzuto, J.M. 1998. Evaluation of antioxidant potential of natural products. Combinatorial chemistry and High Throughput screening **1**: 35-46.

Lee, S.M., Na, M.K., An, R.B., Min, B.S. and Lee, H.K. 2003. Antioxidant activity of two phloroglucinol derivatives from *Dryopteris crassirhizoma*. Biol Pharm Bull. **26(9)**: 1354-1356.

Lee, W.S., Kim, J.-R., Han, J.M., Jang, K.C., Sok, D.-E. and Jeong, T.-S. 2006. Antioxidant activities of abietane diterpenoids isolated from *Torreya nucifera* leaves. J. Agric. Food Chem. **54(15)**: 5369-5374. doi: 10.1021/jf.060896c

Lee, W.S., Kim, J.-R., Im, K.R., Cho, K.H., Sok, D.E. and Jeong, T.S. 2005. Antioxidant effects of diaryl-hepatanoid derivatives from *Alnus japonica* on humn LDL oxidation. Planta Med. **71(4)**: 295-299.

Lee, Y., Howard, L.R. and Villalon, B. 2006. Flavonoids and antioxidant activity of fresh pepper (*Capsicum annuum*) cultivars. J. Food Sci. **60**(3): 473 – 476.

Lee, Y.Y., Jang, D.S., Jin, J.L., Yun-Choi, H.S. 2005. Anti-platelet aggregating and anti-oxidative activities of 11-O-(4′-O-methylgalloyl)-bergenin, a new compound isolated from *Crassula* cv. *himaturi*. Planta Medica **71(8)**: 776-777.

Leelapornpisid, P., Chansakaow, S., Chaiyasu, C. and Wogwataananukul 2007. Antioxidant activity of some volatile oils and absolutes from Thai aromatic plants. ISHS Acta Horticulturae 786.

Leewanich, P., Worachartcheewan, A., Prachayasittikul, S. and Prachayasittikul, V. 2011. Anticancer and antioxidative activities of *Stephania venosa*. Eur.J. Sci. Res. **51(2)**: 150-156.

Leleno, A., Tachibanaa, S. and Itoh, K. 2009. *In vitro* antioxidative activities and polyphenol content of *Eugenia polyantha* Wight grown in Indonesia. Pak. J. Biol. Sci **12(24)**: 1564-1570.

Lelpo, M. T. L., Sole, P. D., Basile, A., Moscatiella, E. L., Costaldo, R., Cobianchi, C. and Vuotto, M. L. 1998. Antioxidant properties of *Lunularia cruciata* (Bryophyte) extract. Immunopharmacol. Immunotoxicol. **20**: 555-556.

Lemos, T.L.G., Machado, L.L., Souza, J.S.N., Fonseca, A.M., Maia, J.L. and Pessoa, D.L. 2006. Antioxidant, ichthytoxicity and brine shrimp lethality tests of *Magonia glabrata*. Fitoterapia **77**: 443-445.

Lenucci, M.S., Cadinu, D., Taurino, M, Piro, G and Dalesandro G. 2006. Antioxidant composition in cherry and high-pigment tomato cultivars. J. Agric. Food Chem. **54(7)**: 2606-2613.

Leontowicz, H., Gorinstein, S., Lojek, A., Leontowicz, M., Ciz, M., Soliva-Fortuny, R., Park, Y-S, Jung, S-T, Trakhtenberg, S. and Martin-Belloso, O. 2002. Comparative content of some bioactive compounds in apple, peaches and pears and their influence on lipids and antioxidant capacity in rats. J. Nutr. Biochem. **13**: 603-610.

Leskovac, A., Joksic, G., Jankovic, T., Savikin, K. and Menkovic, N. 2007. Radioprotective properties of the phytochemically characterized extracts of *Crategus monogyna*, *Cornus mas* and *Gentianella austriaca* on human lymphocytes *in vitro*. Planta Medica **73(11)**: 1169-1175.

Lestario, L.N., Stepahanli, S. and Timotius, K.H. 2008. Antioxidant activity and total phenolic content of red sea weed (*Gracilaria verrucosa* L.). J. Teknologi dan Industri Pangan **19(2)**:

Le Tutor, B. and Guedon, D. 1992. Antioxidative activities of *Olea europaea* leaves and related phenolic compounds. Phytochem. **31**: 1173-1178.

Letteron, P., Labbe, G., Degott, C., Berson, A., Fromenty, B., Delaforge, M., Larry, D. and Pessayre, D. 1990. Mechanism of the protective effects of silymarin against carbon tetrachloride-induced lipid peroxidation and hepatotoxicity in mice. Evidence that silymarin acts both as an inhibitor or metabolic activation and as a chain-breaking antioxidant. Biochem. Pharmacol. **39(12)**: 2027-2034.

Li, C., Du, H., Wang, L., Shu, Q., Zheng, Y. Xu, Y, Zhang, J., Zhang, J., Yang, R. and Ge, Y. 2009. Flavonoid composition and antioxidant activity of tree peony (*Paeonia* section Moutan) yellow flowers. J. Agric. Food Chem. **57(18)**: 8496-8503.

Li, C. and Xie, B. 2000. Evaluation of the antioxidant and pro-oxidant effects of tea catechin oxypolymers. J. Agric. Food Chem. **48**: 6362-6366.

Li, C.Y., Lee, E-J. and Wu, T.S. 2004. Antityrosinase principles and constituents of the petals of *Crocus sativus*. J. Nat. Prod. **67(3)**: 437-440.

Li, H., Wang, X., Li, Y., Li, P. and Wang, H. 2009. Polyphenolic compounds and antioxidant properties of selected China wines. Food Chem. **112**: 454-460.

Li, H., Qiu, N., Ding, H. and Yao, R. 2008. Polyphenols contents and antioxidant capacity of 68 Chinese herbals suitable for medical or food uses. Food Res. Int. **41**: 36-370.

Li, H., Zhang, M. and Ma, G. 2011. Radical scavenging activity of flavonoids from *Trollius chinensis* Bunge. Nutrition. **27(10)**: 1061-1065.

Li, H.B., Cheng, K.W., Wong, C.C., Fan, K.W., Chen, F. and Jiang, Y. 2007. Evlauation of antioxidant capacity and total phenolic content of different fractions of selected microalgae. Food Chem. **102**: 771-776.

Li, H.-B., Jian, Y., Wong, C.-C., Cheng, K.-W. and Chen, F. 2007. Evaluation of two methods for the extraction of antioxidants from medicinal plants. Analytical and Bioanalytical chemistry **388(2)**: 483-488.

Li, H.-B., Li, D., Zhang, Y., Gan, R.-Y., Song, F.-L. and Chen, F. 2008. Antioxidant properties of Chinese medicinal plants. In: Gupta, S.D. (ed.). Reactive Oxygen species and antioxidants in Higher Plants. Science Publishers, Enfield, U.S.A., Chapter 15, pp 331-362.

Li, H-B., Wong, C-C., Cheng, K-W. and Chen, F. 2008. Antioxidant properties *in vitro* and total phenolic contents in methanol extracts from medicinal plants. LWT-Food Science and Technology **41(3)**: 385-390. doi: 10.1016/j.lwt.2007.03.011

Li, J.W., Ding, S.D. and Ding, X.L. 2005. Comparison of antioxidant capacities of extracts from five cultivars of Chinese jujube. Proc. Biochem. **40**: 3607-3613.

Li, K., Li, X.M., Ji, N.Y. and Wang, B.G. 2007. Natural bromophenols from the marine red alga *Polysiphonia urceolata* (Rhodomelaceae): structural elucidation and DPPH radical-scavenging activity. Bioorg. Med. Chem. **15(21)**: 6627-31.

Li *et al.*, 2001. Antioxidation activity of natural *Cordyceps sinensis* and cultured *Cordyceps* mycelia. Phytomedicine **8(3)**: 207-212.

Li, N., Li, X., Zhang, Y., Wang, T. and Xiao, W. 2008. Free radical scavengers, antioxidants and aldose reductase inhibitors from *Camptosorus sibiricus* Rupr. Z. Naturforsch. C. **63(1-2)**: 66-68.

Li, S.Y. and Li, S.P. 2009. Antioxidant activities of essential oil of *Curcuma longa* and *Curcuma wenyujin*. Int. J. Essent. Oil Ther. **3**: 31-34.

Li., W., Friel, J. and Beta, T. 2010. An evaluation of the antioxidant properties and aroma quality of infant cereals. Food Chem. **121(4)**: 1005-1102.

Li, W., Gao, Y., Zhao, J. and Wang, Q. 2007. Phenolic, flavonoid and luteolin ester content and antioxidant activity of 11 cultivars of Chinese marigold. J. Agric. Food Chem. **55(21)**: 8478-8484.

Li, X.L., Zhou, A.G. and Han, Y. 2006. Anti-oxidation and anti-microorganism activities of purification polysaccharide from *Lygodium japonicum in vitro*. Carbohydr. Polym. **66**: 34-42.

Li, X.M. 2007. Protective effect of *Lycium barbarum* polysaccharides on streptozotocin-induced oxidative stress in rats. Int. J. Biol Macromol. **40**: 461-465.

Li, X.M., Li, X.L. and Zhou, A.G. 2007. Evaluation of antioxidant activity of the polysaccharides extracted from *Lycium barbarum* fruits *in vitro*. Eur. Polym. J. **43**: 488-497.

Li, X.-Y. *et al.*, 2010. Study on flavonoids extraction process and antioxidation function in *Limonium bicolor*. J. Anhui Agr. Sci. DOI: CNKI: SUN: ANHY.0.2010-16-143.

Li, Y., Guo, C., Yang, J., Wei,., Xu, J. and Cheng, S. 2006. Evaluation of antioxidant properties of pomegranate peel extract in comparison with pomegranate pulp extract. Food Chem. **96**: 254-260.

Li, Y., Qian, Z.J., Ryu, B., Lee, S.H., Kim, M.M. and Kim. S.K. 2009b. Chemical components and its antioxidant properties *in vitro*: An edible marine brown alga, *Ecklonia cava*. Bioorg. Med. Chem. **17**: 1963-1973.

Li, Y., Wang, C.L., Wang, Y.J., Guo, S.X., Yang, J.S., Chen, X.M. and Xiao, P.G. 2009a. Three new bibenzyl derivatives from *Dendrobium candidum*. Chem. Pharm. Bull. **57(2)**: 218-219.

Li, Y., Wang, C.L., Wang, F.F., Guo, S.X., Yang, J.S. and Xiao, P. 2009b. Four new bibenzyl derivatives from *Dendrobium candidum*. Chem. Pharm. Bull. **57(9)**: 997-999.

Li, Z.-X., Zheng, S.Y. and Xiang, J.L. 2011. Research on changes of antioxidant activity of *Hovenia dulcis* Thunb. Vinegar during fermentation process. Food Sci. **32(7)**:

Liang, D., Zhou, Q., Gong, W., Wang, Y., Nie, Z., He, H., Li, J., Wu, J., Wu, C. and Zhang, J. 2011. Studies on the antioxidant and hepatoprotective activities of polysaccharides from *Talinum triangulare*. J. Ethnopharmacol. **136(2)**: 316-321.

Liao, H., Banbury, L.K. and Leach, D.N. 2008. Antioxidant activiry of 45 Chinese herbs and the relationship with their TCM characteristics. Evid. Based Complem. Alternat. Med. **5**: 429-434.

Lim, B.O., Choi, S.H., Kim, E.K., Lee, S.J., Je, J.Y., Jeon, Y.J., Kim, B., Park, S.H., Moon, S.H., Jeon, B.T., Lee, K.H., Park, T.K. and Park, P.J. 2008. Antioxidant activity of enzymatic extracts from *Stellaria dichotoma*. J. Med. Fod **11(4)**: 723-732.

Lim, B.O., Yu, B.P., Kim, S.C. and Park, D.K. 1999. The antioxidative effect of ganhuangenin against lipid peroxidation. Phytother. Res. **13(6)**: 479-483.

Lim, C.S., Jin, D.Q., Sung, J.Y., Lee, J.H., Choi, H.G., Ha, I. and Han, J.S. 2006. Antioxidant and anti-inflammatory activities of the methanolic extract of *Neorhodomela aculeata* in hippocampal and microglial cells. Biol Pharm. Bull. **29(6)**: 1212-1216.

Lim, H.K., Yoo, E.S., Moon, J.Y., Jeon, Y.J. and Cho, S.K. 2006. Antioxidant activity of extracts from Dangyuja (*Citrus grandis* Osbeck) fruits produced in Jeju Island. Food Sci. Biotech. **15**: 312-316.

Lim, J.P., Song, Y.C., Kim, J.W., Ku, C.H., Eun, J.S., Leem, K,H. and Kim, D.K.2002. Free radical scavengers from the heartwood of *Juniperus chinensis*. Arch Pharm Res. **25(4)**: 449-452

Lim, K.T., Hu, C. and Kitts, D.D. 2001. Antioxidant activity of *Rhus verniciflua* Strokes ethanol extracts. Food and Chemical Toxicol. **39**: 229.

Lim, S.N., Cheung, P.C.K., Ooi, V.E.C. and Ang, P.O. 2002. Evaluation of antioxidative activity of extracts from a brown seaweed, *Sargassum siliquastrum*. J. Agric. Food Chem. **50**: 3862-3866.

Lim, Y.Y. and Murtijaya, J. 2007. Antioxidant properties of *Phyllanthus amarus* extracts as affected by different drying methods. Food Sci. Tech. **40**: 1664-1669.

Lima, C.F. Andrade, P.B., Seabra, R.M., Fernandes-Ferreira, M. and Pereira-Wilson, C. 2005. The drinking of *Salvia officinalis* infusion improves liver antioxidant status in mice and rats. J. Ethnopharmacol. **97(2)**: 383-389.

Limaye, A.,S., Deore, G.B., Shinde, B.M. and Laware, S.L. 2010. Assessment of *Adiantum trapeziforme* L. for antioxidant activities. Asian J. Exp. Biol. Spl. 75-80.

Lin, C.C., Cheng, H.Y., Yang, C.M. and Lin, T.C. 2002. Antioxidant and antiviral activities of *Euphorbia thymifolia* L. J. Biomed. Sci. **9**: 656-664.

Lin, C.C. and Huang, P.C. 2002. Antioxidant and hepatoprotective effects of *Acanthopanax senticosus*. Phytother. Res. **14**: 489-494.

Lin, C.-C., Wu, S.-J., Chang, C.-H. and Ng, L.-T. 2003. Antioxidant activity of *Cinnamomum cassia*. Phytother. Res. **17(7)**: 726-730.

Lin, C.C., Wu, S.J. Wang, J.S., Yang, J.J. and Chang, C.-H. 2001. Evaluation of the antioxidant activity of legumes. Pharm. Biol. **39(4)**: 300–304.

Lin, C.C., Yen, F.L., Hsu, F.F. and Lin, J.M. 2000. Anti-hypercholesterolaemia, antioxidant activity and free radical scavenger effects of traditional Chinese medicine prescriptions used for stroke. J. Pharm. Pharmcaol. **52(11)**: 1387-1393.

Lin, C.H., Li, C.Y. and Wu, T.S. 2004. A novel phenylpropenoyl sulfonic acid and a new chlorophyll from leaves of *Petasites formosanus* Kitamura. Chem. Pharmaceutical Bull. **52(9)**: 1151-1152.

Lin, E.-S., Chou, H.-J., Kuo, P.-L. and Huang, Y.-C. 2010. Antioxidant and antiproliferative activities of methanolic extracts of *Perilla frutescens*. J. Med. Plants Res. **4(6)**: 477-483.

Lin, F.L., Wu, S.J., Lee, S.C. and Ng, L.T. 2009. Antioxdant, antioedema and analgesic activities of *Andrographis paniculata* extracts and their active constituent andrographiolide. Phytother. Res. **23(7)**: 958-964.

Lin, J.M., Lin, C.C., Chen, M.F., Ujiie, T. and Takata, A. 1995a. Radical scavenging and antihepatotoxic activity of *Ganoderma formosanum*, *G. lucidum* and *G. neo-japonicum*. J. Ethnopharmacol. **47**: 33-41.

Lin, J.M., Lin, C.C., Chen, M.F., Ujiie, T. and Takata, A. 1995b. Scavenging effects of *Mallotus repandus* on active oxygen species. J. Ethnopharmacol. **46**: 175-181.

Lin, L.C., Chiou, W.F. and Chou, C.J. 2004. Phenylpropanoid glycosides from *Orobanche caerulescens*. Planta Medica **70(1)**: 50-53.

Lin, S.Y., Liu, H.Y., Lu, Y.L. and Hou, W.C. 2005. Antioxidant activities of mucilages from different Taiwanese Yam cultivars. Bot. Bull. Acad. Sin. **46**: 183-188.

Lin, Y.L., Juan, I.M., Chen, Y.L., Liang, Y.C. and Lin, J.K. 1996. Composition of polyphenols in fresh leaves and associations of their oxygen-radical absorbing capacity with antiproliferative actions in fibroblast cells. J. Agric. Food Chem. **44**: 1387-1394.

Linberg-Madson, H. and Bertelsen, G. 1995. Spices as antioxidants. Trends Food Sci. Technol 6: 271-277.

Lindsey, K.L., Motsei, A.K. and Jäger, M.L. 2002. Screening of South African food plants for antioxidant activity. J. Food Sci. **67(6)**: 2129-2131.

Linnert, A., Latha, P.G., Gincy, M.M., Anuja, G.I., Suja, S.R., Shyamal, S., Shine, V.J., Sini, S., Shikha, P., Dan, M. and Rajasekharan, S. 2010. Anti-inflammatory, analgisc and anti-lipid peroxidative effects of *Rhaphidophora pertusa* (Roxb.) Schott. and *Epipremnum pinnaum* (Linn.) Engl. aerial parts. Indian J. Nat. Prod. Resourc. **1(1)**: 5-10.

Lionis, C., Faresjo, A., Skoula, M., Kapsokefalou, M. and Faresjo, T. 1998. Antioxidant effects of herbs in Crete. Lancet **352**: 1987-1988.

Lirdprapamongkol, K., Mahidol, C., Thongnest, S., Prawat, H., Ruchirawat, S., Srisomsap, C., Surait, R., Punyarit, P. and Svasti, J. 2003. Anti-metastatic effects of aqueous extract of *Helixanthera parasitica*. J. Ethnopharmacol. **86(2-3)**: 253-256.

Lissy, K.P., Simon, T.K. and Latha, M.S. 2006. Antioxidant potential of *Sida retusa*, *Urena lobata* and *Triumfetta rhomboidea*. Ancient Sci. Life **25(3&4)**.

Liu, C., Zhong, S.M., Chen, R.Y., Wu, Y. and Zhu, X.J. 2009. Two new compounds from the dried tender stems of *Cinnamomum cassia*. J. Asan Nat. Prod. Res. **11(9-10)**: 845-849.

Liu, C.-H., Yen, M.-H., Tsang, S.-F., Gan, K.-H., Hsu, H.-Y. and Lin, C.-N. 2010. Antioxidant triterpenoids from the stems of *Momordica charantia*. Food Chem. **118(3)**: 751-756.

Liu, C.L., Chen, Y.S., Yang, J.H.and Chiang, B.H.2008. Antioxidant activity of tartary (*Fagopyrum tataricum* (L.) Gaertn.) and common (*Fagopyrum esculentum* Moench) buckwheat sprouts. J. Agric. Food Chem. **56**: 173-178.

Liu, C.L., Chen, Y.S., Yang, J.H., Chiang, B.H. and Hsu, C.K. 2007. Trace element water improves the antioxidant activity of Buckwheat (*Fagopyrum esculentum* Moench) sprouts. J. Agric. Food Chem. **55(22)**: 8934-8940.

Liu, C.Z., Murch, S.J., El-Demerdash, M. and Saxena, P.K. 2004. *Artemisia judaica* L.: micropropagation and antioxidant activity. J. Biotechnol. **110 (1)**: 63-71.

Liu, F. and Ng, T.-B. 2000. Antioxidative and free radical scavenging activities of selected medicinal herbs. Life Sciences **66**: 725-735.

Liu, F., Ooi, V.E.C. and Chang, S.T. 1997. Free radical scavenging activity of mushroom polysaccharide extracts. Life Sci. **60**: 763-771.

Liu, H., Guo, S.-Y. and Xiao, K.J. 2006. Antioxidant activity of fruits of *Alpinia oxyphylla*. Nat. Prod. Res. Dev. **18(5)**: 768-771.

Liu, H. *et al.*, 2008. Study on the antioxidation of the flowers extracts from 15 ordinary tropical plants. J. Anhui Agr. Sci. DOI: CNKI: SUN: AHNY.0.2008-01-002.

Liu, H., Yang, X., Ren, T., Feng, Y. and Xu, H. 2005. Effects of *Erigeron breviscapus* ethanol extract on neuronal oxidative injury induced by superoxide radical. Fitoterapia **76(7-8)**: 666-670.

Liu, H.Y., Qiu, N.X., Ding, H.H. and Yoo, R.Q. 2008. Polyphenol content and antioxidant capacity of 68 Chinese herbals suitable for medical or food uses. Food Res. Int. **41**: 363-370.

Liu, J. and Mori, A. 1992. Antioxidant and free radical scavenging activities of *Gastrodia elata* Bl. and *Uncaria rhynchophylla* (Miq.) Jacks. Neuropharmacol. **31(12)**: 1287-1298.

Liu, J. and Xiao, P-G. 2006. Recent advances in the study of antioxidative effects of Chinese medicinal plants. Phytother. Res. **8(8)**: 445-451.

Liu, J.K., Hu, L., Dong, Z.J. and Hu, Q. 2004. DPPH radical scavenging activity of ten natural *p*-terphenyl derivatives obtained from three edible mushrooms indigenous to China. Chem Biodivers. **1(4)**: 601-605.

Liu, L., Xie, B., Cao, S., Yang, E., Xu, X. and Guo, S. 2007. A-type procyanidins from *Litchi chinensis* pericarp with antioxidant activity. Food Chem. **105(4)**: 1446-1451.

Liu, L.Y., Chang, L.-Y., Chou, S.-S., Hsiao, Y.-L. and Chien, Y.-W, 2010. Studies on the antioxidant components and activities of the methanol extracts of commercially grown *Hemarocallis fulva* L. (daylily) in Taiwan. J. Food Biochem. Doi: 10.1111/ j.1745-4514.2009.00306x.

Liu, M., Li, X.Q., Weber, C., Lee, C.Y. Brown, J. and Liu, R.H. 2002. Antioxidant and antiproliferative activities of raspberries. J. Agric. Food Chem. **50**: 2926-2930.

Liu, P., Hu, Y., Guo, D.H., Lu, B.R., Rahman, K., Mu, L.H. and Wang, D.X. 2010. Antioxidant activity of oligosaccharide ester extracted from *Polygala tenuifolia* roots in senescence-accelerated mice. Pharm. Bio. **48(7)**: 828-833.

Liu, Q. and Yao, H. 2007. Antioxidant activities of barley seeds extracts. Food Chem. **102**: 732-737.

Liu, S.C., Lin, J. T. Wang, C.K., Chen, H.Y., and Yang, D.J. 2009. Antioxidant properties of various solvent extracts from lychee (*Litchi chinenesis* Sonn.) flowers. Food Chem. **114(2)**: 577-581.

Liu, X., Zhao, M., Wang, J., Yang, B. and Jiang, Y. 2008. Antioxidant activity of methanolic extract of Emblica fruit (*Phyllanthus emblica* L.) from six regions in China. J. Food Comp. Anal. **21**: 219-228.

Liu, X., Zhao, M., Wang, J. and Luo, W. 2009. Antimicrobial and antioxidant activity of emblica extracts obtained by supercritical carbon dioxide extraction and methanol extraction. J. Food Biochem. **33(3)**: 307-330.

Liu, X., Zhao, M., Wang, J., Yang, B. and Jiang, Y. 2008. Antioxidant activity of methanolic extract of emblica fruit (*Phyllanthus emblica* L.) from six regions in China. Journal of food composition and analysis. **21(3)**: 219-228.

Liu, X.H., Hou, D.B., Zhao, N. and Wang, B. 2010. Extraction and antioxidant activity of flavonoids from *Aralia cordata*. Zhong, Yao Cai **33(9)**: 1484-1487.

Liu, X.P., Luan, J.J. and Goldring, C.E. 2009. Comparison of the antioxidant activity amongst *Ginkgo biloba* extract and its main components. Zhong Yao Cai **32(5)**: 736-740.

Liu, Y.H., Liang, H.J., Cheng, H.C., Liu, Y.W. and Hou, W.C. 2006. Comparisons of *in vitro* antioxidant activities of storage proteins in tuber of two *Dioscorea* species. Bot. Studies **47**: 231-237.

Liu, Y.W., Han, C.H., Le, M.H., Hsu, F.L. and Hou, W.C. 2003. Patatin, the tuber storage protein of potato (*Solanum tuberosum* L.) exhibited antioxidant activity *in vitro*. J. Agric. Food Chem. **51**: 4389-4393.

Liyana-Pathiranan, Chandrika, M. and Shahidi, F. 2005. Antioxidant activity of commercial soft and hard wheat (*Triticum aestivum* L.) as affected by gastric pH conditions. J. Agric. Food Chem. **53**: 2433-2440.

Ljubuncic, P., Azaizeh, H., Portnaya, I., Cogan, U., Said, O., Saleh, K.A. and Bomzon, A. 2005. Antioxidant activity and cytotoxicity of eight plants used in Arab traditional medicine in Israel. J. Ethnopharmacol. **99**: 43-47.

Ljubuncic, P., Dawkar, S., Portnaya, I., Coogan, U., Azaizeh, H. and Bomzon, A. 2006. Aqueous extract of *Teucrium polium* possess remarkable antioxidant activity *in vitro*. Evidence Based Coimplement. Alternat. Med. **3**: 329-338.

Ljubuncic, P., Portnaya, I., Coogan, U., Azaizeh, H. and Bomzon, A. 2005. Antioxidant activity of *Crataegus aronia* aqueous extract used in traditional Arab medicine in Israel. J. Ethnopharmacol. **101(1-3)**: 153-161.

Llorach, R., Espín, J.C., Tomás-Barberán, F.A. and Ferreres, F. 2003.Valorization of cauliflower (*Brassica oleracea* L. var. *botrytis*) by-products as a source of antioxidant phenolics. J. Agric. Food Chem. **51(8)**: 2181-2187.

Lo, S.F., Mulabagal, V., Chen, C.L., Kuo, C.L. and Tsay, H.S. 2004a. Bioguided fractionation and isolation of free radical scavenging components from *in vitro* propagated Chinese medicinal plants *Dendrobium tosaense* Makino and *Dendrobium moniliforme* Sw. J. Food Chem. **52(23)**: 6916-6919.

Lo, S.F., Nalawade, S.M., Mulabagal, V., Mathew, S., Chen, C.L. and Tsat, H. 2004b. *In vitro* propagation by asymbiotic seed germination and 1,1-diphenyl-2-picrylhydrazyl (DPPH) radical scavenging activity studies of tissue culture raised plants of three medicinally important species of *Dendrobium*. Biol. Pharmceutical Bull. **27(5)**: 731-735.

Lo Scalzo, **R.**, Iannoccari, T., Summa, C., Morelli, R. and Rapisarda, P. 2003. Effect of thermal treatments on antioxidant and antiradical activity of blood orange juice. Food Chem. **85(1)**: 41-47.

Lobo, V., Pathak, A. and Chandra, N. 2010. Comparative evaluation of antioxidant activity of aqueous and alcoholic extract of *Cassia tora* Linn. leaves. Asian J. Exp. Biol. Sci. **1(4)**: 826-832.

Lock, O., Castillo, P., Doroteo, V. and Rojas, R. 2005. Antioxidant activity *in vitro* of selected Peruvian medicinal plants. ISHS Acta Horticulture 675: III WOCMAP Congress on Medicinal and Aromatic Plants Volume 1: Bioprospecting and Ethnopharmacology.

Lohézic-Le Dévéhat, F., Tomasi, S., Elix, J.A., Bernard, A., Rouaud, I., Uriac, P. and Boustie, J. 2007. Stictic acid derivatives from the lichen *Usnea articulata* and their antioxidant activities. J. Nat. Prod. **70(7)**: 1218-1220.

Loizzo, M.R., Menichini, F., Conforti, F., Tundis, R., Bonesi, M., Saab, A.M., Statti, G.A., de Cindo, B., Houghton, P.J., Menichini, F. and Natale. 2009. Chemical analysis, antioxidant, anti-inflammatory and anticholinesterase activities of *Origanum ehrenbergii* Boiss and *Origanum syriacum* L. essential oils. Food Chem. **117(1)**: 174-180. doi: 10.1016/j.foodchem.2009.03.095.

Loizzo, M.R., Menichini, F., Tundis, R., Bonesi, M., Conforti, F., Nadjafi, F., Statti, G.A., Frega, G.N. and Menichini, F. 2009. *In vitro* biological activity of *Salvia leriifolia* Benth. essential oil relevant to the treatment of Alzhemiers disease. J. Oleo. Sci. **58(8)**: 443-446.

Loizzo, M.R., Tundis, R., Confort, F., Menichini, F., Bonesi, M., Nadjafi, F., Frega, N.G. and Menichini, F. 2010. *Salvia leriifolia* Benth. (Lamiaceae) extract demonstrates *in vitro* antioxidant properties and cholinesterase inhibitory activity. Nutrition Research **30(12)**: 823-830.

Loliger, J. 1991. The use of antioxidants in food. I: Aruoma, O.I. and Halliwell, B. (eds.). Free Radicals and food additives. Taylor and Francis, London, pp. 129-150.

Lombardi-Boccia, G., Lucarini, M., Lanzi, S., Aguzzi, A. and Cappelloni, M. 2004. Nutrients and antioxidant molecules in yellow plums (*Prunus domestica* L.) from conventional and organic productions: A comparative study. J. Agric. Food Chem. **52(1)**: 90-94.

Londhe, J.S., Devasagayam, T.P.A., Yeap Foo, L. and Ghaskadbi, S. 2008. Antioxidant activity of some polyphenols of the medicinal plant *Phyllanthus amarus*. Redox Report **13**: 199-207.

Loots, D.T., Van der Westhuizen, F.H. and Botes, L. 2007. *Aloe ferox* leaf gel phytochemical content, antioxidant capacity, and possible health benefits. J. Agric. Food Chem. **55(17)**: 6891-6896.

Lopes, G.C., Sanches, A.C., Nakamura, C.V., Dias Filho, B.F., Hernandes, L. and Mello, J.C. 2005. Influence of extracts of *Stryphnodendron polyphyllum* Mart. and *Stryphnodendron obovatum* Benth. on the cicatrisation of cutaneous wounds in rats. J. Ethnopharmacol. **99(2)**: 265-272.

Lopez-Arnaldos, T., Zapata, J.M., Calderon, A.A. and Ros Barcelo, A. 1996. Antioxidant activity of lavandin (*Lavandula* x *intermedia*) cell cultures in relation to their rosemarin content. In: Risch, S.J. and Ho, Ch-T. (eds.). Spices, Flavour, Chemistry and antioxidant properties. ACS Symposium series 660. American chemical Society, Washington DC. Pp. 206-218.

Lopez, V., Akerreta, S., Casanova, E., Garcia-Mina, J.M., Cavero, R.Y. and Calvo, M.I. 2008. Screening of Spanish medicinal plants for antioxidant and antifungal activities. Pharmaceutical Biol. **46 (9)**: 602-609.

Lotulung, P.D.N., Minarti, L.B.S., Kardono and Kawanishi, K. 2008. Antioxidant activity of Rhizome *Kaempferia rotunda* L. Pak. J. Biol. Sci. **11(20)**: 2447-2450.

Louli, V., Ragoussis, N. and Magoulas, K. 2004. Recovery of phenolic antioxidants from wine industry by-products. Bioresource Technol. **92**: 201-208.

Lourens, A.C.U., Reddy, D., Baser, K.H.C., Viljoen, A.M. Van Vuuren, S.F. 2004. *In vitro* biological activity and essential oil composition of four indigenous South African *Helichrysum* species. J. Ethnopharmacol. **95**: 253-258.

Lu, F. and Foo, L.Y. 2001. Antioxidant activities of polyphenols from sage (*Salvia officinalis*). Food Chem. **75**: 197-202.

Lu, G.H., Chan, K., Leung, K., Chan, C.L., Zhao, Z.Z. and Jiang, Z.H. 2005. Assay of free ferulic acid and total ferulic acid for quality assessment of *Angelica sinensis*. J. Chromatogr. **68(2)**: 209-219.

Lu, Y. and Foo, L.Y. 1997. Identification and quantification of major polyphenols in apple pomace. Food Chem. **59(2)**: 187-194.

Lu, Y. and Foo, L.Y. 2000. Antioxidant and radical scavenging activities of polyphenols from apple pomace. Food Chem. **68**: 81-85.

Lu, Y., Wang, Q., Melzig, M.F. and Jenett-Siems, K. 2008. Extracts of *Cynomorium songaricum* protect SK-N-SH human neuroblastoma cells against staurosporine-induced apoptosis potentially through their radical scavenging activity. Phytother. Res. **23(2)**: 257-261.

Lu, Y.R. and Foo, L.Y. 2001. Antioxidant activities of polyphenols from sage (*Salvia officinalis*). Food Chem. **75**: 197-202.

Lubag, Jr., A.J.M., Lurena, A.C., Tecson-Mendoza, E.M. 2008. Antioxidants of purple and white greater yam (*Dioscorea alata* L.) varieties from Philippines. Philippine J. Sci. **137(1)**: 61-67.

Lugasi, A., Dworschak, F., Blazovics, A. and Kery, A. 1998. Antioxidant and free radical scavenging properties of squeezed juice from black radish (*Raphanus sativus* L. var. *niger*) root. Phytother. Res. **12**: 502-506.

Lugasi, A., Horvahovich, P. and Dworschák, E.1999. Additional information to the *in vitro* antioxidant activity of *Ginkgo biloba* L. Phytother Res. **13(2)**: 160-162.

Lugasi, A., Hovari, J., Hagymasi, K., Jakoczi, I. and Blazocvics. 2006. Antioxidant properties of a mixture of Lamiaceae plants intended to use as food additive. Acta Alimentaria. **35(1)**: 85-97.

Lui, F. and Ng, T.B. 1999. Antioxidativee and free radical scavenging activities of selected medicinal herbs. Life Sci. **66**: 725-735.

Luis, J.C., Valdes, F., Martin, R., Carmona, A.J. and Dias, J.G. 2006. DPPH radical scavenging activity of two flavonol glycosides from *Aconitum napellus* ssp. *lusitanicum*. Fitoterapia 77: 469-471.

Lukmanul Hakkim, F., Gowri Shankar, C. and Girija, S. 2007. Chemical composition and antioxidant property of holy Basil (*Ocimum sanctum* L.) leaves, stems and inflorescence and their *in vitro* callus cultures. J. Agric. Food Chem. **55(22)**: 9109-9117.

Luo, X.-D., Wu, D.G., Cai, X.H., and Kennelly, E.J. 2006. New antioxidant phenolic glycosides from *Walsura yunnanensis*. Chem. Biodiv. **3(2)**: 224-230.

Luximon-Ramma, A., Bahorun, T., Soobratteee, M.A. and Aruoma, O.I. 2002. Antioxidant activities of phenolic anthocyanidins and flavonoid components in extracts of *Cassia fistula*. J. Agric. Food Chem. **50**: 5042-5047.

Lundgren, L. 1987. Studies on low-molecular phenolic constituents in conifer needles. Dissertation. Sweedish University of Agricultural Sciences, Uppsala, Sweden.

Luo, A., Fan, Y. and Luo, A. 2011. *In vitro* free radicals scavenging activities of polysaccharide from *Polygonum multiflorum* Thunb. J. Med.Plants Res. **5(6)**: 966-972.

Luo, A., Ge, Z., Fan, Y., Luo, A., Chun, Z. and He, X. 2011. *In vitro* and *in vivo* antioxidant activity of a water-soluble polysaccharide from *Dendrobium denneanum*. Molecules **16(2)**: 1579-1592.

Luo, A.X., He, X.J., Zhou, S.D., Fan, Y.J., He, T. and Chun, Z. 2009. *In vitro* antioxidant activities of a water-soluble polysaccharide derived from *Dendrobium nobile* Lindl. extracts. Int. J. Biol. Macromol. **45(4)**: 359-363.

Luo, A.X., He, X.J., Zhou, S.D., Fan, Y.J., Luo, A.S. and Chun, Z. 2010. Purification, composition analysis and antioxidant activity of the polysaccharides from *Dendrobium nobile* Lindl. Carbohydr. Polym. 79: 1014-1019.

Luo, D. 2008. Identification of structure and antioxidant activity of a fraction of polysaccharide purified from *Dioscorea nipponica*. Carbohydrate Polymers, **71 (4)**: 544-549.

Luo, J., Tian, C., Xu, J. and Sun, Y. 2009. Studies on the antioxidant activity and phenolic compounds of enzyme-assisted water extracts from Du-zhong (*Eucommia ulmoides* Oliv.) leaves. J. Enzyme Inhib. Med. Che. **24(6)**: 1280-1287.

Luo, M., Kannar, K., Wahlqvist, M.L. and O'Brien, R.C. 1997. Inhibition of LDL oxidation by green tea extract (comment, letter). Lancet 349: 360-361.

Luo, Q., Cai, Y., Sun, M. and Corke, H. 2004. Hypoglycemic and hypolipidemic effects and antioxidant activity of fruit extracts from *Lycium barbarum*. Life Sci. **76(2)**: 137-149.

Luo, X.D., Basile, M.J. and Kennelly. 2002. Popyphenolic antioxidants from the fruits of *Chrysophyllum canito* L (Star apple). J. Agric. Food Chem. **50**: 1379-1382.

Luo, Y., Cai, Q., Sun, M. and Corke, H. 2004. Antioxidant activity and phenolic compounds of 112 traditional Chinese medicinal plants associated with anticancer. Life Science **74**: 2157-2184.

Luqman, S., Kumar, R., Kausik, S., Srivastava, S., Darokar, M.P. and Khanauja, S.P.S. 2009. Antioxidant potential of the root of *Vetiveria zizanioides* (L) Nash. Indian J. Biochem. and Biophysics **46**: 122-125.

Lv L., Gu X., Tang J., Ho C.-T. 2007. Antioxidant activity of stilbene glycoside from *Polygonum multiflorum* Thunb. *in vivo*. Food Chem. **104(4)**: 1678-1681.

Ly, T.N., Shimoyamada, M. and Yamauchi, R. 2006. Isolation and characterization of Rosmarinic acid oligomers in *Celastrus hindsii* Benth. leaves and their antioxidant activity. J. Agric. Food Chem. **54(11)**: 3786-3793. DOI: 10.1021/jf052743f.

Lyle, N., Gomes, A., Sur, T., Munshi, S., Paul, S., Chatterjee, S. and Bhattacharjee, D. 2009a,b. The role of antioxidant properties of *Nardostachys jatamansi* in alleviation of the symptoms of the chronic fatigue syndrome. Behav. Brain Res. **202(2)**: 285-290.

Lyle, N., Bhattacharya, D., Sur, T., Munshi, S., Paul, S., Chatterjee, S. and Gomes, A. 2009. Stress modulating antioxidant effect of *Nardostachys jatamansi*. Indian J. Biochem Biophys. **46(1)**: 93-98.

Ma, C.J., Kim, S.H., Lee, K.Y., Oh, T., Kim, S., Sung, S.H. and Kim, Y.C. 2009. ESP-102, a combined extract of *Angelica gigas*, *Saururus chinensis* and *Schizandra chinensis*, protects against glutamate-induced toxicity in primary cultures of rat cortical cells. Phytother. Res. **23(11)**: 1587-1591.

Ma, X.M., Liu, Y. and Shi, Y.P. 2007. Phenolic derivatives with free-radical scavenging activities from *Ixeridium gracile* (DC.) Shih. Chem. Biodivers. **4(9)**: 2172-2181.

MacLean, D., Murr, D.P. and DeEll, J.R. 2003. A modified total oxyradical scavenging capacity assay for antioxidants in plant tissues. Postharvest Biology and Technology 29: 183-194.

Madan, S., Singh, G.N., Kumar, Y. and Kohli, K. 2010. Phytochemical analysis and free radical scavenging activity of *Flemingia strobilifera* (Linn.) R.Br. Res. J. Pharmacol. Biol. Chem. Sci. **1(4)**: 183-190.

Madhavan, V., Shah, P., Murali, A. and Yoganarasimhan, S.N. 2010. *In vitro* and *in vivo* antioxidant activity studies on the roots of *Toddalia asiatica* (L.) Lam. (Rutaceae). Asian J. Trad. Med. **5(5)**: 188-198.

Madhavi, D.L., Deshpande, S.S., Salunkhe, D.K. 1996. Introduction. In: Madhavi, D.L., Deshpande, S.S., Salunkhe, D.K. (eds.). Food antioxidants: Technological, Toxicological and Health Perspectives, Marcel Dekker, Inc. New York pp 1-4.

Madhujith, T. and Shahidi, F. 2003. Antioxidant activity of Bluberry and other *Vaccinium* species. ACS Symposium series. Vol. 871. Chapter 12. pp. 149-160.

Madrau, M.A., Piscopo, A., Sanguinetti, A.M., Del Caro, A., Poiana, M., Rom, F.V. and Piga, A. 2009. Effect of drying temperature on polyphenolic content and antioxidant activity of apricots. European Food Res. Techn. **228(3)**: 441-448.

Madsen, H.L. and Bertelsen, G. 1995. Spices as antioxidants. Trends in Food Sci. Technol. **6**: 271-277.

Maduka, H.C. and Okoye, Z.S. 2002. The effect of *Sacoglottis gabonensis* stem bark, a Nigerian alcoholic beverage additive, on the natural antioxidant defences during 2,4-diphenyl hydrazine-induced membrane peroxidation *in vivo*. Vasc. Pharmacol. **39(1-2)**: 81- 21-31.

Maeda, G., Takara, K., Wada, K., Oki, T., Masuda, M., Ichiba, T., Chuda, Y., Ono, H. and Suda, I. 2006. Evaluation of antioxidant activity of vegetables from Okinawa prefecture and determination of some antioxidative compounds. Food Sci. Technol. Res. **12(1)**: 8-14.

Maffei Facino, R., Carini, M., Aldini, G., Bombardelli, E., Morazzoni, P. and Morelli, R. 1994. Free radicals scavenging action and anti-enzyme activities of procyanidines from *Vitis vinifera*. A mechanism for their capillary protective action. Arzneimittelforschung. **44 (5)**: 592-601.

Magali, C., Gary, F.W. and Elizabeth, I.O. 2008. Determination of the antioxidant capacity of culinary herbs subjected to various cooking and storage processes using the ABTS radical cation assay. Plant Foods Hum. Nutr. **63**: 47-52.

Magwa, M.L., Gundidza, M., Gweru, N., Humphrey, G. 2006. Chemical composition and biological activities of essential oil from the leaves of *Sesuvium portulacastrum*. J. Ethnopharmacol. **103(1)**: 85-89.

Mahaboob, M., Rahman, M.F. and Grover, P. 2005. Acute oral effects of *Acorus calamus* on lipid peroxidation and glutathione levels in different organs of rat. Toxicol. Intern. **12(1)**: 51.

Mahabusarakam, W., Nuangnaowarat, W. and Taylor, W.C. 2006. Xanthone derivatives from *Cratoxylum cochinchinense* roots. Phytochem. **67(5)**: 470-474.

Mahakundakorn, P., Tohda, M., Murakami, Y., Matsumoto, K. and Watanabe, H. 2004. Antioxidant and free radical-scavenging activity of Choto-san and its related constituents. Biol. Pharmaceutical Bull. **27(1)**: 38-46.

Maharana, L., Kar, D.M., Pattnaik, S., Sawale, J. and Kalia, A.N. 2010. *In vitro* antioxidant activity of aqueous leaf extract of *Solanum nigrum* L. Pharmacology online **3**: 333-345.

Maharana, L., Pattnaik, S., Kar, D.M., Sahu, P.K. and Si, S.C. 2010. Antioxidant activity of aqueous extract of aerial parts of *Mollugo pentaphylla* Linn. – An *in vitro* study. Pharmacology online **3**: 982-997.

Mahesh, T. and Menon, V.P. 2004. Quercetin allievates oxidative stress in streptozotocin-induced diabetic rats. Phytother. Res. **18(2)**: 123-127.

Mahmoud, A.H., H.M.Motawa, Wahba, H.E. and Ebrahim, A.Y. 2005. Study of some antioxidant parameters in mice livers affected with *Urtica pilulifera* extracts. Egyptian J. Hosptal Med. **21**: 33-42.

Mahour, K. and Saxena, P.N. 2010. *In vitro* antioxidant status of aqueous extract of Indian ginseng D. and *Panax ginseng* D. Pharmacology online **1**: 118-123.

Maiga, A., Malterud, K.E., Diallo, D. and Paulsen, B.S. 2006. Antioxidant and 15-lipoxygenase inhibitory activities of the Malian medicinal plants *Diospyros abyssinica* (Hiern) F. White (Ebenaceae), *Lannea velutina* A.Rich (Anacardiaceae) and *Crossopteryx febrifuga* (Afzel) Benth. (Rubiaceae). J. Ethnopharmacol. **104(1-2)**: 132-137.

Mainar, A.M., Langa, E., Berrueco, B., Maestro, C. an Urieta, J.S. 2008. Antioxidant activity of supercritical extracts of *Populus* buds. www.isasf.net/fileadmin/files/Docs/Barcelona/./PDF/./P_NP_24.pdf -

Maioli, M.A., Alves, L.C., Campanini, A.L., Lima, M.C., Dorta, D.J., Groppo, M., Cavalheiro, A.J., Curti, C. and Mingatto, F.E. 2010. Iron-chelaating-mediated antioxidant activity of *Plectranthus barbatus* extract on mitochondria. Food Chem. **122(1)**: 203-208.

Maisuthisakul, P. 2004. Effect of sample preparation methods and extraction methods and extraction time on yield and antioxidant activity from Kradonbok (*Careya sphaerica* Roxb.) leaves. Proc. 42nd Kasetsart University Annual Conference, Bangkok.

Maisuthisakul, P., Gordon, M.H., Pongsawatmanit, R. and Suttajit, M. 2007. Enhancing the oxidative stability of rice crackers by addition of the ethanolic extract from *Cratoxylum formosum* Dyer. Asia-Pacific J. Clinical Nutr. **16**: 37-42.

Maisuthisakul, P. and Pasuk, S. 2008. Antiradical scavenging activity and polyphenolic compounds extracted from Thai mango seed kernels. As. J.Food Ag.-Ind. **1(2)**: 87-91.

Maisuthisakul, P., Pasuk, S. and Ritthiruangdej, P. 2008. Relationship of antioxidant properties and chemical composition of some Thai plants. J. Food Composition Anal. **21**: 229-240.

Maisuthisakul, P., Pongsawatmanit, R. and Gordon, M.H. 2006. Antioxidant properties of Teaw (*Cratoxylum formosum* Dyer) extract in soybean oil and emulsions. J. Agric. Food Chem. **54**: 2719-2725.

Maisuthisakul, P., Pongsawatmanit, R. and Gordon, M.H. 2007a. Assessment of phenolic content and free radical scavenging capacity of some Thai indigenous plants. Food Chem. **100**: 1409-1418.

Maisuthisakul, P., Pongsawatmanit, R. and Gordon, M.H. 2007 b. Characterization of the phytochemicals and antioxidant properties of extracts from Teaw (*Cratoxylum formosum* Dyer). Food Chem. **100**: 1620-1629.

Maisuthisakul, P.,Trakoontivakorn, K., Suttajit, M., Pongsawatmanit, R. 2005. Comparison of Antioxidant activity methods of Thai Indigenous vegetable leaf extracts. Proc. 6th Agro-Industrial Conference.

Majeed, M., Badmaev, V., Shivakumar, u., and Rajendran, R. 1995. Curcuminoid – Antioxidant phytonutrients. Nutriscience Publishers, Inc., Piscataway, New Jersey.

Majhenic, L., Skerget, M. and Knez, Z. 2007. Antioxidant and antimicrobial activity of guarana seed extracts. Food Chem. **104**: 1258-1268.

Mak, D.H.F., Chiu, P.Y., Poon, M.K.T., Ng, T.T.L., Chung, Y.K., Lam, B.Y.H., Du, Y. and Ko, K.M. 2004. A yang-promoting Chinese herbal suppository preparation enhances the antioxidant status of red cells in male human subjects. Phytother. Res. **18(7)**: 525-530.

Makari, H.K., Haraprasad, N., Patil, H. and Ravikumar, S. 2008. *In vitro* antioxidant activity of the hexane extracts of *Cordia wallichi* and *Celastrus paniculata*. The Internet J. Aesthetic and Antiaging Medicine **1**: 1-10.

Makasci, A.A., Mammadov, R., Dusen, O. and Isik, H.I. 2010. Antimicrobial and antioxidant activities of medicinal plant species *Ornithogalum alpigenum* Stapf from Turkey. J. Med. Plants Res. **4(16)**: 1637-1642.

Makris, D.P., Boskou, G. and Andrikopoulos, N.K. 2007. Polyphenolic content and *in vitro* antioxidant characteristics of wine and other agrifood solid waste extracts. J. Food Compos. Anal. **20**: 125-132.

Maksimovi, Z 2005. Polyphenol contents and antioxidant activity of *Maydis stigma* extracts. Bioresour. Technol. **96 (8)**: 873-877.

Maksimovi, Z., Petrovi, M., Pavlovi, M, Kovacevi, N and Kuki, J. 2007. Antioxidant activity of *Filipendula hexapetala*. Fitoterapia **78(3)**: 265-267.

Malaviya, N., Jain, S., Jain, A., Jain, S. and Gurjar, R. 2010. Evaluation of *in vitro* antioxidant potential of aqueous extract of *Trapa natans* L. fruits. Acta Poloniae Pharmaceutica – Drug Res. **67(4)**: 391-396.

Maldonado, P.D., River-Cruz, I., Mata, R. and Pedraza-Chaverri, J. 2005. Antioxidant activity of A-type proanthocyanidins from *Geranium niveum* (Geraniaceae). J. Agric. Food Chem. **53(6)**: 1996-2001.

Malenci, D., Gasic, O., Popovi, M. and Boza, P. 2000. Screening for antioxidant properties of *Salvia reflexa* Hornem. Phytother. Res. **14(7)**: 546-548.

Malenci, D., Maksimovi, Z., Popovi, M. and Miladinovi, J. 2008. Polyphenol contents and antioxidant activity of soybean seed extracts. Bioresour. Technol. **99 (14)**: 6688-6691.

Malenci, D., Popovi, M. and Miladinovi, J. 2007. Phenolic content and antioxidant properties of soybean (*Glycine max* (L.) Merr.) seeds. Molecules. **12 (3)**: 576-581.

Mamadalieva, N.Z., El-Readi, M.Z., Janibekov, A.A., Tahrani, A., and Wink, A. 2011. Phytoecdysteroids of *Silene guntensis* and their *in vitro* cytotoxic and antioxidant activity. Z. Naturforsch.C. **66(5-6)**: 215-224.

Mammadov, R., Dusen, O., Uysal, D. and Kose, E. 2009. Antioxidant and antimicrobial activities of extracts from tubers and leaves of *Colchicum balansae*. J. Med. Plants Res. **3(10)**: 767-770.

Mammadov, R., Makasci-Afacan, A., Uysal-Demir, D. and Girk, C. 2010. Determination of antioxidant activities of different *Urginea maritima* (L.) Baker plant extracts. Iran. J. Chem. Chem. Eng. **29(3)**: 47-53.

Mamyrbekova-Beko, J.A., Konan, K.M., Bekro, Y.A., Djie Bi, M.G., Zomi Bi, J.J., Mambo, V. and Bona, B.B. 2008. Phytocompounds of the extracts of four medicinal plants of Cote D'Ivoire and assessment of their potential antioxidant by Thin Layer Chromatography. European J. Scientific Res. **24(2)**: 219-229.

Manda, H., Rao, B.K., Yashwant, Kutty, G.N., and Swarnkar, S.K. 2009. Antioxidant, anti-inflammatory and antipyretic activities of ethyl acetate fraction of ethanolic extract of *Schrebera swietenioides* Roxb. root. Int. J. Toxicol. Pharmacol. Res. **1(1)**: 7-11.

Mandal, A., Bishayee, A. and Chatterjee, M. 1996. Restoration of antioxidant balance by *Trianthema portulacastrum* in carbon tetrachloride-induced hepatocellular injury in mice. Phytomedicine **3**(Supp.l.): 48.

Mandal, P., Misra, T.K. and Ghosal, M. 2009. Free-radical scavenging activity and phytochemical analysis in the leaf and stem of *Drymaria diandra* Blume. Int. J. Integrative Biol. **7(2)**: 80-84.

Mancini-Filho, J., Van-Koiij, A., Mancini, D.A., Cozzolino, F.F, and Torres, R. P. 1998 Antioxidant activity of cinnamon (*Cinnamomum zeylanicum* Breyne) extracts. Boll Chim. Farm. **137 (11)**: 443-447.

Manea, S., Mazilu, E., Ristea, C., Setnic, S., Cretu, I., Neagu, M. Ivopol, G. and Popescu, M. 2006. Comparative determination of certain antioxidnts in sea buckthorn (*Hippophae rhamnoides*) leaves, fruits, juice, oil and oil extraction residu. Farmacia **54(4)**: 97-103.

Mangathayaru, K., Sravan, K., Praveen K. Reddy, A., Munikumar, R., Swetha, B. and Umamaheswara Reddy, C. 2007. *In vitro* antioxidant studies of the aerial parts of *Origanum majoranum* Linn. and *Artemisia sieversiana* Ehrh. Pharmacognosy Mag. **3 (10)**: 90-94.

Manigauha, A., Ali, H., and Umamaheshwari, M. 2009. Antioxidant activity of ethanolic extract of *Piper betel*. J. Pharm. Res. **2(3)**: 491-494.

Manoj, G., Venkataraman, L.V. and Srinivas, L. 1992. Antioxidant properties of Spirulina (*Spirulina platensis* Seshadri et Bai). Spirulina MCRC **48**: 154-158.

Manonmani, G., Anbarasi, K., Balakrishna, K., Veluchamy, G. and Shyamala Devi, C.S. 2002. Effect of *Terminalia arjuna* on the antioxidant defense system in alloxan induced diabetes in rats. Biomedicine **22(3-4)**: 52-61.

Manonmani, G., Bhavapriya, V., Kalpana, S. Govindasamy, S. and Apparanantham, T. 2005. Antioxidant activity of *Cassia fistula* Linn. flowers in alloxan induced diabetic rats. J. Ethnopharmacol. **97 (1)**: 39-42.

Manosroi, A., Masahiki, A.B.E. and Manosroi, J. 1999. Comparison of antioxidant activity of extracts from the seeds of white pepper (*Piper nigrum* Linn.) to commercial antioxidants in 2 per cent hydroquinone cream. J. Cosmet. Sci. **50**: 221-229.

Manosroi, A., Saraphanchotiwitthaya, A and Manosroi, J. 2005. *In vitro* immunomodulatory effect of *Pouteria cambodiana* (Pierre ex Dubard) Baehni extract. J. Ethnopharmacol. **101(1-3)**: 90-94.

Mantena, S.K., Jagadish, Badduri, S.R., Siripurapu, K.B. and Unnikrishnan, M.K. 2003. *In vitro* evaluation of antioxidant properties of *Cocos nucifera* Linn. water. Nahrung **47(2)**: 126-131.

Mantle, D., Eddeb, F. and Pickering, A.T. 2000. Comparison on relative antioxidant activities of British medicinal plant species *in vitro*. J. Ethnopharmacol. **72 (1-2)**: 47-51.

Mantle, D., Wilkins, R.M., Gok, M.A. 2003. Comparison of antioxidant activity in commercial *Ginkgo biloba* preparations. J. Altern. Complement Med. **9(5)**: 625-629.

Marcel, K.K., Akhanovna, M.J., Yves-Alain, B., Gabin, D.B.M. and Jerome, Z.B.T. 2011. *In vitro* antioxidant activities of total flavonoids extracts from leaves and stems of *Adenia lobata* (Jacq.) Engl. (Passifloraceae). J. Pharmacogn. Phytother. **3(1)**: 8-12.

Marcocci, L., Maguire, J.J., Droy-Lefaix, M.T. and Parkes, L. 1994a. The nitric oxide scavenging effect of *Ginkgo biloba* extract EGb 761. Biochem. Biophys. Res. Commun. **201**: 748-755.

Marcocci, L., Packer, L., Droy-Lepaiz, M.T., Sekaki, A., Gardes-Albert, M. 1994b. Antioxidant action of *Ginkgo biloba* extract. Meth. Enzymol. **234**: 462-465.

Marianova, E.M. and Yanishelieva, N.V, 1997. Antioxidant activity of extracts from selected species of the family Lamiaceae in sunflower oil. Food Chem. **58**: 245-248.

Mariassova, M. 2006. Antioxidant activity of some herbal extracts in rapeseed and sunflower oils. J. Food Nutr. Res. **45(3)**: 104109.

Maridonneau, I., Braquet, P. and Garay, R.P. 1982. In: L. Frakas, M. Gabor, F. Kallay and H. Wagner (eds): Flavonoids in Bioflavonoids. Elsevier, Amsterdam pp. 427-436.

Marimuthu, P., Wu, C.L., Chand, H.-T. and Chang, S.-T. 2008. Antioxidant activity of the ethanolic extract from the bark of *Chaemaecyparis obtusa* var. *formosana*. J. Sci. Food Agric. **88**: 1400-1405.

Mariod, A.A., Ibrahim, R.M., Ismail, M. and Ismail, N. 2010. Antioxidant activities of phenolic rich fractions (PRFs) obtained from black mahlab (*Monechma ciliatum*) and white mahlab (*Prunus mahaleb*)seed cakes. Food Chem. **118(1)**: 120-127.

Mariutti, L. R. B., Barreto G. P. M., Bragagnolo, N., and Mercadante, A. Z. 2008. Free radical scavenging activity of ethanolic extracts from herbs and spices commercialized in Brazil. Braz. Arch. Biol. Technol. **51(6)**: 1225-1232.

Marja, P.K., Anu, I.H., Hiekki, J.V., Jussi-Pekka, R., Kalevi, P., Tytti, S.K. and Marina, H. 1999. Antioxidant activity of plant extracts containing phenolic compounds. J. Agric. Food Chem. 54: 1151-1157.

Marocci, L., Maguire, J.J. and Droy-Lefaix, M.T. 1994. The nitric oxide scavenging properties of *Ginkgo biloba* extract EGb761. Biochem. Biophysic. Res. Comm. **15**: 748-755.

Marocci, L., Packer, L. and Droy-Lefaix, M.T. 1994. Antioxidant action of *Ginkgo biloba* extract EGb761. Methods Enzymol. **234**: 462-475.

Maroo, J., Ghosh, A., Mathur, R., Vasu, V.T. and Gupta, S. 2003. Antidiabetic efficacy of *Enicostemma littorale* methanol extract in alloxan-induced diabetic rats. Pharmaceutical Biol. **41(5)**: 388-391.

Marongin, B., Porcedda, S., Piras, A., Rosa, A., Deina, M.and Dessi, M.A. 2004. Antioxidant activity of supercritical extract of *Melissa officinalis* subsp. *officinalis* and *Melissa officinalis* subsp. *inodora*. Phytother. Res. **18(10)**: 789-792.

Marquele, F.D., Di Mambro, V.M., Georgetti, S.R., Casagrande, R., Valim, Y.M., and Fonseca M.J. 2005. Assessment of the antioxidant activities of Brazilian extracts of *propolis* alone and in topical pharmaceutical formulations. J. Pharm. Biomed. Anal. **39 (3-4)**: 455-462.

Martinez, G., Delgado, R., Parez, G., Garrido, G., Selles, A.J.N. and Leon, O.S. 2000. Evaluation of *in vitro* antioxidant activity of *Mangifera indica* L. extract (Vimang). Phytother. Res. **14**: 424-427.

Martinez-Tome, M., Jimenez, A., Ruggieri, S., Frega, N., Strabbioli, R. and Murcia, M. 2001. Antioxidant properties of Mediterranean spices compared with common food additives. J. Food Protection **64**: 1412-149.

Martinez-Tome, M., Murcia, M., Frega, N., Ruggieri, S., Jimenez, A.M., Roses, F. and Parras, P. 2004. Evaluation of antioxidant capacity of cereal brans. J. Agric. Food Chem. **52(15)**: 4690-4699.

Maruta, Y., Kawabata, J. and Niki, R. 1995. Antioxidative caffeoylquinic acid derivatives in the roots of burdock (*Arctium lappa* L.). J. Agric. Food Chem.**43(10)**: 2592-2595.

Maruthappana, V. and Shreeb, K.S. 2010. *In vitro* and *in vivo* antioxidant activity of *Phyllanthus reticulatus.* Drug Invention Today **2(6)**:

Marwah, R.G., Fatope, M.O., Al Mahrooqi, R.S., Varma, G.B., Al Abadi, H. and Al-Burtamani, K.S. 2007. Antioxidant capacity of some edible and wound healing plants in Oman. Food Chem. **101**: 465-470.

Mary, N.K., Achutan, C.R., Babu, B.H. and Padikkala, J. 2003. *In vitro* antioxidant and antithrombotic activity of *Hemidesmus indicus* (L.) R.Br. J. Ethnopharmacol. **87(2-3)**: 187-191.

Marx, J.L. 1985. Oxygen free radicals linked to many diseases. Science **235**: 529-531.

Masahiro, T., Risa, M., Harutaka, Y. and Kazuhiro, C. 1996. Novel antioxidant isolated from *Perilla frutescens* Britton var. *crispa* (Thunb.) Biosci. Biotechnol. Biochem. **60**: 1093-1095.

Masaki, H., Sakaki, S., Atsumi, T. and Sakurai, H. 1995. Active oxygen scavenging activity of plant extracts. Biol. Pharmac. Bull. **18(1)**: 162-166.

Masao, H., Yang, X.W., Miyashiro, H. and Nabma, T. 1993. Inhibitory effects of monomeric and dimeric phenyl propanoids from mice on lipid peroxidation *in vivo* and *in vitro*. Phytother. Res. **7**: 395-401.

Masoko, P. and Eloff, J.N. 2007. Screening of 24 South African Combretum and 6 *Terminalia* (Combretaceae) species for antioxidant activities. Afr. J. Tradit. Complement. Altern. Med. **4(2)**: 231-239

Masteikova, R., Muselik, J., Bernatoniene, J. and Bernatoniene, R. 2007. Antioxidative activity of Ginkgo, Echinacea, and Ginseng tinctures. Medicina (Kaunas). **43(4)**: 306-309.

Masuda, T., Inaba, Y., Maekawa, T., Takeda, Y., Yamaguchi, H., Nakamoto, K., Kuninaga, H., Nishizato, S. and Nonaka, A. 2003. Simple detection method of powerful antiradical compounds in the raw extract of plants and its application for the identification of antiradical plant constituents. J. Agric. Food Chem. **51**: 1831-1838.

Masuda, T., Isobe, J., Jitoe, A. and Nakatani, N. 1992. Antioxidant curcuminoids from rhizomes of *Curcuma xanthorrhiza*. Phytochem. **31(10)**: 3645-3647.

Masuda, T., Yonemori., Oyama, Y., Takeda, Y., Tanaka, T. and Andoh, T. 1999. Evaluation of the antioxidant activity of environmental plants: activity of leaf extracts from seashore plants. J. Agric. Food Chem. **47**: 1749-1754.

Masuda, Y., Kikuzaki, H., Hisamoto, M. and Nakatani, N. 2004. Antioxidant properties of gingerol related compounds from ginger. BioFactors **21**: 293-296.

Masuoka, N., Isobe, T. and Kubo, I. 2006. Antioxidants from *Rabdosia japonica*. Phytother. Res. **20(3)**: 206-213.

Masuoka, C., Yokoi, K., Komatsu, H., Kinjo, J., Nohara, T. and Ono, M. 2007. Two novel antioxidant ortho-benzoyloxyphenyl acetic acid derivatives from the fruit of *Vaccinium uliginossum*. Food Sci Tech. Res. **13(3)**: 215-220.

Mateos, R.M., Leon, A.M., Sandalio, L.M., Gomez, M., Rio, L.A. and Palma, J.M. 2003. Peroxisomes from pepper fruits (*Capsicum annuum* L.): purification, characterisation and antioxidant activity. J. Plant Physiol. **160 (12)**: 1507-1516.

Materska, M. and Perucka, I. 2005. Antioxidant activity of the main phenolic compounds isolated from hot pepper fruit (*Capsicum annuum* L.) J. Agric. Food Chem. **53**: 1750-1756.

Mates, J.M., Perez-Gomez, C., Nunez de Castro, I. 1999. Antioxidant enzymes and human diseases. Clin. Biochem. 32: 595-603.

Mathew, G.M. and Sasikumar, J.M. 2007. Antioxidant activity of *Pseudarthria viscida*. Indian J. Pharm. Sci. **69**: 581-582.

Mathew, J., Latha, P.G. and George, V. 2004. Antioxidant flavonoids from *Vitex peduncularis*. J. Trop. Med. Plants **5(2)**: 217-222.

Mathew, S. and Abraham E.T. 2004. *In vitro* antioxidant activity and scavenging effects of *Cinnamomum verum* leaf extract assayed by different methodologies. Food. Chem. Toxicol. **44**: 198-206.

Mathew, S. and Kuttan, G. 1997. Antioxidant activity of *Tinospora cordifolia* and its usefulness in the amelioration of cyclophosphamide induced toxicity, J. Exp. Clin. Cancer Res. **16**: 407-410.

Mathiesen, L., Malterud, K.E, and Sund, R.B. 1995. Antioxidant activity of fruit exudate and C-methylated dihydrochalcones from *Myrica gale*. Planta Med. **61 (6)**: 515-518.

Mathisen, E., Diallo, D., Andersen, O.M. and Malterud, K.E. 2002. Antioxidants from the bark of *Burkea africana*, an African medicinal plant. Phytother. Res. **16(2)**: 148-153.

Mathur, A., Verma, S.K., Singh, S.K., Prasad, G.B.K.S. and Dua, V.K. 2010. Phytochemical investigation and antioxidant activities of some plants of Uttarakhand. IJPI's J. Pharmacogn. Herbal Formulations **1(1)**: 1-7.

Mathur, A., Verma, S.K., Purohit, Gupta, V., Dua, V.K., Prasad, G.B.K.S., Mathur, D., Singh, S.K. and Singh, S. 2010. Evaluation of *in vitro* antimicrobial and antioxidant activities of peel and pulp of some citrus fruits. J. Biotech. Biotherapeutics **1**: 1-17.

Mathur, A., Verma, S.K., Singh, S.K., Prasad, G.B.K.S. and Dua, V.K. 2010. Phytochemical investigation and *in vitro* antioxidant activities of some plants of Uttarakhand. J. Phrmacogn. Herbal Formulations **1**: 1-7.

Matkowski, A. and Piotrowska, M. 2006. Antioxidant and free radical scavenging activities of some medicinal plants from Lamiaceae. Fitoterapia **77**: 346-353.

Matkowski, A., Zielinska, S., Oszmianski, J. and Lamer-Zarawska, E. 2008. Antioxidant activity of extracts from leaves and roots of *Salvia miltiorrhiza* Bunge, *S. przewalskii* Maxim. and *S. verticillata* L. Bioresource Tech. **99(16)**: 7892-7896.

Matook, S.M. 2005. Antioxidant activities of water-soluble polysaccharides from Buncan (*Citrus grandis* Osbeck) fruit flavedo tissues. Pak. J. Biol. Sci. **8**: 1472-1477.

Matsingou, T.C., Kapsokefalou, M. and Salifoglou, A. 2001. Aqueous infusions of Mediterranean herbs exhibit antioxidant activity towards iron promoted oxidation of phospholipids, linoleic acid and deoxyribose. Free Radic. Res. **35**: 593-605.

Matsingou, T.C., Kapsokefalou, M. and Salifoglou, A. 2008. *In vitro* antioxidant activity of black tea and Mediterranean herb infusions toward iron under simultated gastrointestinal conditions. J. Food Sci. **65(6)**: 1060-1065.

Matsingou, T.C., Petrakiss, N., Kapsokefalou, M. and Salifoglou, A. 2003. Antioxidant activity of organic extracts from aqueous infusions of sage. J. Agric. Food Sci. **51**: 6696-6701.

Matsumoto, S., Hanai, T., Matsui, T., Oka, M., Tanaka, M. and Uemura, H. 2010. Eviprostat suppresses urinary oxidative stress in a rabbit model of partial bladder outlet obstruction and in patients with benign prostrate hyperplasia. Phytother. Res. **24(2)**: 301-303.

Matsumoto, M., Hara, H., Chiji, H. and Kasai, T. 2004. Gastroprotective effect of red pigments in black chokeberry fruit (*Aronia melanocarpa* Elliot) on actue gastric hemorrhagic lesions in rats. J. Agric. Food Chem. **52(8)**: 2226-2229.

Matsumoto, T., Moriguchi, R. and Yamada, H. 2005. J. Pharm. Pharmacol **56**: 537.

Mattei, R., Barros, M.P., Galvao, S.M.P., Bechara, E.J.H. and Carlini, E.L.A. 2001. *Heteropteris aphrodisiaca* O. Machado: Effects of extract BST 0298 on the oxidative stress of young and old rat brains. Phytother. Res. **15(7)**: 604-607.

Matthaus, B. 2002. Antioxidant activity of extracts obtained from residues of different oil seeds. Agric. Food Chem. **12**: 3444-3452.

Mattiazzi, M., Vijayvergia, C., Caiewski, C.D., De Vivo, D.C., Lenaz, G., Wiedmam, M. and Mambredi, G. 2004. The mt DNA T8993G (NARP) mutation results in an impairment of oxidative phosphorylation that can be improved by antioxidants. Hum. Mol. Genet. **13**: 868-879.

Mau, J.-L. Ko, P.-T. and Chyau, C.-C. 2003. Aroma characterization and antioxidant activity of supercritical carbon dioxide extracts from *Terminalia catappa* leaves. Food Research International. **36 (1)**: 97-104.

Mau, J.L., Lai E. Y.C., Chien-Chou, N., Chi-Huarng W., Charng-Cherng C. and Chyau. 2003. Composition and antioxidant activity of the essential oil from *Curcuma zedoaria*. Food Chem. **82(4)**: 583-591.

Mau, J.L., Lin, H.C. and Chen, C.c. 2002. Antioxidant properties of several medicinal mushrooms. J. Agric. Food Chem. **50(10)**: 2926-2930.

Mau, J.L., Chang, C.N., Huang, S.J. and Chen, C.C. 2004. Antioxidant properties of methanolic extracts from *Grifola frondosa*, *Morchella esculenta* and *Termitomyces albuminous* mycelia. Food Chem. **87**: 111-118.

Mavelli, I., Rossi, L., Authouri, F., Braquet, P. and Rotilio, G. 1983. In: G. Cohen and R.A.Greenwald (eds.) Oxyradicals Their Scavenger Cyst. Elsevier, New York.

Mavi, A., Terzi, Z., Özgen, U., Yildirin, A. and Coskun, M. 2004. Antioxidant properties of some medicinal plants – *Pragos ferulacea* (Apiaceae), *Sedum sempervioides* (Crassulaceae), *Malva neglecta* (Malvaceae), *Cruciata taurica* (Rubiaceae), *Rosa pimpinellifolia* (Rosaceae), *Galium verum* subsp. *verum* (Rubiaceae), *Urtica dioica* (Urticaceae). Biol. Pharm. Bull. **27(5)**: 702-705.

Mazumder, A., Saha, B.P., Basu S.P. and Mazumder, R. 2003. Preliminary assay on the antioxidant activity of *Maydis stigma* extract. Fitoterapia, **74**: 144-147.

Mazumder, M.E.H. and Rahman, S. 2008. Pharmacological evaluation of Bangladeshi medicinal plants for antioxidant activity. Pharm. Biol. **46(10-11)**: 704-709.

Mazumder, P.M., Sasmal, D. and Nambi, E.A. 2008. Antiulcerogenic and antioxidant effects of *Coccinia grandis* (Linn.) Voigt leaves on aspirin-induced gastric ulcer in rats. Nat. Prod. Radiance 7(1): 15-18.

Mazumder, U.K., Gupta, M. and Rajeswar, Y. 2005. Antihyperglycemic effect and antioxidant potential of *Phyllanthus niruri* (Euphorbiaceae) in streptozotocin induced diabetic rats. European Bull. Drug Res. **13(1)**: 15-23.

Mazza, G. and Oomah, B.-D. 2000. Chemistry, pharmacology and clinical applications of St, John's wort and *Ginkgo biloba*. In: Mazza, G. and Oomah, B.-D. (eds.). Herbs, Botanicals and Teas. Technomic Publishing Co. Inc., Lancaster, PA, USA, pp 131-176.

Mcanlis, G.T., Mceneny, J., Pearce, Young, I.S. 1999. Absorption and antioxidant effects of quercetin from onions, in man. Eur. J. Clin. Nutr. **53**: 92-96.

McCune, L.M. 1999. Antioxidants in Canadian boreal forest: indigenous medicinal plant treatments in relation to non-insulin dependent diabetes mellitus. Ph.D. thesis, McGill University.

McCune, L.M. and Johns, T. 2002. Antioxidant activity in medicinal plants associated with the symptoms of diabetes mellitus used by the indigenous peoples of the North American Boreal forest. J. Ethnopharmacol. **82(2-3)**: 197-205.

McCune, L.M. and Johns, T. 2007. Antioxidant activity relates to plant part, life form and growing condition in some diabetes remedies. J. Ethnopharmacol. **112(3)**: 461-469.

McDonald, S., Prenzler, P.D., Antolovich, M. and Robards, K. 2001. Phenolic content and antioxidant activity of olive oil extracts. Food Chem. **73**: 73-84.

McGaw, L.J., Steenkamp, V. and Eloff, J.N. 2007. Evaluation of Athrixia bush tea for cytotoxicity, antioxidant activity, caffeine content and presence of pyrrolizidine alkaloids. J. Ethnopharmacol. **110**: 16-22.

Mclnerney, J.K., Seccafien, C.A., Stewart, C.M. and Bird, A.R. 2007. Effects of high pressure processing on antioxidant activity and total carotenoid content and availability, in vegetables. Innov. Food Sci. Technol. **8(4)**: 543-548.

McKay, D.L. and Blumberg, J.B. 2006. A review of the bioactivity of south African herbal teas: rooibos (*Aspalathus linearis*) and honeybush (*Cyclopia intermedia*). Phytother. Res. **21(1)**: 1-16.

Mechergui, K., Coelho, J.A., Serra, M.C., Lamine, S.B., Boukhchina, S. and Khouja, M.L. 2010. Essential oils of *Origanum vulgare* L. subsp. *glandulosum* (Desf.) Ietswaart from Tunisia: chemical composition and antioxidant activity. J. Sci. Food Agri. **90**: 1745-1749.

Medina, A.L., Lucero, M.E., Holguin, F.O., Estell, R.E., Posakony, J.J., Simon, J and O'Connell M.A. 2005. Composition and antimicrobial activity of *Anemopsis californica* leaf oil. J. Agric. Food Chem. **53(22)**: 8694-8698.

Meenakshi, S., Manicka, G.D., Tamilmozhi, S., Arumugam, M. and Balasubramanian, T. 2009. Total flavonoid and *in vitro* antioxidant activity of two seaweeds of Rameshwaram coast. Global J. Pharmacol. **3**: 59-62.

Meera, R., Devi, P., Kameshwari, B., Madhumita, B. and Merlin, N.J. 2009. Antioxidant and hepatoprotective activities of *Ocimum basilicum* Linn. and *Trigonella foenum-graecum*, against H_2O_2 and CCl_4 induced hepatotoxicity in goat liver. Indian J. Exp. Bio. **47(7)**: 584-590.

Meera, R., Ramaseshu, K.V., Devi, P. and Kameshwari. 2008. Anti-inflammatory activity and antioxidant activity of *Rimelia reticulata*. Biomed. **3(1)**: 270-273.

Megala, J. and Geetha, A. 2010a. Free radical-scavenging and H+, K+-ATPase inhibition activities of *Pithecellobium dulce*. Food Chem. **121(4)**: 1120-1128.

Megala, J. and Geetha, A. 2010b. Gastroprotective and antioxidant effects of hydroalcoholic fruit extract of *Pithecellobium dulce* on ethanol induced gastric ulcer in rats. Pharmacology online **2**:

Meghashri, S., Vijay Kumar, H., and Gopal, S. 2010. Antioxidant properties of a novel flavonoid from leaves of *Leucas aspera*. Food Chem. **122(1)**: 105-110.

Megias, C., Pastor-Cavada, E., Torres-Fuentes, C., Giron-Calle, J., Alaiz, M., Juan, R., Pastor, J. and Vioque, J. 2009. Chelating, antioxidant and antiproliferative activity of *Vicia sativa* polyphenol extracts. Eur. Food Res. Tech. **230(2)**: 353-359.

Mehdipour, S., Yasa, N., Dehgan, G., Khorasani, R., Mohammadirad, A., Rahimi, R., Abdollahi, M. 2006. Antioxidant potentials of Iranian *Carica papaya* juice *in vitro* and *in vivo* are comparable to alpha-tocopherol. Phytother. Res. **20(7)**: 591-594.

Mehta, B.K., Pandit, V. and Gupta, M. 2009. New principles from seeds of *Nigella sativa*. Nat. Prod. Res. **23(2)**: 138-148.

Mehta, R.G. and Pezzuto, J.M. 2002. Discovery of cancer preventive agents from natural products: from plants to prevention. Curr. Oncol. Rep. **4(6)**: 478-486.

Mehta, R.L., Zayas, J.F. and Yang, s.S. 1994. Ajowan as a source of natural lipid antioxidant. J. Agric. Food Chem. **42**: 1420-1422.

Mehta, S., Kamboj, P., Faujdar, S., Sawale, J. and Kalia, A.N. 2010. *In vitro* antioxidant activity of *Cassia occidentalis* seeds. Pharmacology online **3**: 217-224.

Memnune, S., Yildiz, H., Gungor, N., Centin, B., Eser, Z. and Ercisli, S. 2009. Total phenolic content, antioxidant and antimicrobial activities of some medicinal plants. Pak. J. Phar. Sci. **22(1)**: 102-106.

Mendiola, J.A, Marín, F.R., Hernández, S.F., Arredondo, B.O., Señoráns, F.J., Ibañez, E. and Reglero, G. 2005 Characterization via liquid chromatography coupled to diode array detector and tandem mass spectrometry of supercritical fluid antioxidant extracts of *Spirulina platensis* microalga. J. Sep. Sci. **28 (9-10)**: 1031-1038.

Mendoza, E.M.T., Osawa, T., Nakayama, T., Laurena, A.C. and Kawakishi, S. 1994. Search for new natural antioxidants in selected tropical plant food materials. In: Uritani, I., Garcia, V.V., Mendoza, V.V. (eds.). Post Harvest Biochemistry of plant food materials in the tropics. Japan Scientific societies press, Tokyo, pp. 73-82.

Menezes, P.R., Schwarz, E.A. and Santos, C.A.M. 2004. *In vitro* antioxidant activity of species collected in Parana. Fitoterapia **75**: 398-400.

Meng. L., Lozano, Y.F., Gaydou, E.M. and Li, B. 2009. Antioxidant activities of polyphenols extracted from *Perilla frutescens* varieties. Molecules **14**: 133-140. doi: 10.3390/molecules14010133.

Mensah, A.Y., Houghton, P.J., Akyirem, G.N., Fleischer, T.C., Mensah, M.L., Sarpong, K. and Adosraku, R. 2004. Evaluation of the antioxidant and free radical scavenging properties of *Secamone afzelii* Roem. Phytother. Res. **18(12)**: 1031-1032.

Mensor, L.I., Menezes, F.S., Leiton, A.S., Reis, A.S., Santos, T.C., Coube, C.S. and Leitao, S.G. 2001. Screening of Brazilian plant extracts for antioxidant activity by the use of DPPH free radical method. Phytother. Res. **15 (2)**: 127-130.

Meot-Duras, L., Floch, G.L. and Magne, C. 2008. Radical scavenging antioxidant and antimicrobial activities of halophytic species. J. Ethnopharmacol. **116(2)**: 258-262.

Meral, G. and Konyahoglu, S. 2003. Comparison of total flavonoids and phenol contents and antioxidant capacities of three *Hypericum* L. species growing in Turkey. Acta Pharmaceutica Turcia **45**: 183-185.

Mercan, N., Duru, M.E., Turkoglu, A., Gezer, K., Kivrak, I. and Turkoglu, H. 2006. Antioxidant and antimicrobial properties of ethanolic extract from *Lepista nuda* (Bull.) Cooke. Annals of Microbiol. **56 (4)**: 339-344.

Merlani, M., Barbakadze, V., Amiranashvili, L., Gogilashvili, L., Yannakopulou, E., Papadopoulos, K. and Chankvetadze, B. 2010. Enantioselective synthesis and antioxidant activity of 3(-(3,4-dihydroxyphenyl)-glyceric acid-basic monomeric moiety of a biologically active polyether from *Symphytum asperum* and *S. caucasicum*. Chirality **22(8)**: 717-725.

Metzger, B.T., Robbins, M.G. and Barnes, D.M. 2010. Longitudinal expression of antioxidant phytochemicals in buckwheat (*Fagopyrum esculentum* Moench.). J. Herbs Spices Med. Plants **16(2)**: 106-118.

Meunier, M.T., Durox, E. and Bastide, P. 1989. Plant Medicine. Phytother. **23**: 267.

Meyer, A.S., Donovan, J.L., Pearson, D.A., Waterhoue, A.I. and Frankel, E.N. 1997. Inhibition of human low-density lipoprotein oxidation in relation to composition of phenolic antioxidants in grapes. J. Agric. Food Chem. **45**: 1638-1643.

Meyer, A.S., Donovan, J.L., Pearson, D.A., Waterhouse, A.I. and Frankel, E.N. 1998. Fruit hydroxycinnamic acids inhibit human low-density lipoprotein oxidation *in vitro*. J. Agric. Food Chem. **46**: 1783-1787.

Meyer, B., Schneider, W. and Elstner, E.F. 1995. Antioxidative properties of alcoholic extracts from *Fraxinus excelsior, Populus tremula* and *Solidago virgaurea*. Arzneimittel-Forschung **45(2)**: 174-176.

Meyers, K.J., Watkins, C.B., Pritts, M.P. and Liu, R.H. 2003. Antioxidant and antiproliferative activities of strawberries. J. Agric. Food Chem. **51(23)**: 6887-6892.

Mezadri, T., illano, D., Fernandez-Pachon, M.S., Garcia-Parirlla, M.C. and Tronscoso, A.M.2008. Antioxidant compounds and antioxidant activity in acerola (*Malpighia emarginata* DC.) fruits and derivatives. J. Food Compos. Anal. **21(4)**: 282-290.

Michael, H.N., Awad, H.M., El-Sayeed, N.H. and Pare, P.W. 2010. Chemical and antioxidant investigations: Norfolk pine needles (*Araucaria excelsa*). Pharm. Biol. **48(5)**: 534-538.

Middleton, P., Stewart, F., Al-Qahtaani, S., Egan, P., O'Rourke, C., Abdulrahman, A., Byres, M., Middleton, M. Kumarasamy, Y., Shoeb, M., Nahar, L., Delazar, L. and Sarker, S.D. 2005. Antioxidant, antibacterial activities and general toxicity of *Alnus glutinosa*, *Fraxinus excelsior* and *Papaver rhoeas*. Iranian J. Pharm. Res. **2**: 81-86.

Mighri, H., Hajloui, H., Akrout, A., Najjaa, H. and Neffati, M. 2010. Antimicrobial and antioxidant activities of *Artemisia herba-alba* essential oil cultivated in Tunisian arid zone. C.R. Chim. **13**: 380-386.

Miguel, G., Simones, M., Figueiredo, A.C., Barroso, J.G., Pedro, L.G. and Carvalho, L. 2004. Composition and antioxidant aactivities of the essential oils of *Thymus caespititius*, *Thymus camphoratus* and *Thymus mastichina*. Food Chem. **86**: 183-188.

Miguel, M.G. 2010. Antioxidant and anti-inflammatory activities of essential oils: A short review. Molecules **15**: 9252-9287. doi: 10.3390/molecules 15129252.

Miguel, M.G., Cruz, C., Faleiro, L., Simones, M.T.F., Figueiredo, A.C., Barrosa, J.G. and Pedro, L.G. 2010. *Foeniculum vulgare* essential oils: Chemical composition, antioxidant and antimicrobial activities. Nat. Prod. Comm. **5(2)**: 319-328.

Mikhaeil, B.R., Badria, F.A., Matooq, G.T., Amer, M.A.A. 2004. Antioxidant and immunomodulatory constituents of Henna leaves. Z. Naturforsch. **59**: 468-476.

Mikre, W., Rohloff, J. and Hymete, A. 2007. Volatile constituents and antioxidant activity of essential oils obtained from important aromatic plants of Ethiopia. J. Essential Oil Bearing Plants **10(6)**: 465-474.

Miladi, S. and Damak, M. 2008. *In vitro* antioxidant activities of *Aloe vera* leaf skin extracts. J. de la Soc. Chimiqu de tunisie **10**: 101-109.

Mildner-Szkudlarz, S., Zawirska-Wojtasiak, R., Obuchowski, W. and Goslinski, M. 2009. Evaluation of antioxidant activity of green tea extract and its effect on the biscuits lipid fraction oxidative stability. J. Food Sci. **74(8)**: S 362 – S 370.

Miliauskas, G., van Beek, T.A., Venskutonis, P.R., Linssen, J.P.H. and de Waard, P. 2004. Antioxidative activity of *Geranium macrorrhizum*. Eur. Food Res. Technol. **218**: 253-261.

Miliauskas, G., van Beek, T.A., Venskutonis, P.R., Linssen, J.P.H., de Waard, P., Sudholter, E.J.R. 2009. Antioxidant activity of *Potentilla fruticosa*. J.Sci. Food Agric. **84(15)**: 1997-2009.

Miliauskas, G., va Beok, T.A., de Waard, P., Venskutonis, B.P. and Sucholter, F.J.R. 2005. Identification of radical scavenging compounds in *Rhaponticum carthamoides* by means of LC-DAD-SPE-NMR. J. Nat. Prod. **68(2)**: 168-172.

Miliauskas, G., Venskutonis, P.R. and van Beek, T.A. 2004. Screening of radical scavenging activity of some medicinal and aromatic plant extracts. Food Chem. **85(2)**: 231-237.

Miller, H.E., Rigelhof, F., Marquart, L., Prakash, A. and Kanter, M. 2000. Antioxidant content of whole grain breakfast cereals, fruits and vegetables. J. Am. College of Nutr. **19**: 1-8.

Miller, N.J., Castelluccio, C., Tijburg, L.and Rice-Evans, C. 1996. The antioxidant properties of the aflavins and their gallate esters-Radical scavengers or metal chelators? FEBS Lett. **392**: 40-44.

Miller, N.J. and Rice-Evans, C.A. 1997. Factors influencing the antioxidant activity determined by ABTS+ radical cation assay. Free Radic. Res. **26**: 195-199.

Miller, N.J. and Rice-Evans, C.A. 1997.The relative contributions of ascorbic acid and phenolic antioxidants to the total antioxidant activity of orange and apple fruit juice and black currant drink. Food Chem. **60**: 331-337.

Miller, N.J., Rice-Evans, C., Davies, M.J., Gopinathan, V. and Milner, A. 1993. A novel method for measuring antioxidant capacity and its application to monitoring the antioxidant status in premature neonates. Clin. Sci. **84**: 407-412.

Milovanovic, M., Picuric-Jovanovic, K., Vucelic-Radovic, B. and Vrbaski, Z. 1996. Antioxidant effects of flavonoids of *Anthriscus sylvestris* in lard. J. Amer. Oil Chemists' Soc. **73(6)**: 773-776.

Mimica-Dukic, N., Bozin, B., Sokovic, M., Mihajlovic, B. and Matavulj. 2003. Antimicrobial and antioxidant activities of three *Mentha* species essential oil. Plant Med. **69**: 413-419.

Mimica-Dukic, N., Bozin, B., Sokovic, M., Simin, N. 2004. Antimicrobial and Antioxidant activities of *Melissa officinalis* L. (Laminaceae) essential oil J. Agric. Food Chem. **52(9)**: 2485-2489.

Mimica-Dukic, N., Bugarin, N., Grbovi, S., Miti-Culafi, D., Vukovi-Gaci, B., Orci, D., Jovin, E. and Couladis, M. 2010. Essential oil of *Myrtus communis* L. as a potential antioxidant and antimutagenic agents. Molecules **15(4)**: 2759-2770.

Mimica-Dukic, N., Krstic, L. and Boza, P. 2005. Effect of *Solanum* species (*Solanum nigrum* L. and *Solanum dulcamara* L.) on lipid peroxidation in lecitine liposome. Oxid. Commun. **28**: 536-546.

Mimica-Dukic, N., Simin, N. Cvejic, J., Jovin, E, Orcic, D. and Bozin, B. 2008. Phenolic compounds in field horsetail (*Equisetum arvense* L.) as natural antioxidants. Molecules **13**: 1455-1464.

Min, B.S., Na, M.K., Oh, S.R., Ahn, K.S., Jeong, G.S., Li, G., Lee, S.K., Joung, H. and Lee, H.K. 2004. New furofuran and butyrolactone lignans with antioxidant activity from the stem bark of *Styrax japonica*. J. Nat. Prod. **67(12)**: 1980-1984.

Miniyar, P.B., Chitre, T.S., Deuskar, H.J., Karve, S.S. and Jain, K.S. 2008. Antioxidant activity of ethyl acetate extract of *Aquilaria agallocha* on nitrate-induced methaeglobin formation. Int. J. Green Pharmacy **2(2)**: 116-117.

Miquel, J., Bernd, A., Sempere, J.M. and Diaz-Alperi, R.A. 2002. The curcuma antioxidants: pharmacological effects and prospects future clinical use. A review. Arch. Gerontol. Geriatr. **34**: 37-46.

Mira, L., Silva, M. and Manso, C.F. 1994. Scavenging reactive oxygen species by silibinin dihemisuccinate. Biochem. Pharmacol. **48(4)**: 753-759.

Mirada, M.S., Cintra, R.G., Barros, S.M. and Mancini-Filho, J. 1998. Antioxidant activity of the microalga *Spirulina maxima*. Braz. J. Med. Biol. Res. **31**: 1075-1079.

Mirunalini, S. and Karthiswaran, K. 2010. Evaluation of the antioxidant potential of *Pergularia daemia* Forsk. in experimental oral carcinogenesis and oral carcinoma cell line. 2nd Intern. Seminar on Medicinal plants and Herbal products. S.V.University, Tirupati, pp. 123-124.

Mirunalini, S., Kumaraguruparan, R., Subapriya, R. and Nagini, S. 2004. Garlic oil enhances hepatic and blood antioxidants during hamster buccal pouch carcinogenesis. Pharmaceutical Biol. **42(3)**: 240-245.

Mirzaei, A., Akbartabar, M., Sadeghi, H. and Sharifi, B. 2010. The antioxidant activities and total phenolic of *Artemisia maritima*, *Achillea millefolium* and *Matricaria recutita*. Archives of SID.

Mishima, S., Saito, K., Maruyama, H., Inoue, M. and Yamashita, T. 2004. Antioxidant and immuno-enhancing effects of *Echinacea purpurea*. Biol. Pharmaceutical Bull. **27(7)**: 1004-1009.

Mishra, A., Bapat, M. M., Tilak J. C. and Devasagayam, T.P.A. 2006. Antioxidant activity of *Garcinia indica* (kokam) and its syrup. Curr. Sci. **91(1)**: 90-92.

Mishra, J., Srivastava, R.K., Shukla, S.V. and Raghav, C.S. 2007. Antioxidants in aromatic and medicinal plants. Science Tech Entrepreneur. Pp1-16.

Misik, V., Bezáková, L., Máleková, L. and Kostálová, D. 1995. Lipoxygenase inhibition and antioxidant properties of protoberberine and aporphine alkaloids isolated from *Mahonia aquifolium*. Planta Med. **61(4)**: 372-373.

Misra, L. and Wagner, H. 2007. Extraction of bioactive principles from *Mucuna pruriens* seeds. Indian J. Biochem. Biophys. **44(1)**: 56-60.

Mitchell, R.J., Sasidharan, S., Vijayarathna, S. and Ramanathan, S. 2009. Antioxidant properties of *Ganoderma boninense*. J. Trop. Med. Plants 10(1): 23-26.

Mitic, M.N., Obradovic, M.V., Grahovac, Z.B. and Pavlovic, A.N. 2010. Antioxidant capacities and phenolic levels of different varieties of Serbian white wines. Molecules **15(3)**: 2016-2027.

Miura, K., Kikuzaki, H. and Nakatani, N. 2002. Antioxidant activity of chemical components from sage (*Salvia officinalis* L.) and Thyme (*Thymus vulgaris* L.) measured by the oil stability index method. J. Agric. Food Chem. **50**: 1845-1851.

Miura, K. and Nakatani, N. 1989. Antioxidative activity cr flavonoids from thyme (*Thymus vulgaris* L.). Agric. Biol. Chem. **53**: 3043-3045.

Miura, S., Watanabe, J., Tomita, T., Sano, M. and Tomita, I. 1994. The inhibitory effects of tea polyphenols (flavan-3-ol derivatives) on Cu^{2+} mediated oxidative modification of low density lipoprotein. Biol. Pharm. Bull. **17**: 1567-1572.

Miyajima, Y., Kikuzaki, H., Hisamoto, M. and Nikatani, N. 2004. Antioxidative polyphenols from berries of *Pimenta dioica*. Biofactors **21(1-4)**: 301-303.

Miyake, Y., Ito, C., Itoigawa, M. and Osawa, T. 2009. Antioxidants produced by *Eurotium herbariorum* of filamentous fungi used for the manufacture of Karebushi, dried Bonito (Katsuobushi). Biosci. Biotech. Biochem. **73(6)**: 1323-1327.

Mkaddem, M., Bouajila, J., Ennajar, M., Lebrihi, A., Mathieu, F. and Romdhane, M. 2009. Chemical composition and antimicrobial and antioxidant activities of *Mentha* (*longifolia* L. and *viridis*) essential oils. J. Food Sci. **74(7)**: 358-363.

Modak, B., Rojas, M., Torres, R., Rodilla, J. and Luebert, F. 2007. Antioxidant activity of a new aromatic geranyl derivative of the resinous exudates from *Heliotropium glutinosum* Phil. Molecules **12**: 1057-1063.

Modak, B., Torres, R., Lissi, E. and Monache, F.D. 2003. Antioxidant capacity of flavonoids and a new arylphenol of the resinous exudates from *Heliotropium sinuatum*. Nat. Prod. Res. **17(6)**: 403-407.

Mohammad, S. 2011. Antioxidant activity of kiwifruit (*Actinidia chinensis*). Pharmacology online **1**: 758-764.

Mohamed, H., Lajis, N.H., Abas, F., Ali, A.M., Sukari, M.A., Kikuzaki, H. and Nakatani, N. 2005. Antioxidative constituents of *Etlingera elatior*. J. Nat. Prod. **68(2)**: 285-288.

Mohamed, T.K. 2004. Chemical constituents and antioxidant activity of *Citrus paradisi* (star-ruby red grapefruit) and *Citrus sinensis* (blood sweet orange) Egyptian cultivars. Asian J. Chem. **16(3-4)**: 1753-1764.

Mohammed, A.F., Martina, W., Gudrum Sr, and Ulrike, L. 2007. Antioxidant, antimicrobial and cytotoxic activities of selected medicinal plants from Yemen. J. Ethnopharmacol. **111(3)**: 657-666.

Mohan, S., Abdul, A.B., Wahab, S.I.A., Al-Zubairi A.S., Elhassan, M.M. and Yousif, M. 2008. Antibacterial and antioxidant activities of *Typhonium flagelliforme* (Lodd.) Blume Tuber. Am. J. Biochem. Biotechnol., **4**: 402-407. DOI: 10.3844/ajbbsp.2008.402.407.

Mohana Rao G.M., Venkateswararao, C., Rawat, A.K.S., Pushpangadan, P. and Shirwaikar, A. 2005. Antioxidant and antihepatotoxic activities of *Hemidesmus indicus* R. Br. Acta Pharmaceutica Turcica **47**: 107 – 113.

Mohandas, J., Marshall, J.J., Duggin, G.G., Horvath, J.S. and Tiller, D. 1984. Differential distribution of glutathione and glutathione related enzymes in rabbit kidney: possible implications in analgesic neuropathy. Cancer Res. **44**: 5086–91.

Mohd-Esa, N., Hern, F.S., Ismail, A. and Ye, C.L. 2010. Antioxidant activity in different parts of roselle (*Hibiscus sabdariffa* L.) extracts and potential exploitation of the seeds. Food Chem. **122(4)**: 1055-1060.

Mohsen, S.M. and Ammar, A.S.M. 2009. Total phenolic contents and antioxidant activity of corn tassel extracts. Food Chem. **112(3)**: 595-598.

Mokbel, M. S. and Suganuma, T. 2006. Antioxidant and antimicrobial activities of the methanol extracts from pummelo (*Citrus grandis* Osbeck) fruit albedo tissues. European Food Res. and Technol. **224**: 39–47.

Molyneux Philip. 2004. The use of the stable free radical diphenylpicrylhydrazyl (DPPH) for estimating antioxidant activity. Songklanakarin J. Sci. Technol. **26(2)**: 211-219.

Momeni, J., Ntchatchoua, W.P.D., Akam, F.M.T. and Ngassoum, M.B. 2010. Antioxidant activities of some Cameroonian plants extracts used in the treatment of intestinal and infectious diseases. Indian J. Pharm. Sci. **72(1)**: 140-144. doi: 10.4103/0250-474x62236.

Momin, R.A., and Nair, M.G. 2002. Antioxidant, cyclooxygenase and topoisomerase inhibitory compounds from *Apium graveolens* Linn. seeds. Phytomedicine. **9 (4)**: 312-318.

Monagas, M., Garrido, I, Lebron Aguilar, R., Bartolome, B. and Gomez Cordoves, C. 2007. Almond (*Prunus dulcis* (Mill.) D.A. Webb) skin as a potential source of bioactive polyphenols. J. Agric. Food Chem. **55(21)**: 8498-8507.

Mondal, S.K., Chakraborty, G., Gupta, M. and Mazumder, U.K 2006. *In vitro* antioxidant activity of *Diospyros malabarica* Kostel bark. Indian J. Exp. Biol. **44 (1)**: 39-44.

Mongelli, E., Desmarchelier, C., Rodsiguez, T.J., Coussio, J. and Ciccia, G. 1997. *In vitro* antioxidant and cytotoxic activity of extracts of *Baccharis cordifolia* DC. J. Ethnopharmacol. **58**: 157-163.

Monteiro, M.V.B., de Melo Leite, A.K.R., Bertini, L.M., de Morais, S.M. and Nunes-Pinheiro, D.C.S. 2007. Topical anti-inflammatory, gastroprotective and antioxidant effects of the essential oil of *Lippia sidoides* Cham. Leaves. J. Ethnopharmacol. **111(2)**: 378-382.

Montoro, P., Tuberoso, C.I.G., Piacente, S., Perrone, A., De Feo, V., Cabras, P. and Pizza, **C.** 2006. Stability and antioxidant activity of polyphenols in extracts of *Myrtus communis* L. berries. J. Pharmaceutical and Biomedical Analysis. **41 (5)**: 1614-1619.

Mooi, J.L.Y., Lajis, N.H., Ali, A.M., Sukari, M.A., Rahman, A.A., Othman, A.G., Kikuzaki, H. and Nakatani, N. 2003. Antioxidant and antitumor promoting activities of the flavonoids from *Hedychium thyrsiforme*. Pharmaceutical Biol. 41(7): 506-511.

Moon, H.I, Min, B.S., Lee H. K., and Zee. O.P. 2002. Antioxidant compounds of *Oryza sativa* L. Korean J. Pharmacogn. **33**: 173-176.

Mori, J., Iwashima, M., Wakasugi, H., Saito, H., Matsunaga, T., Ogasawa, M., Takahashi, S., Suzuki, H. and Hayashi, T. 2005. New Plastoquinones isolated from the brown alga, *Sargassum micracanthum*. Chem. Pharm. Bull. **53(9)**: 1159-1163.

Morquio, A., Rivera-Megret, F. and Dajas, F. 2005. Photoprotection by topical application of *Achyrocline satureioides* ('Marcela'). Phytother. Res. **19(6)**: 486-490.

Morrison, J.F. and Twumasi, S.K. 2010. Comparative studies on the *in vitro* antioxidant properties of methanolic and hydro-ethanolic leafy extracts from eight edible leafy vegetables of Ghana. Afr. J. Biotech **9(32)**: 5177-5184.

Morteza-Semnani, K., Saeedi, M. and Nozadi, Z. 2007. Comparison of antioxidant activity of green tea extract to commercial antioxidant in 2 per cent hydroquinone cream. J. Herbs, Spices Med. Plants **13(1)**: 1-9.

Morteza-Semnani, K., Saeedi, M. and Shanavaz, B. 2003. Comparison of antioxidant activity of extract from roots of liquorice (*Glycyrrhiza glabra* L.) to commercial antioxidants in 2 per cent hydroquinone cream. J. Cosmet. Sci. **54(6)**: 551-558.

Mosaddak, M.A., Banbury, L., Forster, P., Booth, R., Larkham, J., Leach, D. and Waterman, P.G. 2004. Screening of some Australian Flacourtiaceae species for *in vitro* antioxidant, cytotoxic and antimicrobial activity. Phytomedicine **11(5)**: 461-466.

Mosquera, O.M., Correa, M., Buitrago, D.C. and Nino, J. 2007. Antioxidant activity of twenty five plants from Colombian biodiversity. Mem. Inst. Oswaldo Cruz. Rio de Janeiro **102(5)**: 631-634.

Mosquera, O.M., Correa, M., Nino, J. 2009. Antioxidant activity of plant extracts from Colombian flora. Revista Brasileira de Farmacognosia 19(2): doi: 10.1059/S0102-695 x 2009000300008.

Mossa, A.T.H. and Nawwar, G.A.2011. Free radical scavenging and antiacetylcholinesterase activities of *Origanum majorana* L. essential oil. Hum. Exp. Toxicol. Doi: 10.1177/0960327110391686.

Mostafa, M., Nahar, N., Mosihuzzaman, M., Makhmoor, T., Choudhary, M.I. and Rahman, A.U. 2007. Free radical scavenging phenylethanoid glycosides from *Leucas indica* Linn. Nat. Prod. Res. **21(4)**: 354-361.

Mota. C.S., Freita, R.B., Athayede, M.L., Boligon, A.A., Augusti, P.R., Somacol, S., Rocha, M.P. and Bauermann, L.F. 2011. Effect of *Vernonia cognata* on oxidative damage induced by ethanol in rats. Hum. Exp. Toxicol. **30(7)**: 675-684. doi: 10.1177/0960327110377646.

Motalleb, G., Hanachi, P., Kua, S.H., Fauziah, O. and Asmah, R. 2005. Evaluation of phenolic content and Total antioxidant activity in *Berberis vulgaris* fruit extract. J. Biol. Sci. **5(5)**: 648-653.

Motamed, S.M. and Naghibi, F. 2010. Antioxidant activity of some edible plants of the Turkmen Sahra region in northern Iran. Food Chem. **119(4)**: 1637-1642.

Mothana, R.A.A., Abdo, S.A.A., Hasson, S., Althawab, F.M.N., Alagbari, S.A.Z. and Lindequist, U. 2010. Antimicrobial, antioxidant and cytotoxic activities and phytochemical screening of some Yemeni medicinal plants. Evidence-Based Compl. Alt. Med. **7(3)**: 323-330.

Mothana, R.A., Al-Rehaily, A.J. and Schultze, W. 2010. Chemical analysis and biological activity of the essential oils of two endemic Soqotri *Commiphora* species. Molecules **15**: 689-698.

Mothana, R.A., Lindequist, U., Gruenert, R., and Bednarski, P.J. 2009. Studies on the *in vitro* anticancer, antimicrobial and antioxidative potentials of selected Yemeni medicinal plants from the island Soqotra. BMC Compl. Alt. Med. **9(2)**: DOI: 10.1186/1472-6882-9-7.

Mounnissmy, V.M., Kavimani, S., Balu, V. and Darlin Quine S. 2008. Effect of eethanol extract of *Cansjera rheedii* J. Gmelin (Opiliaceae) on Hepatotoxicity. J. Pharmacol. Toxicol. **3**: 158-162. doi: 10.3923/jpt.2008.158.162.

Moure, A., Cruz, J.M., Franco, D., Dominiquez, J.M., Sinero, J., Dominiquez, H., Nunej, M.J. and Paraja, J.C. 2001. Natural antioxidants from residual sources. Food Chem. **72**: 145-171. doi: 10.1016/S0308-8146(00)00223-5.

Moure, A., Franco, D., Sineiro, J., Dominguez, H., Nunez, M.J. and Lema, J.M. 2001. Antioxidant activity of extracts from *Gevuina avellana* and *Rosa rubiginosa* defatted seeds. Food Res. Intern. **34(2-3)**: 103-109.

Moure, A., Franco, D., Sineiro, J., Domínguez, H., Núñez, M.J. and Lema, J.M. 2000. Evaluation of extracts from *Gevuina avellana* hulls as antioxidants. J. Agric. Food Chem. **48 (9)**: 3890-3897.

Moussa, A.M., Emam, A.M., Diab, Y.M., Mahmoud, M.E. and Mahmoud, A.S. 2011. Evaluation of antioxidant potential of 124 Egyptian plants with emphasis on the action of *Punica granatum* leaf extract on rats. Int. Food Res. J. **18**: 535-542.

Moussaieff, A., Fride, E., Amar, Z., Lev, E, Steinberg, D., Gallily, R. and Mechoulam, R. 2005. Jerusalem balsam: From the Franciscan monastery in the old city of Jerusalem to Martindale 33. J. Ethnopharmacol. **101(1-3)**: 16-26.

Mowla, S.B., Thomson, J.A., Farrant, J.M. and Mundress, S.G. 2002. A novel stree-induced antioxidant enzyme identified from the resurrection plant *Xerophyta viscosa* Baker. Planta **215(5)**: 716-726.

Moyer, R., Hummer, K., Wrolstad, R.E. and Finn, C. 2000. Antioxidant compounds in diverse *Ribes* and *Rubus* germplasm. ISHS Acta Horticulturae 585: VIII International *Rubus* and *Ribes* Symposium. P.1-2.

Moyer, R.A., Hummer, K.E., Finn, C.E., Frei, B., Wrolstad, R.E. 2002. Anthocyanins, phenolics, and antioxidant capacity in diverse small fruits: *Vaccinium, Rubus,* and *Ribes*. J. Agric. Food Chem. **50(3)**: 519-525.

Moyo, M., Ndhlala, A.R., Finnie, J.F. and van Staden, J. 2010. Phenolic composition, antioxidant and acetylcholinesterase inhibitory activities of *Sclerocarya birrea* and *Harpephyllum caffrum* (Anacardiaceae) extracts. Food Chem. **122(1)**: 69-76 doi: 10.1016/f.foodchem.2010.03.130.

Mrudula, G., Mallikharjuna Rao, P., Jayaveera, K.N., Arshad, M.D. Rehaman, S.A. 2010. Antistress and antioxidant effects of *Prunella vulgaris* leaves. Pharmacology online **2**: 952-962.

Muanda, F., Kone, D., Dicko, A., Soulimani, R. and Younos, C. 2011. Phytochemical composition and antioxidant capacity of three Malian medicinal plants. Evid. Based Compl. Alternat. Med. Doi: 10.1093/ecam/nep109.

Muchuweti, M., Kativu, E., Mupure, C.H., Chidewe, C., Ndhalala, A.R. and Benhura, M.A.N. 2007. Phenolic composition and antioxidant properties of some spices. Amer. J. Food Tech. **5**: 414-420.

Muhammad, I., Takamatsu, S., Walker, L.A., Mossa, J.S., Fong, H.H.S. and El-Feraly, F.S. 2003. Cytotoxic and antioxidant activities of alkylated benzoquinones from *Maesa lanceolata*. Phytother. Res. **17(8)**: 887-891.

Mukherjee, S., Jagtap, S.D., Kuvalekar, A.A., Kale, Y.B., Kulkarni, O.P., Harsulkar, A.M. and Pawar, P.K. 2010. Demonstration of the potential of *Hibiscus cannabinus* Linn. flowers to manage oxidative stress, bone related disorders and free radical induced DNA damage. Indian J. Nat. Prod. Resour. **1 (3)**: 322-327.

Muller, K. and Ziereis, K. 1994. The antipsoriatic *Mahonia aquifolium* and its active constituents; I. Pro- and antioxidant properties and inhibition of 5-lipoxygenase. Planta Medica **60(5)**: 421-424.

Muller-Hocker, J. 1989. Cytochrome-c-oxidase deficient cardiomyocytes in the human hear – an age related phenomenon. A histochemical ultracytochemical study. Amer. J. Pathol. **134**: 1167-1173.

Mulyaningsih, S., Youns,., El-Readi, M.Z., Ashour, M.L., Nibret, E., Sporer, F., Hermann, F., Reiching, J. and Wink, M. 2010. Biologically activity of the essential oil of *Kadsura longipedunculata* (Schisandraceae) and its major components. J. Pharm. Pharmacol. **62(8)**: 1037-1044.

Munasinghe, T.C., Seneviratne, C.K., Thabrew, M.I. and Abeysekere, A.M. 2001. Antiradical and antilipoperoxidative effects of some plant extract used by Sri Lankan traditional medical practitioners for cardioprotection. Phytother. Res. **15(6)**: 519-523.

Munim, A., Negishi, O., Ozawa, T. 2003. Antioxidant compounds from *Crotalaria sessiliflora*. Biosci. Biotech. Biochem. **67**: 410-414.

Murakami, N., Murakami, T., Kadoya, M., Matsuda, H., Yamahara, J. and Yoshikawa, M. 1996. New hypoglycaemic constituents in 'gymnemic acid" from *Gymnema sylvestre*. Chem. Phrm. Bull. **44**: 469-471.

Murali, A., Ashok, P., Madhavan, V. and Raju, A. 2010. *In vitro* and *in vivo* antioxidant activity studies on the leaves of *Smilax zeylanica* L. (Smilacaceae). J. Pharm. Res. 3(10): 2427-2430.

Murali, A., Purnima, A. and Madhavan, V. 2010. Antioxidant activity of roots of *Hemidesmus indicus* var. *pubescens* – an *in vitro* study. Pharmacology online **3**: 121-129.

Murcia, M., Egea, I., Romojaro, F., Parras, P., Jimenez, A. and Martinez-Tome, M. 2004. Antioxidant evaluation in dessert spices compared with common food additives. Influence of irradiation procedure. J. Agric. Food Chem. **52**: 1872-1881.

Murcia, M.A., Jimenez-Monreal, A.M., Garcia-Diz, L., Carmona, M., Maggi, I. and Martinez-Tome, M. 2009. Antioxidant activity of minimally processed (in modified atmospheres), dehydrated and ready-to-eat vegetables. Food Chem. Toxicol. **47(8)**: 2103-2110.

Murcia, M.A., Jimenez Antonia, M. and and Martinez-Tome, M. 2009. Vegetables antioxidant losses during industrial processing and refrigerated storage. Food Res. Int. **42(8)**: 1046-1056.

Murias, M., Jodynis-Liebert, Matlawska, I. and Bylka, W. 2005. Antioxidant activity of isocytisoside and extracts of *Aquilegia vulgaris*. Fitoterapia **76**: 476-480.

Muriel, P., Garciapina, T., Perez-Alvarez, V. and Mourelle, M. 1992. Silymarin protects against paracetamol-induced lipid peroxidation and liver damage. J. Applied Toxicol. **12(6)**: 439-442.

Murkovic, M., Abuja, P.M., Bergmann, A.R., Zirngast, A., Adam, U., Winklhoefer-Roob, B.M. and Toplak, H. 2004. Effects of elderberry juice on fasting and postprandial serum lipids and low-density lipoprotein oxidation in healthy volunteers: A randomized, double-blind placebo-controlled study. Eur. J. Clinical Nutr. **58**: 244-249.

Murtaza, I., Kamli, A., Khan, F.A., Ahmad, I., Mustaq, D. and Bhat, B.A. 2007. Effect of storage on phenolic profiles and antioxidant activity of sweet cherry (*Prunus avium* L.) varieties of Kashmir valley. Agric. Biol. Res. **23(2)**: 130-137.

Murthy, G.P., Mamtharani, D.R., Tejas, T.S. and Suarlikerimath, N.M. 2011. Phytochemical analysis, *in vitro* anti-bacterial and antioxidant activities of wild onion sps. Intern. J. Pharma Biosci. **2(3)**: 230-237.

Murthy, K.N., Rajasekaran, T., Giridhar, P. and Ravishankar, G.A. 2006. Antioxidant property of *Decalepis hamiltonii* Wight and Arn. Indian J. Exp. Biol. **44 (10)**: 832-837.

Murthy, K.N., Vanitha, A., Swamy M. M. and Ravishankar, G.A. 2003. Antioxidant and antimicrobial activity of *Cissus quadrangularis* L. J. Medicinal Food. **6(2)**: 99-105.

Murthy, A.C., Vanitha, J., Rajesh, M., Mahadeva Swamy, P.R. and Ravishankar, G.A. 2005. *In vivo* antioxidant activity of carotenoids from *Dunaliella salina* – a green microalga. Life Sci. **76**: 1381.

Murzakhmetova, M., Moldakarimov, Tamcheva, L., Abarova, S and Serkedjieva, J. 2008. Antioxidant and prooxidant properties of a polyphenol-rich extract from *Geranium sanguineum* L. *in vitro* and *in vivo*. Phytother. Res. **22(6)**: 746-751.

Mutee, A.F., Salhimi, S.M., Yam, M.F., Lim, C.P., Abdullah, G.Z., Ameer, O.Z. Abdulkarim, M.F. and Asmawi, M.Z. 2010. *In vitro* anti-inflammatory and *in vitro* antioxidant activities of *Peperomia pellucida*. Int. J. Pharmacol. **6**: 686-690.

Muthukumaran, P., Pattabhiraman, K. and Kalaiyarasan, P. 2010. Hepatoprotective and antioxidant activity of *Mimosa pudica* on carbon tetrachloride –induced hepatic damage in rats. Int. J. Curr. Res. **10**: 46-53.

Myagmar, B.E. and Aniya, Y. 2000. Free radical scavenging action of medicinal herbs from Mongolia. Phytomedicine **7(3)**: 221-229.

Myagmar, B.E., Shinno, E, Ichiba, T. and Aniya, Y. 2004. Antioxidant activity of medicinal herb *Rhodococcum vitis-idaea* on galactosamine-induced liver injury in rats. Phytomedicine **11(5)**: 416-423.

Myer, B., Schneider, W. and Elstner, E.F. 1995. Antioxidative properties of alcoholic extracts from *Fraxinus excelsior*, *Populus tremula* and *Solidago virgaurea*. Arzeimittel Forschung **45(2)**: 174-176.

Naaz, F., Javed, S. and Abdin, M.Z. 207. Hepatoprotective effect of ethanolic extract of *Phyllanthus amarus* Schum. et Thonn. on aflatoxin B1-induced liver damage in mice. J. Ethnopharmacol. **113(3)**: 503-509.

Nabavi, S.M. 2009. Antioxidant activity of *Vicia canescens*. Pharmacology online **3**: 688-694.

Nabavi, S.M. 2010. Nitric oxide radical scavenging potential of some Elburz medicinal plants. Afr. J. Biotech. **9(32)**: 5212-5217.

Naderi, G. A., Seddigheh, A., Sarraf-Zadegan, N., Oroojy, H. and Afshin-Nia, F. 2004. Antioxidant activity of three extracts of *Morus nigra*. Phytother. Res. **18(5)**: 365-369.

Nadjet, D. Hania, B. Dina, A. Nassima, C. and Djebbar, A. 2008. Antioxidant activity of extracts of three Algerian medicinal plants: *Clematis flammula, Populus nigra* and *Ulmus*. Comparative Biochemistry and Physiology, Part A, **150 (3)**: 182-182.

Nagai, t., Myoda, T. and Nagashima, T. 2004. Antioxidative activities of water extract and ethanol extract from field horsetail (tsushi) *Equisetum arvense* L. Food Chem. **9**: 389-394.

Nagalapur, S.K. and Paramjyothi, S. 2010. *In vitro* antioxidant activity of *Launaea pinnatifida* Cass. Leaves. Bioscan **5(1)**: 105-108.

Nagasathya, A. and Thajuddin, N. 2008. Antioxidant property of Hypersaline Cyanobacterium *Phormidium tenue* (KMD 33). Int. J. Pharmacol. 1-5.

Nagashree. R., Latha, R., Anand, C. and Karthikeyan, V. 2009. Effect of *Hibiscus esculentus* fruit on arsenic induced oxidative changes in albino rats. J. Nat. Remedies **9(2)**: 170-176.

Nagata, H., Takekoshi, S., Takagi, T., Honma, T. and Watanabe, K. 1999. Antioxidative action of flavonoids, quercetin and catechin, mediated by the activation of glutathione peroxidase. Tokai J. Exp. Clin. Med. **24(1)**: 1-11.

Nagatsu, A., Sugitani, T., Mori, Y., Okuyama, H., Sakakibara, J. and Mizukami H. 2004. Antioxidants from rape (*Brassica campestris* var. *japonica* Hara.) oil cake. Nat Prod Res. **18(3)**: 231-239.

Nagavani, V. and Raghava Rao, T. 2010. Evaluation of antioxidant potential and qualitative analysis of major polyphenols by RP-HPLC in *Nymphaea nouchali* Burm flowers. Int. J. Pharm. Pharm Sci. **2**: Suppl. 4: 98-104.

Nagavani, V., Raghava Rao, K.V., Ravi Kiran, Ch. and Raghava Rao, T. 2010. *In vitro* screening of *Nyctanthes arbor-tristis* flowers for antioxidant activity and identification of polyphenols by RP-HPLC. Pharmacology online **2**: 57-78.

Nagendra Prasad, K., Xie, H., Hao, J., Yang, B., Qiu, S., Wei, X., Chen, F and Jiang, Y. 2010. Antioxidant and anticancer activities of 8-hydroxypsoralen isolated from wampee [*Clausena lansium* (Lour.) Skeels] peel. Food Chemistry **118(1)**: 62-66.

Nagesh, K. S., Nayaka, M. A.H., Dharmesh, S. M., Shanthamma, C. and Pullaiah, T. 2008. *In vitro* propagation and antioxidant activity of *Curculigo orchioides* Gaetrn. J. Trop. Med. Plants **9(2)**: 404-410.

Nagulendran, K.R., Velavan, S., Mahesh, R. and Begum, H.V. 2007. *In vitro* antioxidant activity and total polyphenolic content of *Cyperus rotundus* rhizomes. E. J. Chem. **4(3)**: 440-449.

Nagy, M., Spilkova, J., Vrchovska, V., Kontsekova, Z., Sersen, F., Mucaji, P and Grancai, D. 2006. Free radical scavenging activity of different extracts and some constituents from the leaves of *Ligustrum vulgare* and *L. delavayanum*. Fitoterapia **77**: 395-397.

Naidoo, V., Chikoto, H., Bekker, L.C. and Eloff, J.N. 2006. Antioxidant compounds in *Rhoicissus tridentata* extracts may explain their antibacterial activity. South Afr., J. Sci. **102(5-6)**: 198-200.

Naidoo, V., McGaw, L.J., Bisschop, S.P.R., Duncan, N. and Eloff, J.N. 2008. The value of plant extracts with antioxidant activity in attenuating coccidiosis in broiler chickens. Veterinary parasitology **153**: 214-219. doi: 10.1016/j/vetpar.2008.02.013.

Naidu, M.M., Sulochanamma, G, Sampathu, S.R. and Srinivas, P. 2008. Studies on extraction and antioxidant potential of green coffee. Food Chem. **107(1)**: 377-384.

Naik, G.H., Priyadarsini, K.I., Bhagirathi, R.G., Mishra, B., Mishra, K.P., Banavalikar, M.M. and Mohan, H. 2005. *In vitro* antioxidant studies and free radical reactions of triphala, an ayurvedic formulation and its constituents. Phytother. Res. **19 (7)**: 582-586.

Naik, G. H. Priyadarsini, K.I. and Hari Mohan. 2005. Evaluating the antioxidant activity of different plant extracts and herbal formulations. Res. Chem. Intermed. **13(1)**: 145-151.

Naik, G.H., Priyadarsini, K.I., Naik, D.B., Gangabhagirathi, R. and Mohan, H. 2004. Studies on he aqueous extract of *Terminalia chebula* as a potent antioxidant and a probable radioprotector. Phytomedicine **11(6)**: 530-538.

Naik, G.H., Priyadarsini, K.I., Satav, J.G., Banavalikar, M.M., Sohoni, D.P., Biyani, M.K. and Mohan, H. 2003. Comparative antioxidant activity of individual herbal components used in Ayurveda medicine. Phytocheem. **63(1)**: 97-104.

Naik, S.R. and Panda, V.S. 2007. Antioxidant and hepatoprotective effects of *Ginkgo biloba* phytosomes in carbon tetrachloride-induced liver injury in rodents. Liver Int. **27(3)**: 393-399.

Naik, S.R., Pilgaonkar, V.W. and Panda, V.S. 2006. Evaluation of antioxidant activity of *Ginkgo biloba* phytosomes in rat brain. Phytother Res. **20(11)**: 1013-1016.

Naik, Y., Jayaram, S., Nayaka, M.A.H., Lakshman and Dharmesh, S.M. 2007. Gastroprotective effect of swallow root (*Decalepis hamiltonii*) extract: possible involvement of H^+-K^+-APTas inhibition and antioxidative mechanism. J. Ethnopharmacol. **112(1)**: 173-179.

Nainwal., P., Bhagla, A. and Nanda, D. 2010. Study on antioxidant potential and wound healing activity on the aqueous extracts of fruits of *Garcinia mangostana*. IJPI's J. Pharmacognosy and Herbal Formulations **1(1)**: 1-5.

Nair, S., Nagar, R. and Gupta, R. 1998. Antioxidant phenolics and flavonoids in common Indian foods. J. Assoc. Physicians (India) **46**: 708-710.

Nair, V., Dairam, A., Aqboon, Arnason, J.T., Foster. B.C. and Konfer, I. 2007. Investigation of the antioxidant activity of African potato (*Hypoxis hemerocallidea*). J. Agric. Food Chem. **55(5)**: 1707-1711.

Najmen, D., Talla, E., Mbafer, J.T., Formum, Z.T. and Kamyani, A. 2003. Antiinflammatory activity of ericrystagallin, a pterocarpene from *Erythrina mildbraedii*. European J. Pharmacol. **468(1)**: 67-74.

Nakajima, Y., Sato, Y. and Konishi, T. 2007. Antioxidant small phenolic ingredients in *Inonotus obliquus* (Persoon) Pilat (Chaga). Chem. Pharm. Bull. **55(8)**: 1222-1226.

Nakamura, Y., Ohto, Y., Murakami, A. and Ohigashi, H. 1998. Superoxide scavenging activity of rosmarinic acid from *Perilla frutescens* Britton var. *acuta* f. *viridis*. J. Agric. Food Chem. **46**: 4545-4550.

Nakatani, N. 1997. Antioxidants from spices and herbs. In: Shahidi, (ed.). Natural antioxidants: Chemistry, Health Effects and applications. AOCS Press, Champaign, IL, pp 64-75.

Nalewajo-Sielwoniuk, E., Nazaruk, J., Antypink, E. and Kojlo, A. 2008. Determination of phenolic compounds and their antioxidant activity in *Erigeron acris* L. extracts and pharmaceutical formulation by flow injection analysis with inhibited chemiluminiscent detection. J. Pharm. Biomed. Anal. **48(3)**: 579-586. doi: 10.1016/j.jpba,2008.05.026.

Nam, S.H. and Kang, M.Y. 2004. Antioxidant activity of 13 medicinal plants. Pharmaceutical Biol. **42(6)**: 409-415.

Namdeo, A.G., Sharma, A., Mahadik, K.R. 2010. Antioxidant activity of methanolic extracts of *Nothapodytes nimmoniana* (J.Graham) Mabberly. Pharmacology online **1**: 148-159.

Nanditha, B. and Prabhasankar, P. 2009. Antioxidants in bakery products: A Review. Critical Reviews in Food Sci. Nutr. **49(1)**: 1-27.

Nanditha, B.R., Jena, B.S. and Prabhasankar, P. 2009. Influence of natural antioxidants and their carry-through property in biscuit processing. J. Sci. Food Agric. **89(2)**: 288-298.

Nanjo, F., Goto, K., Seto, R., Suzuki, M., Sakai, M. and Hara, Y. 1996. Scavenging effects of tea catechins and their derivatives on 1,1-diphenyl-2-picrylhydrazyl radical. Free Radical Biol. Med. **21**: 895-902.

Naphade, S.S., Khadabadi, S.S., Deora, Sl., Jagtap, N.S. and Hadke, S.P. 2009. Antioxidant activity of different extracts of plant *Tricholepis glaberrima* DC. (Astraceae). Internationa Journal of PharmTech Res. **1(3)**: 502-505.

Nara, K., Yamaguchi, A., Maeda, N. and Koga, H. 2009. Antioxidative activity of water soluble polysaccharide in pumpkin fruits (*Cucurbita maxima* Duchesne). Biosc. Biotech. Biochem. **73(6)**: 1416-1418.

Narasimhan, S., Govindarajan, R., Vijayakumar, M. and Mehrotra, S. 2005. Free radical scavenging potential of *Chlorophytum tuberosum* Baker J. Ethnopharmacol. **104 (3)**: 423-425.

Narayan, S., Devi, R.S., Srinivasan, P. and Devi, C.S.S. 2005. *Pterocarpus santalinus*: a traditional herbal drug as a protectant against ibuprofen induced gastric ulcers. Phytother. Res. **19(11)**: 958-962. doi: 10.1002/ptr. 1764.

Narayanaswamy, N. and Balakrishnan, K.P. 2011. Evaluation of some medicinal plants for their antioxidant properties. Int. J. PharmTech Res. **3(1)**: 381-385.

Nardi, G.M., Felippi, R., Dalbo, S, Siqueira-Junior, J.M., Arruda, D.C., Delle Monache, F., Timbola, A.K., Pizzolatti, M.G., Ckless, K. and Ribeiro-do-Valle, R.M. 2003. Anti-inflammatory and antioxidant effects of *Croton celtidifolius* bark. Phytomedicine **10**: 702-705.

Nariman, K. BadrEl-Din and Stanley T. Omaye. 2007. Concentration-dependent antioxidant activity of conjugated linoleic acid and α-tocopherol in corn oil. J. Sci. Food Afgric. **87(14)**: 2715-2720.

Narvaez-Mastache, J.M., Soto, C. and Delgado, G. 2007. Antioxidant evaluation of *Eysenhardtia* species (Fabaceae): Relay synthesis of 3-O-acetyl-11α, 12α-epoxy-oleanan-28, 13betolide isolated from *E. platycarpa* and its protective effect in experimental diabetes. Biol. Pharm. Bull. **30(8)**: 1503-1510.

Nascimento, M.A., Silva, A.K., Franca, L.C.B., Quignard, E.L.J., Lopez, J.A. and Almeida, M.G. 2006. *Turnera ulmifolia* L. (Turneraceae): Preliminary study of its antioxidant activity. Bioresource Technol. **97(12)**: 1387-1391. doi: 10.1016/j.niortech.2005.07.009

Nascimento, N.C. do, Fragoso, V., Moura, D.J., de Silva, A.C.R., Fett-Neto, A.G. and Saffi, J. 2007. Antioxidant and antimutagenic effects of the crude foliar extract and the alkaloid brachycerine of *Psychotria brachyceras*. Env. Mutagenesis **48**: 728-734.

Natarajan, K.S., Narasimhan, M., Shanmugasundaram, K.R. and Shanmugasundaram, E.R.B. 2006. Antioxidant activity of a salt-spice-herbal mixture against free radical induction. J. Ethnopharmacol. **105**: 76-83.

Naval, M.V., Gomez-Serranillos, M.P., Cartero, M.E. and Villar, A.M. 2007. Neuroprotective effects of a ginseng (*Panax ginseng*) root extract on astrocytes primary culture. J. Ethnopharmacol. **112(2)**: 262-270.

Navarro, M.C., Montilla, M.P., Cabo, M.M., Galisteo, M., Caceras, A., Morales, C. and Berger, I. 2003. Antibacterial, antiprotozoal and antioxidant activity of five plants used in Izabal for infectious diseases. Phytother. Res. **17**: 325-329.

Nazari, A.S. Dias, S.A., da Costa, W.F., Bersani-Amada, C.A., Vidotti, G.J., de Souza, M.C. and Sarragiotto, M.H. 2006. Anti-inflammatory and antioxidant activities of *Randia hebecarpa* and major constituents. Pharm. Biol. **44(1)**: 7-9. doi: 10.1080/13880200500596

Nazaruk, J., Czechowska, S.K., Markiewicz, R. and Borawska, M.H. 2008. Polyphenolic compounds and *in vitro* antimicrobial and antioxidant activity of aqueous extracts from leaves of some *Cirsium* species. J. Nat. Prod. Res. **22(18)**: 1583-1588.

Nazemiyeh, H., Bahadori, F., Delazar, A., Ay, M., Topcu, G., Nahar, L., Majinda, R.R. and Sarker, S.D. 2008. Antioxidant phenolic compounds from the leaves of *Erica arborea* (Ericaceae). Nat. Prod. Res. **22(16)**: 1385-1392.

Nazif, N.M. 2007. Phytochemical and antioxidant activity of *Spathodea campanulata* P.Beauvois growing in Egypt., Nat. Prod. Sci. **13(1)**: 11-16.

Ndidi, E. and Akeem, O.A. 2011. Antioxidant enzyme activity during post harvest deterioration of Cassava (*Manihot esculenta*) root tubers. Int. J. Agric. Biol. **13**: 455-457.

Neeli, G.S., Karki, S.S., Patil, S., Pattan, S.R. and Hariprasad, C.K. 2008. *In vitro* antioxidant activity of flowers of *Hibiscus rosa-sinensis* Linn. Biomed. **3(1)**: 279-282.

Negrao, C., Tommasi, L. and Miceli, A. 2003. Phenolic compounds and antioxidant activity from red grape marc extracts. Bioresource technology. **87**: 41-44.

Nejla, B.H.-B.E. and Moncef, C.M. 2006. Antioxidant activity of *Origanum majorana* L. oil from Tunisia. J. Essential Oil Bearing Plants **9(1)**: 88-92.

Nema, N.K., Maity, N., Sarkar, B. and Mukherjee, P.K. 2011. *Cucumis sativus* fruit-potential antioxidant, antihyaluronidase, and anti-elastase agent. Arch. Dermatol. Res. **303(4)**: 247-252. Doi: 10.1007/s00403-010-1103y.

Nencini, C., Cavallo, F., Capasso, A., Franchi, G.G., Giorgio, G. and Micheli, L. 2007. Evaluation of antioxidant properties of *Allium* species growing wild in Italy. Phytother. Res. **21**: 874-878. doi: 10.1002/ptr.2168.

Nencini, C., Gorgi, G. and Micheli, L. 2007. Protective effect of silymarin on oxidative stress in rat brain. Phytomedicine **14(2-3)**: 129-135.

Neo, Y.-P., Ariffin, A., Tan, C.-O. and Tan, Y.-A. 2010. Phenolic acid analysis and antioxidant activity assessment of oil palm (*E. guineensis*) fruit extracts. Food Chem. **122(1)**: 353-359.

Neogy, S., Das, S., Mahapatra, S.K., Mandal, N. and Roy, S. 2008. Ameliotary effect of *Andrographis paniculata* Nees on liver, kidney, heart, lung and spleen during nicotine induced oxidative stress. Environ. Toxicol. Pharmacol. **25(3)**: 321-328.

Nergard, C.S., Diallo, D., Inngjerdingen, K., Michaelsen, T.E., Matsumoto, T., Kiyohara, H., Yamada, H. and Paulsen, B.S. 2005. Medicinal use of *Cochlospermum tinctorium* in Mali, anti-ulcer, radical scavenging and immunomodulating activities of polymers in the aqueous extract of the roots. J. Ethnopharmacol. **96**: 255-269.

Ness, A.R. and Powles, J.W. 1999. The role of diet, fruit and vegetables and antioxidants in the aetiology of stroke. J. Cardiovascular Risk **6**: 229-234.

Netzel, M., Netzel, G., Tian, Q., Schwartz, S. and Konczak, I. 2007. Native Australian fruits – A novel source of antioxidants for food. Innovative Food Science and Emerging Technologies 8: 339-346.

Netzel, M., Strass, G., Kaul, C., Bitsch, I., Dietrich, H. and Bitch, R. 2002. *In vivo* antioxidative capacity of a composite berry juice. Food Res. Int. **35**: 213-216.

Nevin, K.G. and Vijayammal, P.L. 2005. Effect of *Aerva lanata* against hepatotoxicity of carbon tetrachloride in rats. Environmental Toxicol. Pharmacol. **20**: 471-477.

Ng, P.C., Kong, Y.C., Ko, K.M., So, C.M. and Yick, P.K. 1992. Antioxidant activity of *Ganoderma lucidum*: Protective effects on carbon tetrachloride-induced hepatotoxicity. ISHS Acta Horticulturae 332. WOCMAP 1 – Medicinal and Aromatic Plants Conference.

Ng, T.B., Gao, W., Li, L., Niu, S.M., Zhao, L., Liu, J. Shi, L.S., Fu, M. and Liu, F. 2005. Rose (*Rosa rugosa*) flower extract increases the activity of antioxidant enzymes and their gene expression and reduces lipid peroxidation. Biochem. Cell Biol. **83(1)**: 78-85. doi: 10.1139/004-100.

Ng, T.B., Liu, F. and Wang, Z.T. 2000. Antioxidative activity of natural products from plants. Life Sci. **66(8)**: 709-723.

Ng, T.B., Liu, F. and Wang, Z.T. 2004. The antioxidant effects of aqueous and organic extracts of *Panax quinquefolium, Panax notoginseeng, Codonopsis pilosula, Pseudostellaria heterophylla* and *Glehnia littoralis*. J. Ethnopharmacol. **93(2-3)**: 285-288. doi: 10.1016/j.jep.2004.03.040

Ngouela, S., Lenta, B.N., Noungoue, D.T., Ngopupayo, J., Boyom, F.F., Tsamo, E., Gut, J., Rosenthal, P.J. and Connolly, J.D. 2006. Antiplasmodial and antioxidant activities of constituents of the seed shells of *Symphonia globulifera* Linn.f. Phytochem. **67(3)**: 302-306.

Nguelefack-Mbuyo, E.P., Dimo, T., Nguelefack, T.B., Azebaze, A.G.B., Dongmo, A.B., Kamtchouing, P. and Kamanyi, A. 2010. *In vitro* antioxidant activity of extracts and coumarins from the stem bark of *Mammea africana* Sabine. J. Complementery and Integrative medicine **7(1)**: DOI: 1.2202/1553-3840.1447.

Nhukarume, L., Chikwambi, Z., Muchuweti, M. and Chipurura, B. 2010. Phenolic content and antioxidant capacities of *Parinari curatelifolia, Strychnos spinosa* and *Adansonia digitata*. J. Food Biochem. **34** supplement S1: 207-221. Doi: 10.1111/j.1745-4514.2009.00325.

Niiforovi, N., Mihailovi, V., Maskovi, P., Soluji, S., Stojkovi, A. and Pavlovi Muratspahi, D. 2010. Antioxidant activity of selected plant species; potential new sources of natural antioxidants. Food Chem. Toxicol. **48(11)**: 125-130.

Nickavar, B. and Abolhasani, F.A. 2009. Screening of antioxidant properties of seven Umbelliferae fruits from Iran. Pak. J. Pharm. Sci. **22(1)**: 30-35.

Nickavar, B., Alinaghi, A. and Kamalinejad, M. 2008. Evaluation of the antioxidant properties of five *Mentha* species. Iranian J. Pharm. Res. **7(3)**: 203-209.

Nickavar, B., Kamalinejad, M. and Izadpanah, H. 2007. *In vitro* free radical scavenging activity of five *Salvia* species. Pak. J. Pharm. Sci. **20(4)**: 291-294.

Nidhiya, I.S.R., Pai, K.S.R. and Mallikharjuna Rao, C. 2010. Freee radical scavenging activity of extracts of young proproots of *Ficus benghalensis*. Pharmacology online **2**: 299-307.

Niggeweg, R., Michael, A.J. and Martin, C. 2004. Engineering plants with increased levels of the antioxidant chlorogenic acid. Nature Biotechnol. **22**: 746-754.

Niki, E. 2002. Antioxidant activity: Are we measuring it correctly? Nutrition **18**: 524-525.

Niki, E. and Noguchi, N. 2000. Evaluation of antioxidant capacity. What capacity is being measured by which method? IUBMB Life **50**: 323-329.

Nikolova, M. and Dzhurmanski, A. 2009. Evaluation of free radical scavenging capacity of extracts from cultivated plants. Biotechnol. Biotechnol. XI Anniversary Scientific Conference Special Edition On line pp. 109-111.

Nile, S.H. and Khobrgade, C.N. 2011. Phytochemical analysis, antioxidant and xanthine oxidase inhibitory activity of *Tephrosia purpurea* Linn. root extract. Indian J. Nat. Prod. Res. **2(1)**: 52-58.

Nilnakara, S., Chienchan, N. and Devahastin, S. 2009. Production of antioxidant dietary fibre powder from cabbage outer leaves. Food and Bioproducts Process **87(4)**: 301-307.

Nilsson, J., Pillai, D., Onning, G., Persson, C., Nilsson, A. and Akesson, B. 2005. Comparison of the ABTS and FRAP methods to assess the total antioxidant capacity in extracts of fruit and vegetables. Mol. Nutr. Food Res. 49: 239-246.

Nilsson, J., Stegmark, R and Akesson, B. 2004. Total antioxidant capacity in different pea (*Pisum sativum*) varieties after blanching and freezing. Food Chem. **86**: 501-507.

Ninfali, P., Mea, G., Giorgini, S. Rocchi, M. and Bacchiocca, M. 2005. Antioxidant capacity of vegetables, spices and dressings relevant to nutrition. Br. J. Nutr. **93**: 257-266.

Ningappa, M.B. Dinesha, R. and Srinivas, L. 2008. Antioxidant and free radical scavenging activities of polyphenol-enriched curry leaf (*Murraya koenigii* L.) extracts. Food Chem. **106(2)**: 720-728.

Niranjan, A. and Tewari, S.K. 2008. Phytochemical composition and antioxidant potential of *Desmodium gangeticum* (Linn.) DC. Nat. Prod. Radiance **7(1)**: 35-39.

Nish, P., Abdul Nazir, P. and Jayamurthy, P. 2009. A comparative study on antioxidant activities of different varieties of *Solanum melongena*. Food Chem. Toxicol. **47(10)**: 2640-2644.

Nissen, H.P., Patzelt-Wenczler, R. and Hose, S. 1996. Scavenger and antioxidant properties of special chamomile extracts. Phytomedicine **3** (Suppl.): 241.

Nithya, N. and U.Saraswathi. 2010. *In vitro* antioxidant and antibacterial efficacy of *Feronia elephantum* Correa fruit. Indian J. Nat. Prod. Resources **1(3)**: 301-305.

Nitzsche, A., Tokalov, S.V., Gutzeit, H.O. and Ludwig-Müller, J. 2004. Chemical and biological characterization of cinnamic acid derivatives from cell cultures of lavender (*Lavandula officinalis*) induced by stress and jasmonic acid. J. Agric. Food Chem. **52 (10)**: 2915-23.

Niwano, Y., Saito, K., Yoshizaki, F., Kohno, M. and Ozawa, T. 2011. Extensive screening for herbal extracts with potent antioxidant properties. J. Clin. Biochem. Nutr. **48(1)**: 78-84.doi: 10.3164/jcbn.11-013FR.

Njamen, D., Mbafor, J.T., Formum, Z.T., Kamanyi, A., Mbanya, J.C., Reclo, M.C., Giner, R.M., Manez, S. and Rios, J.L. 2004. Anti-inflammatory activities of two flavanones, sigmoidin A and sigmoidin B, from *Erythrina sigmoidea*. Planta Medica **70(2)**: 104-107.

Nkosi, C.Z., Opoku, A.R. and Terblanche, S.E. 2006. Antioxidative effects of pumpkin seed (*Cucurbita pepo*) protein isolate in CCl_4-induced liver injury in low-protein fed rats. Phytother. Res. **24**:

Nobre Junior, H.V., Maia, F.D., De Oliveira, R.A., Bandeira, M.A.M., Pessoa, C.D.O., Moraes, M.O., Cunha, G.M.A. and Viana, G.S.B. 2007. Neuroprotective actions of tannins from *Myracrodendruon urundeuva* on 6-hydroxydopamine-induced neuronal cell death. Int. J. Fruit Sci. **7(2)**: 41-57.

Noda, Y., Anzai, K., Mori, A., Kohno, M., Shinmel, M. and Packer, L. 1997. Hydroxyl and superoxide anion radical scavenging activities of natural source antioxidants using the computerized JES-FR30 ESR spectrophotometer system. Biochem. Mol. Biol. Int. **42**: 35-44.

Noda, Y., Kneyuki, T., Igarashi, K., Mori, A. and Packer, L. 2000. Antioxidant activity of nasunin, an anthocyanin in egg plant peels. Toxicol. **148 (2-3)**: 119-123.

Nor, F. M., Mohamed, S., Idris, N. A. and Ismail, R. 2008. Antioxidative properties of *Pandanus amaryllifolius* leaf extracts in accelerated oxidation and deep frying studies. Food Chem. **110(2)**: 319-327. doi: 10.1016/j.foodchem.2008.02.004

Noriham, A., Babji, A.S. and Aminah, A. 2004. Determination of antioxidative activities of selected Malaysian plant extracts. ASEAN Food J. 13: 193-199.

Nsimba, R., Kikuzaki, H. and Konishi, Y. 2008. Antioxidant activity of various extracts and fractions of *Chenopodium quinoa* and *Amaranthus* spp. seeds. Food Chem. **106**: 760-766.

Nuengchamnong, N., de Jong, C.F., Bruyneel, B., Niessen, W.M., Irth, H. and Ingkaninan, K. 2005. HPLC coupled on-line to ESI-MS and a DPPH-based assay for the rapid identification of anti-oxidants in *Butea superba*. Phytochem. Anal. **16 (6)**: 422-428.

Nuengchamnong, N. and Ingkaninan, K.2010. On-line HPLC-MS-DPPH assay for the analysis of phenolic antioxidant compounds in fruit wine: *Antidesma thwaitesianum* Muell. Food Chem. **118(1)**: 147-152.

Nuria E. R., Rubén F.G.L., Francisco. I.P. J. Cynthia A.N.B., José-Alberto, G.I. 2007. Effect of pressure cooking on the antioxidant activity of extracts from three common bean (*Phaseolus vulgaris* L.) cultivars. Food chem. **100(1)**: 31-35.

Nuutila, A.M., Pupponen-Pimia, R., Arni, M. and Oksman Caldentey, K.M. 2003. Comparison of antioxidant activities of onion and garlic extracts by inhibition of lipid peroxidation and radical scavenging activity. Food Chem. **81**: 485-493.

Nwanjo, H.U. 2006. Antioxidant activity of the exudates from *Aloe barbadensis* leaves in diabetic rats. Biokemistri **18(2)**: 77-81.

Nwanjo, H.U. 2008. Aqueous extract of *Tridax procumbens* leaves: Effect of lipid peroxidative stress and antioxidant status in chloroquine-induced hepatotoxicity. J. Herbs Spices Med. Plants **14(3-4)**: 154-165.

Nwankwo, J.O., Tanhnteng, J.G. and Emerole, G.O. 2000. Inhibition of aflatoxin B_1 genotoxicity in human liver-derived HepG2 cells by kolaviron biflavonoids and molecular mechanisms of action. Eur. J. Canc. Prev. **9**: 351–361.

Nwanna, E.E. and Oboh, G. 2007. Antioxidant and hepatoprotective properties of polyphenol extracts from *Telfaria occidentalis* (fluted pumpkin) leaves on acetaminophen induced liver damage. Pak. J. Bio. Sci. **10(16)**: 2682-2687.

Nyligira, E., Viljoen, A.M., van Heerden, F.R., Van Zyl, R.L., Van Vuuren, S.F. and Steenkamp, P.A. 2008. Phytochemistry and *in vitro* pharmacological activities of South African *Vitex* (Verbenaceae) species. J. Ethnopharmacol. **119(3)**: 680-685. doi: 10.1016/j.jep.2008.07.004

Oboh, G. 2005a. Effect of blanching on the antioxidant property of some tropical green leafy vegetables. Lebensm Wiss Technol. **38**: 513-517.

Oboh, G. 2005b. Hepatoprotective property of ethanolic and aqueous extracts of fluted pumpkin (*Telfaria occidentalis*) leaves against garlic-induced oxidative stress. J. Med. Food **8(4)**: 560-563.

Oboh, G. 2006a. Antioxidant properties of some commonly consumed and underutilized tropical legumes. Eur. Food Res. Tech. **224(1)**: 61-65.

Oboh, G. 2006b. Nutritive value, antioxidant and antimicrobial properties of *Struchium sparganophora* leaves. J. Med. Food. **9(2)**: 276-280.

Oboh, G., Ademiluyi, A.O. and Akindahunsi, A.A. 2009. Changes in polyphenols distribution and antioxidant activity during fermentation of some underutilized legumes. Food sci. Technol. Intern. **15(1)**: 41-46. doi: 10.1177/1082013208101022.

Oboh, G. and Akindahunsi, A.A. 2004. Change in the ascorbic acid, total phenol and antioxidant activity of some sun-dried green leafy vegetables in Nigeria. Nutr. Health **18**: 29-36.

Oboh, G., Akinyeni, A.J. and Ademiluyi, A.O. 2012. Antioxidant and inhibitory effect of red ginger (*Zingiber officinale* var. *rubra*) and white ginger (*Zingiber officinale* Roscoe) on Fe(2+) reduced lipid peroxidation in rat brain *in vitro*. Exp. Toxicol. Pathol. **64(1-2)**: 31-36. Doi: 10.1016/j.etp.2010.06.002.

Oboh, G., Ekperigin, M.M. and Akindahunsi, A.A. 2007. Coagulants modulate the antioxidant properties and hypocholesterolemic effect of tofu (coagulated soymilk). Nutr. Health **18(4)**: 369-381.

Oboh, G. and Henle, T. 2009. Antioxidant and inhibitory effects of aqueous extracts of *Salvia officinalis* leaves on pro-oxidant-induced lipid peroxidation in brain and liver *in vitro*. J. Med. Food **12(1)**: 77-84.

Oboh, G., Puntel, R.L. and Rocha, J.B.T. 2007. Hot pepper (*Capsicum annuum* Tepin and *Capsicum chinense* Habanero) prevents Fe^{2+} induced lipid peroxidation in brain – *in vitro*. Food Chem. **102**: 178-185.

Oboh, G., Raddatz, H. and Henle, T. 2008. Antioxidant properties of polar and non-polar extracts of some tropical green leafy vegetables. J. Sci. Food Agric. **88**: 2486-2492.

Oboh, G. and Rocha, J.B. 2007a. Antioxidants in foods: a new challenge for food processors. Leading Edge Antioxidants Research, Nova Science Publishers Inc. New York, US, pp. 35-64.

Oboh, G. and Rocha, J.B. 2007b. Polyphenols in red pepper (*Capsicum annuum* var. *aviculare* (Tepin) and their protective effect on some pro-oxidants induced lipid peroxidation in brain and liver. Eur. Food Res. Technol. **225**: 239-247.

O'Brien, P., Carrasao,-Pozo, C. and Speisky, H. 2006. Boldine and its antioxidant or health promoting properties. Chem. Biol. Interact. **159(1)**: 1-17.

Ochani, P.C. and D'Mello, P. 2009. Antioxidant and antihyperlipidemic activity of *Hibiscus sabdariffa* leaves and calyces extracts in rats. Indian J. Exp. Bio. **47(4)**: 276-282.

Ochi, T., Takaishi, Y., Kogure, K. and Yamayuti, I. 2003. Antioxidant activity of a new capsaicin derivative from *Capsicum annuum*. J. Nat. Prod. **66(8)**: 1094-1096.

Odabasoglu, F., Aslan, A., Cakir, A., Suleyman, H., Karagoz, Y., Bayir, Y. and Halici, M. 2005. Antioxidant activity, reducing power and total phenolic content of some lichen species. Fitoterapia **76(2)**: 216-219.

Odabasoglu, F., Aslan, A., Cakir, A., Suleyman, H., Karagoz, Y., Halici, M., Bayir, Y. 2004. Comparison of antioxidant activity and phenolic content of three lichen species. Phytother Res. **18(11)**: 938-41.

Odabasoglu, F., Cakir, A., Suleyman, H., Aslan, A., Bayir, Y., Halici, M. and Kazaz, C. 2006. Gastroprotective antioxidant effects of usnic acid on indomethacin induced gastric ulcer in rats. J. Ethnopharmacol. **103(1)**: 59-65.

Odriozola-Serrano, I., Soliva-Fortuny, R. and Olag, M.B. 2008. Antioxidant properties and shelf-life extension of fresh-cut tomatoes stored at different temperatures. J. Sci. Food Agric. **88(15)**: 2606-2614.

Odukoya, O.A., Ilori, O.O., Sofidiya, M.O., Aniunon, O.A., Lawal, B.M. and Tadi, I.O. 2005. Antioxidant activity of Nigerian dietary spices. Electr. J. Environ. Agr. Food Chem. **4(6)**: 1086-1093.

Odukoya, O.A., Inya-Agha, S.I., Sofidiya, M.O. and Ilori, O.O. 2007. Antioxidant activity of selected Nigerian leafy vegetables. Am. J. Food Technol. **2(3)**: 169-175. DOI: 10.3923/ajft.2007.169.175.

Offord, E.A., Guilot, F., Aeschenbach, R., Lolinger, J and Pfeifer, A.M.A. 1997. Antioxdant and biological properties of rosemary components: implications for food and health. In: Shahidi, F. (ed.) Natural antioxidants, Chemistry, health effects and applications. AOCS Press. Champaign, IL. Pp. 88-96.

Ogbunugafor, H., Sofidiya, O., Okpuzor, J., Kemdilim, M., Anajekwe, B. and Ekechi, A. 2010. Effects of extracts of *Hymenocardia acida* Tul. (Hymenocardiaceae) on rats. J. Amer. Sci. **6(2)**: 143- 146.

Oh, H., Kim, D.H., Cho, J.H. and Kim, Y.C. 2004. Hepatoprotective and free radical scavenging activities of phenolic petrosins and flavonoids isolated from *Equisetum arvense*. J. Ethnopharmacol. **95**: 421-424.

Oh, H., Ko, E.K., Jun, J.Y., Oh, M.H., Park, S.U., Kang, K.H., Lee, H.S. and Kim, Y.C. 2002. Hepatoprotective and free radical scavenging activities of prenylflavonoids, coumarin and stilbene from *Morus alba*. Planta Med. **68(10)**: 932-934.

Oh, H.M., Kang, Y.J., Lee, Y.S., Park, M.K. Kim, H.J., Seo, H.G., Lee, J.H. and Chang, K.C. 2006. Protein kinase G-dependent heme oxygenase-1 induction by *Agastache rugosa* leaf extract protects RAW264.7 cells from hydrogen peroxide-injury. J. Ethnopharmacol. **103(2)**: 229-235.

Ohmori, R., Iwamoto, T., Tago, M., Takeo, T., Unno, T., Itakura, H. and Kondo, K. 2005 Antioxidant activity of teas (*Camellia sinensis*), tochu (*Eucommia ulmoides*), *Gymnema sylvestre*, Japanese mugwort (*Artemisia princeps*), and barley (*Hordeum vulgare*) against free radicals and LDL oxidation. Lipids. **40(8)**: 849-853.

Ohsugi, W.Z., Fan, K., Hase, K., Xiong, Q., Tezuka, Y.Komatsu, K., Namba, T., Saitoh, K., Tazawa, K. and Kadota, S. 1999. Active-oxygen scavenging activity of traditional nourishing-tonic herbal medicines and active constituents of *Rhodiola sacra*. J. Ethnopharmacol. **67**: 111-119.

Ohtani, I.I., Gotoh, N., Tanaka, J., Higa, T., Gyamfi, M.A. and Aniya, Y. 2000. Thonningianins A and B – a new antioxidants from the African medicinal herb *Thonningia sanguinea*. J. Nat. Prod. **63(5)**: 676-679.

Ojha, S.K., Nandave, M., Kumari, S. and Arya, D.S. 2009. Antioxidant activity of *Andrographis paniculata* in Ischemic Myocardium of rats. Global J. Pharmacol. **3(3)**: 154-157.

Okada, Y. and Okada, M. 1998. Scavenging effect of soluble proteins in broad beans on free radicals and active oxygen species. J. Agric. Food Chem. **46**: 401-406.

Okamura, H., Mimura, A., Yakou, Y., Niwano, M. and Takahara, Y.,1993. Antioxidant activity of tannins and flavonoids in *Eucalyptus rostrata*. Phytochem. **33(3)**: 557-561.

Okasaka, M., Takaishi, Y., Kogure, K., Fukuzawa, K., Shibata, H., Higuti, T., Honda, G., Ito, M., Kodzhimatov, O.K. and Ashurmetov. 2004. New stilbene derivatives from *Calligonum leucocladum*. J. Nat. Prod. **67(6)**: 1044-1046.

Oke, J.M. and Hamburger, M.O. 2002. Screening of some Nigerian medicinal plants for antioxidant activity using 2, 2, Diphenyl-Picryl-Hydrazyl radical. African J. Biomedical Res. **5**: 77-79.

Okeola, V.O., Adaramoye, O.A., Nneji, C.M., Falade, C.O., Farombi, E.O. and Ademowo, O.G. 2010. Antimalarial and antioxidant activities of methanolic extract of *Nigella sativa* seeds (black cumin) in mice infected with *Plasmodium yoelli nigeriensis*. Parasitol. Res. 108(6): 1507-1512. doi: 10.1007/s00436-010-2204-4.

Oki, T., Masuda, M., Kobayashi, M., Nishiba, Y., Furuta, S, Suda, I. and Sato, T. 2001. Radical scavenging activity of fried chips made from purple-fleshed sweet potato. Nippon Shokuhin Kagaku Kogaku Kaishi **48**: 926-932 (in Japanese).

Oki, T., Masuda, M., Kobayashi, M., Nishiba, Y., Furuta, S., Suda, I. and Sato, T. 2002. Polymeric procyanidins as radical-scavenging components in red-hulled rice. J. Agric. Food Chem. **50**: 7524-7529.

Oktay, M., Gülçin, O. and Küfreviolu, Ö.I. 2003. Determination of in vitro antioxidant activity of fennel (*Foeniculum vulgare*) seed extracts. Lebens-Wiss Technol. **36**: 263-271.

Okusa, P.N. and Penge, O. 2007. Direct and indirect antimicrobial effects and antioxidant activity of *Cordia gilletii* De Wild. (Boraginaceae). J. Ethnopharmacol. **112(3)**: 476-481.

Olaleye, M.T., Ogunboye, A.A., Akinmoladun, A.C. and Akindahunshi, A.A. 2010. Antioxidant activity and hepatoprotective property of leaf extracts of *Boerhavia diffusa* Linn. against Acetaminophen-induced liver damage in rats. Food Chem. Toxicol. **48(8-9)**: 2200-2205. doi: 10.1016/j.fct.2010.05.047.

Olaleye, M.T. and Rocha, B.T. 2007. Commonly used tropical medicinal plants exhibit distinct *in vitro* antioxidant activities against hepatotoxins in rat liver. Expl Toxicol Pathol **58(6)**: 433-438.

Olaleye, M.T. and Rocha, B.T. 2008. Acetaminophen-induced liver damage in mice: Effects of some medicinal plants on the oxidative defence system. Exp. Toxicol. Pathol. **59(5)**: 319-327.

Olas, B., Saluk-Juszczak, J., Pawlaczyk, I., Nowak, P., Kolodziejczyk, J., Gancarz, R. and Wachowicz, B. 2006. Antioxidant and antiaggregatory effects of an extract from *Conyza canadensis* on blood platelets **17(6)**: 354-360.

Olaleye, M.T., Akinmoladun, A.C., and Akindahunsi. 2006. Antioxidant properties of *Myristica fragrans* (Houtt.) and its effect on selected organs of albino rats. Afr. J. Biotechnol. **5(13)**: 1274-1278.

Oliveira, I., Coelho, V., Baltasar, R., Pereira, J.A. and Baptista, P. 2009. Scavenging capacity of strawberry tree (*Arbutus unedo* L.) leaves on free radicals. Food Chem. Toxicol. **47(7)**: 1507-1511.

Oliveira, I., Sousa, A., Ferreira, I.C.F.R., Bento, A., Estevinho, L. and Pereira, J.A. 2008. Total phenols, antioxidant potential and antimicrobial activity of walnut (*Juglans regia* L.) green husks. Food Chem. Toxicol. **46**: 2326-2331.

Oliveira, I., Sousa, A., Morais, J.S., Ferreira, I.C.F.R., Bento, A., Estevinho, L. and Pereira, J.A. 2008. Chemical composition, and antioxidant and antimicrobial activities of three hazelnut (*Corylus avellana* L.) cultivars. Food Chemic. Toxicol. **46(5):** 1801-1807.

Oliveira, I., Sousa, A., Valentao, P., Andrade, P.B., Ferreira, I.C.F.R., Ferreres, F., Bento, A, Seabra, R., Estevinho, L. and Pereira, J.A. 2007. Hazel (*Corylus avellana* L.) leaves as source of antimicrobial and antioxidative compounds. Food Chem. **105(3):** 1018-1025.

Oliveira, O.M., Vellosa, J.C., Fernandes, A.S., Buffa-Filho, W., Hakime-Silva, R.A., Furlan, M. and Brunetti, I.L. 2007. Antioxidant activity of *Agaricus blazei*. Fitoterapia **78(3):** 263-264.

Oloyede, G.K. and Farombi, O.E. 2010. Antioxidant properties of *Crinum ornatum* bulb extract. World. J. Chem. **5(1):** 132-136.

Oloyede, G.K., Oke, M.J., Raji, Y. and Olugbade, A.T. 2010b. Antioxidant and anticonvulsant alkaloids in *Crinum ornatum* bulb extract. World J. Chem. **5(1):** 26-31.

Oloyede, G.K., Onocha, P.A., Soyinka, J., Oguntokun, O. and Thonda, E. 2010a. Phytochemical screening, antimicrobial and antioxidant activities of four Nigerian Medicinal plants. Ann. Biol.Res. **1:** 114-120.

Olszewska, M.A. and Michel, P. 2009. Antioxidant activity of inflorescences, leaves and fruits of three *Sorbus* species in relation to their polyphenolic composition. Nat. Prod. Res. **23:** 1507-1521.

Oluwaseun, A.A. and Ganiyu, O. 2008. Antioxidant properties of methanolic extracts of mistletoes (*Viscum album*) from cocoa and cashew trees in Nigeria. African J. Biotech. **7(17):** 3138-3142.

Oluwole, F.S., Ayo, J.A., Omolaso, B.O., Emikpe, B.O. and Adesanwo, J.K. 2008. Methanolic extract of *Tetracera potatoria*, an antiulcer agent increases gastric mucus secretion and endogenous antioxidants. Nigerian J. Physiol. Sci. **23(1-2):** 79-83.

Omanakutty, M. 2007. Medicinal properties of Cardamom. Spice India **20(1):** 10-11.

Omar, M.A. 2005. Effects of 2-hydroxy-1,4-naphthaquinone, a natural dye of henna, on aldehyde oxidase activity in guinea pigs. J. Med. Sci **5(3):** 163-168.

Omonkhua, A.A. and Onoagbe, I.O. 2008. Effects of *Irvingea grandiflora*, *Urena lobata* and *Carica papaya* on the oxidative status of normal rabbits. The Internet J. Nutr. Wellness **6(2):**

Om Prakash, Singh, R.M., Mathur, S.C. and Singh, G.N. 2007. Quantification of gallic acid by HPLC and antioxidant activity of amla fruits. J. Pharm. Res. **6(3):** 161-162.

Onay-Ucar, E., Karagoz, A. and Arda, N. 2006. Antioxidant activity of *Viscum album* ssp. *album*. Fitoterapia **77:** 556-560.

Ondrojevic, M., Ondrigova, Z. and Kubincova, J. 2009. Isolation of antioxidants from *Alchemilla xanthochlora*. Nova Biotechnologia **9(3)**: 313-318.

Ono, M., Furushawa, C., Ozono, T., Od, K., Ysuda, S., Okawa, M., Kinjo, J., Ikeda, T., Miyashita, H., Yoshimitsu, H. and Nohara, T. 2011. Four new iridoid glucosides from *Ajuga reptans*. Chem. Pharm. Bull. (Tokyo). **59(8)**: 1065-10683 Densi.pharm.or.jp/home/pubpharm/puvieww.asp?p=c110206.

Ono, M., Morinaga, H., Masuoka, C., Ikeda, T., Okawa, M., Kinjo, J. and Nohara, T. 2005. New bisabolane-type sesquiterpenes from the aerial parts of *Lippia dulcis*. Chem. Pharm. Bull. **53(9)**: 1175-1177.

Ono, M., Nishida, Y., Masuoka, C., Li, J.C., Okawa, M., Ikeda, T. and Nohara, T. 2004. Lignan derivatives and a norditerpene from the seeds of *Vitex negundo*. J. Nat. Prod. **67(12)**: 2073-2075.

Onocha, P.A., Oloyede, G.K. and Afolabi, Q.O. 2011. Phytochemical investigation, cytotoxicity and free radical scavenging activities of non-polar fractions of *Acalypha hispida* (leaves and twigs). EXCLI Journal **10**: 1-8.

Oonsivilai, R., Feruzzi, M.G. and Ningsanond, S. 2008. Antioxidant activity and cytotoxicity of Rang Chuet (*Thunbergia laurifolia* Lindl.) extracts. Asian J. Food Agro-Industry **1(2)**: 116-128.

Op de Beck, P., Cartier, G., David, B., Dijoux-Franca, M. and Mariotte, A. 2003. Antioxidant flavonoids and phenolic acids from *Leea guineense* G. Don (Leeaceae). Phytother. Res. **17**: 345-347.

Opoku, A.R., Maseko, N.F. and Terblanche, S.E. 2002. The *in vitro* antioxidative activity of some traditional Zulu medicinal plants. Phytother. Res. **16**. Suppl. 1: S 51-56.

Ordonez, A.A.L., Gomez, D., Vattuone, M.A. and Isla, M.I. 2006. Antioxidant activities of *Sechium edule* (Jacq.)Swartz extracts. Food Chem. **97**: 452-458.

Orhan, D.D., Aslan, M., Sendogdu, N., Ergun, F. and Yesilada, E. 2005. Evaluation of the hypoglycemic effect and antioxidant activity of three *Viscum album* subspecies (European mistletoe) in streptozotocin-diabetic rats. J. Ethnopharmacol. **98(1)**: 95-102.

Orhan, D.D., Mustafa, A., Sendogdu, A., Ergun, F. and Yesilada, F. 2005. Evaluation of hypoglycemic effect and antioxidant activity of three *Viscum album* subspecies in streptozotocin diabetic rats. J. Ethnopharmacol. **98**: 95-102.

Orhan, D.D., Orhan, N., Ergun, E. and Ergun, F.2007. Hepatoprotective effect of *Vitis vinifera* L. leaves on carbon tetrachloride-induced acute liver damage in rats. J. Ethnopharmacol. **112(1)**: 145-151.

Orhan, I., Senol, F.S., Gulpinar, A.R., Kartal, M., Sekeroglu, N., Deveci, M., Kan, Y. and Sener, B. 2009. Acetylcholinesterase inhibitory and antioxidant properties of *Cyclotrichium niveum*, *Thymus praecox* subsp. *caucasicus* var. *caucasicus*, *Echinacea purpurea* and *E. pallida*. Food Chem. Toxicol. **47(6)**: 1304-1310. doi: 10.1016/j.fct.2009.03.004.

Orhan, I.E., Tosun, F., Tamer, U., Duran, A.,Alan, B. and Kok, A.F. 2011. Quantification of genistin and daidzein in two endemic *Genista* species and their antioxidant activity. J. Serb. Chem. Soc. **76**: 1-8.

Orhan, I. and Ustun, O. 2010. Determination of total phenol content; antioxidant activity and acetylcholinesterase inhibition in selected mushrooms from Turkey. J. Food Composition and Analysis. doi: 10.1016/j.jfca.010.11.005.

Ortet, R., Regalado, E.L., Thomas, O.P., pino, J.A., Fernandez, M.D. and Filippi, J.J.2009. Composition and antioxidant properties of the essential oil of the endemic Cape Verdean *Satureja forbesii*. Nat. Prod. Commun. **4(9)**: 1277-1280.

Ortet, R., Thomas, O.P., Regalado, E.L., Pino, J.A., Filippi, J.J. and Fernandez, M.D. 2010. Composition and biological properties of the volatile oil of *Artemisia gorgonum* Webb. Chem. Biodivers. **7(5)**: 1325-1332.

Osakabe, N., Yamaguchi, M., Sanbogi, C., Natsumo, M., Takizawa, T. and Osawa, T. 1998. The antioxidative substances in Cacao liquor. Nutr. Sci. Vit. **44**: 313-321.

Osakabe, N., Yasuda, A., Natsume, M., Sanbongi, C., Kato, Y., Osawa, T. and Yoshikawa, T. 2002. Rosmarinic acid, a major polyphenolic component of *Perilla frutescens*, reduced lipopolysaccharide (LPS)- induced liver injury in D-galactosamine (D-GaiN) sensitized mice. Free Radical Biol. Med. **33**: 798-806.

Osawa, T. and Namaki, M. 1981. Antioxidative activity of three herbal water extracts. Agric. Biol. Chem. **45**: 735-739.

Osawa, T. and Namaki, M. 1983. A novel type antioxidant isolated from leaf wax of *Eucalyptus* leaves. Agric. Biol. Chem. **45**: 735-739.

Oskoueian, E., Abdullah, N., Saad, W.Z., Omar, A.R., Ahmad, S., Kuan, W.B., Zolkifli, N.A., Hendra, R. and Ho, Y.W. 2011. Antioxidant, anti-inflammatory and anticancer activities of methanolic extracts from *Jatropha curcas* Linn. J. Med. Pl. Res. **5(1)**: 49-57.

Osman, H., Rahim, A.A., Isa, N.M. and Bakhir, N.M. 2009. Antioxidant activity and phenolic content of *Paederia foetida* and *Syzygium aqueum*. Molecules **14**: 970-978.

Ossipov, V., Nurmi, K., Loponen, J., Haukioja, E. and Pihlaja, K. 1996. HPLC separation and identification of phenolic compounds from leaves of *Betula pubescens* and *Betula pendula*. J. Chromatogr. **721**: 59-68.

Oszmianski, J., Wojdylo, A., Lamer-Zarawska, E. and Swiader, K. 2007. Antioxidant tannins from Rosaceae plant roots. Food Chem. **100(2)**: 579-583. doi: 10.1016/j.foodchem.2005.09.086.

Othman, A., Ismail, A., Ghani, N.A. and Adenan, I. 2007. Antioxidant capacity and phenolic content of cocoa beans. Food Chem. **100**: 1523-1530.

Othman, A., Jalil, A.M.M., Weng, K.K., Ismail, A., Ghani, N.A. and Adenan, I. 2010. Epicatechin content and antioxidant capacity of cocoa beans from four different countries. Afr. J. Biotech. **9(7)**: 1052-1059.

Ouattara, M.B., Konate, K., Kiendrebeogo, M., Outtara, N., Compaore, M. and Meda, R. 2011. Antibacterial potential and antioxidant activity of polyphenols of *Sesbania grandiflora*. Curr. Res. J. Biol. Sci. **3(4)**: 351-356.

Ouattara, L., Koudou, J., Zongo, C., Barro, N., Savadogo, A., Bassole, I.H.N., Ouattara, A.S. and Traore, A.S. 2010. Antioxidant and antibacterial activities of three species of *Lannea* from Burkina Faso. J. Applied Sci. **11**: 157-162.

Ouyang, M.A., He, Z.D. and Wu, C.L. 2003. Antioxidative activity of glycosides from *Ligustrum sinense*. Nat. Prod. Res. **17(6)**: 381-387.

Owen, R.W., Giacosa, W.E., Hull, R., Hauber, B, Spiegelhalder, B. and Bartsch, H. 2000. The antioxidant/anticancer potential of phenolic compounds isolated from olive oil. European J. Cancer **36**: 1235-1247.

Owolabi, M.A., Jaja, S.I., Oyekanmi, O.O. and Olatunji, O.J. 2008. Evaluation of antioxidant activity and lipid peroxidation of the leaves of *Vernonia amygdalina*. J. Complement. Integr. Med. **5(1)**: article 21. doi: 10.2202/1553-3840.1152.

Oyedemi, S.O., Bradley, G. and Afolayan, A.J. 2010. *In -vitro* and *–vivo* antioxidant activities of aqueous extract of *Strychnos henningsii* Gilg. Afr. J. Pharm. Pharmacol. **4(2)**: 70-78.

Oza, D. and Pandya, H.A. 2007. Scaling of antioxidant activities in *Citrus limon* (L.) Burm.f. In: 17th Annual conference of Indian Association for Angiosperm Taxonomy and International Seminar on "Changing scenario in Angiosperm systematics". Abstract. P. 182.

Ozcan, M. 2003a. Effect of sumac (*Rhus coriaria* L.) extracts on the oxidative stability of peanut oil. J. Medicinal Food **6**: 63-66.

Ozcan, M. 2003b. Antioxidant activities of rosemary, sage and sumac extracts and their combinations on stability of peanut oil. J. Medicinal Food **6**: 267-270.

Özen, T. 2009. Investigation of antioxidant properties of *Nasturtium officinale* (watercress) leaf extracts. Acta Pol. Pharm. **66(2)**: 187-193.

Özen, T. 2010. Antioxidant activity of wild edible plants in the Black Sea Region of Turkey. Grasas Y Aceites 61(1): 86-94. doi: 10.3989/gya075509.

Özen, T., Collu, Z. and Korkmaz, H. 2010. Antioxidant properties of *Urtica pilulifera* root, seed, flower and leaf extract. J. Med. Food **13(5)**: 1224-1231.

Ozer, H., Sökmen, M., Gulluce, M., Adiguzel, A., Sahin, F., Sökmen, A., Kilic, H. and Baris, O. 2007. Chemical composition and antimicrobial and antioxidant activities of the essential oil and methanol extract of *Hippomarathrum microcarpum* (Bieb.) from Turkey. J. Agric. Food Chem. **55(3)**: 937-942.

Ozer, M.S., Sarikurkcu, C., Tepe, B. and Can. S. 2010. Essential oils composition and antioxidant activities of *Alkanna tinctoria* subsp. *tinctoria*. Food Sci. Biotech. 19: 1177-1183.

Özgen, M., Durgan, C., Serne, S. and Kaya, C. 2008. Chemical and antioxidant properties of pomegranate cultivars grown in Mediterranean region of Turkey. Food Chem. **111**: 703-706.

Özgen, M., Mavi, A., Terzi, Z., Yildirim, A., Coskun, M. and Houghton, P.J. 2006. Antioxidant properties of some medicinal Lamiaceae (Labiatae) species. Pharm. Bio. **44**: 107-112.

Özgen, M., Neil Reese, R., Tulio, A.Z., Scheerens, J.C. and Miller, A.R. 2006. Modified 2,2-azino-bis-3-ethylbenzothiazoline-6-sulphonic acids (ABTS) method to measure antioxidant capacity of selected small fruits and comparison to ferric reducing antioxidant power (FRAP) and 2,2′-diphenyl-1-picrylhydrazyl (DPPH) methods. J. Agric. Food Chem. **50**: 6975-6980.

Özgen, M and Scheerens, J.C. 2006. Determination of antioxidant capacity of black and red raspberries using modified TEAC method and discussion of anticancer peroperties. II. National Soft Fruits Symposium, 14-16 September, 2006 pp. 322-327.

Özgen, M., Scheerens, J.C., Reese, R.N. and Miller, R.A. 2010. Total phenolic, anthocyanin contents and antioxidant capacity of select elderberry (*Sambucus canadensis* L.) accessions. Phcog. Mag. **6(23)**: 198-203.

Özgen, M., Serce, S., Gunduz, Z., Yen, F., Kafkas, E. and Paydas, S. 2007. Determining total phenolics and antioxidant activity of selected *Fragaria* genotypes. Asian J. Cehm. **19(7)**: 5573-5581.

Özgen, M., Serce, S. and Kaya, C. 2009. Phytochemical and antioxidant properties of anthocyanin-rich *Morus nigra* and *M. rubra* fruits. Sci. Hort. **119**: 275-279.

Özgen, M., Tulio Jr. A.Z., Chanon, A.M., Janakiraman, N., Reese, R.N. and Miller, A.R. 2006. Phytonutrient accumulation and antioxidant capacity at eight developmental stages of black raspberry fruit. Hort. Sci. **41**: 1082.

Özgen, M., Wyzgoski, F.J., Tulio, A.Z, Gazula, A., Miller, A.R., Scheerens, J.C. Reese, R.N. and Wright, S.R. 2008. Antioxidant capacity and phenolic antioxidants of Midwestern black raspberries grown for direct markets are influenced by production site. Hort. Sci. **43**: 2039-2047.

Özgen, U., Houghton, P.J., Ogundipe, Y. and Coskun, M. 2003. Antioxidant and antimicrobial activities of *Onosma argenteum* and *Rubia peregrina*. Fitoterapia **74**: 682-685.

Özgen, U., Mavi, A., Terzi, Z., Coskum, M. and Yildirin, A. 2004. Antioxidant activities and total phenolic compounds amount of some Asteraceae species. Turkish J. Pharm. Sci. **1(3)**: 203-216.

Özgen, U., Mavi, A., Terzi, Z., Kazaz, C., Asci, A., Kaya, Y. and Secen, H. 2011. Relationship between chemical structure and antioxidant activity of Luteolin and its glycosides isolated from *Thymus sipyleus* subsp. *sipyleus* var. *sipyleus*. Rec. Nat. Prod. **5(1)**: 12-21.

Ozkan, G., Baydar, H. and Erbas, S. 2010. The influence of harvest time on essential oil composition, phenolic constituents and antioxidant properties of Turkish Oregano (*Origanum onites* L.). J. Sci. Food Agric. **90(2)**: 205-209.

Ozkan, G., Kuleasan, H., Celik, S., Gokturk, R.S. and Unal, O. 2007. Screening of Turkish endemic *Teucrium montbretii* subsp. *pamphylicum* extract for antioxidant and antibacterial activities. Food control **18(5)**: 509-512.

Ozkan, G. and Ozcan, M. 2006. Antioxidant tocopherol constituents from some medicinal and aromatic plants. Chemistry of Natural compounds **42 (5)**: 604-605.

Özkan, G, Sagdic, O., Baydar, N.G. and Baydar, H. 2004. Antioxidant and antibacterial activities of *Rosa damascena* flower extracts. Food Sci. Tech. Intern. **10(4)**: 277-281. DOI: 10.1177/1082013204045882.

Ozkan, G., Sagdic, O., Gokturk, R.S., Unal, O. and Albayrak, S. 2010. Study on chemical composition and biological activities of essential oil and extract from *Salvia pisidica*. LWT – Food Sci. Tech. **43**: 186-190.

Ozkan, G., Sagdic, O., Ozcan, M., Ozcelik, H. and Unver, A. 2005. Antioxidant and antibacterial activities of Turkish endemic *Sideritis* extracts. Grasas y Aceitas **56(1)**: 16-20.

Ozkan, G., Sagdic, O. and Ozcelik, H. 2004. Some Turkish endemic herb extracts as antibacterial and antioxidant agents. In: Fourth International congress environmental, micropaleontology, microbiology and meiobenthology pp. 151-154, Isparta, Turkey.

Ozsoy, N., Can, A., Yanardag, R., Akev, N. 2008. Antioxidant activity of *Smilax excelsa* L. leaf extracts. Food Chem. **110 (3)**: 571-583.

Ozsoy, N., Candoken, E. and Akev, N. 2009. Implications for degenerative disorders: Antioxidative activity, total phenols, flavonoids, ascorbic acid, beta-carotene and beta-tocopherol in *Aloe vera*. Oxid Med. Cell Longev. **2(2)**: 99-106.

Ozsoy, N.,Yilmaz, T., Kurt, O., Can, A. and Yanardag, R. 2009. *In vitro* antioxidant activity of *Amaranthus lividus* L. Food Chem. **116(4)**: 867-872.

Ozsoy-Sacan, O., Yanardag, R., Orak, H., Ozgey, Y., Yarat, A. and Tunali, T. 2006. Effects of parsley (*Petroselinum crispum*) extract versus glibornuride on the liver of streptozotocin-induced diabetic rats. J. Ethnopharmacol. **104(1-2)**: 175-181.

Öztürk, M., Bozan, B., Tunalier, Z., Kosar, M. and Baser, K.H.C. 2002. Antioxidant and free radical scavenging activity of *Eruca sativa*. International Conference on Innovations in Food processing Technology and Engineering. Bangkok, Thailand, pp. 291-301.

Öztürk, M., Öztürk, F.A., Duru, M.E. and Topcu, G. 2007. Antioxidant activity of stem and root extracts of Rhubarb (*Rheum ribes*): An edible medicinal plant. Food Chem. **103(2)**: 623-630.

Öztürk, M., Tumen, I., Ugur, A., Aydogmus- Öztürk, F. and Topcu, G. 2011. Evaluation of fruit extracts of six Turkish *Juniperus* species for their antioxidant, anticholinesterase and antimicrobial activities. J. Sci. Food Agric. **92(5)**: 867-876. doi: 10.1002/jsfa.4258.

Pabuccuoglu, A., Kivcak, B., Bas, M and Mert, T. 2003. Antioxidant activity of *Arbutus unedo* leaves. Fitoterapia **74**: 597-599.

Pabuccuoglu, A., Kivcak, B. and Mert, T. 2003. Antioxidant activity of *Laurus nobilis* L. leaves. J. Faculty of Pharmacy of Gaza University **20(1)**: 15-19.

Pachaly, P. and Klein M. 1987. Inhaltsstoffen von *Andromeda polfolia*. Plant Med. **53**: 442-444.

Pacifico, S., D'Abrosca, B., Pascarella, M.T., Letizia, M., Uzzo, P., Piscopo, V. and Fiorentino, A. 2009. Antioxidant efficacy of iridiod and phenylethanoid glycosides from the medicinal plant *Teucrium chamaedris* in cell-free systems. Bioorganic and Medicinal Chem. **17(17)**: 6173-6179.

Packer, L., Rimbach, G. and Virgili, F. 1999. Antioxidant activity and biologic properties of a procyanidin-rich extract from pine (*Pinus maritima*) bark, pycnogenol. Free Rad. Biol. Med. **27(5-6)**: 704-724.

Padhy, B.M., Srivastava, A. and Kumar, V.L. 2007. *Calotropis procera* latex affords protection against carbon tetrachloride induced hepatotoxicity in rats. J. Ethnopharmacol. **113(3)**: 498-502.

Padma, P.R., Bhuvaneswari, V. and Silambuchelvi, K. 1998. The activities of enzymic antioxidants in selected green leaves. Indian J. Nutr. Diet. **35(1)**: 1-3.

Padma, P.R., Sumathi, S. and Aparna, S. 2000. Antioxidant status of the flowers of *Caesalpinia pulcherrima*. Seminar paper. J. Med. Arom. Pl. Sci. **22**: 36-37.

Padmaja, M., Sravanthi, M. and Hemalatha, K.P.J. 2011. Evaluation of antioxidant activity of two Indian medicinal plants. J. Phytol. **3(3)**: 86-91.

Padmaja, M.V., Mrumugal, P., Santhiya, S.T. and Ramesh, A. 2004. Estimation of total antioxidant activity in different parts of *Annona squamosa* Linn. by ABTS/$K_2S_2O_8$ decoloration method. Chemistry Biology Interface: Synergistic New Frontiers, New Delhi, India p. 23-42, November 21-26, 2004.

Padmaja, S. and Raju, T.N. 2004. Antioxidant effect of curcumin in selenium induced cataract of Wistar rats. Indian J. Exp. Biol. **42(6)**: 601-603.

Padmaja, S. and Raju, T.N. 2005. Protective effect of curcumin during selenium induced toxicity on dehydrogenases in hepatic tissue. Indian J. Physiol. Pharmacol. **49(1)**: 111-114.

Padmakumari, K.P., Sasidharan, I. and Sreekumar, M.M. 2011. Composition and antioxidant activity of essential oil of pimento (*Pimenta dioica* (L.) Merr.) from Jamaica. Nat. Prod. Res. **25(2)**: 152-160.

Pagliosa, C.M., Vieira, M.A., Podesta, R., Maraschin, M., Zeni, A.L.B., Amante, E.R. and Amboni, R.D.M.C. 2010. Food Chem. **122(1)**: 173-178.

Pai, S.R., Nimbalkar, M.S., Pawar, N.V., Patil, R.P. and Dixit, G.B. 2010. Seasonal discrepancy in phenolic content and antioxidant properties from bark of *Nothapodytes nimmoniana* (Grah.) Mabb. Int. J. Pharma Biosci. **1(3)**:

Pakutharivu, T. and Suriyavadhana, M. 2010. *In vitro* antioxidant activity of *Entada pursaetha, Toddallia aculeata* and *Ziziphs mauritiana*. Phcog. J. **2(6)**:

Pal, D., and Mitra, S. 2010. A preliminary study on the *in vitro* antioxidant activity of the stems of *Opuntia vulgaris*. J. Adv. Pharm. Tech. Res. **1(2)**: 268-272.

Pal, D.K. and Dutta, S. 2006. Evaluation of the antioxidant activity of the roots and rhizomes of *Cyperus rotundus* L. Indian J. Pharm. Sci. **68**: 256-258.

Pal, D.K., Kumar, S., Chakrabarty, P. and Kumar, M. 2008. A study on the antioxidant activity of *Semecarpus anacardium* L.f. nuts. J. Nat. Remedies **8(2)**: 160-163.

Palani, S., Raja, S., Naresh, R. and Kumar, B.S. 2010. Evaluation of nephroprotective, diuretic and antioxidant activities of *Plectranthus amboinicus* on a acetaminophen – induced nephrotoxic rats. Toxicol. Mech. Methods **20(4)**: 213-221.

Palanisamy, U., Cheng, H.M., Masilamni, T., Subramaniam, T., Ling, L.T., Radhakrishnan. A.K. 2007. Rind of the rambutan, *Nephelium lappaceum*, a potential source of natural antioxidants. Food Chem. **109(1)**: 54-63.

Palanisamy, U.D., Ling, L.T., Maharan, T., Sivapalan, V., Subramaniam, T., Helme, M.H. and Masilamani, T. 2011. Standardized extract of *Syzygium aqueum*: a safe cosmetic ingredient. Int. J. Cosmetic Sci. **33(3)**: 269-275. doi: 10.1111/j.1468-2494.2010.00637.x

Palipoch, S. Jiraungkoorskul, Tansatit, T., Preyavichyapugdee, Jaikua, W. and Kosai, P. 2010. Evaluation of antioxidant properties in leaf extracts of Thai traditional herb, *Thunbergia laurifolia* Lindl. www.sc.mahidol.ac.th/scpa/./poster/poster_32sarawoot.pdf.

Pallauf, K., Rivas-Gonzalo, J.C., del Castillo, M.D., Cano, M.P. and de Pascual-Teresa, S. 2008. Characterization of the antioxidant composition of strawberry tree (*Arbutus unedo* L.) fruits. Journal of Food Composition and Analysis **21(4)**: 273-281.

Pan, H. 1995. Studies on phenolic and terpenoid constituents extracted from bark of birch, spruce and pine in Sweden. Dissertation, Swedish University of Agricultural Sciences, Uppsala, Sweden.

Pan, Y., He, C., Wang, H., Ji, X., Wang, K., Ji, X., Wang, K. and Liu, P. 2010. Antioxidant activity of microwave-assisted extract of *Buddleia officinalis* and its major active component. Food Chem. **121(2)**: 497-502.

Pan, Y., Zhang, X., Wang, H., Linag,., Zhu, J., Li, H., Zhang, Z. and Wu, Q. 2007. Antioxidant potential of ethanolic extract of *Polygonum cuspidatum* and application in peanut oil. Food Chem. **105(4)**: 1518-1524.

Pan, Y., Zhu, J., Wang, H., Zhang, X., Zhang, Y., He, C., Ji, X. and Li, H. 2007. Antioxidant activity of ethanolic extract of *Cortex fraxini* and use in pea nut oil. Food Chem. **103**: 913-918.

Pan, X., Chen, S. and Zhuang, X. 2009. Study on antioxidant activities in extracts from dry *Pinus massoniana* needles by different methods. Science and Technology of Food Industry. DOI: CNKI: SUN: SPKJ.0.2009-03-030.

Pan, Y., Wang, K., Huang, S., Wang, H., Mu, X., He, C., Ji, X., Zhang, J. and Huang, F. 2008. Antioxidant activity of microwave-assisted extract of longan (*Dimocarpus longan* Lour.) peel. Food Chem. **106(3)**: 1264-1270.

Panda, S., Jafri, M., Kar, A. and Mehta, B.K. 2009. Thyroid inhibitory, antiperoxidative and hypoglycemic effects of stigmasterol isolated from *Butea monosperma*. Fitoterapia **80(2)**: 123-126.

Panda, S.P., Haldar, P.K., Bera, S., Adhikari, S. and Kandar, C.C. 2010. Antidiabetic and antioxidant activity of *Swietenia mahagoni* in streptozotocin-induced diabetic rats. Pharm. Bio. **48(9)**: 974-979.

Pandey, A.M. and Kulkarni, Y. 2010. Evaluation of antioxidant activity of *Gmelina arborea* extracts by *in vitro* techniques. Pharmacology online **2**: 805-811.

Pandey, M.M., Govindarajan, R., Rawat, A.K. and Pushpangadan, P. 2005. Free radical scavenging potential of *Saussurea costus*. Acta Pharm. **55(3)**: 297-304.

Pandey, P., Mehta, A., Hajra, S., John, J. and Mehta, P. 2011. Antioxidant property, total phenolic content and inhibition of α-amylase activity of *Ruta graveolens* L. leaves extract. J. Pharm. Res. **4(6)**:

Pandey, S., Sah, S.P., Sah, M.L. and Mishra, D. 2010. An antioxidant potential of hydromethanolic extract of *Urtica parviflora* Roxb. J. Basic Clin. Pharm. **1(3)**: 191-195.

Pandey, S., Sharma, M., Chaturvedi, P. and Tripathi, Y.B. 1994. Protective effect of *Rubia cordifolia* on lipid peroxide formation in isolated rat liver homogenate. Indian J. Exp. Biol. **32(3)**: 180-183.

Panigrahi, B.B., Panda, P.K. and Patro, V.J. 2011. Antitumor and *in-vivo* antioxidant activities of *Pandanus odoratissimus* Linn. against Ehrlisch Ascites carcinoma in Swiss albino mice. Int. J.Pharmaceutical Sci. Rev. Res. **8(2)**: 202-208.

Paniwnyk, L., Beaufoy, E., Lorimer, J.P and Mason, T.J. 2001 The extraction of rutin from flower buds of *Sophora japonica*. Ultrason Sonochem. **8 (3)**: 299-301.

Panpipat, W., Suttirak, W. and Chaijan, M. 2010. Free radical scavenging activity and reducing capacity of five Southern Thai indigenous vegetable extracts. Walailak J. Sci. Tech. **7(1)**: 51-60.

Pantelidis, G.E., Vasilakakis, M., Manganaris, G.A. and Diamantidis, G. 2007. Antioxidant capacity, phenol, anthocyanin an ascorbic acid contents in raspberries, blackberries, red currants, gooseberries and Cornelian cherries. Food Chem. **102(3)**: 777-783.

Papagiannopoulos, M., Wollseifen, H.R, Mellenthin, A, Haber B, and Galensa R. 2004. Identification and quantification of polyphenols in carob fruits (*Ceratonia siliqua* L.) and derived products by HPLC-UV-ESI/MSn. J. Agric. Food Chem. **52 (12)**: 3784-91.

Papaioannou, P., Lazari, D., Karioti, A., Souleles, C., Helimann, J., Hadjipavlou-Litina, D. and Skaltsa, H. 2007. Phenolic compounds with antioxidant activity from *Anthemis tinctoria* L. (Asteraceae). N.Naturforsch C. **62(5-6)**: 326-330.

Papas, A.M. (ed.). 1999. Antioxidant status, diet and health. CRC Press, Boca Raton, Florida,

Paramapojn, S. and Gritsanapan, W. 2009. Free radical scavenging activity determination and quantitative analysis of curcuminoids in *Curcumaa zedoaria* rhizome extracts by HPLC method. Curr. Sci. **97(7)**: 1069-1073.

Pareira, D.A., Faria, J., Gaspar, L., Ferreres, F., Valenta, P., Sottomayor, M. and Andrade, P.B. 2010. Exploiting *Catharanthus roseus* roots: source of antioxidants. Food Chem. **121(1)**: 55-61.

Pareira, J.A., Oliveira, I., Sousa, A., Valento, P. and Andrade, P.B.Ferreira, I.C., Ferreres, F., Bento, A., Seabra, R. and Estevinho, L. 2007. Walnut (*Juglans regia* L.) leaves: phenolic compounds, antibacterial activity and antioxidant potential of different cultivars. Food Chem. Toxicol. **45**: 2287-2295.

Parejo, I., Bastida, J., Viladomat, F. and Codina, C. 2005. Acylated quercetagetin glycosides with antioxidant activity from *Tagetes maxima*. Phytochem. **66 (19)**: 2356-2362.

Parejo, I., Caprai, E., Bastida, J., Viladomat, F., Jauregui, O. and Codina, C. 2004. Investigation of *Lepechinia graveolens* for its antioxidant activity and phenolic composition. J. Ethnopharmacol. **94(1)**: 175-184.

Parejo, I., Viladomat, F., Bastida, J., Rosas-Romero, A., Flerlage, N., Burillo, J. and Codina, C. 2002. Comparison between radical scavenging activity and antioxidant activity of six distilled and nondistilled Mediterranean herbs and aromatic plants. J. Agric. Food Chem. **50**: 6882-6890.

Parejo, I., Viladomat, F., Bastida, J., Rosas-Romero, A., Saavedra, G., Murcia, M.A., Jimenez, A.M. and Codina, C. 2005. Investigation of Bolivian plant extracts for their radical scavenging activity and antioxidant activity. Life Sci. **73**: 1667-1881.

Pari, L. and Latha, M. 2002. Antidiabetic activity of *Cassia auriculata* flowers: Effect on lipid peroxidation in streptozotocin diabetes rats. Pharmaceutical Biology **40(7)**: 512-517.

Parihar, M.S. and Hemnani, T. 2003. Phenolic antioxidants attenuate hippocampal neuronal cell damage against kainic acid induced excitotoxicity. J. Biosci. **28**: 121-128.

Parihar, M.S., Manjula, Y., Bano, S. Hemnani, T., Javeri, T. and Prakash, P. 1997. Nicotinamide and α-tocopherol combination partially protects t-butyl hydroperoxide-induced neurotoxicity: Implication for neurodegenerative diseases. Curr. Sci. **73**: 290-293.

Park, B.S., Lee, K.G., Shibamoto, T., Lee, S.E. and Takeoka, G.R. 2003. Antioxidant activity and characterization of volatile constituents of Taheebo (*Tabebuia impetiginosa* Martius ex DC). J. Agric. Food Chem. **51(1)**: 295-300.

Park, D., Jeon, J.H., Kwon, S.-C., Shin, S., Jang, J.Y., Kaim, Y.B. and Joo, S.S. 2009. Antioxidative activities of white rose flower extract and pharmaceutical advantages of its hexane fraction via free radical scavenging effects. Biochem. Cell Biol. **87(6)**: 943-952.

Park, E.J. and Pezzuto, J.M. 2002. Botanical in cancer chemoprevention. Cancer Metastasis Rev. **31(3-4)**: 231-255.

Park, E.J., Zhao, Y.Z., Kim, Y.C. and Sohn, D.H. 2005. Protective effect of (S)-bakuchiol from *Psoralea corylifolia* on rat liver injury *in vitro* and *in vivo*. Planta Med. **71(6)**: 508-513.

Park, H.S., Lim, J.H., Kim, H.J., Choi, H.J. and Lee, I.S. 2007. Antioxidant flavone glycosides from the leaves of *Sasa borealis*. Arch. Pharm. Res. **30(2)**: 161-166.

Park, P.J., Heo, S.J., Park, E.J., Kim, S.K., Byun, H.G., Jeon, B.T. and Jeon, Y.J. 2005. Reactive oxygen scavenging effect of enzymatic extracts from *Sargassum thunbergii*. J Agric Food Chem. **53(17)**: 6666-72.

Park, Y.K. Koo, M.H. Ikegaki M. and Contado. J.L. 1997. Comparison of the flavonoid aglycone contents of *Apis mellifera propolis* from various regions of Brazil. Arquivos de Biol. Technol. **40**: 97–106.

Parker, T.L., Wang, X.H., Pazimino, J. and Engeseth, N.J. 2007. Antioxidant capacity of phenolic content of grapes, sun-dried raisins, and golden raisins and their effect on *ex vivo* serum antioxidant capacity. J. Agric. Food Chem. **55(21)**: 8472-8477.

Parmar, J., Sharma, P., Verma, P., Sharma, P. and Goyal, P.K. 2010. Inhibitory potential of *Rosmarinus officinalis* against DMBA-induced skin papillogenesis and modulation of antioxidant defense system in mice. Pharmacology online **1**:

Parsaeimehr, A., Sargsyan, E. and Javidnia, K. 2010. A comparative study of the antibacterial, antifungal and antioxidant activity and total content of phenolic compounds of cell cultures and wild plants of three endemic species of *Ephedra*. Molecules **15**: 1668-1678; doi: 10.3390/molecules15031668

Pasias, I., Farmaki, *E.G.*, Thomaidis, N.S. and Piperaki, E.A. 2009. Elemental content and total antioxidant activity of *Salvia fruticosa*. Food Anal. Methods **3(3)**: 195-204. doi: 10.1007/s12161-009-9122-z.

Pastene, E. R., Avello, M. Gonzalez, M., Vega, M. Peric, I. Montes, M. 2003. Antioxidant activity of selected Chilean medicinal plants on human low density lipoprotein oxidation. Biometals **16**: 240.

Pastrana-Bonilla, E., Akoh, C.C., Sellapppan, S. and Krewer, G. 2003. Phenolic content and antioxidant capacity of Muscardine grapes. J. Agric. Food Chem. **51**: 5497-5503.

Patel, A., Patel, A., Patel, A. and Patel, N.M. 2010. Determination of polyphenols and free radical scavenging activity of *Tephrosia purpurea* Linn. leaves (Leguminosae). Phcog. Res. **2(3)**: 152-158.

Patel, M.A., Pattel, P.K., Kansara Monang, U. and Patel, A.P. 2010. *In vitro* free radical scavenging activity of *Ficus benghalensis* and *Ficus racemosa* Linn. leaf extracts. Pharmacology online **1**: 950-957.

Patel, N. J. and Patel, J. 2010. *In vitro* antioxidant activity of various fractions of *Achyranthes aspera*. Pharmacology online **2**: 681-688.

Patel, R.S. 1979. Antioxidant role of Curry (*Murraya koenigii*) and betel (*Piper betel*) leaves in Ghee. J. Food Sci. Techno. **16**: 158-160.

Patel, S.S., Shah, R.S. and Goyal, R.K. 2009. Antihyperglycemic, antihyperlipidemic and antioxidant effects of Dihar, a polyherbal ayurvedic formulation in streptozotocin induced diabetic rats. Indian J. Exp. Bio. **47(7)**: 564-570.

Pathak, N.L., Patel, N.J., Kasture, S.B., Jivani, N.P., Bhalodia, Y.S. and Malaviyaa, S.V. 20110. Free radical scavenging activity of *Albizia lebbeck* methanolic extract in arthritic rats. Int. J. Pharma Res. Dev. Online **1**: 1-8.

Patil, R.P., Nimbalkar, M.S., Jadhav, U.U., Dawkar, V.V. and Govindwar, S.P. 2010. Antiaflaxogenic and antioxidant activity of an essential oil from *Ageratum conyzoides* L. J. Sci. Food Agri. **90**: 608-614.

Patil, S., Anarthe, S., Jadhav, R. and Surana, S. 2011. Evaluation of anti-inflammatory activity and antioxidant activity of Indian Mistletoe, the hemiparasite *Dendrophthoe falcata* L.f. (Loranthaceae). Iranian J. Pharm. Res. **10(2)**: 253-259.

Patnibul, S., Puangmalai, S., Moung-on, A. and Sriprang, S. 2008. Screening for antioxidant activity in eighteen local North-Eastern vegetables using silica gel Thin layer chromatography followed by spraying with DPPH. NU Science Journ. **5(1)**: 1-6.

Patra, J.K., Mohapatra, A.D., Rath, S.K., Dhal, N.K. and Thatoi, H. 2009. Screening of antioxidant and antifilarial activity of leaf extracts of *Excoecaria agallocha* L. Int. J. Integr. Biol. **7(1)**: 9-15.

Patthamakanokporn, O., Puwastien, P., Nitithamyong, A. and Sirichakwal, P.P. 2008. Changes of antioxidant activity and total phenolic compounds during storage of selected fruits. J. Food Compos. Anal. **21(3)**: 241-248.

Patras, A. Breunton, N., Pieve, S.D., Butler, F. and Diwney, G. 2009. Effect of thermal and high pressure processing on antioxidant activity and instrumental colour of tomato and carrot purees. Innov Food Sci. Emerg. Technol. **10(1)**: 16-22.

Patricia F. Leal, Mara, E.M. Braga, Daisy N. Sato, Joao, E. Carvalho, Marcia O.M.Marques and Angela M.Meireles, A. 2003. Functional properties of spice extracts obtained via super critical fluid extraction. J. Agric. Food Chem. **51**: 2520-2525.

Patro, B.S., Bauri, A.K., Mishra, S. and Chattopadhyay, S. 2005. Antioxidant activity of *Myristica malabarica* extracts and their constituents. J. Agric. Food Chem. **53 (17)**: 6912-6918.

Patro, B.S., Chintalwar, G.J. and Chattopadhyay, S. 2005. Antioxidant activities of *Swertia decussata* xanthones. Nat. Prod. Res. **19(4)**: 347-354.

Pattanayak, S.P. and Sunitha, P. 2008. Wound healing, antimicrobial and antioxidant potential of *Dendrophthoe falcata* (L.f.) Ettingsh. J. Ethnopharmacol. **120 (2)**: 241. 247.

Pauletti, P.M., Castro-Gamboa, I., Siqueira Silva, D.H., Young, M.C.M. Tomazela, D.M., Eberlin, M.N., da Silva Bolzani, V. 2003. New antioxidant Cglucosylxanthones from the stems of *Arabidaea samydoides* J. Nat. Prod. **66(10)**: 1384-1387.

Pavel, S., Borivoj, K. and Vlastimil, K. 2006. Determination of total content of phenolic compounds and their antioxidant activity in vegetables, evaluation of spectrophotometric methods. J. Agric. Food Chem. **54**: 607-616.

Pawar, C.R., Mutha, R.E., Landge, A.D., adhav, R.B. and Surana, S.J. 2009. Antioxidant and cytotoxic activities of *Caesalpinia pulcherrima* wood. Indian J. Biochem. Biophys. **46**: 198-200.

Pawar, R.S. and Bhutani, K.K. 2004. Madhucosides A and B, protobasic acid glycosides from *Madhuca indica* with inhibitory activity on free radical release from phagocytes. J. Nat. Prod. **67(4)**: 668-671.

Pawar, S.S., Khater, S.H. and Sawant, S.D. 2009. Hepatoprotective activity of aqueous extracts of *Dodonaea viscosa* in paracetamol and CCl_4 induced albino rats. Adv. Pharmaol Toxicol. **10(3)**: 107-110.

Pbuccuoglu, A., Kivcak, B., Bas, M. and Mert, T. 2003. Antioxidant activity of *Arbutus unedo* leaves. Fitoterapia **74**: 597-599.

Pearson, D., Frankel, E., Aeschbach, R., Jb, G. 1997. Inhibition of endothelial cell-mediated oxidation of low-density lipoprotein by rosemary and plant phenolics. J. Agric. Food Chem. **45**: 578-582.

Pedraza-chaverri, J., Tapia, E., Medina-Campos, O.N., de los Angeles, Granados, M. and Franco, M. 1998. Garlic prevents hypertension induced by chronic inhibitin of nitric oxide synthesis. Life Sci. **62**: 71-77.

Pedreschi, R., Cisneros-Zevallos, L. 2006. Antimutagenic and antioxidant properties of phenolic fractions from Andean purple corn (*Zea mays* L.). J. Agric. Food Chem. **54**: 4557-4567.

Pegg, R.B., Amarowics, R. and Nazk, M. 2004. Antioxidant activity of Polyphenols from a Bearberry leaf (*Arctostaphylos uva-ursi* (L.) Sprengel) extract in meat system. In: Shahidi, F. and Naczk, M. (eds.) Phenolics in Food and Neutraceuticals. CRC Press, Boca Raton. Pp. 67-82.

Pellati, F., Benvenuti, S., Magro, L., Melegari, M. and Soragni, F. 2004. Analysis of phenolic compounds and radical scavenging activity of *Echinacea* spp. J. Pharm. Biomed. Anal. **35(2)**: 289-301.

Pellati, F., Benvenuti, S., Melegari, M. and Lasseigne, T. 2005. Variability in the composition of antioxidant compounds in *Echinacea* species by HPLC. Phytochem. Anal. **16(2)**: 77-85.

Pellegrini, N., Chiavaro, E., Gardana, C., Mazzeo, T., Contino, D., Gallo, M., Riso, P., Fogliano, V. and Porrini, M. 2010. Effect of different cooking methods on color, phytochemical concentration, and antioxidant capacity of raw and frozen *Brassica* vegetables. J. Agric. Food Chem. **58(7)**: 4310-4321.

Pellegrini, N., Re, R., Yang, M. and Rice-Evans, C. 1999. Screening of dietary carotenoids and carotenoid-rich fruit extracts for antioxidant activities applying

2,2′-azinobis(3-ethylenebenzothiazoline-6-sulfonic acid) radical cation decolorization assay. In: L.Packer (ed.): Methods in Enzymology: Oxidants and antioxidants. Academic press, New York, pp. 379-389.

Pellegrini, N., Simonetti, P., Gardana, C., Brenna, O., Brighenti, F. and Pietta, P. 2000. Polyphenol content and total antioxidant activity of *Vini Novelli* (young red wines). J. Agric. Food chem. **48(3)**: 732-735.

Penarrieta, J.M., Alvarado, J.A., Akesson, B., Bergenstahl, B. 2005. Total antioxidant capacity in Andean food species from Bolivia. Revista Bolivia Quimica **22(1)**: 89-93.

Pendry, B., Busia, K. and Bell, C.M. 2005. Phytochemical evaluation of selected antioxidant-containing medicinal plants for use in the preparation of a herbal formula – a preliminary study. CABI Abstract.

Peng, Z.F., Strack, D., Baumert, A., Subramaniam, R., Goh, N.K., Chia, T.F., Tan, S.N. and Chia, L. 2003. Antioxidant flavonoids from leaves of *Polygonum hydropiper* L. Phytochem. **62(2)**: 219-228.

Pereira, J.A., Oliveira, I., Sousa, A., Valentao, P., Andrade, P.B., Ferreira, I.C.F.R., Ferreres, F. and Estevinho, L. 2007. Walnut (*Juglans regia* L.) leaves: Phenolic compounds, antibacterial activity and antioxidant potential of different extracts. Food and Chem. Toxicol. **45 (11)**: 2287-2295.

Perez, R.M., Vargas, G.R., Martinez, S.F.J. and Cordova, R.I. 2004. Antioxidant and free radical scavenging activities of 5,7,3′-trihydroxy-3,6,4′-trimethoxyflavone from *Brickellia veronicaefolia*. Phytother. Res. **18(5)**: 428-430.

Perucka, I. and Materska, M. 2001. Phenylalanine Ammonia lyase and antioxidant activities of lipophilic function of fresh pepper fruits *Capsicum annuum* L. Innov. Food Sci. Emerg. Tech. **2**: 189-192.

Peschel, W., Sanchez-Rabaneda, F., Diekmann, W., Plescher, A., Gartzia, I., Jiminez, D., Lamuela-Rave. 2006. An industrial approach in the search of natural antioxiants from vegetable and fruit wastes. Food Chem. **97(1)**: 137-150.

Peteros, N.P. and Uy, M.M. 2010. Antioxidant and cytotoxic activities and phytochemical screening of four Philippine medicinal plants. J. Med. Plants Res. **4(5)**: 407-414.

Peterson, D.M., Emmons, C.L. and Hibbs, A.H. 2001. Phenolic antioxidants and antioxidant activity in pearling fractions of oat groats. J. Cereal Sci. **33**: 97-103.

Petrova, A.P., Krasnov, E.A., Saprykina, E.V., Subbotina, Y.A. and Ermilova, E.V. 2009. Chemical composition of couch grass and studies of its antioxidant activity in allergic contact dermatitis. Pharm. Chem. Journ. **43(1)**: 48-50.

Petrus, A.J.A., 2011. Foliar phenolic antioxidant constitution of *Mallotus stenanthus*. Asian J. Chem. **23(2)**: 857-862.

Petrus, A.J.A., Bhuvaneshwari, N. and Alain, J.A.L. 2011. Antioxidative constitution of *Mukia maderaspatana* (Linn.) M. Roem leaves. Indian J. Nat. Prod. Res. **2(1)**: 34-43.

Phadungukit, M., Rattarom, R. and Rattana, S. 2010. Phytochemical screening, antioxidant, antibacterial and cytotoxic activities of *Knema angstifolia* extract. J. Med. Plants Res. **4(13)**: 1269-1272..

Phansawan, B. and Poungbangpho, S. 2007. Antioxidant capacities of *Pueraria mirifica, Stevia rebaudiana* Bertoni, *Curcuma longa* Linn., *Andrographis paniculata* (Burm.f.) Nees and *Cassia alata* Linn. for the development of dietary supplement. Kasaetsart J. (Nat. Sci.) **41**: 548-554.

Phillips, K.M., Carlsen, M.H. and Blomhoff, R. 2009. Total antioxidant content of alternatives to refined sugar. J. Amer Diet Assoc. **109(1)**: 64-71.

Phuwapraisirisan, P., Udomchotphuet, S., Surapinit, S. and Tip-Pyang, S. 2006. Antioxidant xanthones from *Cratoxylum cochinchinense*. Nat. Prod. Res. **20(14)**: 1332-1337.

Piacente, S., Montoro, P., Oleszek, W., Pizza, C. 2004. *Yucca schidigera* bark: phenolic constituents and antioxidant activity. J. Nat. Prod. **67(5)**: 882-885.

Piao, M.S., Kim, M.-R., Gun, L.D., Park, Y, Hahm, K.-S, Moon, Y.-H. and Woo, E.R. 2003. Antioxidative constituents from *Buddleja officinalis*. Arch. Pharm. Res. **26(6)**: 453-457.

Piccinelli, A.L., Arana, S., Caceras, A., Bianca, R.E.V.Sorrentino, R. and Rastrelli, L. 2004. New lignans from the roots of *Valeriana prinophylla* with antioxidative and vasorelaxant activities. J. Nat. Prod. **67(7)**: 1135-1140.

Picerno, P., Mencherini, T., Lauro, M.R., Barbato, F. and Aquino, R. 2003. Phenolic constituents and antioxidant properties of *Xanthosoma violaceum* leaves. J. Agric. Food Chem. **51(22)**: 6423-6428.

Pieme, C.A., Penlap, V.N., Ngogang, J. and Costache, M. 2010. *In vitro* cytotoxicity and antioxidant activities of five medicinal plants of Malvaceae family from Cameroon. Environ. Oxicol. Pharmacol. **29(3)**: 223-228.

Pieroni, A., Janiak, V., Durr, C.M., Ludeke, S., Trachsel, E. and Heinrich, M. 2002. *In vitro* antioxidant activity of non-cultivated vegetables of ethnic Albanians in southern Italy. Phytother. Res. **16**: 467-473.

Pietta, P., Simonetti, P. and Mauri, P. 1998. Antioxidant activity of selected medicinal plants. J. Agric. Food. Chem. **46 (11)**: 4487-4490.

Pietta, P.G. 2000. Flavonoids as antioxidants. J. Nat. Prod. **63**: 1035-1042.

Pihlanto, A., Akkanen, S. and Korhonen, H. J. 2008. ACE-inhibitory and antioxidant properties of potato (*Solanum tuberosum*). Food Chem. **109 (1)**: 104-112.

Pilarski, R., Zielinski, H., Ciesiolka, D. and Gulewicz, K. 2006. Antioxidant activity of ethanolic and aqueous extracts of *Uncaria tomentosa* (Willd.) DC. J. Ethnopharmacol. **104(1-2)**: 18-23.

Pinero Estrada, J.E., Bermejo Bescoc, P. and Villar del Fresno, A.M. 2001. Antioxidant activity of different fractions of *Spirulina platensis* protein extracts. Il. Pharmaco. **56**: 497.

Pino, J.A., Regalado, E.L., Rodriguez, J.L. and Fernandez, M.D. 2010. Phytochemical analysis and *in vitro* free-radical –scavenging activities of the essential oils from leaf and fruit of *Melaleuca leucandra* L. Chem. Biodivers. **7(9)**: 2281-2288.

Pistelli, L., Noccioli, C., Martera, M., Giamperi, L., Bucchini, A., Fraternale, D. and Ricci, D. 2006. Antioxidant flavonol glycosides from *Dorycnium hirsutum*. Chem. Nat. Compounds **42(3)**: 281-284.

Pizzale, L., Botrolomeazzi, R., Vichi, S., Uberegga, E. and Conte, L.S. 2002. Antioxidant activity of sage (*Salvia officinalis* and *S. fruticosa*), and Oregano (*Origanum onites* and *O. intercedens*) extracts related to their phenolic compounds content. J. Sci. Food Agric. **82**: 1645-1651.

Pljevljakusic, D., Savikin, K., Jankovic, T., Zdunic, G., Ristic, M., Godjevac, D and Konic-Ristic, A. 2011. Chemical properties of the cultivated *Sideritis raeseri* Boiss. et Heldr. subsp. *raeseri*. Food Chem. **124(1)**: 226-233.

Poli, F., Muzzoli, M., Sacchetti, G., Tassinato, G., Lazzarin, R. and Bruni, A. 2003. Antioxidant activity of supercritical CO_2 extracts of *Helichrysum italicum*. Pharmaceutical biol. **41(5)**: 379-383.

Policegoudra, R.S., Abiraj, K., Gowda, D.C. and Aradhya, S.M. 2007. Isolation and characterization of antioxidant and antibacterial compound from mango ginger (*Curcuma amada* Roxb.) rhizome. J. Chromatography. **852 (1)**: 40-48.

Politeo, O., Jukic, M. and Milos, M. 2010. Comparison of chemical composition and antioxidant activity of glycosidically bound and free volatiles from clove (*Eugenia caryophyllata* Thunb.). J. Food Biochem. **34**: 129-141.

Poltanov, E.A., Shikov, A.N., Dorman, H.J.D., Pozharitskaya, O.N., Makarov, V.G., Tikhov, V. and Hiltunen, R. 2009. Chemical and antioxidant evaluation of Indian gooseberry (*Emblica officinalis* Gaertn, syn. *Phyllanthus emblica* L.) supplements. Phytother. Res. **23(9)**: 1309-1315.

Poltereit, O. 1997. Antioxidants and free-radical scavengers of natural origin. Currant Org. Chem.**1**: 415-440.

Pongnaravane, B., Goto, M., Sasaki, M., Anekpankul, T, Pavasant, P., Shotipruk, A. 2006. Extraction of anthraquinones from roots of *Morinda citrifolia* by pressurized hot water: antioxidant activity of extracts. The Journal of Supercritical Fluids. **37 (3)**: 390-396.

Pongsawatmanit, R. and Maisuthisakul, P. 2004. Characterization of the antioxidant activity of *Leucaena glauca* Benth. extract. Proc. 10th World Congress on Clinical Nutr., Phuket, Thailand. 152.

Popov, I.N. and Lewin, G. 1996. Photochemiluminescent detection of antiradical activity: IV: Testing of lipid soluble antioxidants. J. Biochem. Biophys. Methods **31**: 1-8.

Popovic, M., Kaurinovic, B., Trivic, S., Mimica-Dukic, N., Bursac, M. 2006. Effect of celery (*Apium graveolens*) extracts on some biochemical parameters of oxidative stress in mice treated with carbon tetrachloride. Phytother. Res. **20**: 531-537.

Porwal, P. 2010. Comparative *in vitro* antioxidant activity of *Pongamia pinnata* Linn. leaves extracts and isolated compound. Int. J. Pharm. Biol. Arch. **1**:

Potchoo, Y., Guisson, I.P., Lompo, M., Sakie, E. and Yaro, B. 2008. Antioxidant activity of aqueous methanol and ethyl acetate extract of leaves of *Annona senegalensis* Pers. From Togo versus the one originated from Burkina Faso. Int. J. Pharmacol. Pp. 1-5.

Poudel, P.R., Tamura, H., Kataoka, I. and Mochioka, R. 2008. Phenolic compounds and antioxidant activities of skins and seeds of five wild grapes and two hybrids native to Japan. J. Food Compos. Analysi **21(8)**: 622-625.

Pourmorad, F., Hosseinimehr, S.J. and Shahabimajd, N. 2006. Antioxidant activity, phenol and flavonoid contents of some selected Iranian medicinal plants. African J. Biotech **5(11)**: 1142-1145.

Pouvreau, J.B., Morançais, M., Taran, F., Rosa, P., Dufossé, L., Guérard, F., Pin, S., Fleurence, J. and Pondaven, P. 2008. Antioxidant and free radical scavenging properties of marennine, a blue-green polyphenolic pigment from the diatom *Haslea ostrearia* (Gaillon/Bory) Simonsen responsible for the natural greening of cultured oysters. J. Agric. Food Chem. **56(15)**: 6278-86.

Povichit, N., Phrutivorapongkul, A., Suttajit, M., Chaiyasut, C. and Leelapornpisid, P. 2010. Phenolic content and *in vitro* inhibitory effects on oxidation and protein glycation of some Thai medicinal plants. Pak. J. Pharm. Sci. **23(4)**: 403-408.

Povilaitye, V., Cuverlier, M.E. and Berst, C. 2001. Antioxidant properties of Moldavian dragonhead (*Dracocephalum moldavica* L.). J. Food Lipids **26**: 233-243.

Pownall, T.L., Udenigwe, C.C. and Aluko, R.E. 2010. Aminoacid composition and antioxidant properties of pea seed (*Pisum sativum* L) enzymatic protein hydrolysate fractions. J. Agric. Food Chem. **58(8)**: 4712-4718. doi: 10.1021/jf904456r.

Pozharitskaya, O.N., Ivanova, S.A., Shikov, A.N. and Makarov, V.G. 2007a. Separation and evaluation of free radical-scavenging activity of phenol components of *Emblica officinalis* extracts by using and DPTLC-DPPH method. J. Sep. Sci. **30**: 1250-1254.

Pozharitskaya, O.N., Ivanova, S.A., Shikov, A.N., Makarov, V.G. and Galambosi, B. 2007b. Separation and evaluation of free radical-scavenging activity of phenol components of green, brown, and black leaves of *Bergenia crassifolia* by using DPTLC-DPPH method. J. Sep. Sci. **30**: 2447-2451.

Prabakan, M., Anandan, R. and Devaki, T. 2000. Protective effect of *Hemidesmus indicus* against rifampicin and sioniazid-induced hepatotoxicity in rats. Fitoterapia **71**: 55-59.

Prabhakar, K.R., Veeresh, P., Vipan, K., Sudheer, M., Priyadarshini, K.I., Satish, R.B.S.S. and Unnikrishnan, M.K. 2006. Bioactivity-guided fractionation of *Coronopus didymus*: A free radical scavenging perspective. Phytomedicine **13(8)**: 591-595.

Prabhakar, K.R., Veerapur, V.P., Bansal, P., Parihar, V.K., Reddy, K.M., Bhagath Kumar, P., Priyadarsini, K.I. and Unnikrishnan, M.K. 2007. Antioxidant and radioprotective effect of the active fractions of *Pilea microphylla* L. ethanolic extract. Chem. Biol. Interact. **165(1)**: 22-32.

Prakash, D., Singh, B.N. and Upadhyay Garima. 2007a. Antioxidant and free radical scavenging activities of phenols from onion (*Allium cepa*). Food Chem. **102(4)**: 1389-1393.

Prakash, D., Suri, S., Upadhyay, G., Singh, B.N. 2007b. Total phenol, antioxidant and free radical scavenging activities of some medicinal plants. Int. J Food Sci. Nutr. **58**: 18-28.

Prakash, D., Upadhyay, G., Singh, B.N., Singh, H.B., 2007c. Antioxidant and free radical-scavenging activities of seeds and agri-wastes of some varieties of soybean (*Glycine max*). Food Chem. **104**: 783-790.

Prakash, D., Upadhyay, G. Singh, B.N., Dhakarey, R., Kumar, S., Singh, K.K. 2007d. Free radical scavenging activity of Himalayan Rhododendrons. Curr. Sci. **92(4)**: 526-532.

Prasad, K.N., Yang, B., Dong, X., Jiang, G., Zhang, H., Xie, H. and Jiang, Y. 2009. Flavonoid contents and antioxidant activities from *Cinnamomum* species. Innovative Food Sci. Emerg. Tech. **10(4)**: 627-632.

Prasad, N.K., Divakar, S., Shivamurthy, G.R. and Aradhya, S.M. 2005. Isolation of a free radical scavenging antioxidant from water spinach (*Ipomoea aquatica* Forsk.). J. Sci. Food Agric. **85**: 1461-1468.

Prasanna Anjaneya Reddy, L., Bhakshu, L.Md., Venkata Ratnam, K., Veeranjaneya Reddy, L., Venkata Raju, R,.R. and Reddy, B.N. 2011. *In vitro* susceptibility, antioxidant activity and chemical constituents of essential oil of *Glossocardia bosvallea*. National Seminar on Trends in Plant Sciences, Sri Krishnadevaraya University, Abstract, pp. 46.

Pratt, D.E. 1992. Natural antioxidants from plant material. In: Hang, M., Ho, C. and C.Lee (eds.): Phenolic compounds in foods and their effects on health II: Antioxidants and cancer prevention I. ACS Symposium series 507, American Chemical Society, Washington DC, pp. 54.

Pratt, D.E. and Miller, E.E. 1984. A flavonoid antioxidant in Spanish peanuts (*Arachis hypogaea*). J. Amer. Oil. Chem. Soc. **61**: 1064-1067.

Preethi, K.C., Kuttan, G. and Kuttan, R. 2006. Antioxidant potential of an extract of *Calendula officinalis* flowers *in vitro*. Pharm. Biol. **44(9)**: 691-697. doi: 10.1080/13880200601009149.

Premalatha, B., Muthulakshmi, V., Vijayalakshmi, T. and Sachdanandam, P. 1997. *Semecarpus anacardium* nut extract induced changes in enzymatic antioxidants studied in aflatoxin B1 caused hepatocellular carcinoma bearing Wistar rats. Int. J. Pharmacogn. **35(2)**: 161-166.

Prince, P.S.M. and Venugopala Menon, P. 1999. Antioxidant activity of *Tinospora cordifolia* roots in experimental diabetes. J. Ethnopharmacol. **65**: 277-281.

Prieto, P., Pineda, M. and Aguilar, M. 1999. Spectrophotometric quantitation of antioxidant capacity through the formation of a phorsphomolybdenum complex: Specific application to the determination of vitamin E. Anal. Biochem. **269**: 337-341.

Prior, R.L. and Cao, G. 1999. Antioxidant capacity and polyphenolic components of teas: Implications for altering *in vivo* antioxidant status. Proc. Soc. Exp. Biol. Med. **220**: 255-261.

Prior, R.L. and Cao, G. 2000. Antioxidant phytochemicals in fruits and vegetables: diet and health implications. Hort. Sci. **35**: 588-592.

Prior, R.L., Cao, G., Martin, A., Sofic, E., Mcewen, J., O'Brien, C. Lischner, N., Ehlenfeldt, M., Kalt, W., Krewer, G. and Mainland, C.M. 1998. Antioxidant capacity as influenced by total phenolic and anthocyanin content, maturity and variety of *Vaccinium* species. J. Agric. Food Chem. **46**: 2686-2693.

Prior, R.L., Cao, G., Prior, R.L. and Cao, G. 2000. Analysis of botanicals and dietary supplements for antioxidant capacity: a review. J AOAC Int. **83(4)**: 950-956.

Prior, R.L., Wu, X. and Schaich, K. 2005. Standardized methods for the determination of antioxidant capacity and phenolics in foods and dietary supplements. J. Agric. Food Chem. **53(10)**: 4290-4302.

Pripdeevech, P. and Chukeairote, E. 2010. Chemical compositions, antifungal and antioxdidant activities of essential oil and various extracts of *Melodorum fruticosum* L. flowers. Food Chem. Toxicol. **48 (10)**: 2754-2758.

Priya, C.L., Kumar, G., Karthik, L. and Bhaskara Rao, K.V. 2010. Antioxidant activity of *Achyranthes aspera* Linn. stem extracts. Pharmacology online **2**: 228-237.

Priya, T.T., Sabu, M.C. and Jolly, C.I. 2008. Free radical scavenging and anti-inflammatory properties of *Lagerstroemia speciosa* L. Inflammatopharmacol. **16(4)**: 182-187.

Proestos, C., Boziaris, I.S., Nychas, G.J.E. and Komaitis, M. 2006. Analysis of flavonoids and phenolic acids in Greek aromatic plants: Investigation of their antioxidant capacity and antimicrobial activity. Food Chem. **95(4)**: 664-671.

Prosper-Cabral, N.B., Agbor, G.A., Oben, J.E. and Ngogang, J.Y. 2007. Phytochemical studies and antioxidant properties of four medicinal plants used in Cameroon. Afr. J. Tradit. Complement Altern. Med. **4(4)**: 495-500.

Proteggente, A.R., Pannala, A.S., Paganga, G., Van Buren, L., Wagner, E., Wiseman, S., Van de Put, F., Dacombe, C. and Rice-Evans, C. 2002. The antioxidant activity of regularly consumed fruit and vegetables reflects their phenolics and vitamin C composition. Free Radicals Research **36**: 217-233.

Protogeras, J., Oreopolou, V. and Tzia, C. 1008. Natural antioxidants from *Salvia triloba* for vegetable oils. Riv. Ital. Sostanze Grasse **75(11)**: 507-509.

Psotova, J., Kolar, M., Sousek, J., Svagera, Z., Vicar, J. and Ulrichova, J. 2003. Biological activities of *Prunella vulgaris* extract. Phytother. Res. **17**: 1082-1087.

Pulido, R., Bravo, L. and Saura-Calixto, F. 2000. Antioxidant activity of dietary polyphenols as determined by a modified ferric reducing/antioxidant power assay. J. Agric. Food Chem. **48(8)**: 3396-3402.

Pullela, S.V., Takamatsu, S., Khan, S. and Khan, I.A. 2005. Isolation of lignans and biological activity studies of *Ephedra viridis*. Planta Medica **71(8)**: 789-971.

Punitha, S.R., Bhat, N., Rajendran, K. and Shirwaikar, A. 2005a. Antioxidant activity of aqueous extract of *Coscinium fenestratum* in STZ-nicotinamide induced diabetic rats. Nat. Prod. Sci. **11(3)**: 155-159.

Punitha, R. and Manoharan, S. 2006. Antihyperglycemic and antiperoxidative effects of *Pongamia pinnata* (Linn.) Pierre flowers in alloxan induced diabetic rats. J. Ethnopharmacol. **105**: 39-46.

Punitha, S.R., Rajendran, K., Shirwaikar, A. and Shirwaikar, A. 2005. Alcoholic stem extract of *Coscinium fenestratum* regulates carbohydrate metabolism and improves antioxidant status in Streptozotocin-Nicotinamide induced diabetic rats. Evidence Based Complement. Alternat. Med. **2(3)**: 375-381.

Punzalan, G.A.N. Concepcion, G.P. Dator, R.P. Mazahery, A.R. and Jacinto, S.D.2006. Antioxidant activity of flavonoids and other polyphenols isolated from *Annona squamosa* Linn. leaf extracts. EJC Supplements: **4 (1)**: 58-58.

Puri, A., Srivastava, A.K., Singhal, B., Mishra, S.K., Srivastava, S. and Lakshmi, V. 2010. Antidislipidemic and antioxidant activity of *Pinus roxburghii* needles. Med. Chem. Res. Doi: 10.1007/s00044-010-9424-0.

Puttaraju, N.G., Venkateshaiah, S.U., Dharmesh, S.M., Urs, S.M. and Somasundaram, R. 2006. Antioxidant activity of indigenous edible mushrooms. J. Agric Food Chem. **54(26)**: 9764-72.

Puvanendran, S., Wickramasinghe, A., Karunaratne, D.N., Carr, G., Wijesundara, D.S.A., Anderson, R. and Karunaratne, V. 2008. Antioxidant constituents from *Xylopia championii*. Pharmaceutical Biol. **46(5)**: 352-355.

Puvanendran, S., Wickramasinghe, A., Karunaratne, D.N., Wijesundara, D.S.A. and Karunaratne, V. 2007. Screening of antioxidant activity of some endemic Annonaceae plant extracts. Proc. Peradeniya Univ. Res. Sessions, Sri Lanka **12**: 57-59.

Pyo, Y. H., Lee, T. C., Logendra, L. and Rosen, R.T. 2004. Antioxidant activity and phenolic compounds of Swiss chard (*Beta vulgaris* subspecies *cycla*) extracts. Food Chem. **85 (1)**: 19-26.

Qi, H., Liu, X., Ma, J., Zhang, Q. and Li, Z. 2010. *In vitro* antioxidant activity of acetylated derivatives of polysaccharide extracted from *Ulva pertusa* (Chlorophyta). J. Med. Plants Res. **4(23)**: 2445-2451.

Qi, H., Zhang, Q., Zhao, T., Chen, R., Zhang, H., Niu, X. and Li, Z. 2005a. Antioxidant activity of different sulfate content derivatives of polysaccharide extracted from *Ulva pertusa* (Chlorophyta) *in vitro*. Int. J. Biol. Macromol. **37(4)**: 195-9.

Qi, H., Zhang, Q., Zhao, T., Hu, R., Zhang, K. and Li, Z. 2006. *In vitro* antioxidant activity of acetylated and benzoylated derivatives of polysaccharide extracted from *Ulva pertusa* (Chlorophyta). Bioorg. Med. Chem. Lett. **16(9)**: 2441-2445.

Qi, H., Zhao, T., Zhang, Q., Li, Z., Zhao, Z. and Xing, R. 2005b. Antioxidant activity of different molecular weight sulfated polysaccharide from *Ulva pertusa* Kjellm. (Chlorophyta). J. Appl. Phycol. **17**: 527-534.

Qi, X.Y., Chen, W.J., Zhang, L.Q. and Xie, B.J. 2008. Magrosides extract from *Siraitia grosvenori* scavenges free radicals *in vitro* and lowers oxidative stress, serum glucose, and lipid levels in alloxan-induced diabetic mice. Nutr. Res. **28(4)**: 278-284.

Qingming, Y., Xianhui, P., Weibao, K., Hong, Y, Yidan, S, Li, Z., Yanan, Z, Yuling, Y, Lan, D and Guoan, L. 2010. Antioxidant activities of malt extract from barley (*Hordeum vulgare* L.) toward various oxidative stress *in vitro* and *in vivo*. Food Chem. **118(1)**: 84-89.

Qiu, Y., Liu, Q and beta, T. 2010. Antioxidant properties of commercial wild rice and analysis of soluble and insoluble phenolic acids. Food Chem. **121(1)**: 140-147.

Que, F., Mao, L. and Zheng, X. 2007. *In vitro* and *vivo* antioxidant activities of daylily flowers and the involvement of phenolic compounds. Asia Pacific J. Clin. Nutr. **16** Suppl. 1: 196-203.

Quettier-Delau, C., Gressier, B., Vasseur, J., Dine, T., Brunet, C., Luyckx, M., Cazin, M., Cazin, J.C., Bailleul, F and Trotin, F. 2000. Phenolic compounds and antioxidant activities of buckwheat (*Fagopyrum esculentum* Moench) hulls and flour. J. Ethnopharmcol. **72**: 35-42.

Rababah, T.M., Hettiarachchy, N.S. and Horax, R. 2004. Total phenolics and antioxidant activities of fenugreek, green tea, black tea, grape seed, ginger, rosemary, gotu kola, and ginkgo extracts, vitamin E, and tert-butylhydroquinone. J. Agric. Food Chem. **52(16)**: 5183-5186.

Racková, L., Majeková, M., Kostálová, D. and Stefek, M. 2004. Antiradical and antioxidant activities of alkaloids isolated from *Mahonia aquifolium*. Structural aspects. Bioorg. Med. Chem. **12(17)**: 4709-4715.

Racková, L., Oblozinsky, M., Kostálová, D., Kettmann, V. and Bezakova, L., 2007. Free radical scavenging activity and lipoxygenase inhibition of *Mahonia aquifolium* extract and isoquinoline alkaloids. J. Inflamm. **4 (1)**: 15 -18 doi: 10.1186/1476-9255-4-15.

Radhika, T., Mahender, P., Venkatesham, A., Anreddy, R.N.R, Narasimha Reddy, Y., Krishna, D.R., Christopher, T and Sadanandam, A. 2010. Antioxidant and analgesic activities of *Pterocarpus marsupium* Roxb. J. Herbs Spices Med. Plants **16(1)**: 63-68.

Rafat, A., Philip, K. and Muniandy, S. 2010. Antioxidant potential and content of phenolic compounds in ethanolic extracts of selected parts of *Andrographis paniculata*. J. Med. Plants Res. **4(3)**: 197-202.

Ragaee, S., El-Sayeed, M., Abdel, A. and Noaman, M. 2006. Antioxidant activity and nutrient composition of selected cereals for food use. Food Chem. **98**: 32-38.

Raghu, K.L., Ramesh, C.K., Srinivasa, T.R. and Jamuna, K.S. 2010. DPPH scavenging and reducing power properties in common vegetables. Res. J. Pharm. Biol. Chem. Sci. **1(4)**: 399-406.

Ragini, V., Prasad, K.V.S.R.G. and Bharathi, K. 2011. Antidiabetic and antioxidant activity of *Shorea tumbaggaia* Roxb. Int. J. Innovative Phar. Res. **2(2)**: 113-121.

Rahimi, R., Nikfar, S., Larijani, B. and Abdollahi, M. 2005. A review on the role of antioxidants in the management of diabetes and its complications. Biomed. Pharmacother. **59**: 365-373.

Rahman, A., Kim, E.L. and Kang, C. 2009. Antibacterial and antioxidant properties of *Ailanthus altissima* Swingle leaves extract to reduce food borne pathogens and spoiling bacteria. J. Food Safety **29(4)**: 499-510.

Rahman, A., Rahman, M.M., Sheik, M.I., Rahman, M.M., Shaihidi, S.M. and Alam, M.F. 2008. Free radical scavenging activity and phenolic content of *Cassia sophera*. Afr. J. Biotech. **7(10)**:

Rahman, A., Mizaton, H.H., Kassim, M., Mohsin, H.F. and Wahab, I.A., 2010. The antioxidant properties of *Pandanus sanderi* leaves extracts, and their triterpenoid compositions. www.biotek.gov.my/

Rahman, M.A.A. and Moon, S.S. 2007. Antioxidant polyphenol glycosides from the plant *Draba nemerosa*. Bull Korean Chem. Soc. **28(5)**: 827-831.

Rahman, M.H. 2011. Antioxidant, analgesic and toxic potentiality of *Stephania japonica* (Thunb.) Miers leaf. Int. J. Pharmacol. **7(2)**: 257-262.

Rahman, M.S., Sadhu, S.K. and Hasan, C.M. 2007. Preliminary antinociceptive, antioxidant and cytotoxic activities of *Leucas aspera* root. Fitoterapia **78(7-8)**: 552-555.

Rahmani, M., Hashim, N.M., Sukari, M.A., Ee, G.C.L., Ali, A.M., Alitheen, N.B. and Go, R. 2009. Antioxidant, cytotoxic and antibacterial activities of thirteen species of *Artocarpus* (Moraceae). J. Ned. Aromatic Plant Sci. **31(2)**: 142-146.

Rai, S., Wahile, A., Mukherjee, K., Saha, B.P. and Mukherjee, P.K. 2006. Antioxidant activity of *Nelumbo nucifera* (sacred lotus) seeds. J. Ethnopharmacol. **104 (3)**: 322-327.

Raihan, A.Z., Chowdhury, A.K.A., Rabbani, G.H., Marni, F., Ali, M.S., Nahar, L. and Sarker, S.D. 2009. Effect of aqueous extracts of black and green teas in arsenic-induced toxicity in rabbits. Phytother. Res. **23(1)**: 1603-1608.

Raj, K.J. and Shalini, K. 1999. Flavonoids – A review of biological activities. Indian Drugs **36**: 668-676.

Raja, S., Ahmed, K.F., Kumar, V., Mukherjee, K., Bandhyopadhyay, A. and Mukherjee, P.K. 2007. Antioxidant effect of *Cytisus scoparius* against carbon tetrachloride treated liver injury in rats. J. Ethnopharmacol. **109(1)**: 41-47.

Rajani, G.P. and Ashok, P. 2009. *In vitro* antioxidant and antihyperlipidemic activities of *Bauhinia variegata* Linn. Indian J. Pharmacol. **41(5)**: 227-232.

Rajarajan, T., Sathishkumar, G., Sharmilabanu, G. and Kumar, G. 2010. Free radical scavenging activity of *Plumbago zeylanica* L. on acetaminophen induced hepatic damage in rats. Indian J. Applied Pure Biol **25(1)**: 131-136.

Rajaram, K. and Suresh Kumar, P. 2011. *In vitro* antioxidant and antidiabetic activity of *Tephrosia tinctoria* Pers: an endemic medicinal plant of South India. J. Pharm. Res. **4(3)**:

Rajesh, L. and Mehta Joseph, F. 1994. Ajowan as a source of natural lipid antioxidants. Agric. Food Chem. **42**: 1420-1422.

Rajesh, N.G. and Latha, M.S. 2004. Protective activity of *Glycyrrhiza glabra* L. on carbon tetrachloride-induced peroxidative damage. Indian J. Pharmacol. **36**: 284-287.

Rajeshwar, Y., Gupta, M. and Mazumdar, U.K. 2005. Antitumor activity and *in vivo* antioxidant status of *Mucuna pruriens* seeds against Ehrlich Ascites Carcinoma in Swiss Albino mice. **4**: 46-53.

Rajeshwar, Y., Senthil Kumar, G.P., Gupta, M. and Mazumder, U.K. 2005. Studies on *in vitro* antioxidant activities of methanol extract of *Mucuna pruriens* (Fabaceae) seeds. Eur. Bull. Drug Res. **13(1)**: 31-39.

Rajkapoor, B., Jayakar, B., Murugesh, N. and Sakthisekaran, D. 2006. Chemoprevention and cytotoxic effect of *Bauhinia variegata* against N-nitrosodiethylamine induced liver tumors and human cancer cell lines. J. Ethnopharmacol. **104(3)**: 407-409.

Rajkumar, V., Guha, G. and Ashok Kumar, R. 2010a. *In vitro* antioxidant and anti-cancer potentials of *Rheum emodi* rhizome extracts. Evidence Based Compl. Alt. Med. Doi: 10.1093/ecam/neq048.

Rajkumar, V., Guha, G. and Ashok Kumar, R. 2010b. Therapeutic potential of *Acalypha fruticosa*. Food Chem. Toxicol. **48(6)**: 1709-1713.

Rajkumar, V., Guha, G. and Ashok Kumar, R. 2011. Antioxidant and neo-plastic activities of *Picrorrhiza kurroa* extracts. Food Chem. Toxicol. **49(2)**: 363-369.

Rajkumar, V., Guha, G., Mathew, L. and Ashok Kumar, R. 2010c. Curative potential of *Acorus calamus* rhizome extracts against oxidative stress to biomolecules. J. Pharm. Res. **3(4)**: 763-768.

Rajkumar, V., Guha, G., Ashok Kumar, R. and Mathew, L. 2010d. Evaluation of antioxidant activities of *Bergenia ciliata* rhizome. Rec. Nat. Prod. **4(1)**: 38-48.

Raju, B.L., Lin, S.J., Hou, W.C., Lai, Z.Y., Liu, P.C. and Hsu, F.L. 2004. Antioxidant iridoid glucosides from *Wendlandia formosana*. Nat. Prod. Res. **18(4)**: 357-364.

Rakesh, S.U., Patil, P.R. and Salunkhe, V.R. 2010. Free radical scavenging activity of hydroalcoholic extracts of dry flowers of *Nymphaea stellata* Willd. Int. J. Pharm. Biosci. **1(2)**: 1-9.

Rakotoarison, D.A., Greisser, B., Trotin, F., Dine, T., Brunct, C., Luyckx, M., Cazin, M. and Cazin, J.C. 1996. Effects of proanthocyanidin oligomers from *Cola nitida* in the scavenging of reactive oxygen species. Phytomedicine **3**(suppl. 1): 243.

Rakotoarison, D.A., Greisser, B., Trotin, F., Dine, T., Brunct, C., Luyckx, M., Vasseur, J., Cazin, M. and Cazin, J.C. 1996. Polyphenolics from callus and cell suspension cultures of *Crataegus monogyna*: Extraction, chemical analysis and evaluation of antioxidant properties. Phytomedicine **3**(Suppl.): 244.

Ram, H.N.A., Shreedhara, C.S., Gajera, F.P. and Zanwar, S.B. 2008a. Antioxidant studies of aqueous extract of *Phyllanthus reticulatus* Poir. Pharmacology online **1**: 351-364.

Ram, H.N.A., Shreedhara, C.S., Gajera, F.P. and Zanwar, S.B. 2008b. *In vitro* free radical scavenging potential of methanolic extract of entire plant of *Phyllanthus reticulatus* Poir. Pharmacology online **2**: 440-451.

Ramadan, Md. F., Lothar W. Kroh, and, Jörg-T. Mörsel. 2003. Radical scavenging activity of black cumin (*Nigella sativa* L.), coriander (*Coriandrum sativum* L.), and niger (*Guizotia abyssinica* Cass.) crude seed oils and oil fractions. *J. Agric. Food Chem.* *51* **(24)**: 6961-6969.

Ramadan, Md. F., Kinni, S.G., Rajanna, L.N., Seetharam, Y.N., Seshagiri, M. and Morsel Jorg-homas. 2009. Fatty acids, bioactive lipids and radical scavenging activity of *Celastrus paniculatus* Willd. seed oil. Scient Hortic. **123(1)**: 104-109.

Ramakrishna, D., Vidyasagar, G., Pavan Kumar, K., Madhusudana Reddy, I. and Atyam, A.V.S.S.G. 2011. *In-vitro* antioxidant activity of *Triumfetta pilosa* Roth. Der Pharmacia Lettre **3(3)**: 220-224.

Ramalakshmi, S. and Muthuchelian, K. 2010. Antioxidant and cytotoxic potential of crude ethanolic extract of *Tabebuia rosea* (Bertol) DC. leaves against human colon carcinoma (COLO320 DM) cells *in vitro*. 2nd Intern. Seminar on Medicinal Plants and Herbal Products, S.V.University, Tirupati, Abstract, pp. 145.

Ramalakshmi, K., Rahath Kubra, I. and Jagan Mohan Rao, L. 2008. Antioxidant potential of low-grade coffee beans. Food Res. Int. **41 (1)**: 96-103.

Ramamoorthy, P.K. and Bono, A. 2007. Antioxidant activity, total phenolic and flavonoids content of *Morinda citrifolia* fruit extracts from various extraction processes. J. Engineering Sci. and Technol. **2(1)**: 70-80.

Raman, C., Rajesh, A., Raj K., Ashok, S., Jagdish P., Surendar, S., Ravinder, S., Pankaj, C., Sandeep S., Gurpreet K., Sharma R. K., Puri, S. C. Dar, K. L., Geeta, H., Gupta V. K. and Azi, G. Q. 2005. Antioxidant activity of fractionated extracts of rhizomes of high-altitude *Podophyllum hexandrum*: Role in radiation protection. Molecular and cellular biochemistry. **273(1)**: 193-208.

Raman, P., DeWitt, D.L. and Nair, M.G. 2007. Lipid peroxidation and cyclooxygenase enzyme inhibitory activities of acidic aqueous extracts of some dietary supplements. Phytother. Res. **22(2)**: 204-212.

Ramanathan, N., Ochi, H. and Takeuchi, M. 1997. Antioxidant defense system in vegetable extracts. In: Shahid, F. (ed.) Natural antioxidants, Health effects and applications, AOCS Press, Champaign, IL, pp76-87.

Ramasamy, D., Meena, R., Khan, S.A., Mageswari, S., Sultana, G. and Siddiqi, M.K. 2010. Evaluation of *in vitro* antioxidant activity of the rhizome of *Alpinia officinarum* Hance. Hippocratic J. Unani Med. **5(1)**: 77-81.

Ramirez-Mares, M.V., Sanchez-Burgos, J.A. and Hernandez-Carlos, B. 2010. Antioxidant, antimicrobial and antitopoisomerase screening of the stem bark extracts of *Ardisia compressa*. Pak. J. Nutr. **9(4)**: 307-313.

Ramos, A., Visozo, A., Piloto, J., Garcia, A., Rodriguez, C.A., Rivero, R. 2003. Screening of antimutagenicity via antioxidant activity in Cuban medicinal plants. J. Ethnopharmacol. **87(2-3)**: 241-246.

Ramteke, A., Hasnu, S., Borah, S. and Das, R.K. 2007. *In vitro* study of DPPH radical-scavenging activity of leaf extracts of *Phlogacanthus tubiflorus* Nees. J. Appl. Biosc. Biotechnol. **3(1)**: 74-77.

Rana, M.G., Katbamna, R.V., Padhya, A.A., Dudhrejia, A.D., Jivani, N.P. and Sheth, N.R. 2010. *In vitro* antioxidant and free radical scavenging studies of alcoholic extract of *Medicago sativa* L. Rom. J. Biol. Plant Biol. **55(1)**: 15-22.

Rana, P., Vadhera, S. and Soni, G. 2004. *In vivo* antioxidant potential of rice bran oil (RBO) in albino rats. Indian J. Physiol. Pharmacol. **48(4)**: 428-436.

Rana, V.S. and Rawat, M.S.M. 2005. A new xanthone glycoside and antioxidant constituents from the rhizome of *Swertia speciosa*. Biochem. Biodiv. **2(10)**: 1310-1315. doi: 10.1002/cbdv.200590102

Ranawat, L., Bhatt, J. and Patel, J. 2010. Hepatoprotective activity of ethanolic extracts of bark of *Zanthoxylum armatum* DC in CCl_4 induced hepatic damage in rats. J. Ethnopharmacol. **127(3)**: 777-780.

Ranjbar, A., Khorami, S., Safarabadi, M., Shahmoradi, A., Malekirad, A.A., Vakilian, K., Mandegary, A. and Abdollahi, A. 2006. Antioxidant activity of Iranian *Echium amoenum* Fisch. et C.A.Mey. lower decoction in humans: A cross-sectional before/after clinical trial. Evidence Based Complement. Alternat. Med. **3(4)**: 469-473.

Rao, A.D., Devi, K.N. and Thyagaraju, K. 1998. Isolation of antioxidation of principle from *Azadirachta* seed kernels: determination of its role on plant lipoxygenases. J. Enzyme Inhib. **14**: 85.

Rao, A.R., Sarada, R., Baskaran, V., Ravishankar, G.A. 2006. Antioxidant activity of *Botryococcus braunii* extract elucidated *in vitro* models. J Agric Food Chem. **54(13)**: 4593-4599.

Rao, B.S.S., Shanbhoge, R., Upadhya, D., Jagetia, G.C., Adiga, S.K., Kumar, P., Guruprasad, K. and Gayathri, P. 2006. Antioxidant, anticlastogenic and radioprotective effect of *Coleus aromaticus* on Chinesee hamster fibroblast cells (V79) exposed to gamma radiation. Mutagenesis **21(4)**: 237-242. doi: 10.1093/mutage/gel023.

Rao, K.S., Chaudhury, P.K. and Pradhan, A. 2010. Evaluation of antioxidant activities and total phenolic content of *Chromolaena odorata*. Food Chem. Toxicol. **48(2)**: 729-732.

Rao, L.J.M., Ramalakshmi, K., Borse, B.B., Raghavan, B. 2007. Antioxidant and radical-scavenging carbazole alkaloids from the oleoresin of curry leaf (*Murraya koenigii* Spreng.). Food Chem. **100 (2)**: 742-747.

Rao, R.S.P. and Muralikrishna, G. 2006. Water soluble feruloyl arabinoxylans from rice and ragi: Changes upon malting and their consequence on antioxidant activity. Phytochem. **67(1)**: 91-99.

Raphael, K.R., Sabu, M.C. and Ramadasan, K. 2002. Hypoglycemic effect of methanol extract of *Phyllanthus amarus* Schum. et Thonn. On alloxan induced diabetes mellitus in rats and its relation with antioxidant potential. Indian J. Exp. Biol. **40**: 905-909.

Rathee, J.S., Patro, B.S., Mula, S., Gamre, S. and Chattopadhyay, Y. 2006. Antioxidant activity of *Piper betel* leaf extract and its constituents. J. Agric. Food Chem. **54(24)**: 9046-9054.

Rathee P, Rathee, S., Rathee, D. and Rathee, D. 2010. *In vitro* antioxidant studies and total phenolic content of *Ficus religiosa* fruit extract. Pharmacology online **2**: 737-744.

Ratheesh, M., Shyni, G.L. and Helen, A. 2009. Methanolic extract of *Ruta graveolens* L. inhibits inflammation and oxidative stress in adjuvant induced model of arthities in rats. Inflammopharmacology **17(2)**: 100-105. doi: 10.1007/s10787-009-8044-0.

Rathi, M.A., Meenakshi, P., Gurukumar, D., Arul Raj, C., Thirumoorthi, L. and Gopalakrishnan, V.K. 2010. Potential antioxidant and antiproliferative activities of *Alysicarpus vaginalis* (L.) DC. J. Pharma. Res. **3(10)**:

Rathod, N.R., Raghuveer, I., Chitme, H.R. and Chandra, R. 2009. Free radical scavenging activity of *Calotropis gigantea* on streptozotocin-induced diabetic rats. Indian J. Pharm. Sci. **71(6)**: 615-621.

Ratnasooriya, W.D., Jayakody, JRAC, Premakumara, GAS and Ediriweera, E.R.H.S.S. 2005. Antioxidant activity of water extract of *Scoparia dulcis*. Fitoterapia **76(2)**: 220-222.

Rauter, A.P., Martins, A., Lopes, R., Ferreira, J., Serralheiro, L.M., Arauzo, M.E., Borges, C., Justino, J., Silva, F.NV., Gaulart, M., Thomas-Oates, J., Rodrigues, J.A., Edwards, E., Noronha, J.P., Pinto, R. and Mota-Filipe, H. 2009. Bioactivity studies and chemical profile of the antidiabetic plant *Genista tenera*. J. Ethnopharmacol. **122(2)**: 384-393.

Ravi, K., Ramachandran, B. and Subramanian, S. 2004. Protective effect of *Eugenia jambolana* seed kernel on tissue antioxidants in streptozotocin-induced diabetic rats. Biological Pharmaceutical Bull. **27(8)**: 1212-1217.

Ravikumar, V., Shivashangari, K.S. and Devaki, T. 2005. Effect of *Tridax procumbens* on liver antioxidant defense system during lipo-polysaccharide induced hepatitis in D-galactosamine sensitized rats. Mol. Cell Biochem. **269(1-2)**: 131-136.

Ravishankara, M.N., Shrivastava, N., Padh, H. and Rajani, M. 2002. Evaluation of antioxidant properties of root bark of *Hemidesmus indicus* R.Br. (Anantmul). Phytomedicine. **9(2)**: 153-60.

Rawal, A., Muddeswar, M., and Biswas, S. 2004. Effect of *Rubia cordifolia, Fagonia cretica* Linn. and *Tinospora cordifolia* on free radical generation and lipid peroxidation during oxygen-glucose deprivation in rat hippocampal slices. Biochem. Biophys. Res. Communications **324**: 588-596.

Rawat, M.S.M., Rana, V.S. and Pant, J.N. 2004. Antioxidants from the rhizomes of *Swertia speciosa*. IUPAC Internal Conference on Biodiversity and Natural products Chemistry and Medical applications, New Delhi. P-285.

Rawat, S., Bhatt, I.D. and Rawat, R.S. 2011. Total phenolic compounds and antioxidant potential of *Hedychium spicatum* Buch.-Ham. ex D.Don in west Himalayas, India. J.Food. Composition Anal. **24(4-5)**: 574-579.

Raymundo, M.S., Paula, M.M.S., Franco, C. and Fett, R. 2007. Quantitative determination of phenolic antioxidants using voltametric techniques. Food Sci. Tech. **40**: 1133-1139.

Razali, N., Razab, R., Junit, S.M. and Aziz, A.A. 2008. Radical scavenging and reducing properties of extracts of cashew shoots (*Anacardium occidentale*). Food Chem. **111(1)**: 38-44. DOI: 10.1016/j.foodchem.2008.03.024.

Razavi, S.M., Zahri, S., Nazemiyeh, H., Zarrini, G., Mohammadi, S. and Abolghassemi-Fakhri, M. 2009. A furanocoumarin from *Prangos uloptera* roots, biological effects. Nat. Prod. Res. **23(16)**: 1522-1527.

Re, R., Pelligrini, N., Proteggente, A., Pannala, A. Yang, M. and Rice-Evans, C.A. 1999. Antioxidant activity applying an improved ABTS radical cation decolorization assay. Free Radical Biol. Med. **26**: 1231–37.

Rebecca, O.P.S., Boyce, A.N. and Chandran, S. 2010. Pigment identification and antioxidant properties of red dragon fruit (*Hylocereus polyrhizus*). Afr. J. Biotech. **9(10)**: 1450-1454.

Rebecca, O.P.S., Harivaidaran, K.V., Boyce, A.N. and Chandran, S. 2009. Potential natural dye with antioxidant properties from red dragon fruit (*Hylocereus polyrhizus*). ISHS Acta Horticulturae 875:

Rebecca, O.P.S., Zuliana, R., Wijenthiran, K., Boyce, A.N. and Chandran. 2008. A comparative study on selected antioxidants during pollination induced senescence on *Dendrobium sonia* and *Dendrobium savin* White. Asian J.Plant Sci. **7(8)**: 724-729.

Reddy, B. S., Reddy, R. K. K., Naidu, V.G.M., Madhusudhana, K., Agwane, S.B., Ramakrishna, S. and Diwan, P.V. 2008. Evaluation of antimicrobial, antioxidant and wound-healing potentials of *Holoptelea integrifolia*. J. Ethnopharmacol. **115 (2)**: 249-256.

Reddy, C.S.S.S., Subramanyam, M.V.V., Vani, R. and Devi, S.A. 2007. *In vitro* models of oxidative stress in rat erythrocytes: Effect of antioxidant supplements. Toxicol. *In vitro* **21(8)**: 1355-1364.

Reddy, S.V., Tiwari, A.K., Sampath Kumar, U., Rao, R.J. and Rao, J.M. 2005. Free radical scavenging, enzyme inhibitory constituents from antidiabetic ayurvedic medicinal plant *Hydnocarpus wightiana* Blume. Phytother. Res. **189(4)**: 277-281.

Reddy, V., Urooj, A. and Kumar, A. 2005. Evaluation of antioxidant activity of some plant extracts and their application in biscuits. Food Chem. **90**: 317-321.

Rehecho, S., Hidalgo, O., Garcia-Iniguez de Cirano, M., Navarro, I., Astiasaran, I., Ansorena, D. and Yolanda, R. 2011. Chemical composition, mineral content and antioxidant activity of *Verbena officinalis* L. LWT-Food Sci. Technol. **44(4)**: 875-882.

Rehman, Z., Salariya, A. and Habib, F. 2003. Antioxidant activity of ginger extract in sunflower oil. J. Science of Food and Agriculture **83**: 624-629.

Reitman, S. and Frankel. S. 1957. A colorimetric method for the determination of serum glutamic oxalacetic and glutamic pyruvic transaminases. Am. J. Clin.Path. **28**: 56–63.

Rema, J., Mathew, P.A. and Krishnamoorthy, B. 2008. *Garcinia hombroniana* – A lesser known tree spice. Spice India **21(1)**: 28-29.

Ren, B.B., Ma, Y.P., Sheng, Y. and Gao, B. 1995. Protective action of *Lycium barbarium* L. and betaine on lipid peroxidation of RBC membrane induced by hydrogen peroxide. Zhongguo Zhongyao Zazhi **20(5)**: 303-304.

Ren, H., Zhang, W., Shan, C., Jiang, H. and Zhang, J. 2010. Study on antioxidant activity of ethanol extract from the flower of nymphaea hybrid. Food Sci. Technol. Doi: CNKI: SUN: SSPJ.0.2010-03-061.

Repetto, M.G. and Llesuy, S.F. 2002. Antioxidant properties of natural compounds used in popular medicine for gastric ulcers. Braz. J. Med. Biol. Res. **35(35)**: 523-534.

Resmi, C.R., Venukumar, M.R. and Latha, M.S. 2006. Antioxidant activity of *Albizia lebbeck* (Linn.) Benth. in alloxan diabetic rats. Indian J. Physiol. Pharmacol. **50 (3)**: 297-302.

Reys, L.F. and Cisneros-Zevallos, L. 2003. Wounding stress increases the phenolic content antioxidant capacity of purple-flesh potatoes (*Solanum tuberosum* L.). J. Agric. Food Chem. **51**: 5296-5300.

Reyes-Carmona, J., Youseg, G.G., Martinez-Peniche, R.A. and Lila, M.A. 2005. Antioxidant capacity of fruit extracts of blackberry (*Rubus* spp.) produced in different climatic regions. J. Food Sci. **70**: 497-503.

Reynertson, K.A., Yang, H., Jiang, B., Basile, M.J. and Kennelly, E.J. 2008. Quantitative analysis of antiradical phenolic constituents from fourteen edible Myrtaceae fruits. Food Chem. **109(4)**: 883-890. doi: 10.1016/j.foodchem.2008.01.021

Rezende, K.R., Davino, S.C., Barros, S.B. and Kato, M.J. 2005. Antioxidant activity of aryltetralone lignans and derivatives from *Virola sebifera* (Aubl.) Nat. Prod. Res. **19(7)**: 661-666.

Riaz, N., Malik, A., Rehman, A., Ahmed, Z., Muhammad, P., Nawaz, S.A., Siddiqui, J. and Choudhary, M.I. 2004. Lipoxygenase inhibiting and antioxidant oligostilbene and monoterpene galactoside from *Paeona emodi*. Phytochem. **65(8)**: 1129-1135.

Ribeiro, B., Valentão, P., Baptista, P., Seabra, R.M., Andrade, P.B. 2007. Phenolic compounds, organic acids profiles and antioxidative properties of beefsteak fungus (*Fistulina hepatica*). Food Chem. Toxicol. **45(10)**: 1805-13.

Ribeiro, S.M.R., Barbosa, L.C.A., Queiroz, J.H., Knodler, M., Schieber, A. 2008. Phenolic compounds and antioxidant capacity of Brazilian mango (*Mangifera indica* L.) varieties. Food Chem. **110 (3)**: 620-626.

Ricci, D., Fraternale, D., Giamperi, L., Bucchini, A., Epifano, F., Burini, G and Curini, M. 2005. Chemical composition, antimicrobial and antioxidant activity of the essential oil of *Teucrium marum* L. (Lamiaceae). J. Ethnopharmacol. **98(1-2)**: 195-200.

Ricci, D., Giamperi, L., Bucchini, A. and Fraternale, D. 2006. Antioxidant activity of *Punica granatum* fruits. Fitoterapia **77 (4)**: 310-312.

Rice-Evans, C.A. and Miller, N.J. 1994. Total antioxidant status in plasma and body fluids. Methods in Enzymol. **234**: 279-293.

Rice-Evans, C.A. and Miller, N.J. 1998. Structure-antioxidant activity relationships of flavonoids and isoflavonoids, p. 199-219. IN: Rice-Evans and L. Packer (eds.). Flavonoids in health and disease. Marcel Dekker, New York.

Rice-Evans, C.A., Miller, N.J., Bolwell, P.G., Bramley, P.M. and Pridham, J.B. 1995. The relative antioxidant activities of plant-derived polyphenolic flavonoids. Free Radical Res. **22**: 375-383.

Rice-Evans, C.A., Miller, N.J. and Paganga, G. 1996. Structure-antioxidant activity relationships of flavonoids and phenolic acids. Free Radical Biol. Medicine **20**: 933-956.

Rice-Evans, C.A., Miller, N.J. and Paganga, G. 1997. Antioxidant properties of phenolic compounds. Trends in Plant Science **2**: 152-159.

Ripa, F.A., Haque, M. and Bulbul, I.J. 2010. *In vitro* antibacterial, cytotoxic and antioxidant activities of plant *Nephelium longan*. Pak. J. Biol. Sci. **13(1)**: 22-27.

Ripa, F.A., Haque, M. and Imran-Ul-Haque, Md. 2009. *In vitro* antimicrobial, cytotoxic and antioxidant activity of flower extract of *Saccharum spontaneum* Linn. Eur. J. Sci. Res. **30(3)**: 478-483.

Risco, E., Ghia, F., Vila, R., Iglesias, J., Alvarz, E. and Canigueral, S. 2003. Immunomodulatory activity and chemical characterization of Sangre de Drago (Dragon's blood) from *Croton lechleri*. Planta Medica **69(9)**: 785-794.

Rivas- Arreola, M.J., Rocha-Guzman, N.E., Gallegos-Infante, J.A., Gonzalez-Laredo, R.F., Rosales-Castro, M., Bacon, Rong (Tsao) Cao, J.R., Proulx A. and Intriago-Ortega, P. 2010. Antioxidant activity of oak (*Quercus*) leaves infusions against free radicals and their cardioprotective potential. Pak. J. Biol. Sci. **13**: 537-545.

Rivera, P.B. and Pareja, B.P. 201. Flavonoids de *Ephedra americana* (pinco-pinco), Accion biological sibre El sistema immunologico (IgE). Ciencia e Investigacion **4(1)**: 57-77.

Rizvi, S.I. and Jha, R. 2009. Current perspectives in aging research. In: Khan, I.A. and Khanum, A. (eds.) Herbal Therapy for Ageing. Ukaaz Publications, Hyd. pp. 1-37.

Rodriguez-Garcia, I. and Guil-Guerrero, J.L. 2007. Evaluation of the antioxidant activity of three microalgal species for use as dietary supplements and in the preservation of foods. Food Chem. **108(3)**: 1023-1026. doi: 10.1016/j.foodchem.2007.11.059.

Rodriguez-Saona, L.E., Giusti, M.M. and Weolstad, R.E. 1998. Anthocyanin pigment composition of red-fleshed potatoes. J. Food Sci. **63**: 458-465.

Rohini, G., Sabitha, K.E. and hyamala Devi, C.S. 2004. *Bacopa monniera* Linn. extract modulates antioxidant and marker enzyme status in fibrosarcoma bearing rats. Indian J. Exp. Bio. **42(8)**: 776-780.

Rohn, S., Rawel, H.M. and Kroll, J. 2004. Antioxidant activity of protein-bound quercetin. J. Agric. Food Chem. **52(15)**: 4725-4729.

Roja, G., Vikrant, B.H., Sandur, S.K., Sharma, A. and Pushpa, K.K. 2011. Accumulation of vasicine and vasicinone in tissue cultures of *Adhatoda vasica* and evlauation of free radical-scavenging activities of the various crude extracts. Food Chem. **126(3)**: 1033-1038. doi: 10.1016/j.foodchem.2010.11.115

Romani, A., Coinu, R., Carta, S., Pinelli, P., Galardi, C., Vincieri, F.F. and Franconi, F. 2004. Evaluation of antioxidant effect of different extracts of *Myrtus communis* L. Free Radic. Res. **38(1)**: 97-103.

Ronald, L. 1998. Antioxidant capacity and health benefits of fruits and vegetables. NABC meeting in Portland, Oregan.

Rony, K.A., Mathew, J. and Glandy, D. 2008. Antioxidant and antitumor activities of extract of *Phellinus rimosus* (Berk) Pilat. Amala Res. Bull. **28**: 40-45.

Roome, T., Dar, A., Ali, S., Naqvi, S. and Choudhary, M.I. 2008. A study on antioxidant, free radical scavenging, anti-inflammatory and hepatoprotective actions of *Aegiceras corniculatum* (stem) extracts. J. Ethnopharmacol. **118(3)**: 514-521.

Rop, O., Jurikova, T., Mlcek, J., Krmarova, D. and Sengee, Z. 2009. Antioxidant activity and selected nutritional values of plums (*Prunus domestica* L.) typical of the white Carpathian mountains. Scientia Horticulturae **122(4)**: 545-549.

Rop, O., Reznicek, V., Valsikova, M., Jurikova, T., Mlceb, J. and Kramarova, D. 2010. Antioxidant properties of European cranberry bush fruit (*Viburnum opulus* var. *edule*). Molecules **15**: 4467-4477. doi: 10.3390/molecules15064467.

Rosas-Romero, A. and Saavedra, G. 2005. Screening Bolivian plants for antioxidant activity. Pharmaceutical Biol. **43(1)**: 79-86. DOI: 10.1080/13802005909034l7.

Rouanet, J.-M., Decorde, K., Rio, D.D., Auger, C., Borges, G., Cristol, J.-P., Lean, M.E.J. and Crozier, A. 2010. Berry juices, teas, antioxidants and the prevention of atherosclerosis in hamsters. Food Chem. **118(2)**: 266-271.

Roussis, I.G., Lambropoulous, I., Tzimas, P., Gkoulioti, A., Marinos, V., Tsoupeis, D. and Boutaris, I. 2008. Antioxidant activity of some Greek wine phenolic extracts. J. Food Compos. Anal. **21(8)**: 614-621.

Roy, M.K., Takenaka, M. and Isobe, S. 2007. Thermal processing enhances anti-radical activity and reduces pro-oxidant activity in water-soluble fraction of selected *Allium* vegetables. J. Sci. Food Agric. **87(12)**: 2259-2265.

Rozenberg, O., Howell, A. and Aviram, M. 2006. Pomegranate juice sugar fraction reduces macrophage oxidative state, whereas white grape juice fraction increases it. Atherosclerosis **188**: 68-76.

Ruan, Z.P., Zhang, L.L. and Lin, Y.M. 2008. Evaluation of antioxidant activity of *Syzygium cumini* leaves. Molecules **13**: 2545-2556.

Ruberto, G. and Baratta, M.T. 2000. Antioxidant activity of selected essential oil components in two lipid model systems. Food Chem. **69**: 167-174.

Ruberto, G., Baratta, M.T., Biondi, D.M. and Amico, V. 2001. Antioxidant activity of extracts of marine algal genus *Cystoseira* in a micellar model system. J. Appl. Phycol. **13**: 403-407.

Ruberto, G., Baratta, M.T., Deans, S.G. and Dorman, H.J.D. 2000. Antioxidant and antimicrobial activity of *Foeniculum vulgare* and *Crithmum maritimum* essential oils. Planta Medica **66**: 687-693.

Ruby, A.J., Kuttan, G., Babu, K.D., Rajasekharan, K.N. and Kuttan, R. 1995. Anti-tumor and antioxidant activity of natural curcuminoids. Cancer Lett. **94**: 79.

Ruch, R.J., Cheng, S.J. and Klaunig, J.E. 1989. Prevention of cytotoxicity and inhibition of intercellular communication by antioxidant catechins isolated from Chinese green tea. Carcinogen **10**: 1003–1008.

Rudnicki, M., Roberto de Oliveira, M., Pereira, T.V., Reginatto, F.H., Dal-Pizzol, F. and Moreira, J.C.F. 2007a. Antioxidant and antiglycation properties of *Passiflora alata* and *Passiflora edulis* extracts. Food Chem. **100(2)**: 719-724.

Rudnicki, M., Silveira, M.M., Pereira, T.V., Oliveira, M.R., Reginatto, F.H., F.Dal-Pizzol and Moreira, J.C.F. 2007b. Protective effects of *Passiflora alata* extract pretreatment on carbon tetrachloride induced oxidative damage in rats. Food Chem. Toxicol. **45(4)**: 656-661. doi: 10.1016/j.fct.2006.10.022.

Ruiz-Terán, F., Medrano-Martinez, A. and Navarro-Ocana, A. 2008. Antioxidant and free radical scavenging activities of plant extracts used in traditional medicine in Mexico. Afr. J. Biotech. **7(12)**: 1886-1893.

Rukkumani, R., Balasubashini, M.S. and Menon, V.P. 2003. Protective effects of curcumin and photo-irradiated curcumin on circulatory lipids and lipid peroxidation products in alcohol and polyunsaturated fatty acid-induced toxicity. Phytother. Res. **17(8)**: 925-929.

Rupasinghe, V.H.P. and Clegg, S. 2007. Total antioxidant capacity, total phenolic content, mineral elements, and histamine concentrations in wines of different fruit sources. J. Food Compos. Anal. **20**: 133-137.

Russo, A., Acquaviva R., Campisi A., Sorrenti, V, Di Giacomo, C., Virgata, G. and Barcellona, M.L, 2000. Bioflavonoids as antiradicals, antioxidants and DNA cleavage protectors. Cell Biol. Toxicol. **16 (2)**: 91-8.

Russo, A., Cardile, V., Lombardo, L., Vanella, L., Vanella, A. and Garbarino, J.A. 2005. Antioxidant activity and antiproliferative action of methanolic extract of *Geum quellyon* Sweet roots in human tumor cell lines. J. Ethnopharmacol. **100(3)**: 323-332.

Russo, A., Izzo, A.A., Borrelli, F., Renis, M., and Vanella, A. 2003. Free radical scavenging capacity and protective effect of *Bacopa monnieri* L. on DNA damage. Phytother. Res. **17(8)**: 870-875.

Ry, G., Ju, J.H., Park, Y.J., Ryu, S.Y., Choi, B.W. and Lee, B.H. 2002. The radical scavenging effects of stilbene glycosides from *Polygonum multiflorum*. Arch. Pharm. Res. **25(5)**: 636-639.

Saada, H.N., Said, U.Z., Meky, N.H. and Azime, A.S.A.E. 2008. Grape seed extract *Vitis vinifera* protects against radiation-induced oxidative damage and metabolic disorders in rats. Phytother. Res. **23(3)**: 434-438.

Saarinen, N.M., Wärri, A., Mäkelä, S.I., Eckerman, C., Reunanen, M., Ahotupa, M., Salmi, S.M., Franke, A.A., Kangas, L. and Santti, R. 2000. Hydroxymatairesinol, a novel enterolactone precursor with antitumor properties from coniferous tree (*Picea abies*). Nutr. Cancer.**36 (2)**: 207-16.

Sabir, S.M. and Rocha, J.B.T. 2008. Antioxidant and hepatoprotective activity of aqueous extract of *Solanum fastigiatum* (false "Jurubeba") against paracetamol-induced liver damage in mice. J. Ethnopharmacol. **120 (2)**: 226-232.

Sabir, S.M. and Rocha, J.B.T. 2008. Water extractable phytochemicals from *Phyllanthus niruri* exhibit distinct *in vitro* antioxidant and *in vivo* hepatoprotective activity against paracetamol induced liver damage in mice. Food Chem. **111(4)**: 845-851. doi: 10.1016/j.foodchem.2008.04.060.

Sabu, M.C. and Kuttan, R. 2002. Antidiabetic activity and its relationship with their antioxidant property. J. Ethnopharmacol. **81 (2)**: 155-160.

Sabu, M.C. and Kuttan, R. 2003. Antioxidant activity of Indian herbal drugs in rats with alloxan-induced diabetes. Pharmaceutical Biol. **41(7)**: 500-505.

Sabu, M.C. and Kuttan, R. 2004. Antidiabetic activity of *Aegle marmelos* and its relationship with its antioxidant properties. Indian J. Physiol. Pharmacol. **48**: 81-88.

Sabu, M.C. and Kuttan, R. 2009. Antidiabetic and antioxidant activity of *Terminalia belerica* Roxb. Indian J. Exp. Biol. **47**: 270-275.

Sabuncuoglu, S,. Sohretoglu, D. Sakar M.K. and Ozgunes. H. 2008. Antioxidant activity of polyphenolic compounds and extracts from *Geranium purpureum*.Toxicology Letters. **180(1)**: 242-246

Sacchetti, G., Màietti, S., Muzzoli, M., Scaglianti, M., Manfredini, S. Radice M. and Bruni, R. 2005. Comparative evaluation of 11 essential oils of different origin as functional antioxidants, antiradicals and antimicrobials in foods. Food Chem. **91(4)**: 621-632.

Sacchetti, G., Medici, A.,Maietti, S., Radice, M., Muzzoli, M., Manfredini, S., Braccioli, E. and Bruni, R. 2004. Composition and functional properties of the essential oil of Amazonian basil, *Ocimum micranthum* Willd. Labiatae in comparison with commercial essential oils. J. Agric. Food Chem. **52(11)**: 3486-3491.

Sadiq, A.M., Venkatesan, R.S., Kumar, V.S. and Yoganantharaj, G. 2010. Antioxidant effect of *Tinospora cordifolia* stem extract in streptozotocin induced diabetic rats. Bioscan **5(1)**: 93-95.

Saeed, M.K., Shahzadi, I., Ahmed, I., Ahmad, R., Shahzad, K., Ashraf, M., and Viquar-un-Nisa. 2010. Nutritional analysis and antioxidant activity of bitter gourd (*Momordica charantia*) from Pakistan. Pharmacology online **1**: 252-260.

Saei-Dehkordi, S.S., Tajik, H., Moradi, M. and Khalighi-Sigaroodi, F. 2010. Chemical compositon of essential oils of *Zataria multiflora* Boiss. from different parts of Iran and their radical scavenging and antimicrobial activity. Food Chem.Toxicol. **48**: 1562-1567.

Safaei-Ghomi, J., Ebrahimabadi, A.H., Djafari-Bidgoli, Z. and Batooli, H. 2009. GC/MS analysis and *in vitro* antioxidant activity of essential oil and methanol extracts of *Thymus caramanicus* Jalas and its main constituent carvacrol. Food Chem. **115(4)**: 1524-1528.

Safitri, R., Tarigan, P., Freisleben, J., Rumampuk, R.J., Murakami, A. 2003. Antioxidant activity *in vitro* of two aromatic compounds from *Caesalpinia sappan* L. BioFactors **19 (1-2)**: 71-77.

Sagar, R.K., Chawla, R., Arora, R., Singh, S., Krishna, B., Sharma, R.K., Puri, S.C., Singh, P., Rajkumar, Sharma, A.K. *et al.*, 2006. Protection of the hemopoietic system by *Podophyllum hexandrum* against gamma radiation-induced damage. Planta Medica **72(2)**: 114-120.

Sagdic, O., Aksoy, A., Ozkan, G., Ekici, L. and Albayrak, S. 2007. Biological activities of the extracts of two endemic *Sideritis* species in Turkey. Innovative Food Science and Emerging Technologies **9(1)**: 80-84.

Sagdic, O., Aksoy, A. and Ozkan, G. 2006. Evaluation of the antibacterial and antioxidant potentials of gilaburu (*Viburnum opulus* L.) fruit extract. Acta Alimentaria **35(4)**: 487-492.

Saha, K., Lajis, N.H., Israf, D.A., Hamzah, A.S., Khozirah, S., Khamis, S. and Syahida, A. 2004. Evaluation of antioxidant and nitric oxide inhibitory activities of selected Malaysian medicinal plants. J. Ethnopharmacol. **92(2-3)**: 263-267.

Saha, M.R., Alam, M.A., Akter, R., and Jahangir, R. 2008. *In vitro* free radical scavenging activity of *Ixora coccinea* L. Bangladesh J. Pharmacol. **3(2)**: 90-96.

Saha, M.R., Hasan, S.M.R., Akter, R., Hossain, M.M., Alam, M.S. and Mazumder, M.E.H. 2008. *In vitro* free radical scavenging activity of methanol extract of the leaves of *Mimusops elengi* Linn. Bangl. J. Vet. Med. **6(2)**: 197-202.

Sahgal, G., Ramanathan, S., Sasidharan, S., Mordi, M.N., Ismail, S. and Mansor, S.M. 2009. *In vitro* antioxidant and xanthine oxidase inhibitory activities of methanolic *Swietenia mahagoni* seed extracts. Molecules **14**: 4476-4485.

Sahin, F., Gulluce, M., Daferera, D., Sokmen, A., Sokmen, M., Polissiou, M., Agar, G. and Ozer, H. 2004. Biological activities of the essential oils and methanol extract of *Origanum vulgare* ssp. *vulgare* in the Eastern Anatolia region of Turkey. Food Control **15(7)**: 549-557.

Sahreen, S., Khan, M.R. and Khan, R.A. 2010. Evaluation of antioxidant activities of various solvent extracts of *Carissa opaca* fruits. Food Chem. **122(4)**: 1205-1211. doi: 10.1016/j.foodchem. 2010. 03.120.

Sahu, K.G., Khadabadi, S.S. and Bhide, S.S. 2009. Evaluation of *in vitro* antioxidant activity of *Amorphophallus campanulatus* (Roxb.) Blume ex Decne. Int. J. Chem.Sci. **7(3)**: 1553-1562.

Sai Khan, M.S., Howard, L.R. and Miller, J.C. 1995. Antioxidant activity and total phenolics in different genotypes of potato (*Solanum tuberosum* L.) J. Food Sci. **60(2)**: 341-343.

Sai Ram, M., Neetu, D., Deepti, P., Vandana, M., Ilavazhagan, G, Kumar, D. and Selvamurthy, W. 2003. Cytoprotective activity of Amla (*Emblica officinalis*) against chromium (VI) induced oxidative injury in murine macrophages. Phytother. Res. **17(4)**: 430-433.

Sai Ramesh, A., Christopher, J.G., Setty, C.R. and Thankamani, V. 2010. *In vitro* study on antioxidant property of Arjunolic acid. J. Pharmacogn. Herbal Formulations **1**: 1-10.

Saisin, S., Tip-pyang, S. and Phuwapraisirisan, P. 2009. A new antioxidant flavonoid from the lianas of *Gnetum macrostachyum*. Nat. Prod. Res. **23(16)**: 1472-1477.

Saito, S.T., Gosmann, G., Saffi, J., Presser, M., Richter, M.F. and Bergold, A.M. 2007. Characterization of the constituents and antioxidant activity of Brazilian green tea (*Camellia sinensis* var. *assamica* IAC-259 cultivar) extracts. J. Agric. Food Chem. **55 (23)**: 9409-9414.

Sajjad Khan, S., Khanam, S. and Deepak, M. 2006. Antioxidant activity of a new diarylheptanoid from *Zingiber officinale*. Pharmacog. Mag. **2(8)**: 254-257.

Sala, A., Recio, M.D., Giner, R.M., Manez, S., Tournier, H., Schniella, G. and Rios, J.L. 2002. Anti-inflammatory and antioxidant properties of *Helichrysum italicum*. J. Pharmacy Pharmacol. **54(3)**: 365-371.

Salah, N., Miller, N.J., Paganga, G., Tijburg, L., Bolwell, G.P. and Rice Evans, C. 1995. Polyphenolic flavonols as scavengers of aqueous phase radicals and as chain-breaking antioxidants. Arch. Biochem. Biophys. **322(2)**: 339-346.

Salazar, R., Pozos, M.E., Cordero, P., Perez, J., Salinas, M.C. and Waksman, N. 2008. Determination of the antioxidant activity of plants from Northeast Mexico. Pharm. Biol. **46**: 166-170.

Salawu, S.O., Akindahunsi, A.A. and Comuzzo, P. 2006. Chemical composition and *in vitro* antioxidant activities of some Nigerian vegetables. J. Pharmacol. Toxicol. **1(5)**: 429-437.

Salawu, S.O., Akindahunsi, A.A., Sanni, D.M., Decorti, G., Cvorovic, J., Trmaer, F., Passamonti, S., and Mulinacci, M. 2011 Cellular antioxidant activities and cytotoxic properties of ethanolic extracts of four tropical green leafy vegetables. Afr. J. Food Sci. **5(4)**: 267-275.

Saleem A, Ahotupu, M. and Pihlaja, K. 2001. Total phenolics concentration and antioxidant potential of extracts of medicinal plants of Pakistan. Z. Naturforsch. **56c**: 973-978.

Saleem A, Kivelä H, Pihlaja K. 2003. Antioxidant activity of pine bark constituents. Z Naturforsch [C]. **58(5-6)**: 351-354.

Salehi, P., Sonboli, A., and Allahyari, L. 2007. Antibacterial and antioxidant properties of the essential oil and various extracts of *Nepeta ispahanica* from Iran. J. Essential Oil Bearing Plants **10(4)**: 324-331.

Salehi, P., Sonboli, A., and Asghari, B. 2007. Chemical composition of essential oil of *Stachys acerosa* and its antibacterial and antioxidant activities. Chem. Nat. Compounds **43(3)**: 339-341.

Salehi, P., Sonboli, A., Ebrahimi, S.N. and Yousefzadi, M. 2007. Antibacterial and antioxidant activities of the essential oils and various extracts of *Salvia sahendica* in different phonological stages. Chem. Nat. Compounds **43(3)**: 328-330.

Salehi, P., Sonboli, A., Eftekhar, F., Nejad-Ebrahimi, S. and Yousefzadi, M. 2005. Essential oil composition, antibacterial and antioxidant activity of the oil and various extracts of *Ziziphora clinopodioides* subsp. *rigida* (Boiss.) Rech.f from Iran. Biol. Pharm. Bull. **28(10)**: 1892-1896.

Sales, J.M. and Resurreccion, A.V.A. 2010. Maximizing phenolics, antioxidants and sensory acceptance of UV and ultrasound-treated peanuts. LWT-Food Sci. Tech. **43(7)**: 1058-1066.

Salluca, T.G., Penarrieta, J.M., Alvarado, J.A. and Bergenstahl, B. 2008. Determination of total phenolic compounds content and the antioxidant capacity of Andean tubers and roots (Isano, Oca, Ulluco and Arracacha). Revista Boliviana de Quimica **25(1)**: 58- 61.

Salvador, M.J., Ferreir, E.O., Mertens, Talcott, S.U., De Castro, W.V., Butterweck, V., Derendorf, H. and Dias, D.A. 2011. Isolation and quantitative analysis of antioxidant flavonoids from *Alternanthera tenella* Colla. Z. Naturforsch. **61(1-2)**: 19-25.

Salvayre, R., Braquet, P., Perruchot, Th., Douste-Blazy, L. 1982. In: L. Frakas, M. Gabor, F. Kallay and H. Wagner (eds): Flavonoids in Bioflavonoids. Elsevier, Amsterdam pp. 437-442.

Sam, S.K.G., Senthil, K.B., Ramachandran, S., Saravanan, M. and Sridhar, S.K. 2001. Antioxidant and wound healing properties of *Glycyrrhiza glabra* root extract. Indian Drugs **38(7)**: 355-357.

Sambathkumar, R., Sivakumar, T., Shanmuga Sundaram, R., Siva Kumar, P., Nethaji, R., Gupta, M. and Mazumdar, U.K. 2006. Antimicrobial and antioxidant activities of *Careya arborea* Roxb. stem bark. Iranian J. Pharm. Therapeutics **5**: 35-41.

Samojlik, I., Laki, N., Mimica-Dukic, N., Dakovi-Svajcer, K. and Bozin, B. 2010. Antioxidant and hepatoprotective potential of essential oils of coriander (*Coriandrum sativum* L.) and caraway (*Carum carvi* L.) (Apiaceae). J. Agric. Food Chem. **58(15)**: 8848-8853.

Sampathkumar, U., Tiwari A.K., Reddy,S.V., Aparna, P., Rao, R.J., Ali, A.Z and Rao, J.M. 2005. Free-radical scavenging and xanthine oxidase inhibitory constituents from *Stereospermum personatum*. J. Nat. Prod. **68(11)**: 1615-1621.

Samu, Z., Nyiredy, S., Baitz-Gacs, E., Varga, Z., Kurtan, T., Dinya, Z and Antus, S. 2004. Structural elucidation and antioxidant activity (-)-isolandrin isolated from *Silybum marianum* L. Chem. Biodiv. **1(11)**: 1668-1677.

Sanbogi, C., Osakabe, N., Natsume, M., Takizawa, T., Gomi, S. and Osawa, T. 1998. Antioxidative polyphenols isolated from *Theobroma cacao*. J. Agric. Food Chem. **46**: 454-457.

Sanches, A.C.C., Lopes, G.C., Nakamura, C.V., Filho, B.P.D. and de Mello, J.C.P. 2005. Antioxidant and antifungal activities of extracts and condensed tannins from *Stryphnodendron obovatum* Benth. Brazilian J. Pharm. Sci. **41(1)**:

Sancheti, G. and Goyal, P.K. 2006. Effect of *Rosmarinus officinals* in modulating 7,12-dimethylbenz(a)anthracene induced skin tumorogensis in mice. Phytother. Res. **20(11)**: 981-986.

Sanchez, J., Melchor, G., Martinez, G., Escobar, A. and Faure, R. 2006. Antioxidant activity of *Rhizophora mangle* bark. Fitoterapia **77**: 141-143.

Sánchez-Arana, A., Estarrón-Espinosa, M., Obledo-Vázquez, E.N., Camberos, E.P., Silva-Vázquez, R. and Lugo-Cervantes, E. 2010. Antimicrobial and antioxidant activitie of Mexican oregano essential oils (*Lippia graveolens* H.B.K.) with different composition when microcapsulated in β-cyclodextrin. Lett. Appl. Microbiol. **50**: 585-590.

Sanchez-Medina, A., Garcia-Sasa, K., May-Pat, F., Pena-Rodriguez, L.M. 2001. Evaluation of biological activity of crude extracts from plants used in Yucatecan traditional medicine part. Antioxidant, antimicrobial and beta-glucosidase inhibition activities. Phytomedicine **8(2)**: 144-151.

Sanchez-Moreno, C. 2002. Review: methods used to evaluate the free radical scavenging activity in foods and biological systems. J. Food Sci. Tech. Int. **8(3)**: 121-137.

Sanchez-Moreno, C., Larrauri, J.A. and SauraCalixto, F. 1998. A procedure to measure the anti-radical efficiency of polyphenols. J. Sci. Food. Agric. **76**: 270-276.

Sanja, S.D., Sheth, N.R., Parmer, G. and Golwala, D.K. 2009. Antioxidant properties of the methanolic extract of *Ipomoea reniformis*. Intern. J. Pharmacol. Biol. Sci. **3(3)**: 85-93.

Sanjay, U.P., Kumar, G.S., Jayaveera, K.N., Kishore, D.V.K., Ashok Kumar, C.K. and Dhanpal, R. 2007. Antimicrobial, wound healing and antioxidant activities of *Anthocephalus cadamba*. African J. Tradit. Complement. Alt. Med. **4**: 481-487.

Sanmugapriya, E and Venkataraman, 2006. Studies on hepatoprotective and antioxidant actions of *Strychnos potatorum* Linn. seeds on CCl_4 – induced acute hepatic injury in experimental rats. J. Ethnopharmacol. **105**: 154-160.

Santos, A.B.dos, Silva, D.H.S., Bolzani, V.D.S., Santos, L.A., Schmidt, T.M. and Baffa, O. 2009. Antioxidant properties of plant extracts: an EPR and DFT comparative study of the reaction with DPPH, TEMPOL and Spin Trap DMPO. J. Braz. Chem. Soc. **20(8)**: 1483-1492.

Santos-Cervantes, M.E., Ibarra-Zazueta, M.E., Loarca-Piña, G., Paredes-López, O. and Delgado-Vargas, F. 2007. Antioxidant and antimutagenic activities of *Randia echinocarpa* fruit. Plant Foods Hum. Nutr. **62 (2)**: 71-77.

Santos-Gomes, P.C., Seabra, R.M., Andrade, P.B. and Fernandes-Ferreira, M. 2003. Determination of phenolic antioxidant compounds produced by calli and cell suspension of sage (*Salvia officinalis* L.). J. Plant Physiol. **160**: 1025-1032.

Santoso, M., Naka, Y., Angkawidjaja, C., Yamaguchi, T., Motoba, T. and Takamura, H. 2010. Antioxidant and DNA damage prevention activities of the edible parts of *Gnetum gnemon* and their changes up on heat treatment. Food Sci. Tech. Res. **16(6)**: 549-556.

Saranya, P., Geetha, A., Karthikeyan, S.M. and Selvamurthy, N. 2010. The antioxidant and H^+K^+ ATPase inhibitory effect of *Andrographis paniculata* and Andrographiolide – *in vitro* and *in vivo* studies. Pharmacology online **1**: 356-376.

Saranya, P. and Guruswamy, K. 2010. Free radical scavenging aactivity of *Amaranthus blitum*. Pharmacology online **3**: 625-636.

Saraswathy, A., Nandini Devi, S. and Ramaswamy, D. 2008. Antioxidant, heavy metals and elemental analysis of *Holoptelia integrifolia* Planch. Indian J. Pharmaceutical Sci. **70**: 683-686

Saraswathy, A., Ramaswamy, D. and Nandinidevi, S. 2009. *In vitro* antioxidant activity, heavy metal and mineral analysis of aerial part of *Aerva lanata* (Linn.) Juss. ex Schultes. J. Med. Arom. Plant Sci. **31(1)**: 53-56.

Sarikcioglu, S.B., Oner, G. and Tercan, G. 2004. Antioxidant effect of EGb 761 on hydrogen peroxide-induced lipoperoxidation of G-6-PD deficient erythrocytes. Phytother. Res. **18(10)**: 837-840.

Sarikurkcu, C., Arisoy, K., Tepe, B., Cakir, A. Abali, G. and Mete, E. 2009. Studies on the antioxidant activity of essential oil and different solvent extracts of *Vitex agnus-castus* L. fruits from Turkey. Food Chemical Toxicol. **47(10)**: 2479-2483.

Sarikurkcu, C., Ozer, M.S., Eskici, M., Tepe, B., Can, S. and Mete, E. 2010. Essential oil composition and antioxidant activity of *Thymus longicaulis* C. Presl. subsp. *longicaulis* var. *longicaulis*. Food Chemical Toxicol. **48**: 1301-1805.

Sarikurkcu, C., Tepe, B., Daferera, D., Polissiou, M. and Harmandar, M. 2008. Studies on the antioxidant activity of the essential oil and methanol extract of *Marrubium globosum* subsp. *globosum* (Lamiaceae) by three different chemical assays. Bioresource technol. **99(10)**: 4239-4246.

Sarikurkcu, C., Tepe, B., Karslisemiz, D. and Solak, M.H. 2010. Evaluation of metal concentration and antioxidant activity of three edible mushrooms from Mugla, Turkey. Food Chem. Toxicol. **48**: 1230-1233.

Sarikurkcu, C., Tepe, B. and Yamac, M. 2008. Evaluation of the antioxidant activity of four edible mushrooms from the Central Anatolia, Eskisehir - Turkey: *Lactarius deterrimus, Suillus collitinus, Boletus edulis, Xerocomus chrysenteron*. Bioresour Technol. **99(14)**: 6651-6655.

Saritha, V., Anilakumar, K.R. and Khanum, F. 2010. Antioxidant and antibacterial activity of *Aloe vera* gel extracts. Int. J. Pharm. Biol. Arch. **1(4)**: 376-384.

Sarju, N., Ghani, M.A., Hamdam, S. and Samad, A.A. 2011. Antioxidant activity and cytotoxicity of the leaves of *Melastoma decemfidum* Roxb. ex Jack. serisubuh.wordpress.com/2011/02/12/melastoma-decemfidum

Sarker, S.D., Eynon, E., Fok, K., Kumarswamy, Y., Murphy, E.M., Nahar, L., Shaheen, E.M., Shaw, N.M. and Siakalima, M. 2003. Screening the extracts of the seeds of *Achillea millefolium, Angelica sylvestris* and *Phleum pratense* for antibacterial, antioxidant activities nad general toxicity. Oriental Pharmacy and Exp. Medicine **3**: 157-162.

Sarosi, S. and Bernath, J. 2008. The antioxidant properties of *Prunella vulgaris* L. Acta Alimentaria **37(2)**: 293-300.

Sarumathy, K., Vijay, T., Jayakanthi, K., Dhanarajan, M.S. 2010. A protective effect of *Caesalpinia sappan* (CS) on acetaminophen induced nephrotoxicity and oxidative stress in male albino rats. J. Pharmacol. Toxicol. **1**: 1-12.

Sasidharan, S., Aravindran, S., Latha, L.Y., Vijenthi, R., Saravanan, D. and Amutha, S. 2010. *In vitro* antioxidant activity and hepatoprotective effects of *Lentinula edodes* against paracetamol-induced hepatotoxicity. Molecules **15(6)**: 4478-4489.

Sasikumar, J.M. and Doss, P.A. 2006. *In vitro* antioxidant and antibacterial activity of *Rhaphidophora pertusa* stem. Fitoterapia **77**: 605-607.

Sasikumar, J.M., Jinu, U. and Shamna, R. 2009. Antioxidat activity of HPTLC analysis of *Pandanus odorotissimus* L. root. European J. Biol. Sci. **1(2)**: 17-22.

Sasikumar, J.M., Maheshu, V., Aseervatham, G.S. and Darsini, D.T. 2010. *In vitro* antioxidant activity of *Hedyotis corymbosa* (L.) Lam. aerial parts. Indian J. Biochem. Biophys. **47(1)**: 49-52.

Satheesh, M.A. and Pari, L. 2004. Antioxidant effect of *Boerhavia diffusa* L. in tissues of alloxan induced diabetic rats. Indian J. Exo. Biol **42(10)**: 989-992.

Sathiya, M.. and Muthuchelian, K. 2010. Evaluation of antioxidant and antitumor potentials of *Prosopis juliflora* DC. leaves *in vitro*. Pharmacology online **2**: 328-343.

Satish Kumar, T., Shanmugam, S., T. Palvannan and Bharathi Kumar, V.M. 2008. Evaluation of antioxidant properties of *Canthium parviflorum* Lam. leaves. Nat. Prod. Radiance **7(2)**: 122-126.

Satue-Gracia, M.T., Heinonen, M. and Frankel, E.N. 1997. Antioxidant activity of anthocyanins in LDL and lecithin liposome systems. J. Agric. Food Chem. **45**: 3362-3367.

Saura-Calixto, F. and Goni, I. 2006. Antioxidant capacity of the Spanish Mediterranean diet. Food Chem. **94**: 442-447.

Sawa, T., Nakao, M., Akaike, T., Ono, K. and Maeda, H. 1999. Alkylperoxyl radical-scavenging activity of various flavonoids and other phenolic compounds: implications for antitumor-promotor effect of vegetables. J. Agric. Food Chem. **47**: 397-402.

Sawadogo, W.R., Meda, A., Lamien, C.E., Kiendrebeogo, M., Guissou, I.P. and Nacoulma, O.G. 2006. Phenolic content and antioxidant activity of six Acanthaceae from Burkina Faso. J. Biol. Sci. **6 (2)**: 249-252.

Sawai, Y. and Moon, J.H. 2000. NMR analytical approach to clarify the molecular mechanisms of the antioxidative and radical-scavenging activities of antioxidants in tea using 1,1 dipenyl-2-picrylhydrazyl. J. Agric. Food Che. **43**: 27-32.

Sawant, H.P., Joshi, Y.M. and Kadam, V.J. 2010. *In vitro* antioxidant activity of hydroalcoholic extract of leaves of *Jatropha curcas* Linn. J. Pharm. Res. **3(4)**: •

Saxena, R., Venkiah, K., Anitha, P., Venu, L. and Raghunath, M. 2007. Antioxidant activity of commonly consumed plant foods of India: contribution of their phenolic content. Int. J. Food Sci. Nutr. **58**: 250-260.

Scartezzini, P., Antognoni, F., Raggi, M.A., Poli, F. and Sabbioni, C. 2006. Vitamin C content and antioxidant activity of the fruit and of the Ayurvedic preparation of *Emblica officinalis* Gaertn. J. Ethnopharmacol. **104(1-2)**: 113-118.

Scartezzini, P. and Speroni, E. 2000. Review on some plants of Indian Traditional medicine with antioxidant activity. J. Ethnopharmacol. **71 (1-2)**: 23-43.

Schinella, G.R., Tournier, H.A., Prieta, J.M., Mordujovich de Buschiazzo, P. and Rios, J.L. 2002. Antioxidant activity of anti-inflammatory plant extracts. Life Sci. **70(9)**: 1023-1033.

Schlesier, K., Harwat, M., Bohm, V. and Bitsch, R. 2002. Assessment of antioxidant activity using different *in vitro* methods. Free Radical Research **36**: 177-187.

Schmeda-Hirschmann, G., Rodriguez, J.A., Theoduloz C, Astudillo, S.L., Feresin,. G.E. and Tapia, A. 2003. Free-radical scavengers and antioxidants from *Peumus boldus* Mol. ("Boldo"). Free Radic. Res. **37 (4)**: 447-52.

Schmeda-Hirschmann, G., Tapia, A., Theoduloz C, Rodriguez, J.A., Lopez, S. and Feresin, G.E. 2004. Free radical scavengers and antioxidants from *Tagetes mendocina*. Z. Naturforsch C. **59(5-6)**: 345-353.

Schmourlo, G., de Morais-Filho, Z.B., de Oliveira, D.B., Costa, S.S., Mendonca-Filho, R.R., Alviano, C.S. and Miranda, A.L.P. 2007. Antioxidant and antimicrobial activity of edible plants and their potential use as nutraceuticals. ISHS Acta Horticulturae756: 355-368.

Schuetz., K., Carle, R. and Schieber, A. 2006. *Taraxacum* – A review on its phytochemical and pharmacological profile. J. Ethnopharmacol. **107(3)**: 313-323.

Schuldt, E.Z., Farias, M.R., Ribeiro-doValle, R.M. and Ckless, K. 2004. Comparative study of radical scavenger activities of crude extract and fractions from *Cuphea carthagenesis* leaves. Phytomedicine **11(6)**: 523-529.

Schwartz., K., Bertelsen, G., Nissen, L.R. and Gardner, P. 2003. Investigation of plant extracts for the protection of processed foods against lipid oxidation. Comparison of antioxidant assay based on radical scavenging, lipid peroxidation and analysis of the principal antioxidant compounds. European Food Res. Tech. **212**: 319-328.

Sciacco, M., Bonilla, E., Schon, E.A., DiMauro, S., Moraes, C.T. 1994. Distribution of wild-type and common deletion forms of mtDNA in normal and respiration-deficient muscle fibres from patients with mitochondrial myopathy. Human Mol. Genet. **3**: 13-19.

Scott, G. 1997. Antioxidants in Science, Technology, Medicine and Nutrition. Allion Publishing, Chichester.

Seaver, B. and Smith, J.R. 2004. Inhibition of COX isoforms by nutraceuticals. J. Herb. Pharmacother. **4(2)**: 11-18.

Secco, A., Rosa, D.W., Freitas, R.A., Biawatti, M.W., Creczynski-Pasa, T.B. 2010. Antioxidant activity and low cytotoxicity of extracts and isolated compounds from *Araucaria angustifolia* dead bark. Redox Rep. **15(6)**: 234-242.

Sedlak and Lindsay. R.H. 1968. Estimation of total, protein-bound, and nonprotein sulfhydryl groups in tissue with Ellman's reagent. Anal. Biochem. **25**: 192–205.

Seidel, V., Verholle, M., Malard, Y., Tillequin, F., Fruchart, J-C., Duriez, P., Bailleul, F. and Teissier, E. 2000. Phenylpropanoids from *Ballota nigra* L. inhibit *in vitro* LDL peroxidation. Phytother. Res. **14(2)**: 93-98.

Sekretar, S., Schmidt, S. and Vajdak, M. 2004. Screening of antioxidative and antimicrobial activity of selected plant extracts. Chemicke Listy **98**: 685-686.

Sellappan, S. and Akoh, C.C. 2002. Flavonoids and antioxidant capacity of Georgia-grown Vidalia onions. J. Agric. Food Chem. **50**: 5338-5342.

Sellappan, S., Akoh, C.C. and Krewer, K. 2002. Phenolic compounds and antioxidant capacity of Georgia-grown blueberries and blackberries. J Agric. Food Chem. **50**: 2432-2438.

Selloum, L., Sebihi, L., Mekhalfia, A., Mahdadi, R. and Senator, A 1997. Antioxidant activity of *Cleome arabica*. Biochem. Soc. Trans. **25(4)**: S 608.

Selvameena, P., Arumugasamy, K., Baluprakash, T., Kokilavani, R. and Balakrishnan, A. 2010. Antioxidant activity of methanolic extract of *Swertia corymbosa* (Griseb.)

Wight ex Clarke (Gentianaceae) on alloxan induced diabetic rats. Int. Res. J. Pharm. **1(1)**: 392-394.

Selvan, R., Subramanian, L., Gayathri, R. and Angayarkanni, N. 1995. The antioxidant activity of turmeric (*Curcuma longa*). J. Ethnopharmacol. **47**: 59-67

Selvendian, K., Senthilnathan, P., Magesh, V. and Sakthiskaran, D. 2004. Modulatory effect of piperine on mitochondrial antioxidant system in Benzo(a) pyrene-induced experimental lung carcinogenesis. Phytomedicine **11(1)**: 85-89.

Selvi, S. and Chinnaswamy, P. 2007. *In vitro* antioxidant and antilipidperoxidative potential of *Pleurotus florida*. Ancient Sci. Life **24(4)**: 11-17.

Sendogdu, N., Aslan, M, Deliorman Orhan, D., Ergum, F. and Yesilada, E. 2005. Antidiabetic and antioxidant effects of *Vitis vinifera* L. leaves in streptozotocin-diabetic rats. Turkish J. Pharm. Sci. **3(1)**: 7-18.

Sengul, M., Yildiz, H., Gungor, N., Cetin, B., Eser, Z. and Ercisli, S. 2009. Total phenolic content, antioxidant and antimicrobial activities of some medicinal plants. Pak. J. Pharm. Sci. **22**: 102-106.

Sengupta, S., Mukherjee, A., Goswami, R. and Basu, S. 2009. Hypoglycemic activity of the antioxidant saponarin, characterized as alpha-glucosidase inhibitor present in *Tinospora cordifolia*. J. Enzyme Inhib. Med. Chem. **24(3)**: 684-690.

Senthilkumar, K.K. and Rajkapoor, B. 2010. Effect of *Oxalis corniculata* L. extracts on antioxidant enzyme levels in rat brain after induction of seizure by MES and PTZ. Int. J. Biopharmaceutics 1(2): 58-61.

Senthilkumar, N., Badami, S., Cherian, M.M. and Hariharapura, R.C. 2007. Potent *in vitro* cytotoxic and antioxidant activity of *Careya arborea* bark extract. Phytother. Res. **21(5)**: 492-495.

Senthilkumar, N., Badami, S., Dongre S.H. and Bhojraj, S. 2008. Antioxidant and hepatoprotective activity of the methanol extract of *Careya arborea* bark in Ehrilich ascites carcinoma-bearing mice. J. Nat. Med. **62(3)**: 336-339.

Senthilkumar, S., Ahmedjohn, S., Archunan, G. and Manoharan, N. 2008. Antioxidant activity of *Wedelia chinensis* in alloxan induced diabetic rats. Pharmacognosy online **2**: 640-651.

Sentjurc, M., Nemec, M., Connor, H.D. and Abram, V. 2003. Antioxidant activity of *Sempervivum tectorum* and its components. J. Agric. Food Chem. **51(9)**: 2766-2771.

Seo, W.G., Pae, H.O., Oh, G.S., Chai, K.H.. Kwon, T.A., Yun, Y.G. *et al.*, 2001. Inhibitory effect of methanol extract of *Cyperus rotundus* rhizomes on nitric oxide and superoxide productions by murine macrophage cell line, RAW 264.7 cells. J. Ethnopharmacol. **76**: 59-64.

Seraffini, M., Bellocco, R., Wolk, A. and Ekstrom, A.M. 2002. Total antioxidant potential of fruit and vegetables and risk of gastric cancer. Gastroenterology **123(4)**: 985-991.

Serafini, M., Ghiselli, A., and Ferro-Luzzi, A. 1994. Red wine, tea and anti-oxidants. Lancet **344**: 626.

Serafini, M., Mainani, G. and Ferro-Luzzi, A. 1998. Alcohol-free red wine enhances plasma antioxidant capacity in humans. J. Nutr. **128(6)**: 1003-1007.

Serce, S., Ozgen, M., Torun, A.A. and Ercisli, S. 2010. Chemical composition, antioxidant activities and total phenolic content of *Arbutus andrachne* L. (Fam. Ericaceae) (the Greek strawberry tree) fruits frim Turkey. Journal of Food composition and Analysis. **23(6)**: 619-623.

Serra, A.T., Seabra, I.J., Braga, M.E.M., Bronze, M.R., de Sousa, H.C. and Duarte, C.M.M. 2010. Processing cherries (*Prunus avium*) using supercritical fluid technology. Part 1. Recover of extract fractions rich in Bioactive compounds. The Journal of Super critical fluids **55(1)**: 184-191.

Serrano, M., Zapata, P.J., Castillo, S., Guillen, F., Martinez-Romero, D. and Valero, D. 2010. Antioxidant and nutritive constituents during sweet pepper development and ripening are enhanced by nitrophenolate treatments. Food Chem. **118(3)**: 497-503.

Serrilli, A.M., Graziosi, V., Ballero, M., Foddis, C., Serafini, M., Poli, F., Scartzzini, P. and Bianco, A. 2010. Endemic Sardinian plants: the case of *Genista cadasonensis* Valsecchi. Nat. Prod. Res. **24(10)**: 942-947.

Sersen, F., Mucaji, P., Grancai, D. and Nagy, M. 2005. Antioxidative properties of methanol extracts of leaves and fruit of *Ligustrum vulgare* L. Acta Facultatis Pharmaceuticae Comeniae **52**: 204-209.

Sersen, F., Vencel, T. and Annus, J. 2006. Silymarin and its components scavenge phenylglyoxylic ketyl radicals. Fitoterapia **77**: 525-529.

Serteser, A., Kargioglu, M., Gok, V, Bagci, Y, Musa Ozca, M and Arslan, D. 2009. Antioxidant properties of some plants growing wild in Turkey. Grasas Y Aceites **60(2)**: 147-154.

Sethi, A. and Sharma, R.A. 2011. Antioxidant activity with total phenolic constituents from *Aerva tomentosa* Forsk. Int. J. Pharm. Biosci. **2(2)**: 596-603.

Sethi, J., Sood, S., Seth, S. and Talwar, A. 2004. Evaluation of hypoglycemic and antioxidant effect of *Ocimum sanctum*. Indian J. Clin. Biochem. **19(2)**: 152-155.

Shafaghat, A. 2010. Free radical scavenging and antimicrobial activities and GC/MS analysis of essential oils from different parts of *Falcaria vulgaris* from two regions. Nat. Prod. Commun. **5(6)**: 981-984.

Shafazila, T.S., Lee, P.M. and Lee, K.H. 2010. Radical scavenging ativities of extract and solvent-solvent partition fractions from *Dendrobium sonia* "Red Bom" flower. Sci. Soc. Res. (CSSR) pp. 762-765. doi: 10.1109/CSSR.2010.5773886.

Shahat, A.A., Cos, P., Hermans, N., Apers, S., De Bruyne, T., Pieters, L., Berghe, D.V. and Vlietinck, A.J. 2003. Anticomplement and antioxidant activities of new acetylated flavonoid glycosides from *Centaurium spicatum*. Planta Medica **69(12)**: 1153-1156.

Shahidi, F. (ed.) 1996. Natural antioxidants: Chemistry, Health Effects and applications. AOCS Press, Champaign, IL,

Shahidi, F., Alasalvar, C. and Liyana-Pathirana, C.M. 2007. Antioxidant phytochemicals in hazelnut kernel (*Corylus avellana* L.) and hazelnut byproducts. J. Agric. Food Chem. **55(4)**: 1212-1220.

Shahidi, F., De Silva, C. and Amarowwicz, R. 2003. Antioxidant activity of extract of defatted seeds of niger (*Guizotia abyssinica*). J. Amer. Oil Chem. Soc. **80(5)**: 443-450. doi: 10.1007/s11746-003-0718-9

Shahidi, F., Janitha, P.K. and Wanasundara, P.D. 1992. Phenolic antioxidants. Crit. Rev. Food Sci. Nutr. 32: 67-103.

Shahidi, F. and Naczk, M. 2004. Phenolic compounds of major oil seed and plant oils. Pp. 83-130. In: Shahidi, F. and Naczk, M. (eds.) Phenolics in Food and Neutraceuticals. CRC Press, Boca Raton.

Shahidi, F. and Wanasandra, P.K.J.P.D. 1992. Phenolic antioxidants. Food Sci. Nutr. **32**: 67-103.

Shahidi, F., Wettasinghe, M., Amarowicz, R. and Khan, M.A. 20. Antioxidants of Evening primrose. Chapter 25. In: Shahidi, F. and Ho, C.-T. Phytochemicals and Phytopharmaceuticals pp. 278.

Shahwar, D., Shafiq-ur-Rehman, Ahmad, N., Ullah, S. and Raza, M.A. 2010. Antioxidant activities of the selected plants from the family Euphorbiaceae, Lauraceae, Malvaceae and Balsaminaceae. African J. Biotech. **9(7)**: 1086-1096.

Shalini, S., Arunachalam, G., and Sravanakumar, A. 2010. Antidiabetic, antihyperglycemic and antioxidant activities of *Cansjera rheedii* in alloxan induced rats. Pharmacognosy online **2**: 1054-1071.

Shamman, N.A., Kadir, K.A., Rahmat, A., and Ngah, W.Z. 1998. Vitamin C and *Aloe vera* supplementation protects from chemical hepatocarcinogenesis in the rat. Nutrition **14**: 846.

Shams, K.A., Radwan, H.M., Tawfik, W.A., Habib, A.A. and Soliman, A.M.M. 2010. Flavonoid constituents of *Cakile maritima* (Scope) growing in Egypt and their biological activity. Asian J. Chem. **22(5)**: 3981-3988.

Shan, B., Cai, Y.Z., Sun, M. and Corke, H. 2005. Antioxidant capacity of 26 spice extracts and characterization of their phenolic constituents. J. Agric. Food Chem. **53**: 7749-7759.

Shang, Y.F., Kim, S.M., Song, D.G., Pan, C.H., Lee, W.J. and Um, B.H. 2010. Isolation and identification of antioxidant compounds from *Ligularia fischeri*. J. Food Sci. **75(6)**: C530-535.

Shang, Y.Z., Qin, B.W., Cheng, J.J. and Miao, H. 2006. Prevention of oxidative injury by flavonoids from stems and leaves of *Scutellaria baicalensis* Georgi in PC 12 cells. Phytother. Res. **20(1)**: 53-57.

Shanmugasundaram, K.R., Sujatha, R. and Shanmugasundaram, E.R.B. 1994. Amrita Bindu – A salt-spice-herbal health food supplement for the prevention of nitrosamine induced depletion of antioxidants. J. Ethnopharmacol. **42**: 83-93.

Shanmugasundaram P. and Venkataraman S. 2006. Hepatoprotective and antioxidant effects of *Hygrophila auriculata* (K. Schum.) Heine Acanthaceae root extract. J. Ethnopharmacol. **104 (1-2)**: 124-128.

Shanmukha, I., Abubaker, S., Gupt, K., Manjamkhan and Ramachandra Setty, S. 2010. Antioxidant and nephroprotective activity of *Spathodea campanulata* bark against gentamycin induced nephrotoxicity. Pharmacology online **1**: 666-675.

Sharififar, F., Dehghn-Nudeh, G. and Mirtajaldini, M. 2009. Major flavonoids with antioxidant activity from *Teucrium polium* L. Food Chem. **112(4)**: 885-888.

Sharma, A., Upadhyay, J., Jain, A., Kharya, M.D., Namdeo, A. and Mahadik, K.R. 2011. Antioxidant activity of aqueous extract of *Boswellia serrata*. J. Chem. Biol. Phys. Sci. **1**: 60-71.

Sharma, N. and Garg, V. 2009. Antidiabetic and antioxidant potential of ethanolic extract of *Butea monosperma* leaves in alloxan-induced diabetic mice. Indian J. Biochem. Biophys. **46(1)**: 99-105.

Sharma, N.K., Dey, S. and R.Prasad. 2007. *In vitro* antioxidant potential evaluation of *Euphorbia hirta* L. plant. Pharmacology on line **1**: 91-98.

Sharma, N.K., Dey, S. and Prasad, R. 2008. Antioxidants from medicinal plants and their impact on human health. In: Ashwini Kumar and S.K.Sopory (eds.) Recent Advances in Plant Biotechnology and its applications pp. 470-496.

Sharma, N.K. and Prasad, R. 2008a. *In vitro* study on hepatoprotective role of antioxidant saponin from *Aegle marmelos* against CCl_4 mediated oxidative damage to liver. J. Hepatol. (in press).

Sharma, N.K. and Prasad, R. 2008b. Oxidative protein damage and their inhibition by phenolic acid antioxidants from *Euphorbia hirta* leaves. J. Biotechnol. (in press).

Sharma, R.K., Chatterji, S., Rai, D.K., Mehta, S., Rai, P.K., Singh, R.K., Watal, G. and Sharma, B. 2009. Antioxidant activities and phenolic contents of the aqueous extracts of some Indian medicinal plants. J. Med. Plants Res. **3(11)**: 944-948.

Sharma, S., Prashanth Kumar, K.C., Sikarwar, M.S., Patil, M.B. and Kokate, C.K. Antioxidant activity of ethanolic extract of leaves of *Eclipta alba*. Plant Archives **7(2)**: 863-864.

Sharma, U., Sahu, R.K., Roy, A. and Golwala, D.K. 2010. *In vivo* antidiabetic and antioxidant potential of *Stephania hernandifola* in streptozotocin-induced-diabetic rats. J. Young Pharm. **2(3)**: 255-260.

Shaw, C-Y., Chen, C-H., Hsu, C-C., Chen, C-C. and Tsai, Y-C. 2003.Antioxidant properties of scopoletin isolated from *Sinomonium acutum*. Phytother. Res. **17(7)**: 823-825.

She, G.M., Hu, J.B., Zhang, Y.J. and Yang, C.R. 2010. Phenolic constituents from *Rhopalocnemis phalloides* with DPPH radical scavenging activity. Pharm. Biol. **48(1)**: 116-119.

She, G.M., Zhang, Y.J. and Yang, C.R. 2009. Phenolic constituents from *Balanophora laxiflora* with DPPH radical-scavenging activity. Chem. Biodivers. **6(6)**: 875-880.

Sheeba, M.S. and Asha, V.V. 2006. Effect of *Cardiospermum halicacabum* on ethanol-induced gastric ulcers in rats. J. Ethnopharmacol. **106(1)**: 105-110.

Sheela, C.G. and Augusti, K.T. 1995. Antiperoxide effects of S-allyl cysteine sulphoxide isolated from *Allium sativum* Linn. and gugulipid in cholesterol fed rats. Indian J. Exp. Biol. **33(5)**: 337-341.

Sheeja, K. Shihab, P.K. and Kultan, G. 2006. Antioxidant and anti-inflammatory activities of the plant *Andrographis paniculata* Nees. Immunopharmacol. Immunotoxicol. **28(1)**: 129-140.

Shen, C.C., Shen, Y.C., Wang, Y.H., Lin, L.C., Don, M.J., Liou, K.T., Wang, W.Y., Hou, Y.C. and Chang, T.T. 2006. New lanostanes and naphthaquinones isolated from *Antrodia salmonea* and their antioxidative burst activity in human leukocytes. Planta Medica **72(3)**: 199-203.

Sherin, L., Shujaat, S. and Baqai, F. 2010. Antioxidant studies of various fractions of methanolic extract of *Terminali bellerica* Roxb. Asian J. Chem. **22(6)**: 4573-4579.

Shetty, T.K., Satav, J.G. and Nair, C.K.K. 2005. Radiation protection of DNA and membrane *in vitro* by extract of *Hemidesmus indicus*. Phytother. Res. **19(5)**: 387-390.

Shi, G.-F., An, L.J., Jiang, B., Guan, S. and Bao, Y.-M. 2006. Alpinia protocatechuic acid protects against oxidative damage *in vitro* and reduces oxidative stress *in vivo*. Neuroscience letters **403(3)**: 206-210.

Shi, H., Wang, H., Wang, M. and Li, X. 2009. Antioxidant activity and chemical composition of *Torreya grandis* cv. Merrilli seed. Nat. Prod. Commun. **4(11)**: 1565-1570.

Shi, H.M., Wen, J., Jia, C.Q., Jin, W., Zhang, X.F., Yao, Z.R. and Tu, P.F. 2006. Two new phenolic glycosides from the barks of *Hydnocarpus annamensis* and their anti-inflammatory and antioxidant activities. Planta Medica **72(10)**: 948-950.

Shi, J., Gong, J., Liu, J, Wu, X and Zhang, Y. 2009. Antioxidant capacity of extract from edible flowers of *Prunus mume* in China and its active components. LWT-Food Science and Technology **42(2)**: 477-482.

Shieh, D.E., Liu, L.T. and Lin, C.C. 2000. Antioxidant and free radical scavenging effects of baicalein, baicalin and wogonin. Anticancer Res. **20(5A)**: 2861-2865.

Shigenori, K., Masa, T., Yasuyuki, S., Masayo, S., Misun, K. and Tsutomo, N. 2002. Antioxidant activities of polyphenols in carob pods. Agric. Food Chem. **50**: 373-377.

Shikha, P., Latha, P.G., Suja, S.R., Anuja, G.I., Shyamal, S., Shine, V.J., Sini, S., Krishnakumar, N.M. and Rajasekharan, S. 2010. Anti-inflammatory and analgesic activity of *Barringtonia racemosa* Roxb. fruits. Indian J. Nat. prod. Resources **1(3)**: 356-361.

Shikov, A.N., Poltanov, E.A., Dorman, H.J., Makarov, V.G., Tikhonov, V.P., Hiltunen, R. 2006. Chemical composition and *in vitro* antioxidant evaluation of commercial water-soluble willow herb (*Epilobium angustifolium* L.) extracts. J. Agric. Food Chem. **54(10)**: 3617-3624.

Shilova, I.V., Pisareva, S.I., Krasnov, E.A., Brughes, M.A. and Pyak, A.I. 2006. Antioxidant properties of *Bergnia crasssifolia* extract. Pharm. Chem. J. **40(11)**: 620-623. doi: 10.1007/s11094-006-0206/4

Shim, J.U. and Lim, K.T. 2009. Antioxidant activity of glycoprotein isolated from *Geranium sibiricum* Linne. Nat. Prod. Res. **23(4)**: 375-387.

Shimada, K., Fujikawa, K., Yahara, K. and Nakamura. 1992. Antioxidative properties of xanthan on the autoxidation of soybean oil in cyclodextrin emulsion. J. Agric. Food Chem. **40**: 945-948.

Shinde, A.N., Malpathak, N. and Fulzele, D.P. 2010. Determination of isoflavone content and antioxidant activity in *Psoralea corylifolia* L. callus cultures. Food Chem. **118(1)**: 128-132.

Shindo, K., Saito, E., Sekiya, M., Matsui, T. and Koike, Y. 2008. Antioxidative activity of the flower of *Torenia fournieri*. J. Nat. Med. **62(2)**: 247-248. doi: 10.1007/s11418-007-0207-y

Shirwaikar, A., Bhilegaonkar, P.M., Malini, S. and Sharath Kumar, J. 2003. The gastroprotective activity of the ethanol extract of *Ageratum conyzoides*. J. Ethnopharmacol. **86(1)**: 117-121.

Shirwaikar, A., Issac, D. and Malini, S. 2004. Effect of *Aerva lanata* on cisplatin and gentamycin models of acute renal failure. J. Ethnopharmacol. **90**: 81-86.

Shirwaikar, A., Prabhu, K.S. and Punitha, I.S.R. 2006. *In vitro* antioxidant studies of *Sphaeranthus indicus* (Linn.). Indian J. Exp. Biol. 44: 993-996.

Shirwaikar, A., Rajendran, K. and Dinesh Kumar, C. 2004. *In vitro* antioxidant studies of *Annona sqamosa* Linn. leaves. Indian J. Exp. Biol. **42**: 803- 807.

Shirwaikar, A., Setty, M.M., Bommu, P. and Krishnanand, B. 2004. Ethanol extract of *Crataeva nurvala* stem bark reverses cisplatin-induced nephrotoxicity. Pharmaceutical Biol. **42(7)**: 559-564.

Shirwaikar, A., Shirwaikar, A., Rajendran, K. and Punitha, I.S.R. 2006. *In vitro* antioxidant studies in the benzyl tetra isoquinoline alkaloid berberine. Bio. Pharma Bull. **29(9)**: 1906-1910.

Shirwaikar, A. and Somasekhar, A.P.2003. Anti-inflammatory activity and free radical scavenging of *Aristolochia bracteolata* Lam. Indian J. Pharm. Sci. **65**: 67-69

Shirwaikar, A., Somasekhar, A.P., Udupa, A.L., Udupa, S.L. and Somasekhar, S. 2003. Wound healing studies of *Aristolochia bracteolata* Lam. with supportive action of antioxidant enzymes. Phytomedicine **10**: 558-562.

Shivaprasad, H.N., Kharya, M.D. and Rana, A.C. 2008. Antioxidant and adaptogenic effect of an herbal preparation triphala. J. Nat. Remedies **8(1)**: 82-88.

Shi-Yu, M., Li,.L., Ji-Liang, W., Jiang-Qi, Y. and De-Sheng, L. 2004. Antioxidant effects of extract of solution of bamboo leaves in mice. Chinesee J. Cli. Pharmacol. Ther. 5-25.

Shobhana, S. and Naidu, K. 2000. Antioxidant activity of selected Indian Spices. Prostaglandins, Leucotrienes and Essential Fatty acids **62**: 107-110.

Shoeb, M., MacManus, S.M., Kumaraswamy, Y., Jaspars, M., Nahar, L., Thoo Lin, P.K., Nazemiyeh, H. and Sarker, S.D. 2006. Americanin, a bioactive dibenzylbutyrolactone lignan, from the seeds of *Centaurea americana*. Phytochem. **67(21)**: 2370-2375.

Shon, M.Y., Choi, S.D., Kahng, G.G., Nam, S.H. and Sung, N.J. 2004. Antimutagenic, antioxidant and free radical scavenging activity of ethyl acetate extracts from white, yellow and red onions. Food Chem. Toxicol. **42**: 659-666.

Shon, M.Y., Kim, T.H. and Sung, N.J. 2003. Antioxidants and free radical scavenging activity of *Phellinus baumii* (*Phellinus* of Hymenochaetaceae) extracts. Food Chem. **82**: 593-597.

Shreedhara, C.S., Aswatham, H.N., Zanwar, S.B. and Falguni, G.P. 2009. Free radical scavenging activity of aqueous root extract of *Argyreia nervosa* (Burm.f.) Boj. (Convolvulaceae). J. Nat. Remedies **9(2)**: 216-223.

Shreedhara, C.S. and Vaidya, V.P. 2006. Screening of *Momordica dioica* for hepatoprotective, antioxidant, and anti-inflammatory activities. Nat. Prod. Sci. **12(3)**: 157-161.

Shreejayan and Rao, M.N.A. 1994. Curcuminoids as potent inhibitors of lipid peroxidation. J. Pharm. Pharmacol. 46: 1013-1016.

Shrififar, F., Yassa, N. and Shafiee, A. 2003. Antioxidant activity of *Otostegia persica* (Labiatae) and its constituents. Iranian J. Pharmaceutical Res. **2**: 235-239.

Shrikumar, S., Sukumar, M. and Ravi, T.K. 200. *In vitro* antioxidant activity in various extracts of *Curculigo orchioides* Gaertn. (Fam.: Amaryllidaceae). J. Nat. Remedies **7(2)**: 184-188.

Shui, G. and Leong, L.P. 2006. Residue from star fruit as vegetable source for functional food ingredients and antioxidant neutraceuticals. Food Chem. **97(2)**: 277-284.

Shui, G. and Peng, L.L. 2004. An improved method for the analysis of major antioxidants of *Hibiscus esculentus* Linn. J. Chromatogr. A. **1048 (1)**: 17-24.

Shukla, R., Gupta, S., Gambhir, J.K., Prabhu, K.M. and Murthy, P.S. 2004. Antioxidant effect of aqueous extract of the bark of *Ficus bengalensis* in hypercholesterolaemic rabbits. J. Ethnopharmacol. **92(1)**: 47-51.

Shukla, S., Mehta, A., Bajpai, V.K. and Shukla, S. 2009. *In vitro* antioxidant activity and total phenolic content of ethanolic leaf extract of *Stevia rebaudiana* Bert. Food Chem. Toxicol. **47(9)**: 2338-2343.

Shukla, S., Mehta, A., John, J., Singh, S., Mehta, P and Vyas, S.P. 2009. Antioxidant activity and total phenolic content of ethanolic extract of *Caesalpinia bonducella* seeds. Food and Chemical Toxicol. **47(8)**: 1848-1851.

Shukla, V., Vashistha M. and Singh, S.N. 2009. Evaluation of antioxidant profile and activity of Amalaki (*Emblica officinalis*), Spirulina and wheat grass. Indian Journal of Clinical Biochem. **24(1)**: 70-75.

Shyam, R., Singh, S.N., Vats, P., Singh, V.K., Bajaj, R., Singh, S.B. and Banerjee, P.K. 2007. Wheat grass supplementation decreases oxidative stress in healthy subjects. A comparative study with spirulina. J. Alternat. Comp. Med. **13**: 789-791.

Shyamala, B.N., Naidu, M.M., Sulochanamma, G. and Srinivas, P. 2007. Studies on the antioxidant activities of natural *Vanilla* extract and its constituent compounds through *in vitro* models. J. Agric. Food Chem. **55 (19)**: 7738-43.

Shyamala, M.P., Venukumar, M.R. and Latha, M.S. 2003. Antioxidant potential of the *Syzygium aromaticum* (Gaertn.) Linn. (cloves) in rats fed with high fat diet. Indian J. Pharmacol. **35**: 99-103.

Shyamasundar, K.V., Singh, B., Thakur, R.S., Hussain, A., Kiso, Y. and Hikino, H. 1985. Antihepatoprotective principles of *Phyllanthus niruri* herbs. J. Ethnopharmacol. **14**: 41-44.

Shylesh, B.S. and Padikkala, J. 1999. Antioxidant and anti-inflammatory activity of *Emilia sonchifolia*. Fitoterapia **70 (3)**: 275-278.

Shyur, L-F., Tsung, J-H., Chen, J-H., Chiu, C-Y. and Lo, C-P. 2005. Antioxidants properties of extracts from medicinal plants popularly used in Taiwan. Intern. J. Applied Sci. Engg. **3(3)**: 195-202.

Si, C.-L., Xu, J., Kim, J.-K., Bae, Y.-S., Liu, P.T. and Liu, Z. 2011. Antioxidant properties and structural analysis of phenolic glucosides from bark of *Populus ussuriensis* Kom. Wood Sci. Tech. **45(1)**: 5-13. doi: 10.1007/s00226-009-0286x

Siatka, T. and Kasparova, M. 2010. Seasonal variation in total phenolic and flavonoid contents and DPPH scavenging activity of *Bellis perennis*. Molecules **15(12)**: 9450-9461.

Sidduraju, M.N. and Dharmesh, S.M. 2007. Inhibition of gastric H+, K+: -: ATPase and *Helicobacter pylori* growth by phenolic antioxidants of *Curcuma amada*. J. Agric. Food Chem. **55(18)**: 7377-7386.

Siddhuraju, P. and Becker, K. 2003. Antioxidant properties of various solvent extracts of total phenolic constituents from three different agroclimatic origins of drumstick tree (*Moringa oleifera* Lam.) leaves. J. Agric. Food Chem. **51 (8)**: 2144-55.

Siddhuraju, P. and Becker, K. 2003. Studies on antioxidant activities of Mucuna seed (*Mucuna pruriens* var. *utilis*) extract and various non-protein aminoacids through *in vitro* models. J. Sci. Food Agric. 83(14): 1517-1524. doi: 10.1002/jsfa.1587.

Siddhuraju, P. and Manian, S. 2007. The antioxidant activity and free radical-scavenging capacity of dietary phenolic extracts from horse gram (*Macrotyloma uniflorum* (Lam.) Verdc.) seeds. Food Chem. **105(3)**: 950-958.

Siddhuraju, P., Mohan, P.S. and Becker, K. 2002. Studies on the antioxidant activity of Indian Laburnum (*Cassia fistula* L.): a preliminary assessment of crude extracts. Food Chem. **79 (1)**: 61-67.

Siebenhandl, S., Grausgruber, H., Pellegrini, N., Rio, D., Fogliano, V., Pernice, R. and Berghofer, E. 2007. Phytochemical profile of main antioxidants in different fractions of purple and blue wheat and black barley. J. Agric. Food Chem. **55(21)**: 8541-8547.

Sieliwanowicz, B. 1998. Antioxidant properties of beer phenols and their potential nutritional effects. Przem. Ferm. Owoc.-Warz. **4**: 9-11.

Silava, B.A., Ferreres,., Malva, J.O. and Dias, A.C.P. 2005 Phytochemical and antioxidant characterization of *Hypericum perforatum* alcoholic extracts. Food Chem. **90**: 157-167.

Silva, B.M., Andrade, P.B., Valentao, P., Ferreres, F., Seabra, R.M. and Ferreira, M.A. 2004. Quince (*Cydonia oblonga* Miller) fruit (pulp, peel and seed) and jam: antioxidant activity. J. Agric. Food Chem. **52(15)**: 4705-4712.

Silva, C.G., Herdeiro, R.S., Mathias, C.J., Panek, A.D., Silverira, C.S., Rodrigues, V.P., Renno, M.N., Falcao, D.Q., Cerqueira, D.M., Minto, A.B., Nogueira, F.L., Quaresma, C.H., Silva, J.F., Menezes, F.S. and Eleutherio, E.C. 2005. Evaluation of antioxidant activity of Brazilian plants. Pharmacol. Res. **52(3)**: 229-233.

Silva, C.G., Raulino, R.J., Cerqueira, D.M., Mannarino, S.C., Pereira, M.D., Panek, A.D., Silva, J.F.M., Menezes, F.S. and Eleutherio, E.C.A. 2009. *In vitro* and *in vivo* determination of antioxidant activity and mode of action of isoquercetin and *Hyptis fasciculata*. Phytomedicine doi: 10.1016/j.phymed.2008.12.019.

Silva, E.G., Behr, G.A., Zanotto-Filho, A., Lorenzi, R., Pasquali, M.A., Ravazolo, L.G., Bordignon, C.L. Jr, Silva, F.A., Aboy, A.L., Bassani, V.L., Henriques, A.T, Reginatto, F.H, Dal-Pizzol, F. and Moreira, J.C. 2007. Antioxidant activities and free radical scavenging potential of *Bauhinia microstachya* (Caesalpiniaceae) extracts linked to their polyphenol content. Biol. Pharm. Bull. **30 (8)**: 1488-96.

Sim, K.S., Nureshi, A.M.S. and Norhnom, A.W. 2010. Phenolic contents and antioxidant activity of *Pereskia grandifolia* Haw. (Cactaceae). Phcog. Mag. **6(2-3)**: 248-254.

Simic, M., Kundakovic, T. and Kovacevic, N. 2003. Preliminary assay on the antioxidative activity of *Laurus nobilis* extracts. Fitoterapia **74**: 613-616.

Simmler, C., Antheaume, C. and Lobstein, A. 2010. Antioxidant biomarkers from *Vanda caerulea* stems reduce irradiated HaCaT PGE-2 production as a result of COX-2 inhibition. PLoS One **5(10)**: e 13713.

Simoes-Pires, C.A., Queiroz, E.F., Henriques, A.T. and Hostettmann, K. 2005. Isolation and on-line identification of antioxidant compounds from three *Baccharis* species by HPLC-UV-MS/MS with post-column derivatisation. Phytochem. Anal. **16(5)**: 307-314.

Simonetti, P., Piergiorgio, P. and Testolin, G. 1997. Polyphenol content and total antioxidant potential of selected Italian wines. J. Agric. Food Chem. **45(4)**: 1152-1155.

Singh, A. and Handa, S.S. 1995. Hepatoprotective activity of *Apium graveolens* and *Hygrophila auriculata* against paracetamol and tnioacetamide intoxication in rats. J. Ethnopharmacol. **38**: 151-155.

Singh, A., Naidu, P.S., Gupta, S. and Kulkarni, S.K. 2002. Effect of natural and synthetic antioxidants in a mouse model of chronic fatigue syndrome. J. Med. Food. **5**: 211-220.

Singh, A. and Sukla, Y. 1998. Antitumor activity of eiallul sulfide in two-stage mouse skin model of carcinogenesis. Biomed. Environ. Sci. **11**: 258.

Singh, B.N., Prakash, D. and Singh, H.B. 2009. Antioxidant power of *Acacia* species. Herbication.com/herblife/Acacia/Acacia-HEP09.pdf.

Singh, B.N., Singh, B.R., Singh, R.L., Prakash, D., Sarma, B.K. and Singh, H.B. 2009. Antioxidant and anti-quorum sensing activities of green pod of *Acacia nilotica* L. Food Chem. Toxicol. **47(4)**: 778-786.

Singh, G., Kapoor, I.P.S., Singh, P., de Heluani, C.S., de Lampasona, M.P., Catalan, C.A.N. 2008. Chemistry, antioxidant and antimicrobial investigations on essential oil and oleoresins of *Zingiber officinale*. Food and Chemical Toxicology, **46 (10)**: 3295-3302.

Singh, G., Kiran, S., Marimuthu, P., Isidorov, V. and Vinogorova, V. 2007. Antioxidant and antimicrobial activities of essential oil and oleoresins of *Elettaria cardamomum* (seeds and pods) part 60. J. Sci. food Agric. **88(2)**: 280-290.

Singh, G., Marimuthu, P., de Heluani, C.S. and Catalan, C.A. 2006. Antioxidant and biocidal activities of *Carum nigrum* (seed) essential oil, oleoresin, and their selected components. J. Agric. Food Chem. **54(1)**: 174-81.

Singh, G., Marimuthu, P., de Heluani, C.S. and Catalan, C.A. 2005. Antimicrobial and antioxidant potentials of essential oil and acetone extract of *Myristica fragrans* Houtt. (aril part) (part 51). J. Food Sci. **70(2)**: M141-148.

Singh, G., Marimuthu, P., Murali, H.S. Bawa, A.S., Heluani, C.S.D., Lampasona, M.P. and Catalan, C. 2005. Antioxidant and antibacterial potential of essential oils and extracts isolated from various spice materials. J. Food Safety **25**: 130-145.

Singh, G., Maurya, S., Catalan, C. and de Lampasona, M.P. 2004. Chemical constituents, antifungal and antioxidative effects of ajwan essential oil and its acetone extract. J. Agric. Food Chem. **52(11)**: 3292-3296.

Singh, G., Maurya, S., Catalan, C. and de Lampasona, M.P. 2005. Studies on essential oils, Part 42: Chemical, antifungal, antioxidant and sprout suppressant studies on ginger essential oil and its oleoresin. Flavour and Grance Journ. **20(1)**: 1-6.

Singh, G., Maurya, S., Lampasoma, M.P. and Catalan, C. 2006a. Chemical constituents, antimicrobial and antioxidative potential of volatile oil and acetone extract of Star anise fruits. Part 54. J. Sci. Food Agric. **86**: 111-121.

Singh, G., Maurya, S., Lampasoma, M.P. and Catalan, C. 2006b. Studies on essential oils, part 41. Chemical composition, antifungal, antioxidant and sprout suppressant activities of coriander (*Coriandrum sativum*) essential oil and its oleoresin. Flavour and Fragrance Journ. **21(3)**: 472-479.

Singh, H.P., Kaur, S., Mittal, S., Batish, D.R. and Kohli, R.K. 2010. *In vitro* screening of essential oil from young and mature leaves of *Artemisia scoparia* compared to its major constituents for free radical scavenging activity. Food Chem. Toxicol. **48**: 1040-1044.

Singh, H.P., Mittal, S., Kaur, S., Batish, D.R. and Kohli, R.K. 2009. Characterization and antioxidant activity of essential oils from fresh and decaying leaves of *Eucalyptus tereticornis*. J. Agric. Food Chem. **57(15)**: 6962-6966.

Singh, J. and Kakkar, P. 2009. Antihyperglycemic and antioxidant effect of *Berberis aristata* root extract and its role in regulating carbohydrate metabolism in diabetic rats. J. Ethnopharmacol. **123**: in press.

Singh, M., Govindarajan, R., Nath, V., Rawat, A.K.S., Mehrotra, S. 2006. Antimicrobial, wound healing and antioxidant activity of *Plagiochasma appendiculatum* Lehm. et Lind. J. Ethnopharmacol. **107 (1)**: 67-72.

Singh, P., Shivhar, Y. and Patil, U.K. 2009. Free radical scavenging potential of *Chenopodium album* Linn. Adv. Pharmacol. Toxicol. **10(3)**: 111-118.

Singh, P., Shukla, R., Prakash, B., Kumar, A., Singh, S., Mishra, P.K. and Dubey, N.K. 2010. Chemical profile, antifungal, antiaflatoxigenic and antioxidant activity of *Citrus maxima* Burm. and *Citrus sinensis* (L.)Osbeck essential oils and their cyclic monoterpene, DL-limonene. J. Chem. Toxicol. **48**: 1734-1740.

Singh, R. and Arora, S. 2007. Attenuation of free radicals by acetone extract/fractions of *Acacia nilotica* (L.) Willd. ex Del. J. Chinese Clin. Med. **2(4)**:

Singh, R., Jain, S.C. and Jain, R.C. 2009. Antioxidant activity of some medicinally important Arid Zone plants. Asian J. Exp. Sci. **23(1)**: 215-221.

Singh, R., Sharma, R.S., Mishra, V., Seth, N. and Babu, C.R. 2004. Chemical prospecting for medicinal compounds from a hemi-parasitic plant (*Dendrophthoe trigona*) inhbiting Western Ghats of India. IUPAC International conference on Biodiversity and Natural Products Chemistry and Medical applications, New Delhi, p. p-325.

Singh, R., Singh, B., Singh, S., Kumar, N., Kumar, S. and Arora, S. 2008. Anti-free radical activities of kaempferol isolated from *Acacia nilotica* (L.) Willd. ex Del. Toxicol. *In vitro* **22(8)**: 1965-1970.

Singh, R., Singh, N., Saini, B.S. and Rao, H.S. 2007. Hepatoprotective and antioxidant properties of methanolic extract of *Piper nigrum* Linn. in rats. Phcog. Mag. **3(12)**: 251-258.

Singh, R., Singh, N., Saini, B.S. and Rao, H.S. 2007. Hepatoprotective and antioxidant properties of methanolic extract of *Piper nigrum* Linn. in rats. Phcog. Mag. **3(12)**: 251-258.

Singh, R., Singh, N., Saini, B.S. and Rao, H.S. 2008. *In vitro* antioxidant activity of pet ether extract of black pepper. Indian J. Pharmacol. **40 (4)**: 147-151.

Singh, R., Singh, S., Kumar, S. and Arora, S. 2007a. Evaluation of antioxidant potential of ethyl acetate extract/fractions of *Acacia auriculiformis* A. Cunn. Food Chem. Toxicol. **45 (7)**: 1216-1223.

Singh, R., Singh, S., Kumar, S. and Arora, S. 2007b. Studies on antioxidant potential of methanol extract of fractions of *Acacia auriculiformis* A. Cunn. Food Chem. **103(2)**: 505-511.

Singh, R., Singh, S., Kumar, S. and Arora, S. 2007c. Free radical scavenging activity of acetone extract/fractions of *Acacia auriculiformis* A.Cunn. Food Chem. **103(4)**: 1403-1410.

Singh, R.P., Banerjee, S., Kumar, Raveesha, K.A. and Rao, A.R. 2006. *Tinospora cordifolia* induces enzymes of carcinogen/drug metabolism and antioxidant system and inhibits lipid peroxidation in mice. Phytomedicine **13**: 74-84.

Singh, R.P., Chidambara Murthy, K.N. and Jayaprakasha, G.K. 2002. Studies on the antioxidant activity of pomegranate (*Punica granatum*) peel and seed extract using *in vitro* models. J. Agric. Food Chem. **50**: 81-86. Doi: 10.1021/jf010865b

Singh, S., Singh, R., Gupta, P., Singh, P. and Gupta, K.C. 2004. *Morinda citrifolia* – herbs for health. Chemistry Bology Interface: Synergistic New Frontiers, New Delhi, India p25-57, November 21-26, 2004.

Singh R.P., Murthy, K.N.C. and Jayaprakasha, G.K. 2002. Antioxidant activity of pomegranate peel and seed extract. J. Food Chem. **50(1)**: 81-86.

Singhal, S.S., Awasthi, S, Pandya, U., Peper, J.T., Sajini, M.K., Cheng, J.Z. and Awasthi, Y.C. 1999. The effect of curcumin on glutathione-linked enzymes in K562 human leukemia cells. Toxicol Left. **109**: 87.

Singhatong, S., Leelarungrayub, D. and Chaiyasut, C. 2010. Antioxidant and toxicity activities of *Artocarpus lakoocha* Roxb. heartwood extract. J. Med. Plants Res. **4(10)**: 947-953.

Singleton, V.L., Orthofer, R. and Lamuela-Raventos, R.M. 1999. Analysis of total phenols and other oxidation substrates and antioxidants by means of Folin – Ciocalteu reagent. In: Methods in Enzymology, oxidant and Antioxidants, Part A. by L.Packer (ed.). Academic press, San Diego, C.A. vol. 299, pp. 152-178.

Sini, S. and Devi, K.S. 2004. Antioxidant activities of chloroform extract of *Solanum trilobatum*. Pharmaceutical Biol. **42(6)**: 462-466.

Siqueira, I.R., Fochesatto, C., Torres, I.L.S., da Silva, A.L., Nunes, D.S., Elisabetsky, E. and Netto, C.A. 2007. Antioxidant activities of *Ptychopetalum olacoides* ('muirapuama") in mice brain. Phytomedicine **14(11)**: 763-769.

Siriwoharn, T., Wrolstad, R.E., Finn, C.E. and Pereira, C.B. 2004. Influence of cultivar maturity, and sampling on blackberry (*Rubus* L. hybrids) anthocyanins, polyphenolics, and antioxidant properties. J. Agric. Food Chem. **52**: 8021-8030.

Sirmah, P., Dumarcay, S., Masson, E. and Gerardin, P. 2009. Unusual amount of (-)mesquitol from the heartwood of *Prosopis juliflora*. Nat. Prod. Res. **23(2)**: 183-189.

Siswoyo, T.A., Mardianat, E., Lee, K.O. and Hoshokawa, K. 2011. Isolation and characteriztion of antioxidant protein fractions from Melinjo (*Gnetum gnemon*) seeds. J. Agric. Food Chem. 59(10): 5648-5656. doi: 10.1021/jf2000647

Sithisarn, P. and Jarikasem, S. 2010. Antioxidant activity and phenolic content of *Acanthopanax trifoliatus* and *Toddalia asiatica*. Kasetsart J. (Nat.Sci.) **44**: 234-242.

Sithisarn. P., Supabphol, R. and Gritsanapan, W. 2005. Antioxidant activity of Siamese neem tree (VP 1209). J. Ethnopharmacol. **99**: 109-112.

Sithisarn. P., Supabphol, R. and Gritsanapan, W. 2006. Comparison of free radical scavenging activity of Siamese neem tree (*Azadirachta indica* A. Juss var. *siamensis* Valeton) leaf extracts prepared by different methods of extraction. Med. Princ. Pract. **15(3)**: 219-22.

Sivakumar, P., Sambath Kumar, R., Siva Kumar, T., Nethaji, R., Vijayabhaskaran, M., Perumal, P., Bukkaver, A.R., Jayakar, B., Parvathi, K. and Sarasu, C. 2008. Antitumor, and antioxidant activities of *Triumfetta rhomboidea* against Dalton's Ascites Lymphoma bearing Swiss albino mice. Res. J. Med. Sci. **2(4)**: 203-208.

Sivapriya, M. and Leela, S. 2007. Isolation and purification of a novel antioxidant protein from the water extract of Sundakai (*Solanum torvum*) seeds. Food Chem. **104(2)**: 510-517.

Skandrani, I., Boubaker, J., Bhouri, W., Limem, I, Kilani, S., Ben Sghaier, M, Neffati, A., Bouhle, I., Ghedira, K., Chekir-Ghedira, L. 2010. Leaf extracts from *Moricanda arvensis* promote antiproliferation of human cancer cells, induce apoptosis, and enhance antioxidant activity. Drug Chem. Toxicol. **33(1)**: 20-27.

Skandrani, I., Sghaier, M.B., Neffati, A., Boubaker, J., Bouhlel, I., Kilani, S., Mahmoud, A., Ghedira, K. and Chekir-Ghedira, L. 2007. Antigenotoxic and free radical scavenging activities of extracts from *Moricanda arvensis*. Drug Chem. Toxicol. **30(4)**: 361-382.

Skottova, N., Kazdova, L., Oliyarnik, O., Vecera, R., Sobolova, L. and Ulrichova, J. 2004. Phenolics rich extracts from *Silybum marianum* and *Prunella vulgaris* reduce a high sucrose diet induced oxidative stress in hereditary hypertriglyceridemic rats. Pharmacol. Res. **50**: 123-130.

Skottova, N., Vecera, R., Vana, P., Ulrichova, J. and Walterova, D. 2001. Antioxidative activities of polyphenolic extract from *Prunella vulgaris* in rat. Atherosclerosis **2**: 116-117.

Slinkard, K and Singleton, V. L. 1977. Total phenol analyses: automation and comparison with manual methods. Am. J. Ecol. Viticul. **28**: 49–55.

Sloley, B.D., Urichuk, L.J., Tywin, C., Coutts, R.T., Pang, P.K. and Shan, J.J. 2001. Comparison of chemical components and antioxidants capacity of different *Echinacea* species. J. Pharm. Pharmacol. **53(6)**: 849-857.

Slusarczyk, S., Hajnos, M., Skalinska-Wojniak, K. and Matkowski, A. 2008. Antioxidant activity of polyphenols from *Lycopus lucidus* Turcz. Food Chem. **113(1)**: 138-138. doi: 10.1016/j.foodcheem.2008.07.037

Smejkal, K., Holubova, P., Zima, A., Muselik, J. and Dvorska, M. 2007. Antiradical activity of *Paulownia tomentosa* (Scrophulariaceae) extracts. Molecules. **12(6)**: 1210-1219.

Smirnoff, N. 2005. Antioxidants and Reactive oxygen species in plants. Blackwell Publishers.

Soares, A.A. de Souza, C.G.M., Daniel, F.M., Ferrari, G.P., da Costa, S.M.G., Peralta, R.M. 2009. Antioxidant activity and total phenolic content of *Agaricus braziliensis* (*Agaricus blazei* Merril) in two stages of maturity. Food Chem. **112(4)**: 775-778.

Soares, J.R., Dinis, T.C.P., Cunha, A.P. and Ameida, L.M. 1997. Antioxidant activities of some extracts of *Thymus zygis*. Free Radical Res. **26**: 469-478.

Soboli, A., Esmaeili, M.A., Gholipous, A. and Kanani, M.R. 2010. Composition, cytotoxicity and antioxidant activity of the essential oil of *Dracocephalum surmandicum* from Iran. Nat. Prod. Commun. **5(2)**: 341-344.

Sofidiya, M.O., Odukoya, O.A., Afolayan, A.J. and Familoni, O.B. 2009. Phenolic contents, antioxidant and antibacterial activities of *Hymenocardia acida*. Nat. Prod. Res. **23(2)**: 168-177.

Sofidiya, M.O., Odukoya, O.A., Familoni, O.B. and Inya-Agha. 2006. Free radical scavenging activity of some Nigerian medicinal plants. Pak. J. Bio. Sci. **9**: 1438-1441.

Sohal, R.S. and Weindruch, R. 1996. Oxidative stress, calorie restriction and ageing. Science **273**: 59-63.

Sohn, H.Y., Ryu, H.Y., Jang, Y.J., Jang. H.S., Park, Y.M. and Kim, S.Y. 2008. Evaluation of antimicrobial, antithrombin and antioxidant activity of aerial parts of *Saxifraga stolonifera*. Korean J. Microbiol. Biotech. **36(3)**: 195-200.

Sokmen, A., Gulluce, M., Akpulat, H.A., Daferera, D., Tepe, B., Polissopi, Munevver Sokmen, and Sahin, F. 2004. The *in vitro* antimicrobial and antioxidant activities of the essential oils and methanol extracts of endemic *Thymus spathulifolius*. Food Control **15(8)**: 627-634. doi: 10.1016/j.foodcont.2003.10.005.

Sokmen, M., Serkedjieva, J., Dafera, D.S., Gulluce, M., Polissiou, M., Tepe, B., Akpulat, A. and Sahin, F. 2004. *In vitro* antioxidant, antimicrobial and antiviral activities of the essential oil and various extracts from herbal parts and callus cultures of *Origanum acutidens*. J. Agric. Food Chem. **52(11)**: 3309-3312.

Sokmen, A., Sokmen, S., Dafrera, E., Polissiou, M., Candan, F., Unlu, M. and Akpulat, H. A. 2004. The *in vitro* antioxidant and antimicrobial activities of the essential oil and methanol extracts of *Achillea biebersteini* Afran. (Asteraceae). Phytother. Res. **18(6)**: 451-456.

Somova, L.I., Shode, F.O., Ramnanan, P. and Nadar, A. 2003. Antihypertensive, antiatherosclerotic and antioxidant activity of triterpenoids isolated from *Olea europaea*, subspecies *africana* leaves. J. Ethnopharmacol. **84(2-3)**: 299-305.

Sonboli, A., Esmaeli, M.A., Gholipous, A. and Kanani, M.R. 2010. Composition, cytotoxicity and antioxidant activity of the essential oil of *Dracocephalum surmandicum* from Iran. Nat. Prod. Commun. **5(2)**: 341-344.

Song, F., Qi, X., Chen, W., Jia, W., Yao, P., Nussler, A.K., Sun, X. and Liu, L. 2007. Effect of *Momordica grosvenori* on oxidative stress pathways in renal mitochondria of normal and alloxan-induced diabetic mice. Involvement of heme oxygenase-1. Eur. J. Nutr. **46(2)**: 61-69.

Song, F.-L., Gan, R.-Y., Zhang, Y., Xiao, Q., Kuang, L. and Li, H.-B. 2010. Total phenolic contents and antioxidant capacities of selected Chinese medicinal plants. Int. J. Mol. Sci. 11(6): 2362-2372. doi: 10.3390/ijms11062362.

Song, H.S., Park, Y.H., Jung, S.H., Kim, D.P., Jung, Y.H., Lee, M.K. and Moon, K.Y. 2006. Antioxidant activity of extracts from *Smilax china* root. J. Kor. Soc. Food Sci. Nutr.

Song, M.C., Yang, H.J., Bang, M.H., Kim, D.K., Jeong, T.S., Kim, J.P. and Baek, N.I. 2007. Antioxidant and antiaterogenic activity of cis-Hinokiresinol from *Trapa pseudoincisa*. Arch. Pharm. Res. **30(11)**: 1392-1397.

Song, Y.S., Kim, S.H., Sa, J.H., Jin, C., Lim, C.J. and Park, E.H. 2003 Anti-angiogenic, antioxidant and xanthine oxidase inhibition activities of the mushroom *Phellinus linteus*. J. Ethnopharmacol. **88 (1)**: 113-116.

Soni, H., Pandey, H., Phatak, A.K., Nayak, G., Singhai, A.K., Parihar, A., Singh, V. and Rathur, A.S. 2009. Evaluation of antioxidant potential of hydroalcoholic extract of leaves of *Coleus aromaticus*. Adv. Pharmacol Toxicol. **10(1)**: 75-82.

Soni, K., Kumar, P.S. and Saraf, M.N. 2003. Free radical scavenging and antilipid peroxidation acivity of *Tephrosia purpurea* Linn. Indian J. Pharmaceutical Sci. **65**: 27-30.

Soni, K., Kumar, P.S. and Saraf, M.N. 2006. Antioxidant activity of fraction of *Tephrosia purpurea* Linn. Indian J. Pharm. Sci. **68**: 456-60.

Sonogo, R., Vassallo, A., Malafronte, N., Imparaato, S., Russo, A., Dal Piaz, F. 2009. New phenolic glycosides from *Securinega virosa* and their antioxidant activity. Nat. Prod. Commun. **4(12)**: 1645-1650.

Soobrattee, M.A., Bahorun, T., Neergheen, V.S., Googoolye, K. and Aruoma, O.I. 2008. Assessment of the content of phenolics and antioxidant actions of the Rubiaceae, Ebenaceae, Celastraceae, Erythroxylaceae and Sterculiaceae families of Mauritian endemic plants. Toxicology *in vitro* **22(1)**: 45-56.

Soon, Y.Y. and Tan, B.K. 2002. Evaluation of the hypoglycemic and antioxidant activities of *Morinda officinalis* in streptozotocin-induced diabetic rat. Singapore Med. J. **43(2)**: 77-85.

Soong, Y. and Barlow, P.J. 2004. Antioxidant activity and phenolic content of selected fruit seeds. Food Chem. **88**: 411-417.

Soong, Y. and Barlow, P.J. 2006. Quantification of gallic acid and ellagic acid from Longan (*Dimocarpus longan* Lour.) seed and mango (*Mangifera indica* L.) kernel and their effects on antioxidant activity. Food Chem. **97**: 524-530.

Souri, E., Amin, G., Dehmobed-Sharifabadi, A., Nazifi, A. and Farsam, H. 2004a. Antioxidant activity of sixty plants from Iran. Iranian J. Pharmaceutical Res. **3**: 55-59.

Souri, E., Amin, G., Farsam, H. and Andaji, S. 2004b. The antioxidant activity of some commonly used vegetables in Iranien diet. Fitoterapia **75**: 585-588.

Souri, E., Amin, G., Farsam, H. and Barazandeh Tehrani, M. 2008. Screening of antioxidant and phenolic content of 24 medicinal plant extracts. DARU **16(2)**: 83-87.

Souri, E., Amin, G., Farsam, H., Jalalizadeh, H. and Barezi, S. 2008. Screening of thirteen medicinal plant extracts for antioxidant activity. Iranian J. Pharmaceutical Res. **7(2)**: 149-154.

Souri, E., Amin, G., Sharifabadi, A.D., Nazifi, A. and Farsam, H. 2004. Antioxidative activity of sixty plants from Iran. Iranian J. Pharmac. Res. **3**: 55-59.

Souri, E., Farsam, H., Sarkhail, P. and Ebadi, F. 2004. Antioxidant activity of some furanocoumarins isolated from *Heracleum persicum*. Pharm. Biol. **42**: 396-399

Souri, E., Sarkhail, P., Kaymanesh, P., Amini, M. and Farsam, H. 2006. Antioxidant activity of extract and a new isolated Dioxaspiran derivative of *Tripleurospermum disciforme*. Pharm. Biol **43(7)**: 620-623.

Souza, J.G., Tomei, R.R., Kanashir, A., Kabeya, L.M., Azzaloni, A.E.C.S., Dias, D.A., Salvaador, M.J. and Lucisano-valim, Y.M. 2010. Ethanolic crude extract and flavonoids isolated from *Alternanthera maritima*: neutrophil chemiluminescence inhibition and free radical scavenging activity. Z. Naurforsch. **62(5-6)**: 339-347.

Souza, J.S., Machado, L.L., Pessoa, L.D.L., Braz-Filho, R., Overk, C.R., Yao, P., Cordel, G.A. and Lemos, T.L.G. 2005. Pyrrolizidine alkaloids from *Heliotropium indicum*. J. Braz. Chem. Soc. **16(6)**:

Sowndhararajan, K., Joseph, J.M. and Manian, S. 2010. Antioxidant and free radical scavenging activities of Indian Acacias: *Aacacia dealbata* Link. and *A. pennata* (L.) Willd. 20th Annual Conference of IAAT and Intern. Symposium on Taxonomy, Plant Diversity and Conservation. Abstract pp. 265.

Spada, P.D.S., de Souza, G.G.N., Bortolini, G.V., Henriques, J.A.P. and Salvador, M. 2008. Antioxidant and antimutageneic activity of frozen fruits. J. Med. Food **11**: 144-151.

Speroni, E., Cervellati, R., Costa, S., Guerra, M.C., Utan, A., Govoni, P., Berger, A., Müller, A. and Stuppner, H. 2007. Effects of differential extraction of *Verbena officinalis* on rat models of inflammation, cicatrization and gastric damage. Planta Med. **73 (3)**: 227-35.

Spiridon, I., Bodirlau, R. and Teaca, C.-A. 2011. Total phenolic content and antioxidant activity of plants used in traditional Romanian herbal medicine. Central European J. Biol. DOI: 10.2478/s11535-011-0028-6.

Sreedevi, A., Bharathi, K and Prasad, K.V.S.R.G. 2009. Effect of alcoholic extract of *Dichrostachys cinerea* Wight et Arn. against cisplatin-induced nephrotoxicity in rats. Nat. Prod. Radiance **8(1)**: 12-18.

Sreejayan, N. and Rao, M.N.A. 1996. Free radical scavenging activity of Curcuminoids. Drug Res. **46**: 169.

Sreejayan, N. and Rao, M.N.A. 1997. Nitric oxide scavenging by curcuminoids. J. Pharm. Pharmacol. **47**: 105-107.

Sreeramulu, D., Vijayakumar Reddy and Raghunath, M. 2009. Antioxidant activity of commonly consumed cereals, millets, pulses and legumes in India. Indian J. Biochem. Biophys. **46**: 112-115.

Sridhar, C., Subbaraju, G.V., Venkateswarlu, Y. and Venugopal, R.T. 2004. New acylated iridoid glucosides from *Vitex altissima*. J. Nat. Prod. **67(12)**: 2012-2016.

Srikanth, J. and Muralidharan, P. 2010. Evaluation of antioxidant properties of *Sapindus emarginatus* Vahl. Asian J. Exp. Biol. Sci. **1(3)**: 693-699.

Srinivasan, G.V., Ravi, S., Tushar, K.V. and Balachandran, I. 2009. Quantitative determination of Mangiferin from the roots of *Salacia fruticosa* and the antioxidant studies. J. Trop. Med. Plants **10(2)**:

Srinivasan, G.V., Sadashiva, C.T., George, N. and Balachandran, I. 2008. Chemical compounds in *Biophytum reinwardtii* and their antioxidant potential. J. Trop. Med. Plants **9(2)**: 338-342.

Srinivasan, M., Padmanabhan, M. and Prince, P.S. 2005. Effect of *Enicostemma littorale* Blume extract on key carbohydrate metabolic enzymes, lipid peroxides and antioxidants in alloxan-induced diabetic rats. J. Pharmacy and Pharmacol. **57**: 497-503.

Srinivasan, R., Chandrasekar, M.J.N., Nanjan, M.J. and Suresh, B 2007a. Antioxidant activity of *Caesalpinia digyna* root. J. Ethnopharmacol. **113 (2)**: 284-291.

Srinivasan, R., Chandrasekar, M.J.N., Nanjan, M.J. and Suresh, B 2007b. Free radical scavenging activity of *Ipomoea obscura* (L.) Ker-Gawl. J.Nat. Remedies **7(2)**: 184-188.

Srinivas Reddy, B., Kiran Kumar Reddy, R., Purushotham Reddy, B., Ramakrishna, S., and Diwan, P.V. 2008a. Potential *in vitro* antioxidant protective effects of *Soymida febrifuga* on ethanol induced oxidative damage inHep G2 cell. Food Chem. Toxicol. **46(11)**: 3429-3442. doi: 10.1016/j.fct.2008.08.034.

Srinivas Reddy, B., Purushotham Reddy, B., Raghavulu, A.V., Ramakrishna, S., Venkateswarlu, Y. and Diwan, P.V. 2008b. Evaluation of antioxidant and antimicrobial properties of *Soymida febrifuga* leaf extracts. Phytother. Res. **22(7)**: 943-947.

Srivastava, A., Harish, S.R., and Shivanandappa, T. 2006. Antioxidant activity of the roots of *Decalepis hamiltonii* Wight et Arn. LWT - Food Science and Technology, **39 (10)**: 1059-1065.

Srivastava, A. and Shivanandappa, T. 2010. Hepatoprotective effect of the root extract of *Decalepis hamiltonii* against carbon tetrachloride-induced oxidative stress in rats. Food Chem. **118(2)**: 411-417.

Sroka, Z., Rzadkowska-Bodalska, H. and Mazol, I. 1994. Antioxidative effect of extracts from *Erodium cicutarium* L. Zeitschrift fur Naturforschung C. **49(11-12)**: 881-884.

Stadtman, E.R. 2004. Role of oxidant species in aging. Curr. Med. Chem. **11**: 1105-1112.

Štajner, D., Canadanovic-Brunet, J. and Pavlovic, A. 2004. *Allium schoenoprassum* L., as a natural antioxidant. Phytother. Res. **18(7)**: 522-524.

Štajner, D., Igic, R., Popovic, B.M. and Malencic, Dj. 2008. Comparative study of antioxidant properties of wild growing and cultivated *Allium* species. Phytother. Res. **22(1)**: 113-117.

Štajner D., Milic, N., Canadanovic-Brunet, J. 1999. An investigation into the antioxidant activity of *Allium nutans* L. Phytother. Res. **13(4)**: 333-336.

Štajner D., Milic, N., Canadanovic-Brunet, J., Kapor, M., Stajner, M. and Popovic, B.M. 2006. Exploring *Allium* species as a source of potential medicinal agents. Phytother. Res. **20**: 581-584.

Štajner D., Milic-Demarino, N., Canadanovic-Brunet. 2003. Screening for antioxidant properties of Leeks, *Allium sphaerocephalon* L. J. Herbs, Spices Med. Plants **10(3)**: 75-82.

Štajner, D., Milic-Demarino, N., Canadanovic-Brunet, J. and Popovic, M. 2002. Scavenger activity of *Allium psekemense* B. Fedtsch. Phytother. Res. **16(5)**: 484-487.

Štajner, D., Milic-Demarino, N., Canadanovic-Brunet, J., Kapor, A., Stajner, M. and Popovic, B.M. 2006. Screening for antioxidant properties of *Allium giganteum*. Fitoterapia **77**: 268-270.

Štajner, D., Milic, N., Lazic, B., Mimica-Dukic, N. 1998a. Study on antioxidant enzymes in *Allium cepa* L. and *Allium fistulosum* L. Phytother. Res. **12**: 15-17.

Štajner, D., Milic, N., Mimica-Dukic, N., Lazic, B. and Iric, R. 1998b. Antioxidant abilities of cultivated and wild species of garlic. Phytother. Res. **12**: 13-14.

Štajner, D., Mimica-Dukic, N. and Braic, M. 1997. Oxygen radical scavenging activity of Reatival. Fitoterapia **68(3)**: 261-264.

Štajner, D. and Popovic, B.M. 2010. Comparativee study of antioxidant capacity in organs in different *Allium* species. Central European J. Biol. **4(2)**: 224-228. doi: 10.2478/s11535-009-0010-8

Štajner, D., Popovic, B.M. and Boza, P. 2007. Evaluation of willow herb's (*Epilobium angustifolium* L.) antioxidant and radical scavenging capacities. Phytother. Res. **21(12)**: 1242-1245.

Štajner, D., Popovic, B.M., Boza, P. and Kapor, A. 2009. Antioxidant capacity of *Melampyrum barbatum* – weed and medicinal plant. Phytother. Res. **23(7)**: 1006-1010.

Štajner, D., Popovic, B.M., Calic-Dragonsavac, D., Malencic, D. and Zdrakovic-Koraac, S. 2011. Comparative study on *Allium schoenoprasum* cultivated plant and *Allium schoenoprasum* tissue culture organs antioxidant status. Phytother. Res. Doi: 10.1002/ptr.3394.

Štajner, D., Popovic, B.M., Canadanovic-Brunet, J. and Boza, P. 2006. Free radical scavenging activity of three *Equisetum* species (*Equisetum arvense, Equisetum ramosissimum* and *Equisetum telmateia*) from Fruška gora mountain. Fitoterapia. **77 (7-8)**: 601-604.

Štajner, D., Popovic, B.M., Canadanovic-Brunet, J. and Igic, R.S. 2008. Antioxidant and free-radical scavenging activities of *Allium roseum* and *Allium subhirsutum*. Phytother. Res. **22(11)**: 1469-1471.

Štajner, D., Popovic, B.M., Canadanovic-Brunet, J. and Stajner, M. 2008. Antioxidant and scavenger activities of *Allium ursinum*. Fitoterapia **79(4)**: 303-305.

Štajner, D., Popovic, B.M., Kapor, A., Boza, P. and Stajner, M. 2010. Antioxidant and scavenging capacity of *Anacamptis pyrimidalis* L. – Pyrimidal orchid from Vojvodina. Phytother. Res. **24**: 759-763.

Stangeland, T., Remberg, S. F. and Lye. K. A. 2009. Total antioxidant activity in 35 Ugandan fruits and vegetables. Food Chem. **113(1)**: 85-91.

Stanley Mainzen Prince, P., Padmanabhan, M. and Menon, V.P. 2004. Restoration of antioxidant defence by ethanolic *Tinospora cordifolia* root extract in alloxan-induced diabetic liver and kidney. Phytother. Res. **18(9)**: 785-787.

Stanner, S.A., Hughes, J., Kelly, C.N. and Buttriss, J.A. 2004. Review of the epidemiological evidence for the 'antioxidant hypothesis'. Public Health Nutrition **7**: 407-422.

Stanojevi, L., Stanakovi, Nikoli, V. D.and Nikoli, L. 2008. Anti-oxidative and antimicrobial activities of *Hieracium pilosella* L. extracts. J. Serb. Chem. Soc. **73**: 531-540.

Stanojevi, L., Stanakovi, Nikoli, V., Nikoli, L., Risti, D., Canadovic-Brunet, J. and Tumbas, V. 2009. Antioxidant activity and total phenolic and flavonoid contents of *Hieracium pilosella* L. extracts. Sensors **9**: 5702-5714.

Stashenko, E.E., Jaramillo, B.E. and Martínez, J.R. 2004. Comparison of different extraction methods for the analysis of volatile secondary metabolites of *Lippia alba* (Mill.) N.E. Brown, grown in Colombia, and evaluation of its *in vitro* antioxidant activity. J Chromatogr. A. **1025 (1)**: 93-103.

Stasi, L.C.D., Camuesco, D., Nietl, A., Vilegas, W., Zaryeski, A. and Galvez, J. 2004. Intestinal anti-inflammatory activity of Paepalantine an isocoumarin isolated from the capitula of *Paepalanthus bromelioides*, in the trinitrobenzenesulphonic acid model of rat colitis. Planta Medica **70(4)**: 315-320.

Steenkamp, V., Gouws, M.C., Gulumian, M., Elgorashi, E.E. and van Staden, J. 2006. Studies on antibacterial, anti-inflammatory and antioxidant activity of herbal remedies used in the treatment of benigen prostatic hyperplasia and prostates. J. Ethnopharmacol. **103(1)**: 71-75.

Steenkamp, V., Mathivha, V., Gouws, M.C. and van Rensburg, C.E.J. 2004. Studies on antibacterial, antioxidant and fibroblast growth stimulation of wound healing remedies from South Africa. J. Ethnopharmacol. **95**: 353-357.

Stevic, T., Savikin, K., Zdunic, G., Stanojkovic, T., Juranic, Z., Jankovic, T. and Menkovic, N. 2010. Antioxidant, cytotoxic and antimicrobial activity of *Alnus incana* L. ssp. *incana* Moench and *A.viridis* (Chaix.) DC. ssp. *viridis* extracts. J. Med. Food **13(3)**: 700-704. doi: 10.1089/jmf. 2009.0111.

Stoilova, I., Krastanov, A., Stoyanova, A., Denev, P. and Gargova, S. 2007. Antioxidant activity of a ginger extract (*Zingiber officinale*). Food Chem. **102 (3)**: 764-770.

Stojjkovic, S., Petrovic, S., Kukic, J., Dzamic, A., Ristic, M., Milenkovic, M., Glamoclija, J., Sokovic, M. and Stojkovic, D. 2009. Chemical composition and antimicrobial and antioxidant activity of *Seseli rigidum* flower essential oil. Chem. Nat. Compounds **45(2)**: 253-256.

Strack, D., Heilemann, J., Wray, V., Dirks, H. 1989. Structure and accumulation patterns of soluble and insoluble phenolics from Norway spruce needles. Phytochemistry **28**: 2071-2078.

Su, B.N, Pawlus, A. D., Jung, H.A., Keller, W. J., McLaughlin, J. L. and Kinghorn D. A. **2005.** Chemical constituents of the fruits of *Morinda citrifolia* (Noni) and their antioxidant activity. J. Nat. Prod. **68 (4)**: 592-595.

Su, C.R., Kuo, P.C., Wang, M.L., Liou, M.J., Damu, A.G. and Wu, T.S. 2003. Acetophenone derivatives from *Acronychia pedunculata*. J. Nat. Prod. **66(7)**: 990-993.

Su, L., Yin, J.-J., Charles, D., Zhou, K., Moore, J. and Yu, L. 2007.Total phenolic contents, chelating capacities, and radical-scavenging properties of black peppercorn, nutmeg, rosehip, cinnamon and oregano leaf. Food Chem. **100**: 990-997.

Su, M.S. and Silva, J.L. 2006. Antioxidant activity, anthocyanins, and phenolics of rabbiteye blueberry (*Vaccinium ashei*) by-products as affected by fermentation. Food Chem. **97(3)**: 447-451.

Su, X.-Y., Wang, Z.-Y. and Liu, J.R. 2009. *In vitro* and *in vivo* antioxidant activity of *Pinus koraiensis* seed extract containing phenolic compounds. Food Chem. **117(4)**: 681-686. doi: 10.1016/j.foodchem.2009.04.076

Su, Y.L., Chen, R., and Chen, Z.Y. 2004. Studies on antioxidant constituents from black tea. J. Chinese medicinal materials **27**: 732-743.

Su, Y.L., Huang, Y. and Chen, Z.Y.2004. Isolation and elucidation of antioxidant constituents from acetone extract in root of *Scutellaria rehderiana*. China J. Chinese Materia Medica **29**: 863-864.

Su, Y.L., Leung, L.K., Bi, Y.R., Huang, Y. and Chen, Z.Y. 2000. Antioxidant activity of flavonoids isolated from *Scutellaria rehderiana*. J. Am. Chem. Soc. **77**: 807-812.

Suanarunsawat, T., Ayutthya, W.D.N., Songsak, T., Thirawarapan, S. and Poungshompoo, S. 2010. Antioxidant activity and lipid-lowering effect of essential oils extracted from *Ocimum sanctum* L. leaves in rats fed with a high cholesterol diet. J. Clin. Biochem. Nutr. **46**: 52-59.

Subhan, N., Alam, M.A., Ahmed, F., Awal, M.A. Nahar, L. and Sarker, D.S. 2008. *In vitro* antioxidant property of the extract of *Excoecaria agallocha* (Euphorbiaceae). DARU **16(3)**: 149-154.

Subhasree, B., Bhaskar, R., Keerthana R. Laxmi, Susan R. Lijina and Rajasekharan, P.2009. Evaluation of antioxidant potential in selected green leafy vegetables. Food Chem. **115(4)**: 1213-1220.

Suboh, S.M., Bilto, Y.Y. and Aburjai, T.A. 2004. Protective effect of selected medicinal plants against protein degradation, lipid peroxidation and deformability loss of oxidatively stressed human erythrocytes. Phytother. Res. **18**: 280-284.

Subramanian, M., Chintalwar, G.J. and Chattopadhyay, S., 2002. Antioxidant properties of a *Tinospora cordifolia* polysaccharide against iron-mediated lipid damage and γ-ray induced protein damage. Redox Report **7**: 137-143.

Subramanian, P., Sendaresan, S. and Manivasagam, T. 2005. Influence of garlic extract on temporal characteristics of lipid peroxidation products and antioxidants in tumor bearing rats. Pharm. Biol. **43(3)**: 209-218.

Sudati, J.H., Fachinetto, R., Pereira, R.P., Boligon, A.A., Athayde, M.L., Soares, F.A., Vargas, Barbosa, N.B. and Rocha, J.B.T. 2009. *In vitro* antioxidant activity of *Valeriana officinalis* against different neurotoxic agents. Neurochemical Res. **34(8)**: 1372-1379.

Sudheesh, S., Sandhya, C., Koshy, S. and Vijayalakshmi, N.R. 199. Antioxidant activity of flavonoids from *Solanum melongena*. Phytother. Res. **13(5)**: 393-396.

Sudjaroen, Y., Haubner, R., Wurtele, G., Hull, W.E., Erben, G., Spiegelhalder, B., Changbumrung, S., Bartsch, H. and Owen, R.W. 2005. Isolation and structure elucidation of phenolic antioxidants from Tamarind (*Tamarindus indica* L.) seeds and pericarp. Food Chem. Toxicol. **43**: 1673–1682.

Suganthy, N., Kesika, P., Karutha, S., Kasi, P. and Devi, P. 2009. Mangrove plant extracts: radical scavenging activity and the battle against food-borne pathogens. Forsch Komplementmed 16(1) 41-48.

Sugie, S., Okanoto, K., Rahman, K.M.W., Tanaka, T., Kawai, K., Yamahara, J. and Mori, H. 1998. Inhibitory effects of plumbagin and juglone on azoxymethane induced intestinal carcinogenesis in rats. Cancer Letters **127**: 177-183.

Sugiura, M., Ohshima, M., Ogawa, K. and Yano, M. 2006. Chronic administration of Satsuma mandarin fruit (*Citrus unshiu* Marc.) improves oxidative stress in streptozotocin-induced diabetic rat liver. Biol. Parm. Bull. **29(3)**: 588-591.

Suhaj, M. 2006. Spice antioxidants isolation and their antiradical activity: a review. J. Food Compos. Analy. **19(6-7)**: 531-537.

Suhaj, M., Racova, J., Polovka, M. and Brezova, V. 2006. Effect of ±c-irradiation on antioxidant activity of black pepper (*Piper nigrum* L.). Food Chem. **97 (4)**: 696-704.

Suja, K.P., Jaylekshmy, A. and Arumughan, C. 2004. Free radical scavenging behavior of antioxidant compounds of sesame (*Sesamum indicum* L.) in DPPH system. J. Agric. Food Chem. **52(4)**: 912-915.

Sujatha, R. and Srinivas, L. 1995. Modulation of lipid peroxidation by dietary components. Toxicol. *In vitro* **9**: 231-236.

Sukesh, Hidayathulla, S., Haneef, M., Aruneekumar, K. and Chandrasekhar, K.R. 2011. Phytochemical evaluation, antioxidant and antibacterial activity of seed wings of *Hopea ponga* (Dennst.) Mabberly. J. Pharm. Res. **4(8)**:

Suksamran, A., Poomsing, P., Aroonrerk, N., Punjanon, T., Suksamran, S. and Kongkun, S. 2003. Antimycobacterial and antioxidant flavones from *Limonophila geoffrayi*. Arch. Pharm. Res. **26(10)**: 816-820.

Sulaiman, M.M., Bagla, V., Naidoo, V. and Eloff, J.N. 2010. Evaluation of selected South African plant species for antioxidant, antiplatelet and cytotoxic activity. Pharmaceut. Biol. **49**: 1-8

Sulaiman, S.F., Shuang, K.H. and Yusof, S.R. 2010. The antioxidant effect of *Carissa carandas* unripe fruit extracts and fractions. eprints.usm.my/./ The_**Antioxidant**_effects_of_**Carissa**_carandas_unripe_fruit_extracts_and_fractions.pdf

Sulanova, N., Makhmoor, T., Yasin, A., Abilov, Z.A., Omurkamzinova, V.B., Atta-ur-Rahman and Choudhary, M.I. 2004. Isotamarixen – A new antioxidant and prolylendopeptidase-inhibiting triterpenoid from *Tamarix hisida*. Planta Medica **70(1)**: 65-67.

Sultana, S., Parwaiz, S. Iqbal, M. and Athar, M. 1995. Crude extracts of hepatoprotective plants, *Solanum nigrum* and *Cichorium intybus* inhibit free-radical mediated DNA damage. J. Ethnopharmacol. **45**: 189-192.

Suma, P.F. and Urooj, A. 2011. Antioxidant activity of extracts from foxtail millet (*Setaria italica*). J. Food Sci. Tech. doi: 10.1007/s13197-011-0300-9

Sumanth, M. and Rana, A.C. 2006. *In vivo* antioxidant activity of hydroalcoholic extract of *Taraxacum officinale* roots in rats. Indian J. Pharmacol. **38**: 54-55.

Sumino, M., Sekine, T., Ruangrungsi, N., Igarashi, K. and Ikegami, F. 2002. Ardissiphenols and antioxidant principles from the fruits of *Ardisia clorata*. Chem. Pharm. Bull. (Tokyo) **50(11)**: 1484-1487.

Sumino, M., Sekine, T., Ruangrungsi, N. and Ikegami, F. 2001. Ardisphenols A-C, novel antioxidants from the fruits of *Ardisia colorata*. Chem. Pharmceutical Bull. 49(12): 1664-1665.

Sun, J., Chu, Y.F., Wu, X.Z. and Liu, R.H. 2002. Antioxidant and antiproliferative activities of common fruits. J. Agric. Food Chem. **50(25)**: 7449-7454.

Sun, J, He, H, and Xie, B.J. 2004. Novel antioxidant peptides from fermented mushroom *Ganoderma lucidum*. J. Agric. Food Chem. **52(21)**: 6646-6652.

Sun, J., Jiang, Y., Shi, J., Wei, X., Xue, J., Shi, J. and Yi, C. 2010. Antioxidant activities and contents of polyphenol oxidase substrates from pericarp tissues of litchi fruit. Food Chem. **119(2)**: 753-757.

Sun, T. and Ho, C.T. 2005. Antioxidant activities of buckwheat extracts. Food Chem. **90**: 743-749.

Sun, T., Simon, P.W., Tanumihardjo, S.A. 2009. Antioxidant phytochemicals and antioxidant capacity of Biofortified carrots (*Daucus carota* L.) of various colors. J. Agric. Food Chem. **57(10)**: 4142-4147.

Sun, W., Xu, Z., Wang, C., Qu, W. and Lin, C. 2005. Study on antioxidant activity of essential oils and its monomer from *Pelargonium graveolens*. Zhong Yao Cai **28(2)**: 87-89.

Sundaram, R. and Mitra,.K. 2006. Antioxidant activity of *Caesalpinia digyna* in *in vitro* and *in vivo* experimental models. Antiseptic **103(9)**: 522-528.

Sundaram, R. and Mitra, S.K. 2007. Antioxidant activity of ethyl acetate soluble fraction of *Acacia arabica* in rats. Indian J. Pharmacol. **39**: 33-38.

Sundararajan, R., Haja, N.A, Venkatesan, K., Mukherjee, K., Saha, B.P., Bandyopadhyay, A. and Mukherjee, P.K. 2006. *Cytisus scoparius* Link–a natural antioxidant. BMC Complement. Altern. Med. **16**: 6: 8.

Sundararaju, D., Krishnaraju, A.V., Rao, C.V. and Trimurtulu. 2010. Antioxidant activity and brine shrimp lethality of *Punica granatum* fruit rind. Asian J. Chem. **22(2)**: 1628-1630.

Suneetha, B., Pavan Kumar, P., Prasad, K.V.S.R.G., Vidydhara and Sambasiva Rao, K.R.S. 2011. Hepatoprotective and antioxidant activities of methanolic extract of *Mimosa pudica* roots against carbon tetrachloride induced hepatotoxicity in albino rats. Int. J Pharm. **1(1)**: 46-53.

Sunil, C. and Ignacimuthu, S. 2011. *In vitro* and *in vivo* antioxidant activity of *Symplocos cochinchinensis* S. Moore leaves containing phenolic compounds. Food Chem. Toxicol. **49(7)**: 1604-1607. doi: 10.1016/j.fct.2011.04.010

Sunilson, J.A.J., Jayaraj, P., Syam Mohan, M., Kumari, A.A.G. and Varatharajan, R. 2008. Antioxidant and hepatoprotective effect of the roots of *Hibiscus esculentus* Linn. Int. J. Green Pharmacy **2(4)**: 200-203.

Sunilson, J.A.J., Varatharajan, R., Jayaraj, P., John, T., Jisha, J. and Promwichit, P. 2008. Gastroprotective and antioxidant activities of the roots of *Hibiscus aculeatus* Roxb. in rats. Int. J. Pharmacol. **4(4)**: 252-257.

Sunitha, S., Nagaraj, M. and Varalakshmi, P. 2001. Hepatoprotective effect of lupeol and lupeol lineolate on tissue antioxidant defense system in cadmium-induced hepatotoxicity in rats. Fitoterapia **72**: 516-523.

Sunitha, T., Gupta, V. and Bhattacharya, S. 2000. Comparative study of the antioxidant potential of tea with and without additives. Indian J. Physiol Pharmacol. **44**: 215-219.

Sur, P., Chaudhuri, T., Vedasiromani, j.R., Gomes, A. and Ganguly, D.K. 2001. Antiinflammatory and antioxidant property of saponins of tea [*Camellia sinensis* (Linn.)O.Kuntze] root extract. Phytother. Res. **15(2)**: 174-176.

Suresh, K., Sivakumar, K., Vijayanand, M.A., Rajalingam, K. and Rajkamal, G. 2010. Antilipoperoxidative and antioxidant effects of *Zingiber officinale* Roscoe root extract in 7, 12-Dimethyl Benz(A) anthracene induced oral carcinogensis. Pharmacology online **2**: 689-701.

Suresh, R., Saravanakumar and Suganyadevi, P. 2011. Anthocyanins from Indian cassava (*Manihot esculenta* Crantz) and its antioxidant properties. Int. J. Pharm. Sci. Res. **2(7)**: 1819-1828.

Suresh Kumar, P., Sucheta, S., Deepa, V.S., Selvamani, P. and Latha, S. 2008. Antioxidant activity in some selected Indian medicinal plants. African J. Biotechnol. **7(12)**: 1826-1828.

Surh, Y. 1999. Molecular mechanisms of chemopreventive effects of selected dietary and medicinal phenolic substances. Shinlim-dong, Kwanak-gu, Seoul, South Korea. Mut. Res. **428**: 305.

Surh, Y.Z. and Ferguson, L.R. 2003. Dietary and medicinal antimutagens and anticarcinogens: molecular mechanisms and chemopreventive potential. Highlight of a symposium.

Surveswaran, S., Cai, Y.-Z., Corke, H., Sun, M. 2007. Systematic evaluation of natural phenolic antioxidants from 133 Indian medicinal plants. Food Chem. **102(3)**: 938-953.

Surveswaran, S., Cai, Y.-Z., Xing, J., Corke, H. and Sun, M. 2010. Antioxidant properties and principal phenolic phytochemicals of Indian medicinal plants from Asclepiadoideae and Periplocoideae. Nat. Prod. Res. **24(3)**: 206-221.

Sushruta, K. 2002. Antioxidant activity of *Acacia nilotica* bark. Asian J. Chem. **14(1)**: 523-525.

Susithra, E., Mallikharjuna Rao, K., Rameshu, K.V. and Meena, S. 2011. Evaluation of *in vitro* antioxidant activity of isolated compounds of lichen *Usnea undulata*. J. Pharm. Res. **4(2)**:

Suthar, M., Rathore, G.S. and Pareek, A. 2009. Antioxidant and antidiabetic activity of *Helicteres isora* L. fruits. Indian J. Pharm. Sci. **71(6)**: 695-699.

Suzec, S., Mericli, A.H., Houghton, P.J. and Cubukcu, B. 2005. Flavonoids of *Helichrysum compactum* and their antioxidant and antibacterial activity. Fitoterapia **76(2)**: 269-272.

Szollosi, R. and Szollosi Varga, I. 2002. Total antioxidant power in some species of Labiatae (Adaptation of FRAP method). Acta Biol. Szeged. **46**: 125-127.

Tachakittirungrod, S., Ikegami, F. and Okonogi, F. 2007. Antioxidant active principles isolated from *Psidium guajava* grown in Thailand. Scientia Pharmaceutica **75**: 179-193.

Tadhani, M.B., Patel, V.H. and Sushash, R. 2007. *In vitro* antioxidant activities of *Stevia rebaudiana* leaves and callus. J. Food Comp. Anal. **20(3-4)**: 323-329. doi: 10.1016/j.fca.2006.08.004.

Taga, M.S., Miller, E.E. and Pratt, D.E. 1984. Chia seeds as a source of natural lipid antioxidants. J. Amer. Oil Chem. Soc. **61**: 928-931.

Takacsova, M., Pribela, A. and Faktorova, M. 1995. Study of the antioxidant effects of thyme, sage, juniper and oregano. Nahrung **39**: 241-243.

Takagi, T. and Iida, T. 1980. Antioxdant for fats and oils from canary seed: Sterol and triterpene alcohol esters of caffeic acids. J. Am. Oil Chem. Soc. **57**: 326-329.

Takahashi, H., Hirata, S., Minami, H. and Fukuyama, Y. 2001. Triterpene and flavonone glycoside from *Rhododendron simsii*. Phytothemistry **56**: 875-879.

Takahata, Y., Ohnishi-Kameyama, M., Furuta, S., Takahashi, M. and Suda, I. 2001. Highly polymerized procyanidins in brown soybean seed coat with a high radical-scavenging activity. J. Agric. Food Chem. **49**: 5843-5847.

Tamura, H. and Yamagami. 1994. Antioxidative activity of monoacylated and anthocyanins isolated from Muscat Bailey A grape. J. Agric. Food Chem. **42**: 1612-1615.

Tan, S.-K., Osman, H., Wong, K.-C., Boey, P.-L. and Ibrahim, P. 2009. Antimicrobial and antioxidant activities of *Swietenia macrophylla* leaf extracts. Asian J. Food Ag-Ind. **2(2)**: 181-188.

Tanas, S., Odabasoglu, F., Halici, Z., Cakir, A., Aygun, H., Aslam,. and Suleyman, H. 2010. Evaluation of anti-inflammatory and antioxidant activities of *Peltigera rufescens* lichen species in acute and chronic inflammation models. J. Nat. Medicines **64(1)**: 42-49.

Tang, J.S., Gao, H., Wang, C.X., Dal, Y., Bao, L., Ye, W.C. and Yao, X.S. 2009. Antioxidative phenylpropanoid substituted epicatechin glycosides from *Parabarium huaitingii.* Planta Medica **75(15)**: 1586-1590.

Tang, S.Y., Whiteman, M., Peng, Z.F., Jenner, A., Yong, E.L. and Halliwell, B. 2004. Characterization of antioxidant and antiglycation properties and isolation of active ingredients from traditional Chinese medicines. Free Radic. Biol.Med. **36(12)**: 1575-1587.

Tang, X. and Yuan, H.M. 1997. Extraction of *Salvia sclarea* L. oil and its antioxidative activity in foods. Science and Technology of Food Industry **3**: 10-13.

Tanizawa, H., Ohkawa, Y., Takino, Y., Miyase, T., Ueno, A., Kageyama, T. and Hara, S. 1992. Studies on natural antioxidants in *Citrus* species. I. Determination of antioxidative activities of *Citrus* fruits. Chem. Pharm. Bull. **40**: 1940-1942.

Taniazawa,H., Toda, S., Sazuka, Y., Taniyama, T., Hayashi, T., Arichi, S. and Takino, Y. 1984. Natural antioxidants I: Antioxidative components of tea leaf (*Thea sinensis* L.). Chemical Pharmac. Bull. **32**: 2011-2014.

Tanti, B., Buragohain, A.K., Gurun, L., Kakati, D., Das, A.K. and Borah, S.P. 2010. Assessment of antimicrobial and antioxidant activities of *Dendrocnide sinuata* (Blume) Chew leaves – a medicinal plant used by ethnic communities of North East India. Indian J. Nat. Prod. Resource. **1(1)**: 17-21.

Tanwer, B.S., Choudhary, R. and Vijayvergia, R. 2010. *In vitro* and *in vivo* comparative study of primary metabolites and antioxidant activity in *Spilanthes acmella* Murr. Int. J. Biotechnol. Biochem. **6(5)**: 819-825.

Tapia, A., Rodriguez, J., Theoduloz, C., Lopez, S., Feresin, G.E. and Schmeda-Hischmann, G. 2004. Free radical scavengers and antioxidants from *Baccharis grisebachi.* J. Ethnopharmacol. **95**: 155-161.

Tapondjou, L.A., Talat, M., Iqbal, C.M., Lontsi, D., Sondengam, L.C. and Atta ur Rahman. 2005. Bancoensone, a new antioxidant caffeoyl derivative from *Cussonia bancoensis*. Cameroon J. Exp. Biol. **1(1)**: 57-61.

Tarhan, L., Kayali, H. A. and Urek, R. O.2007. *In vitro* antioxidant properties of *Cucurbita pepo* L. male and female flowers extracts. Plant Foods for Human Nutrition. **42**: 49-51.

Tariq, M. and Al-Yahya, M.A. 1983. Effect of vitamin E on endotoxin shock induced changes in cardiac structure, function and metabolism. Toxicon **3**: 449-451.

Taruscio, T.G., Barney, D.L. and Exon, J. 2004. Content and profile of flavonoid an phenolic acid compounds in conjunction with the antioxidant capacity for a variety of Northwest *Vaccinium* berries. J. Agric. Food Chem. **52(10)**: 3169-3176.

Tasdemir, D., Donmez, A.A., Calis, I. and Ruedi, P. 2004. Evaluation of biological activity of Turkish plants. Rapid screening for the antimicrobial, antioxidant and acetylcholinesterase inhibitory potential by TLC bioautographic methods. Pharmaceutical Biol. **42(4-5)**: 374-383.

Tasduq, S.A., Reeta, D., Sandeep, S.P., Johri, R.K. and Qazi. G.N. 2007. Antioxidant and cytoprotective activity of water extract from *Labisia pumila*. Proc. Conference – WHAT Medicine pp. 213.

Tateyama, C., Honma, N., Namiki, K. and Uchiyama, T. 1997. Polyphenol content and antioxidant activity of various flower petals. Nippon Shokuhin Kagaku Kogaku Kaishi **44**: 290-299. (in Japanese).

Tatiya, A.U., A.S.Kulkarni, S.J.Surana and N.D.Bari. 2010. Antioxidant and hypolipidemic effect of *Caralluma adscendens* Roxb. in alloxanized diabetic rats. Int. J. Pharmacol. **6**: 400-406. doi: 10.3923/ijp.2010.400.406.

Tatiya, A.U., Saluja, A.K. and Surana, S.J. 2008. Evaluation of antioxidant and anti-inflammatory activity of stem bark of *Bridelia airy-shawii* in rats. J. Nat. Remedies **8(1)**: 18-23.

Tatiya, A.U., Ushir, Y.V. and Surana, S.J. 2007. Antioxidant, analgesic and antipyretic activity of essential oil from *Anisomeles indica* (Labiatae) on experimental animal model. J. Res. Education in Indian Medicine **13(4)**: 39-48.

Tatlidede, E., Sehrli, O., Velioglu-Ogunc, A., Cetinel, S., Yegen, B.C., Yarat., A., Suleymanoglu, S. and Sener, G. 2009. Resveratrol treatment protects against doxorubicin-induced cardiotoxicity by alleviating oxidative damage. Free Radical Res. **43(3)**: 195-205.

Tawaha, K., Alali, F.Q., Gharaibeh, M., Mohammad, M. and El-Elimat, T. 2007. Antioxidant activity and total phenolic content of selected Jordanian plant species. Food Chem. **104**: 1372-1378.

Tedesco, A. C., Gomes, A. J., Bonilha, J. B. S. and Gonzalez, S. 1997. The antioxidant action and royal jelly, *Polypodium leucotomos* and kojic acid, as scavenger of reactive oxygen species. Photochem. Photobiol. **65**: 30S.

Teffo, L.S., Aderogba, M.A. and Eloff, J.N. 2010. Antibacterial and antioxidant activities of four kaempferol methyl ethers isolated from *Dodonaea viscosa* Jacq. var. *angustifolia* leaf extracts. S Afr. J. Bot. **76**: 25-29.

Teissedre, P. and Waterhouse, A. 2000. Inhibition of oxidation of human low-densitty lipoprotyeins by phenolic substances in different essential oils varieties. J. Agric. Food Chem. **48**: 3801-3805.

Telci, I., Elmastas, M. and Shahin, A. 2009. Chemical composition and antioxidant activity of *Ocimum minimum* essential oils. Chem. Nat. Compounds **45(4)**: 568-570.

Temraz., A. and El-Tantaway, W.H. 2008. Characterization of antioxidant activity of extract from *Artemisia vulgaris*. Pak. J. Pharm. Sci. **21(4)**: 321-326.

Tenpe, C.R., Upaganlawar, A.B., Mane, G.U. and Yeole, P.G. 2008. *In vitro* antioxidant activity of *Pongamia pinnata* Linn. leaf extracts. Biomed. **3(1)**: 66-72.

Tepe, B. 2007. Antioxidant potentials and rosmarinic acid levels of the methanolic extracts of *Salvia virgata* (Jacq.), *Salvia staminea* (Monbret and Aucher ex Bentham) and *Salvia verbenaca* (L.) from Turkey. Bioresource Technol. **99(6)**: 1584-1588.

Tepe, B., Akpulat, H.A.and Sokmen, M. 2011. Evaluation of the chemical composition and antioxidant activity of the essential oils of *Peucedanum longifolium* (Waldst. and Kit.) and *P. palimboides* (Boiss.) from Turkey. Records of Natural Products **5**: 108-116.

Tepe, B., Akpulat, H.A., Sokmen, M., Daferera, D., Yumrutas, O., Aydin, E., Polissiou, M and Sokmen, A. 2006. Screening of the antioxidant and antimicrobial properties of the essential oils of *Pimpinella anisetum* and *Pimpinella flabellifolia*. Food Chem. **97(4)**: 719-724.

Tepe, B., Daferera, D., Sokmen, M., Polissiou, M. and Sokmen, A. 2004. *In vitro* antimicrobial and antioxidant activities of the essential oils and various extracts of *Thymus eigii* M.Zohary et P.H.Davis. J. Agric. Food Chem. **52(5)**: 1132-1137.

Tepe, B., Daferera, D., Sokmen, A., Sokmen, M. and Polissiou, M. 2005. Antimicrobial and antioxidant activities of the essential oil and various extracts of *Salvia tomentosa* Miller (Lamiaceae). Food Chem. **90**: 333-340.

Tepe, B., Daferera, D., Sokmen, M., Polissou, M. and Sokmen, A. 2004. The *in vitro* antioxidant and antimicrobial activities of the essential oil and various extracts of *Origanum syriacum* L. var. *bevanii* (Holmes) Letswart. J. Sci. Food Agric. **84**: 1389-1396.

Tepe, B., Daferera, D., Tepe, A.-S., Polissious, M. and Sokmen, A. 2007. Antioxidant activity of the essential oil and various extracts of *Nepeta flavida* Hub.-Mor. from Turkey. Food Chem. **103(4)**: 1358-1364.

Tepe, B., Degerli, S., Arslan, S., Malatyali, E. and Sarikurkcu, C. 2011. Determination of chemical profile, antioxidant, DNA damage protection and antiamoebic activities of *Teucrium polium* and *Stachys iberica*. Fitoterapia **82**: 237-246.

Tepe, B., Donmez, E., Unlu, M. Candna, F., Daferera, D., Vardar-Unlu, G. Polissiou, M. and Sokmen, A. 2004. Antimicrobial and antioxidative activities of the essential oils and methanol extracts of *Salvia cryptantha* Montbret et Aucher ex Benth. and *Salvia multicaulis* Vahl. Food Chem. **84**: 519-525.

Tepe, B., Eminagaoglu, O., Akpulat, H.A. and Aydin, E. 2007. Antioxidant potentials and rosmarinic acid levels of the methanolic extracts of *Salvia verticillata* L. subsp. *verticillata* and *S. verticillata* subsp. *amasiaca* (Freyn et Bornm.) Bornm. Food Chem. **100**: 985-989.

Tepe, B., Gursoy, N. and Akpulat, H.A.2012. Chemical composition and antioxidant activity of the essential oils of *Salvia palaestina* and *S. ceratophylla*. Record of Natural Products, Basimda.

Tepe, B., Sarikurkcu, C., Berk,., Alim, A. and Akpulat, H.A. 2011. Chemical composition, radical scavenging and antimicrobial activity of the essential oils of *Thymus bovii* and *Thymus hyemalis*. Rec. Nat. Prod. **5**: 208-220.

Tepe, B., Sihoglu Tepe, A., Daferera, D., Polissiou, M. and Sokmen, A. 2007. Chemical composition and antioxidant activity of the essential oil of *Clinopodium vulgare* L. Food Chem. **103**: 766-770.

Tepe, B. and Sokmen, A. 2007. Screening of the antioxidative properties and total phenolic contents of three endemic *Tanacetum* subspecies from Turkish flora. Bioresource Tech. **98**: 3076-3079.

Tepe, B., Sokman, M., Akpulat, H.A and Sokmen, A. 2005. *In vitro* antioxidant activities of the methanol extracts of four *Helichrysum* species from Turkey. Food Chem. **90**: 673-677.

Tepe, B., Sokman, M., Akpulat, H.A and Sokmen, A. 2005. *In vitro* antioxidant activities of the methanol extracts of five *Allium* species from Turkey. Food Chem. **92**: 89-92.

Tepe, B., Sokman, M., Akpulat, H.A., Yumrutas, O. and Sokmen, A. 2006. Screening of antioxidative properties of the methanolic extracts of *Pelargonium endlicherianum* Fenzl., *Verbascum wiedemannianum* Fisch. et Mey., *Sideritis libanotica* Labill. subsp. *linearis* (Bentham) Borm., *Centaurea mucronifera* DC. and *Hieracium cappadocicum* Freyn from Turkish flora. Food Chem. **98(1)**: 9-13. doi: 10.1016/j.foodchem.2005.05.046.

Tepe, B., Sokman, M., Akpulat, H.A. and Sokmen, A. 2006. Screening of antioxidant potentials of six *Salvia* species from Turkey. Food Chem. **95**: 200-204.

Tepe, B., Sökmen, M., Akpulat, H.A., Daferera, D., Polissiou, M. and Sökmen, A. 2005. Antioxidative activity of the essential oils of *Thymus sipyleus* subsp. *sipyleus* var. *sipyleus* and *Thymus sipyleus* subsp. *sipyleus* var. *rosulans*. J. Food Eng. **66(4)**: 447-454.

Tepe, B., Sökmen, M., Sökmen, A., Daferera, D. and Polissiou, M. 2005. Antimicronial and antioxidant activities of the essential oil and various extracts of *Cyclotrichum origanifolium* (Labill.) Manden et Scheng. J. Food Eng. **69**: 335- 342..

Terahima, K., Takaya, Y. and Niwa M. 2002. Powerful antioxidative agents based on Garcinoic acid from *Garcinia kola*. Biorg. Med. Chem. **10**: 1619–1625.

Terao, J., Karasawa, H., Arai, H., Nagao, A., Suzuki, T. and Takama, K. 1993. Peroxyl radical scavenging activity of caffeic acid and its related phenolic compounds in solution. Biosci. Biotechnol. Biochem. **57**: 1204-1205.

Termentzi, A., Kefalas, P. and Kokkalou, E. 2006. Antioxidant activities of various extracts and fractions of *Sorbus domestica* fruits at different maturity stages. Food Chem. **98**: 599-608.

Tesoriere, L., Butera, D., Gentile, C. and Liverea, M.A. 2007. Bioactive components of Caper (*Capparis spinosa* L.) from Sicily and antioxidant effects in a red meat simulated gastric digestion. J. Agric. Food Chem. **55(21)**: 8465-8471.

Tewari, S.K., Niranjan, A. and Lehri, A. 2010. Variations in yield, quality, and antioxidant potential of kalmegh (*Andrographis paniculata* Nees) with soil alkalinity and season. J. Herbs Spices Med. Plants **16(1)**: 41-50.

Thabrew, M.I., Hughes, R.D., McFarlane, I.G. 1998. Antioxidant activity of *Osbeckia aspera*. Phytother. Res., **12**: 288-290.

Thaipong, K., Boonprakob, U., Crosby, K., Cisneros-Zevallos, L. abd Byrne, D.H. 2006. Comparison of ABTS, DPPH, FRAP and ORAC assays for estimating antioxidant activity from guava fruit extracts. J. Food Compos. Anal. **19**: 669-675.

Thakare, V.N., Suralkar, A.A., Deshpande, A.D. and Naik, S.R. 2010. Stem bark extraction of *Ficus benghalensis* Linn. for anti-inflammatory and analgesic activity in animal models. Indian J. Exp. Biol. **48(1)**: 39-45.

Tharakan, B., Dhanasekaran, M. and Manyam, B.V. 2005. Antioxidant and DNA protecting properties of anti fatigue herb *Trichopus zeylanicus*. Phytother. Res. **19**: 669-673.

Thatte, U., Bagadey, S. and Dahanukumar, S. 2000. Modulation of programmed cell death by medicinal plants. Cell Mol. Biol. Noisy le grnd **46**: 199.

Then, A., Szentmihalyi K., Gere A., Jasztrab S. and Szoke, E. 2005. Antioxidant properties of *Myrtilli folium, Phaseoli fructus* sine semnibus and drug mixture extracts. Acta Alimentaria **34**: 169–176.

Then, M., Szentmihalyi, K., Agnes Sarkozi and Varga, I.S. 2003. Examination on antioxidant activity in the greater celandine (*Chelidonium majusL.*) extracts by FRAP method. Acta Biologia Szegediensis **47(1-4)**: 115-117.

Thirugnanasampandan, R. Mahendran, G. and Narmatha Bai, V. 2008. Antioxidant properties of some medicinal Aristolochiaceae species. African J. Biotechnol. **7 (4)**: 357–361.

Thitilertdecha, N., Teerawutgulrag, A., Kiburn, J.D. and Rakarinyatham, N. 2010. Identification of major phenolic compounds from *Nephelium lappaceum* L. and their antioxidant activities. Molecules **15**: 1453-1465.

Thitilertdecha, N., Teerawutgulrag, A. and Rakarinyatham, N. 2008. Antioxidant and antibacterial activities of *Nephelium lappaceum* L. extracts. LWT Food Sci. Tech. **41(10)**: 2029-2035.

Thomas, R., Sah, N. and Sharma, P. 2008. Therapeutic biology of *Jatropha curcas*: A mini review. Curr. Pharm. Biotechnol. **9(4)**: 315-324.

Thongphasuk, P., Suttisri, R., Bavovada, R. and Verpoorte, R. 2004. Antioxidant lignan glucosides from *Strychnos vanprukii*. Fitoterapia **75**: 623-628.

Thuong, P.T., Hung, T.M., Ngoc, T.M., Ha, D.T., Min, B.S., Kwack, S.J., Kang, T.S., Choi, J.S. and Bac, K. 2010. Antioxidant activities of coumarins from Korean medicinal plants and their structure-activity relationships. Phytother. Res. **24(1)**: 101-106.

Thuong, P.T., Na, M.K., Dang, N.H., Huang, T.M., Ky, P.T., Thanh, T.V., Nam, N.H., Thuan, N.D., Sok, D.E. and Bae, K.H. 2006. Antioxidant activities of Vietnamese medicinal plants. Nat. Prod. Sci. **12(1)**: 29-37.

Thuong, P.T., Pokharel, Y.R., Lee, M.Y., Kim, S.K., Bae, K.H., Su, N.D., Oh, W.K. and Kang, K.W. 2009. Dual antioxidative effect of fraxetin isolated from *Fraxinus rhinchophylla*. Biol. Pharm. Bull. **32(9)**: 1527-1532.

Tian, L.L. and White, P.J. 1994. Antioxidant activity of oat extract in soybean and cotton seed oils. J. Am. Oil Chem. Soc. **70**: 1079-1085.

Tibri, A., Sawadogo, R.W. and Ouedraogo, N. 2010. Evaluation of antioxidant activity, total phenolic and flavonoid contents of *Entada africana* organ extracts. Res. J. Med. Sci. **4(2)**: 81-87.

Tieppo, M., Porawsu, M., Salvador, M., Moreira, A.J., Collado, P.S., Gonzalez-Gallego, J. and Marroni, N.P. 2006. *Croton cajucara* Benth. leaf extract scavenges the stable free radical DPPH and protects against oxidative stress induced by paraquat. Biol. Pharm. Bull. **29(1)**: 161-165.

Tiew, P., Ioset, J.R., Kokpol, U., Chavasiri, W. and Hostettmann, K. 2003. Antifungal, antioxidant and larvicidal activities of compounds isolated from the heartwood of *Mansonia gagei*. Phytother. Res. **17(2)**: 190-193.

Tilak, J.C., Adhikari, S. and Devasagayam, T.P.A. 2004. Antioxidant properties of *Plumbago zeylanica* – an Indian medicinal plant and its active ingredient plumbagin. Redox Rep. **9(4)**: 219-227.

Tilak, J.C., Banerjee, M., Mohan, H. and Devasagayam, T.P.A. 2004. Antioxidant availability of turmeric in relation to its medicinal and culinary uses. Phytother. Res. **18(10)**: 798-804.

Tilak, J.C., Devasagayam, T.P.A. and Adhikari, S. 2003. Radioprotective and antioxidant properties of Indian medicinal plant *Terminalia arjuna*. BARC Newsletter **249**: 167-175.

Tilak-Jain, J. 2005. Studies of the antioxidant effects and other beneficial properties of natural compounds from Indian medicinal plants. Ph.D. thesis, Mumbai University.

Tirgar, P.R., Jadhav, P.D., Sheth, D.B. and Desai, T.R. 2010. Therapeutic role of antioxidant properties of *Emblica officinalis* (amla) in streptozotocin induced Type I diabetic rats. Pharmacology online 1:

Tiwari, A.K., Srinivas, P.V., Kumar, S.P. and Rao JM. 2001. Free radical scavenging active components from *Cedrus deodara*. J Agric Food Chem. **49(10)**: 4642-4645.

Tiwari, B.K. and Khosa, R.L. 2009. Hepatoprotective and antioxidant effect of *Sphaeranthus indicus* against acetaminophen induced hepatotoxicity in rats. J. Pharm. Sci. Res. **1(2)**: 26-30.

Tiwari, S. 2010. *Stevia rebaudiana*: A medicinal and nutraceutical plant and sweet gold for diabetic patients. Int. J. Pharm. Life Sci. **1(8)**: 451-457.

Tlili, I., Hdider, C., Lenucci, M.S., Riadh, I., Jebari, H. and Dalessandro, G. 2010. Bioactive compounds and antioxidant activities of different watermelons (*Citrullus lanatuss* (Thunb.) Mansfeld) cultivars as affected by fruit sampling area. J. Food Composition and Analysis (in press). Doi: 10.1016/j/jfca.2010.06.005.

Toda, S., Miyase, T., Arichi, H., Tanizawa, H. and Takino, Y. 1985a. Natural antioxidants II: Antioxidative components isolated from seeds of *Plantago asiatica* Linn. Chem. Pharm. Bull. **33(3)**: 1279-1273.

Toda, S., Miyase, T., Arichi, H., Tanizawa, H. and Takino, Y. 1985b.Natural antioxidants III. Antioxidative components isolated from rhizomes of *Curcuma longa* L. Chem. Pharm. Bull. **33(4)**: 1725-1728.

Toda, S. and Shirataki, Y. 2004. Inhibitory effects of resveratrol oligomers in *Sophora mooracraftiana* on lipid peroxidation by superoxide anion. Pharm. Biol. **42(1)**: 55-58.

Toda, S. and Shirataki, Y. 2005. Antioxidant activity of 6a-hydroxypterocarpan from the root of *Sophora chrysophylla*. Fitoterapia **76**: 728-729.

Tomaturo, T. 1994. A simple screening method for antioxidants and isolation of several antioxidants produced by marine bacteria from fish and stellfish. J. Biosci, Bistech. Biochem. **58(10)**: 1780-1783.

Tomczyk, M. and Latte, K.P. 2009. *Potentilla* – A review of its phytochemical and pharmacological profile. J. Ethnopharmacol. **122(2)**: 184-204.

Tomi, S., Božin, B., Samojlik, I., Milenkovi, M., Mimica-Duki, N. and Petrovi, S. 2010. Effects of *Athamanta turbinth* fruit essential oils on CCl_4-induced hepatic failure in mice and their antioxidant properties. Phytother. Res. **24**: 787-790.

Tommanaro, G., Segura Rodriguez, C.S., Santillana, M., Immirzi, B., De Prisco, R., Nicolaus, B. and Poli, A. 2007. Chemical composition and biotechnological properties of a polysaccharide from the peels and antioxidative content from the pulp of *Passiflora ligularis* fruits. J. Agric. Food Chem. **55(18)**: 7427-7433.

Toor, R.K. and Savage G.P. 2005. Antioxidant activity in different fractions of tomatoes. Food Res. Int. **38 (5)**: 487-494.

Topcu, G., Ay, M., Bilici, A., Sarikurkcu, C., Ozturk, M. and Ulubelen, A. 2007. A new flavone from antioxidant extracts of *Pistacia terebinthus*. Food Chem. **103**: 816-822.

Torras-Claveria, L., Jauregui, O., Bastida, J., Codina, C. and Viladomat, F. 2007. Antioxidant activity and phenolic composition of Lavandin (*Lavandula x intermedia* Emeric ex Loiseleur) waste. J. Agric. Food Chem. **55(21)**: 8436-8443.

Torres, R., Faini, F., Modak, B., Urbina, F., Labbe, C., Guerrero, J. 2006. Antioxidant activity of coumarins and flavonols from the resinous exudates of *Haplopappus multifolius*. Phytochem. **67(10)**: 984-987.

Toshio, F., Kazue, S., Taro, N. and Hiroshi, S. 2003. Antinephritis and radical scavenging activity of phenyl flavonoids. Fitoterapia **74**: 720-740.

Tosun, I. and Ustun, N.S. 2003. An investigation about antioxidant capacity of fruit nectars. Pak. J. Nutr. **2(3)**: 167-169.

Tosun, M., Ercisli, S., Sengul, M., Ozer, H., Polat, T. and Ozturk, E. 2009. Antioxidant properties and total phenolic content of eight *Salvia* species from Turkey. Biol. Res. **42(2)**: 175-181.

Tourino, S., Lazarraga, D., Carreras, A., Lorenzo, S., Ugartondo, V., Mitjans, M., Vinardell, M.P., Julia, L., Cascante, M. and Torres, J.L. 2008. Highly galloylated tannis fractions from witch hazel (*Hamamelis virginiana*) bark: electron transfer capacity, *in vitro* antioxidant activity and effects on skin-related cells. Chem. Res. Toxicol. **21(3)**: 696-704.

Tram Ngoc Ly, Makoto Shimoyamada, Kojo Kato and Ryo Yamauchi. 2004. Antioxidative compounds isolated from the rhizomes of smaller galangal (*Alpinia officinarum* Hance), BioFactors **21(1-4)**: 305-308.

Trifunovic, A. and Larson, N.G. 2008. Mitochondrial dysfunction as a cause of aging. J. Intern. Med. **263(2)**: 167-178.

Tripathi, R. and Kamat, J.P. 2007. Free radical induced damages to rat liver subcellular organelles: inhibition by *Andrographis paniculata* extract. Indian J. Exp. Biol. **45(11)**: 956-967.

Tripathi, V. 2003. Triterpenoids from *Lantana camara* (Fam. Verbenaceae) and other related species. National Symposium on Emerging Trends in Indian Medicinal Plants, Lucknow, p. 52.

Tripathi, Y.B., Chaurasia, S., Tripathi, E., Upadhyay, A. and Dubey. G.P. 1996. *Bacopa monniera* Linn. as an antioxidant: Mechanism of action. Indian J. Exp. Biol. **34**: 523–526.

Tripathi, Y.B., Pandey, E. and Dubey, G.P. 1999. Antioxidant property of *Hypericum perforatum* (L.) of Indian origin and its comparison with established Medhya rasayanas of ayurvedic medicine. Curr. Sci. **76(1)**: 27-29.

Tripathi, Y.B. and Sharma, M. 1998. Comparison of the antioxidant action of the alcoholic extract of *Rubia cordifolia* with rubiadin. Indian J. Biochem. Biophysic. **35**: 313-316.

Tripathi, Y.B., Sharma, M. and Manickam, M. 1997. Rubiadin, a new antioxidant from *Rubia cordifolia*. Indian J. Biochem. Biophys. **34(3)**: 302-306.

Tripathi, Y.B., Tripathi, E. and Upadhya, A.K. 1996. Antiperoxidative property of *Nardostachys jatamanshi*. Indian J. Exp. Biol. **34**: 1150-1151.

Tripathi, Y.B. and Upadhya, A.K. 2002. Effect of the alcohol extract of the seeds of *Mucuna pruriens* on free radicals and oxidative stress in albino rats. Phytother. Res. **16**: 534-538.

Trivedi, N. P. and Rawal, U. M. 2001. Hepatoprotective and antioxidant property of *Andrographis paniculata* Nees in BHC Induced liver damage in Mice. Indian J. Exp. Biol. **39(1)**: 41-46.

Trojakova, L., Reblova, Z., Nguyen, H.T. and Pokorny, J. 2001. Antioxidant activity of rosemary and sage extracts in rape oil. J. Food Lipids **41**: 5181-5190.

Troszynska, A. and Ciska, E. 2002. Phenolic compounds of seed coats of white and coloured varieties of pea (*Pisum sativum* L.) and their total antioxidant activity. Czech. J. Food Sci. **20**: 15-22.

Troszynska, A., Estrella, I., Lopez-Amores, M.L. and Hernandez, T. 2002. Antioxidant activity of pea (*Pisum sativum* L.) seed coat acetone extract. Lebensmittel-Wissenschaft und Technologie **35(2)**: 158-164. doi: 10.1006/fstl.2001.0831.

Trouillas, P., Calliste, C.A., Allais, D.P., Simon, A., Marfak, A., Delage, C. and Duroux, J.L. 2003. Antioxidant, anti-inflammatory and antiproliferative properties of sixteen water plant extracts used in the Limousin countryside as herbal teas. Food Chem. **80(3)**: 399-407.

Trovato, A., Taviano, M.F., Pergolizzi, S., Campolo, L., Pasquale, R.D. and Miceli, N. 2009. *Citrus bergamia* Risso et Poiteau juice protects against renal injury of diet-induced hypercholesterolemia in rats. Phytother. Res. **24(4)**: 514-519.

Trumbeckaite, S., Beneetis, R., Bumblauskiene, L., Burdulis, D., Janulis, V., Toleikis, A., Viskelis, P. and Jackstoss, V. 2011. *Achillea millefolium* L. s.l. herb extract: Antioxidant activity and effect on the rat heart mitochondrial functions. Food Chem. **127(4)**: 1540-1548.

Tsaknis, J. and Lalas, S. 2005. Extraction and identification of natural antioxidant from *Sideritis euboea* (mountain tea). J. Agric. Food Chem. **53 (16)**: 6375-81.

Tsao, A.S., Kim, E.S. and Hong, W.K. 2004. Chemoprevention of cancer. CA Cancer J. Clin. **54**: 150-180.

Tsao, R., Yang, R. and Young, J.C. 2003. Antioxidant isoflavones in Osage orange, *Maclura pomifera* (Raf.) Schneid. J. Agric. Food Chem. **51 (22)**: 6445-6451.

Tseng, T.H., Kao, E.S., Chu,C.Y., Chou, F.P., Lin, Wu, H.W. and Wang, C.J. 1997. Protective effects of dried flower extracts of *Hibiscus sabdariffa* L. against oxidative stress in rat primary hepatocytes. Food Chem Toxicol. **35(12)**: 1159-64.

Tsuda, T., Watanabe, M., Ohsima, K., Yamamoto, A., Kawakishi, S. and Osawa, T. 1994. Antioxidative components isolated from the seed of Tamarind (*Tamarindus indica* L.). J. Agric. Food Chem. **42(12)**: 2671-2674.

Tsushida, T., Suzuki, M. and Kurogi, M. 1994. Evaluation of antioxidant activity of vegetable extracts and determination of some antioxidants. Nippon Shokuhin Kagaku Kogaku Kaish **41**: 611-618 (in Japanese).

Tuberoso, C.I., Kowalczyk, A., Coroneo, V., Russo, M.T., Dessì, S. and Cabras, P. 2005. Chemical composition and antioxidant, antimicrobial, and antifungal activities of the essential oil of *Achillea ligustica*. J. Agric. Food Chem. **53(26)**: 10148-53.

Tuberoso, C.I.G., Montoro, P., Piecentre, S., Corona, G., Deina, M., Dessi, M.A., Pizza, C. and Cabras, P. 2009. Flavonoid characterization and antioxidant activity of hydroalcoholic extracts from *Achillea ligustica* All. J.Pharm. Biomed. Anal. **50(3):** 440-448.

Tunali-Akbay, T., Sener, G., Salvarli, H., Sehirli, O. and Yarat, A. 2007. Protective effects of *Ginkgo biloba* extract against mercury (II)-induced cardiovascular oxidative damage in rats. Phytother. Res. **21(1)**: 26-31.

Tunalier, Z., Kosar, M., Ozturk, N., Baser, K.H.C., Duman, H. and Kirimer, N. 2004. Antioxidant properties and phenolic composition of *Siderites* species. Chem. Nat. Comp. **40**: 206-210.

Tunalier, Z., Kosar, M., Kupeli, E., Calis, I. and Baser, K.H.C. 2007. Antioxidant, anti-inflammatory, anti-nociceptive activities and composition of *Lythrum salicaria* L. extracts. J. Ethnopharmacol. **110**: 539-547.

Tundis, R., Bonesi, M., Deguin, B., Loizzo, M.R., Menichini, F., Conforti, F., Tillequin, F. and Menichini, F. 2009. Cytotoxic activity and inhibitory effect on nitric oxide production of triterpene saponins from the roots of *Physospermum verticillatum* (Waldst et Kit.) (Apiaceae). Bioorg. Med. Chem. **17(13)**: 4542-4547.

Tung, Y.T. and Chang, S.T. 2010. Variation in antioxidant activity of extracts of *Acacia confusa* of different ages. Nat. Prod. Commun. **5(1)**: 73-76.

Tung, Y.T., Wu,J.-J.H., Hsieh, C.-Y., Chen, P.S. and Chang, S.T. 2009a. Free radical-scavenging phytochemicals of hot water extracts of *Acacia confusa* leaves detected by an on line screening method. Food chem. **115**: 1019-1024.

Tung, Y.T., Wu, J.H., Huang, C.Y., Kuo, Y.H. and Chang, S.T. 2009b. Antioxidant activities and phytochemical extracts of *Acacia confusa* bark. Bioresource Technol. **100(1)**: 509-514.

Tung, Y.T., Wu, J.H., Huaang, C.-C., Peng, H.C., Chen, Y.L., Yang, S.C., Chang, S.T. 2009c. Protective effect of *Acacia confusa* bark and its active compound gallic acid against carbon tetrachloride liver injury in rats. Food Chem. Toxicol. doi: 10.1016/jfct.2009.03.0211

Tung, Y.T. Wu, J.H. Kuo, Y.H. and Chang, S.T. 2007. Antioxidant activities of natural phenolic compounds from *Acacia confusa* bark. Bioresource Technol. **98(5)**: 1120-1123.

Tural, S. and Koca, I. 2008. Physico-chemical and antioxidant properties of cormelian cherry fruits (*Cornus mas* L.) grown in Turkey. Sci. Hort. **116**: 362-366.

Türemis, N., Kafkas, E., Kafkas, S., Kürkcüoglu, M. and Baser, K.H.C. 2003. Determination of Aroma compounds in blackberry by GC/MS Analysis. Chem. Matural Compounds 39: 2.

Turkmen, N., Sari, F. and Velioglu, Y. 2005. The effect of cooking methods on total phenols and antioxidant activity of selected green vegetables. Food Chem. **93**: 713-718.

Turkoglu, A., Duru, M.E. and Mercan, N. 2007. Antioxidant and antimicrobial activity of *Russula delica* Fr.: An edible wild mushroom. Eurasian J. Anal. Chem. **2(1)**: 54-67.

Turkoglu, A., Duru, M.E., Mercan, N., Kivrak, I. and Gezer, K. 2007. Antioxidant and antimicrobial activities of *Laetiporus sulphureus* (Bull.) Murril. Food Chem. **101**: 267-273.

Turkoglu, A., Kivrak, I., Mercan, N., Duru, M.E., Gezer, K. and Turkoglu, H. 2006. Antioxidant and antimicrobial activities of *Morchella conica* Pers. African J. Biotech. **5(11)**: 1146-1150.

Turkoglu, S., Turkoglu, I., Kahyaoglu, M. and Celik, S. 2010. Determination of antimicrobial and antioxidant activities of Turkish endemic *Ajuga chamaepitys* (L.) Schreber subsp. *euphratica* P.H. Davis (Lamiaceae). J. Med. Plants Res. **4(3)**: 1260-1268.

Udapurkar, P.P., Heda, V.S., Kale, M.K., Duragkar, N.J. and Bhusari, K.P. 2009. Role of antioxidant activity of *Syzygium cumini* seeds in peptic ulcer induced normal and diabetic rats. International J. Pharmacol. Boil. Sci. **3(3)**: 95-100.

Uddin, S.J., Shilpi, J.A., Byres, M., Middleton, M., Shoeb, M., Nahar, L. and Sarker, S.D. 2007. Swarnalin and cis-swarnalin, two new tetrahydrofuran derivatives with free radical scavenging activity, from the aerial parts of *Cuscuta reflexa*. Nat. Prod. Res. **21**: 663-668.

Udegbunam, R.I., Asuzu, U.I., Kene, R.O.C., Udegbunam, S.O. and Chinaka Nwaehujor, 2011. Anti-Nociceptive, Anti-Inflammatory and Anti-Oxidant Effects of the Methanol leaf Extract of *Sterculia tragacantha* Lindl. J. Pharmacol. Toxicol. **6**: 516-524. **DOI**: 10.3923/jpt.2011.516.524.

Ueda, Y., Komatsu, M. and Hiramatsu, M. 1996. Free radical scavenging activity of the Japanese herbal medicine toki-shakuyaku-san (TJ-23) and its effect on superoxide dismutase activity, lipid peroxides, glutamate, and monoamine metabolites in aged rat brain. Neurochem. Res. 21(8): 909-914.

Ugochukwu, N.H. and Babady, N.E. 2002. Antioxidant effects of *Gongronema latifolium* in hepatocytes of rats on non-insulin dependent diabetes mellitus. Fitoterapia **73(7-7)**: 612-618.

Ugochukwu, N.H. and Cobourne, M.K. 2003. Modification of renal oxidative stress and lipid peroxidation in streptozotocin-induced diabetic rats treated with extracts from *Gongronema latifolium* leaves. Clinica Chimica Acta **336**: 73-81.

Ugur, A., Duru, M.E., Ceylan, O., Sarac, N., Varol, O. and Kivrak, I. 2009. Chemical composition, antimicrobial and antioxidant activities of *Centaurea ensiformis* Hub-Mor. (Asteraceae), a species endemic to Mugla (Turkey). Nat. Prod. Res. **23(2)**: 149-167.

Umadevi, P. 2001. Radioprotective, anticarcinogenic and antioxidant properties of the Indian holy basil, *Ocimum sanctum* (Tulsi), Indian J. Exp. Biol. **39(3)**: 185-190.

Umamaheswari, M., Asokkumar, K., Rathidevi, R., Sivashanmugam, A.T., Subhadradevi, V. and Ravi, T.K. 2007. Antiulcer and *in vitro* antioxidant activities of *Jasminum grandiflorum* L. J. Ethnopharmacol. **110 (3)**: 464-470.

Unno, T., Sakane, I., Masumizu, T., Kohno, M. and Kakuda, T. 1998. Antioxidative activity of water extracts of *Lagerstroemia speciosa* leaves. Biosci. Biotech. Biochem. **61(10)**: 1772-1774.

Unno, T., Tamemoto, K., Yahabe, F. and Kakuda, T. 2003. Urinary excretion of 5-(3′, 4′-dihydroxymethyl)-gamma-valerolactone, a ring-fission metabolite of (–)-epicatechin, in rats and its *in vitro* antioxidant activity. J. Agric. Food Chem. **51(23)**: 6893-6898.

Upadhya, S., Shanbhag, K.K., Suneetha G., Naidu, B.M., and Upadhya, S. 2004. A study of hypoglycemic and antioxidant activity of *Aegle marmelos* in alloxan induced diabetic rats. Indian J. Physiol. Pharmacol. **48 (4)**: 476–480.

Urquiaga, I. and Leighton, F. 2000. Plant polyphenol antioxidants and oxidative stress. Biol. Res. **33**: 55-64.

Usenik, V., Fabcic, J. and Stampar, F. 2008. Sugars, organic acids, phenolic composition and antioxidant activity of sweet cherry (*Prunus avium* L.). Food Chem. **107 (1)**: 185-192.

Usha Nagalakshmi, R. 2011. Morphogenetic and antimicrobial studies of *Phyllanthus acidus* (L.) Skeels and *Euphorbia hypericifolia* L.; antioxidant and hepatoprotective activity of *Phyllanthus acidus* fruits. Ph.D. thesis, Sri Venkateswara University, Tirupati.

Usoh, F.I., Akpan, E.J., Etim, E.O., Farombi, E.O. 2005. Antioxidant actions of dried flower of *Hibiscus sabdariffa* L. on sodium arsenite-induced oxidative stress. Pak. J. Nutr. **4**: 135-141.

Vadiraja Bhat, B. and Madhyastha, K.M. 2000. C-phycocyanin: A potent peroxyl radical scavenger *in vivo* and *in vitro*. Biochem. Biophys. Res. Commun. **275**: 20.

Vadlapudi, V. and Naidu, K.C. 2009. Evaluation of antioxidant potential of selected mangrove plants. J. Pharmacy Res. **2(11)**: 1742-1745.

Vadlapudi, V. and Naidu, K.C. 2010. *In vitro* bioevaluation of some Indian medicinal plants. Drug Invention Today **2(1)**: 65-68.

Vaghasiya, Y. and Chanda, S. 2010. Antimicrobial and free radical scavenging activity of different solvent extracts of *Mangifera indica* L. seeds. Res. J. Microbiol. **5**: 1207-1212.

Vagi, E., Rapavi, E. Hadolin, M., Vasarhelyine Peredi, K., Balazas, A. Blazovics, A. and Simandi, B. 2005. Phenolic and triterpenoid antioxidants from *Origanum majorana* L. herb and extracts obtained with different solvents. J. Agric. Food Chem. **53**: 17-21.

Vaillant, F., Perez, A., Davila, I., Dornier, M. and Reynes, M. 2005. Colorant and antioxidant properties of red pitahaya (*Hylocereus* sp.). Fruits. **60**: 1-7.

Valamarthi, S. and Harathi, P.B. 2009. Evaluating and antioxidant activity of the medicinal plant *Cynodon dactylon*. J. Ecotoxicol. Environ. Monitoring **19(2)**: 143-147.

Valentão, P. Fernandes, E., Carvalho, F., Andrade, P.B., Seabra, R.M. and de Lourdes Bastos, M. 2001. Antioxidant activity of *Centaurium erythraea* infusion evidenced by its superoxide radical scavenging and xanthine oxidase inhibitory activity. J. Agric. Food Chem. **49(7)**: 3476-3479.

Valentão, P. Fernandes, E., Carvalho, F., Andrade, P.B., Seabra, R.M. and de Lourdes Bastos, M. 2002a. Studies on the antioxidant activity of *Lippia citriodora* infusion: Scavenging effect on superoxide radical, hydroxyl radical and hypochlorous acid. Biol. Pharm. Bull. **25(10)**: 1324-1327.

Valentão, P. Fernandes, E., Carvalho, F., Andrade, P.B., Seabra, R.M. and Bastos, M.L. 2002b. Antioxidative properties of cardoon (*Cynara cardunculus* L.) infusion against superoxide radical, hydroxyl radical and hypochlorous acid. J. Agric. Food Chem. **50(17)**: 4989-4993.

Valentão, P. Fernandes, E., Carvalho, F., Andrade, P.B., Seabra, R.M. and Bastos, M.L. 2002c. Antioxidant activity of *Hypericum androsaemum* infusion: scavenging activity against superoxide radical, hydroxyl radical and hypochlorous acid. Biol. Pharm. Bull. **25(10)**: 1320-1323.

Valentão, P. Fernandes, E., Carvalho, F., Andrade, P.B., Seabra, R.M. and Bastos, M.L. 2003. Hydroxyl radical and hypochlorous acid scavenging activity of small centaury (*Centaureum erythraea*) infusion. A comparative study with green tea (*Camellia sinensis*). Phytomedicine **10(6-7)**: 517-522.

Valentová, K., Sersen, F. and Ulrichová, J. 2005. Radical scavenging and anti-lipoperoxidative activities of *Smallanthus sonchifolius* leaf extracts. J. Agric. Food Chem. **53(14)**: 5577-5582.

Valenzula, A., Aspillaga, M. Vial, S. and Guerra, R. 1989. Selectivity of silymarin on the increase of the glutathione content in different tissues of the rat. Planta Med. **55**: 420-422.

Valyova, M., Hadjimitova, V., Stoyanov, S., Ganeva, Y., Traykov, T. and Petkov, I. 2009. Radical scavenger and antioxidant activities of extracts and fractions from Bulgarian *Ononis spinosa* L. and GC-MS analysis of Ethanol extract. The Internet J. Alternative Med. **7(2)**: ispub.com

Van Acker, S.A.B.E., Van den Vijgh, W.J.F. and Bast, F. 1996. Structural aspects of antioxidant activity of flavonoids. Free Rad. Bio. Med. **20(3)**: 331-342.

Vanden, B.R., Haenen, G.R., Vanden B.H. and Bast, A.A. 1999. Applicability of an improved trolox equivalent antioxidant capacity (TEAC) assay for evaluation of antioxidant capacity measurements of mixtures. Food Chem. **66**: 511–517.

Vander Jagt, T.J., Ghattas, R., Vander Jagt, D.J., Crossey, M. and Glow, R.H. 2002. Comparison of the total antioxidant content of 30 widely used medicinal plants of New Mexico. Life Sci. **70**: 1035-1040.

Van Gadow, A., Joubert, E. and Hansmann, C.F. 1997. Comparison of the antioxidant activity of aspalathin with that of other plant phenols of rooibos tea (*Aspalathus linearis*), α-tocopherol, BHT and BHA. J. Agric. Food Chem. **45**: 632-638.

Vani, T. Rajani, M., Sarkar, S. and Shishoo, C.J. 1997. Antioxidant properties of the Ayurvedic formulation Triphala and its constituents. Int. J. Pharmacog. **35**: 313–17.

Vardar Unlu, G., Candan, F., Sokemen, A., Daferera, D., Polissiou, M., Sokmen, M., Domex, E. and Tepe, B. 2003. Antimicrobial and antioxidant activity of the essential oil and methanol extracts of *Thymus pectinatus* Fish. et Mey. var. *pectinatus* (Lamiaceae). J. Agric. Food Chem. **51**: 63-67.

Vasconcellos, C.L.C., Vitorio, K.C.D., Andrade, P.A., Cambui, E.V.E., Lira, A.E., Cavalcante, C.H., Estevan, C.S., Antoniolli, A.R. and Thomazzi, S.M. 2009. Antinociceptive, anti-inflammatory, and antioxidant properties of *Phoradendron piperoides* leaves. Pharmaeutical Biol. **47(7)**: 645-652.

Vasu, V.T., Modi, H., Thaikoottathil, J.V. and Gupta, S. 2005. Hypolipidemic and antioxidant effect of *Enicostemma littorale* Blume aqueous extract in cholesterol fed rats. J. Ethnopharmacol. **101(1-3)**: 277-282.

Vaya, J., Belinky, P.A. and Aviram, M. 1997. Antioxidant constituents from licorice roots: isolation, structure elucidation and antioxidant capacity towards LDL peroxidation. Free Radical Biol. Med. **23(2)**: 302-313.

Vayalil, P.K. 2002. Antioxidant and antimutagenic properties of aqueous extract of date fruit (*Phoenix dactylifera* L. Arecaceae). J. Agric. Food Chem. **50**: 610-617.

Veerapur, V.P., Prabhakar, K.R., Parihar, V.K., Kandadi, M.R., Ramakrishna, S, Mishra, B. Satish Rao, B.S., Srinivasan, K.K., Priyadarsini, K.I. and Unnikrishnan, M.K. 2007. *Ficus racemosa* stem bark extract: A potent antioxidant and a probable natural radioprotector. Evid. Based Complement Altern. Med. **3**: 205-208.

Veeresh, B., Veeresh Babu, S.V., Eswara Prasad, M., Biradar, S.S. and Patil, D.D. 2009. Protective effect of methanolic extract of *Euphorbia antiquorum* on haloperidol-induced catalepsy in albino mice. Int. J. Pharmacol. Biol. Sci. **3(3)**: 55-59.

Veeru, P., Kishor, M.P. and Meenakshi, M. 2009. Screening of medicinal plants extracts for antioxidant activity. J. Med. Plants Res. **3(8)**: 608-612.

Velavan, S., Nagulendran, K., Mahesh, R. and Hazeena Begum, V. 2002. *In vitro* antioxidant activity of *Asparagus racemosus* root. Phcog. Mag. **3(9)**: 26-33.

Velazquez, E., Tournier, H.A., de Buschiazzo, P.M., Saavedra, G. and Schinella, G.R. 2003. Antioxidant activity of Paraguayan plant extracts. Fitoterapia **74**: 91-97.

Velioglu, Y.S., Mazza, G., Gao, L. and Oomah, B.D. 1998. Antioxidant activity and total phenolics in selected fruits, vegetables and grain products. J. Agric. Food Chem. **46**: 4113-4117.

Vellosa, J.C.R., Khalil, N.M., Formenton, V.A.F., Ximenes, V.F., Fonseca, L.M., Furlan, M., Brunetti, I.L. and Oliveira, O.M.M.F. 2006. Antioxidant activity of *Maytenus ilicifolia* root bark. Fitoterapia **77**: 243-244.

Venkatswaran, S. and Pari, L. 2002. Effect of *Coccinia indica* on blood glucose, insulin and hepatic key enzymes in experimental diabetes. Pharm. Biol. **40**: 165-170.

Venkatswaran, S. and Pari, L. 2003. Effect of *Coccinia indica* leaf extract on plasma antioxidants in streptozotocin induced experimental diabetes in rats. Phytother. Res. **17(6)**: 605-608.

Venskutonis, P.R., Dvaranauskaite, A and Labokas, J. 2007. Radical scavenging activity and composition of raspberry (*Rubus idaeus*) leaves from different locations in Lithuania. Fitoterapia **78**: 162-165.

Venskutonis, P.R., Skemaite, M. and O. Ragazinskiene. 2007. Radical scavenging capacity of *Agrimonia eupatoria* and *Agrimonia procera*. Fitoterapia **78**: 166-168.

Venskutonis, P.R., Van Beek, T.A., Pukalskas, A., Miliauskas, G., Dapkevicius, A. 2004. Radical scavengers from some plants grown in Lithuania. IUPAC International Conference on Biodiversity and Natural Products Chemistry and Medicinal Applications, New Delhi, p.-401, 26-31.

Venukumar, M.R. and Latha, M.S. 2002a. Antioxidant activity of *Curculigo orchioides* in carbon tetrachloride induced hepatopathy in rats. Indian J. Clin. Biochem. **17**: 80-87.

Venukumar, M.R. and Latha, M.S. 2002b. Antioxidant effect of *Coscinium fenestratum* in carbon tetrachloride treated rats. Indian J. Physiol. Pharmacol. **34**: 269-273.

Verma, N., Behera, B.C., Makhija, U. 2008. Antioxidant and hepatoprotective activity of a lichen *Usnea ghattensis in vitro.* Appl Biochem. Biotechnol. **151(2-3)**: 167-81.

Verma, N. and Vinayak, M. 2008. Antioxidant action of *Andrographis paniculata* on lymphoma A. Mol. Bio. Rep. **35**: 535-540.

Verma, N. and Vinayak, M. 2009. *Semecarpus anacardium* nut extract promotes the antioxidant defence system and inhibits anaerobic metabolism during development of lymphoma. Biosci. Rep. **29(3)**: 151-164.

Verma, R., Sisodia, R. and Bhatia, A.L. 2002. Dietary supplementation of Spinach improves learning and memory in mice after radiation induced oxidative stress. Indian J. Gerontol. **18**: 18-33.

Vesela, D., Kubinova, R., Muselik, J., Zemlicka, M and Suchy, V. 2004. Antioxidative and EROD activities of osajin and pomiferin. Fitoterapia **75**: 209-211.

Veselova, M.V., Fedoreev, S.A., Vasilevskaya, N.A., Denisneko, V.A. and Gerasimenio, A.V. 2007. Antioxidant activity of polyphenols from the far-east plant *Taxus cuspidata*. Pharmaceutical Chem. J. **41(2)**: 88-93. doi: 10.1007/s11094.

Vianna, H.R. Cortes, S.F., Ferreira, A.J., Capettinim, L.S.A., Almeida, A.R., Massensini, A.R. and Lemos, V.S. 2006. Antiarrhythmogenic and antioxidant effect of the flavonoid dioclein in a model of cardiac ischaemia/reperfusion. Planta Med. **72(4)**: 300-303.

Vicas, S., Prokisch, J., Rugina, D. and Socaciu, C. 2009. Hydrophilic and Lipophilic antioxidant activities of Mistletoe (*Viscum album*) as determined by FRAP method. Notulae Botanicae Horti Agrobotanici ClujNapoca **37(2)**: 112-116.

Vidyadhar, S., Sheela, T., Shiva Kumar Reddy, L., Gopal, T.K., Chamundeswari, D., Saidulu, A. and Uma Maheswara Reddy, C. 2010. *In vitro* antioxidant activity of chloroform extract of aerial parts of *Securinega leucopyrus* (Willd.) Muell. Der Pharmacia Lettre **2(6)**: 252-256.

Vijaya Bharathi, R., Kumudhaveni, B., Jayashree, Suseela, L. and Thirumal, M. 2010. Antioxidant and wound healing studies on different extracts of *Stereospermum colais* leaf. Int. J. Res. Pharm. Sci. **1(4)**: 435-439.

Vijayabhaskaran, M., Badkhal, A.K., Babu, G., Sivakumar, P., Perumal, P., Sivakumar, T., Sampathkumar, R. and Jayakar, B. 2010. Antioxidant activity and antioxidant status of *Symplocos racemosa* Roxb. against Ehrlich Ascites carcinoma in Swiss Albino mice. Res. J. Pharm. Biol. Chem. Sci. 1(3): 306-314.

Vijayalakshmi, M.A. and Sarada, N.C. 2008. Screening of *Costus speciosus* extracts for antioxidant activity. Fitoterpia **79(3)**: 197-198.

Vijayakumar, M., Govindarajan, R., Rao, G.M.M., Rao, C.V., Shirwaikar, A., Mehrotra, S. and Pushpangadan, P. 2006. Action of *Hygrophila auriculata* against streptozotocin-induced oxidative stress. J. Ethnopharmacol. **104(3)**: 356-361.

Vijayakumar, M., Govindarajan, R., Shirwaikar, A., Vivek Kumar, Rawat, A.K.S., Mehrotra, S. and Pushpangadan, P. 2005. Free radical scavenging and lipid peroxidation inhibition potential of *Hygrophila auriculata*. Nat. Prod. Sci. **11(1)**: 22-26.

Vijayakumar, M., Govindarajan, R., Singh, D.P. and Rawat, A.K.S. 2005. Review of some important plants with antioxidant and other biological activities. Phcog. Mag. **1(4)**: 129-135.

Vijayakumar, R.S., Surya, D. and Nalini, N. 2004. Antioxidant efficacy of black pepper (*Piper nigrum* L.) and piperine in rats with high fat diet induced oxidative stress. Redox Rep. **9(2)**: 105-110.

Vijayakumar, S., Prasannakumar, G. and Vijayalakshmi, N.R. 2008. Antioxidant activity of banana flavonoids. Fitoterapia, **79 (4)**: .279-282.

Vijayalakshmi, M.A. and Sarada, N.C. 2008. Screening of *Costus speciosus* extracts for antioxidant activity. Fitoterapia **79(3)**: 197-198. doi: 10.1016/j.fitote.2007.11.014

Vijayavel, K., Anbuselvam, C. and Balasubramanian, M.P. 2007. Antioxidant effect of the marine algae *Chlorella vulgaris* against naphthalene-induced oxidative stress in the albino rats. Mol Cell Biochem. **303(1-2)**: 39-44.

Vikas, V.T., Hiren, M., Jyothi T. V. and Sarita, G. 2005. Hypolipidaemic and antioxidant effect of *Enicostemma littorale* Blume aqueous extract in cholesterol fed rats. J. Ethnopharmacol. **101(2)**: 277-282.

Villareal, A., Stoecker, B.J., Garcia, C., Garcia, K., Rios, R., Gonzales, C., Mandadi, K., Faraji, B., Patil, B.S., Deyhim, F. 2007. Cranberry juice improved antioxidant status without affecting bone quality in orchidectomized male rats. Phytomedicine **14(12)**: 815-820.

Vinholes, J., Griosso, C., Andrade, P.B., Gil-Izquierdo, A., Valentao, P., de Pinho, P.G. and Ferreres, F. 2011. *In vitro* studies to assess the antidiabetic, anticholinesterase and antioxidant potential of *Spergularia rubra*. Food Chem. **129(2)**: 454-462.

Vinson, J.A., Dabbagh, Y.A., Serry, M.M. and Jang, J. 1995. Plant flavonoids, especially tea flavonols, are powerful antioxidants using an *in vitro* oxidation model for heart disease. J. Agric. Food Chem. **43(11)**: 2800-2802.

Vinson, J.A., Hao, Y., Su, X. and Zubik, L. 1998. Phenol antioxidant quantity and quality in foods: vegetables. J. Agric. Food Chem. **46**: 3630-3634.

Vinson, J.A., Liang,.Q. *et al.,* 2002. Polyphenol antioxidants in citrus juices *in vitro* and *in vivo* studies relevant to heart disease. Adv. Exp. Med. Biol. **505**: 113-122.

Visavadiya, N. P. and Narasimhacharya, R.L. 2005. Hypolipidemic and antioxidant activities of *Asparagus racemosus* in hypercholesteremic rats. Indian J. Pharmacol. **37**: 376-80.

Vishalakshi Devi, D. and Urooj, A. 2010. Nutrient profile and antioxidant components of *Costus speciosus* Sm. and *Costus igneus* Nak. Indian J. Nat. Prod. Resourc. **1(1)**: 116-118.

Vishwakarma, S.L., Rajani, M., Bagul, M.S. and Goyal, R.K. 2004. A rapid method for the isolation of swertiamarin from *Enicostemma littorale*. Pharm. Biol. **42**: 400-403.

Visweswara Rao, P. and Dhanajaya Naidu, M. 2010. *Rhinacanthus nasutus*: a plant with potential activity in radical scavenging capacity. Current trends in Biotechnology and Pharmacy **4(3)**: 791-794.

Vitalini, S., Beretta, G., Iriti, M., Orsenigo, S., Basilico, N., Dall'Acqua, S., Lorizzi, M. and Fico, G. 2011. Phenolic compounds from *Achillea millefolium* L. and their bioactivity. Acta Biochimica Polonica **58(2)**: 203-209.

Vitalini, S., Grande, S., Visioli, F., Agradi, D., Fico, G. and Tome, F. 2006. Antioxidant activity of wild plants collected in Valsesia, an alpine region of Northern Italy. Phytother. Res. **20(7)**: 576-580.

Viturro, C., Molina, A. and Schmeda-Hischmann, G. 1999. Free radical scavengers from *Mutisia friesiana* (Asteraceae) and *Sanicula graveolens* (Apiaceae). Phytother. Res. **13**: 422-424.

Viuda-Martos, M., Navajas, Y.R., Zapata, E.S., Fernandez-Lopez, J., Perez-Alvarez, J.A. 2010. Antioxidant activity of essential oils of five spice plants widely used in a Mediterranean diet. Flavour and Fragrance Journal **25(1)**: 13-19.

Vivek Kumar, R., Kumar, S., Shashidhara, S., Anitha, S. and Manjula, M. 2010. Antioxidant and antimicrobial activities of various extracts of *Michelia champaca* Linn. flowers. 2nd Intern. Seminar on Medicinal plants and Herbal Products, S.V.Univ., Tirupati, Abstract pp. 115.

Vivek Kumar and Sharma, S.K. 2006. Antioxidant studies on some plants: A review. Hamdard Medicus **49(4)**: 25-36.

Viveros-Valdez, E., Rivas-Morales, C., Carranza-Rosales, P., Mendoza, S. and Schmeda-Hirschmann, G. 2008. Free radical scavengers from the Mexican Herbal tea "Poleo" (*Hedeoma drummondii*). Z. Naturforsch **63c**: 341-346.

Vivot, E., de Dios, Muñoz, J, del Carmen, Cruañes, M., Cruañes, M.J., Tapia, A., Hirschmann,G.S., Martínez, E., Di Sapio, O., Gattuso, M. and Zacchino, S. 2001. Inhibitory activity of xanthine-oxidase and superoxide scavenger properties of *Inga verna* subsp. *affinis*. Its morphological and micrographic characteristics. J. Ethnopharmacol. **76(1)**: 65-71.

Voravuthikunchai, S.P., Kanchanapoom, T., Sawangjaroen, N. and Hutadilok-Towatana, N. 2010. Antioxidant, antibacterial and antigiardial activity of *Walsura robusta* Roxb. Nat. Prod. Res. **24(9)**: 813-824.

Vouldoukis, I., Lacan, D., Kamate, C., Coste, P., Calenda, A., Mazier, D., Conti, M. and Dugas, B. 2004. Antioxidant and anti-inflammatory properties of a *Cucumis melo* L C. extract rich in superoxide dismutase activity. J. Ethnopharmacol. **94 (1)**: 67-75.

Vrchovska, V., Spilkova, J., Valentao, P., Sousa, C., Andrade, P.B. and Seabra, R.M. 2007. Antioxidative properties and phytochemical composition of *Ballota nigra* infusion. Food Chem. **105(4)**: 1396-1403.

Vuda, M., D'Souza, R., Upadhya, S., Kumar, V., Rao, N., Kumar, V., Boillat, C. and Mungli, P. 2011. Hepatoprotective and antioxidant activity of aqueous extract of *Hybanthus enneaspermus* against CCl_4 induced liver injury in rats. Exp. Toxicol. Path. (epub ahead of print).

Wada, L. and Ou, B. 2002. Antioxidant activity and phenolic content of Oregan cranberries. J. Agric. Food Chem. **50**: 3495-3500.

Wagensteen, H., Samuelsen, BA., Malterude, E.K. 2004. Antioxidant activity in extracts from coriander. Food Chem. **88**: 293-297.

Wahab, K.A., Stanslas, J., Abu Bakar, A., Promwichit, P., Aziz, S., Marzuki, A. and Wah, O.K. 2001. Cytotoxic and Antioxidant properties of wood tissues of three Malaysian Timber species. J. Trop. Med. Plants **3**: 163-168.

Wahile, A., Mukherjee, K., Venkatesan, K., Saha, B.P. and Mukherjee, P.K. 2007. Antioxidant potentials of *Hypericum hookerianum* (Family: Hypericaceae) on CCl_4 induced hepatotoxicity in rats. Oriental Pharmacy and Experimental medicine **7(1)**: 85-93.

Walia, S., Singh, M., Kaur, C., Kumar, R. and Joshi, S. 2010. Antioxidant composition of red and orange cultivars of tomatoes (*Solanum lycopersicon* L.): A comparative evaluation. J. Plant Biochem. Biotech. **19(1)**: 95-97.

Wall, M.M. 2010. Functional lipid characteristics, oxidative stability and antioxidant activity of macadamia nut (*Macadamia integrifolia*) cultivars. Food chem. **121(4)**: 1103-1108.

Walton,.J. and Brown, D.E. 1999. Chemicals from plants: Perspectives on plant secondary products. Imperial college press, London.

Wang, B.Y. 2009. *In vitro* antioxidative activities of extract and semi-purified fractions of the marine red alga, *Rhodomela confervoides* (Rhodomelaceae). Food Chem. **113**: 1101-1105.

Wang, G.J., Lin, S.Y., Wu, W.C. an Hou, W.C. 2007. DPPH radical scavenging semicarbazide-sensitive amineoxidase inhibitory and cytotoxic activities of *Taiwanofungus camphoratus* (Chang-Chin). Biosci. Biotechnol. Biochem. **71**: 1873-1878.

Wang, H., Cao, G. and Prior, R.L. 1996. Total antioxidant capacity of fruits. J. Agric.Food Chem. **44**: 701-705.

Wang, H., Cao, G. and Prior, R.L. 1997. Oxygen radical absorbing capacity of anthocyanins. J. Agric. Food. Chem. **45**: 304-309.

Wang, H., Gan, D., Zhang, X. and Pan, Y. 2010. Antioxidant capacity of the extracts from pulp of *Osmanthus fragrans* and its components. LWT- Food Sci. Tech. **43(2)**: 319-325.

Wang, H., Nair, M.G., Straburg, G.M., Booren, A.M. and Gray, J.I. 1999. Antioxidant polyphenols from tart cherries (*Prunus cerasus*). J. Agric. Food Chem. 47: 840-844.

Wang, H., Sun, Y., Ye, W.C., Xiong, F., Wu, J.J. and Yang, C.H. and Zhao, S.X. 2004. Antioxidative phynylethanoid and phenolic glycosides from *Picrorhiza scrophulariiflora*. Chem. Pharm. Bull. **52(5)**: 615-617.

Wang, H.C., Hu, Z.-Q., Wang, Y., Chen, H.B., and Huang, X.-M. 2011. Phenolic compounds and the antioxidant activities in litchi pericarp; Difference among cultivars. Scientia Horticulturae **129(4)**: 784-789.

Wang, H.M., Lo, W.L., Huang, L.Y., Wang, Y.D. and Chen C.Y. 2010. Chemical constituents from the leaves of *Michelia alba*. Nat. Prod. Res. **24(5)**: 398-406.

Wang, J., Yuao, X., Jin, Z., Tian, Y. and Song, H. 2007. Free radical and reactive oxygen species scavenging activities of peanut skin extracts. Food Chem. **104(1)**: 242-250. doi: 10.1016/j.foodchem.2006.11.035

Wang, J., Zhang, Q., Zhang, Z., Li, Z. 2008. Antioxidant activity of sulfated polysaccharide fractions extracted from *Laminaria japonica*. Int. J. Biol. Macromol. **42(2)**: 127-32.

Wang, K., Pan, Y., Wang, H., Zhang, Y., Lei, Q., Zhu, Z., Li, H. and Liang, M. 2009. Antioxidant activities of *Liquidamber formosana* Hance leaf extract. Med. Chem. Res. **19(2)**: 166-176.

Wang, K.-J., Yang, C.-R. and Zhang, Y.-J. 2007. Phenolic antioxidants from Chinese toon (fresh young leaves and shoots of *Toona sinensis*). Food Chem. **101**: 365-371.

Wang, K.J., Zhang, Y.J.and Yang, C.R. 2005. Antioxidant phenolic compounds from rhizomes of *Polygonum paleaceum*. J. Ethnopharmacol. **96(3)**: 483-487.

Wang, K.J., Zhang, Y.J.and Yang, C.R. 2006. New phenolic constituents from *Balanophora polyandra* with radical-scavenging activity. Chem. Biodivers. **3(12)**: 13171324.

Wang, M., Li, J., Rangarajan, M., Shao, Y., La Voie, E.J., Huang, T. and Ho, C. 1998. Antioxidative phenolic compounds from Sage (*Salvia officinalis*). J. Agr. Food Chem. **46**: 4869-4873.

Wang, M., Tsao, R., Zhang, S., Dong, Z., Yang, R., Gong, J. and Pei, Y. 2006. Antioxidant activity, mutagenicity/anti-mutagenicity, and clastogenicity/anti-clastogenicity of lutein from marigold. Food and Chemical Toxicology. **44 (9)**: 1522-1529.

Wang, M.Y., Lutfiyya, M.N., Weidenbacher-Hoper, V., Anderson, G., Su, S.X. and West. B.J. 2009. Antioxidant activity of Noni juice in heavy smokers. Cem. Central Journ. **3**: 130. doi: 10.1186/1752-153x-3-13.

Wang, M.Y. and Su, C. 2001 Cancer preventive effect of *Morinda citrifolia* (Noni). Amer. N.Y. Acad. Sci. **952**: 161-168.

Wang, R.F., Xie, W.D., Zhang, Z., Xing, D.M., Ding, Y., Wang, W. and Du, L.J. 2004. Bioactive compounds from the seeds of *Punica granatum* (Pomegranate). J. Nat. Prod. **67(12)**: 2096-2098.

Wang, S.I. and Lin, H.S. 2000. Antioxidant activity in fruits and leaves of blackberry, raspberry and strawberry varies with cultivar and developmental stage. J. Agric. Food Chem. **48**: 140-146.

Wang, S.Y. and Ballington, J.R. 2007. Free radical scavenging capacity and antioxidant enzyme activity in deerberry (*Vaccinium stamineum* L.) LWT **40**: 1352-1361.

Wang, S.-Y., Chang, H.-N., Lin, K.-T., Lo, C.-P., Yang, N.S. and Shyur, L.-F. 2003. Antioxidant properties and phytochemical characteristics of extracts from *Lactuca indica*. J. Agric. Food Chem. **51**: 1506-1512.

Wang, S.Y. and Jio, H. 2000. Scavenging capacity of berry crops on superoxide radicals, hydrogen peroxide, hydroxyl radicals and singlet oxygen. J. Agric. Food Chem. **48**: 5677-5684.

Wang, S.-Y., Kuo, Y.-H., Chang, H.-N., Tsay, H.-S., Lin, K.-F., Yang, N.S. and Shyur, L.-F. 2002. Profiling and characterization of antioxidant activities in *Anoectochilus formosanus* Hayata. J. Agric. Food Chem. **50**: 1859-1865.

Wang, S.-Y. and Lin, H.S. 2000. Antioxidant activity in fruits and leaves of blackberry, raspberry, and strawberry varies with cultivar and developmental stage. J. Agr. Food Chem. **48**: 140-146.

Wang, S.-Y. and Lin, H.S. 2003. Compost as a soil supplement increases the level of antioxidant compounds and oxygen radical absorbance capacity in strawberries. J. Agric. Food Chem. **51(23)**: 6844-6850.

Wang, S.-Y., Wu, J.-H., Cheng, S.-S., Lo, C.-P., Chang, H.-N., Shyur, L.-F. and Chang, S.-T. 2004. Antioxidant activity of extracts from *Calocedrus formosana* leaf, bark and heartwood. J. Wood Sci. **50**: 422-426.

Wang, S.-Y., Wu, J.-H., Shyur, L.-F., Kuo, Y.-H. and Chang, S.-T. 2002. Antioxidant activity of Abietane-Type Diterpenes from Heartwood of *Taiwania cryptomerioides* Hayata. Holzforschung, **56**: 487-492.

Wang, T. C., Tim, C., Lo, S.C., Yang, C. C. 2007. Free radical-scavenging activity of aqueous extract of *Pteris multifida* Poiret. Fitoterapia. **78(3)**: 248-249.

Wang, W., Bostic, T.R. and Gu, L. 2010. Antioxidant capacities, procyanidins and pigments in avocados of different strains and cultivars. Food Chem. **122(4)**: 1193-1198. doi: 10.1016/j.foodchem.2010.03.114.

Wang, X.J. and Feng, P. 2000. Antioxidant activity of Qizhu tang. Acta Pharmacol. Sin. **21(12)**: 1141-1144.

Wang, X.-W., Zhang, H.P., Chen, F., Wang, X and Wen, W.Y. 2010. *In vitro* antioxidant and antimicrobial activities of lignan compounds from *Gynostemma pentaphyllum*. **31(13)**: 154-157 (in Chinese).

Wang, X.Y., Shi, H.M., Zhang, L. and Li, X.B. 2009. A new chalcone glycoside, a new tetrahydrofuranoid lignan, and antioxidative constituents from the stems and leaves of *Viburnum propinquum*. Planta Medica **75(11)**: 1262-1265.

Wang, Z. and Luo, D. 2007. Antioxidant activities of different fractions of polysaccharide purified from *Gynostemma pentaphyllum* Makino. Carbohydrate Polymers **68(1)**: 54-58.

Wang, Z.Y., Cheng, S.J. and Zhou, Z.C. 1989. Antimutagenic activity of green tea polyphenols. Mut. Res. **223**: 273-285.

Wangensteen, H., Dang, H.C., Uddin, S.J., Alamgir, M. and Malterud, K.E. 2009. Antioxidant and antimicrobial effects of the mangrove tree *Heritiera fomes*. Nat. Prod. Commun. **4(3)**: 371-376.

Wangensteen, H., Miron, A., Alamgir, M., Rajia, S., Samuelsen, A.B. and Malterud, K.E. 2006. Antioxidant and 15-lipoxygenase inhibitory activity of rotenoids, isoflavones and phenolic glycosides from *Sarcolobus globosus*. Fitoterapia **77**: 290-295.

Wangensteen, H., Samuelsen, A.B. and Malterud, K.E. 2004. Antioxidant activity in extracts from coriander. Food Chem. **88 (2)**: 293-297.

Wannes, W.A., Mhamdi, B., Sriti, J., Jenia, M.B., Ouchikh, O., Hamdaoni, G., Kchouk, M.E. and Marzouk, B. 2010. Antioxidant activities of the essential oils and methanol extracts from the myrtle (*Myrtus communis* var. *italica*) leaf, stem and flower. Food Chem. Toxicol. **48**: 1362-1370.

Wansi, J.D., Wandji, J., Mbaze Meva'A, L., Kamdem Waffo, A.F., Ranjit, R., Khan, S.N., Asma, A., Iqbal, C.M., Lallemand, M.C., Tillequin, F. *et al.*, 2006. Alpha-glucosidase inhibitory and antioxidant acridone alkaloids from the stem bark of *Oriciopsis glaberrima* Engl. (Rutaceae). Chem. Pharm. Bull. **54(3)**: 292-296.

Wanyo, P., Siriamornpun, S. and Meeso, N. 2011. Improvement of quality and antioxidant properties of dried mulberry leaves with combined far-infrared radiation and air convection in Thai tea process. Food and Bioproducts processing **89**: 22-30.

Waszkiewicz-Robak, B., Rusaczonek, A. and Swiderski, F. 2005. Characteristic of antioxidant properties of leaf teas. Annales Universaitatis Mariae Curie-Sklodowska **16**: 169-172.

Watanabe T, Nakajima Y, Konishi T. 2008. *In vitro* and *in vivo* anti-oxidant activity of hot water extract of basidiomycetes-X, newly identified edible fungus. Biol. Pharm. Bull. **31(1)**: 111-117.

Wattenberg, L.W. 1996. Chemoprevention of cancer. Preventive Med. **25**: 44-45.

Wazir, D., Ahmad, S., Muse, R., Mahmood, M. and Shukor, M.Y. 2011. Antioxidant activities of different parts of *Gnetum gnemon* L. J. Plant Biochem. Biotech. **20(2)**: 234-240. doi: 10.1007/s13562-011-0051-8

Wei, A. and Shibamoto, T. 2007. Antioxidant activities and volatile constituents of various essential oils. J. Agric. Food Chem. **55(5)**: 1737-1742.

Wei, Q., Chuan, Z., Nan, Q., Huiyuan, Z. and Hongquan, D. 2010. Identification of trans-tiliroside as active principle with anti-hyperglycemic, anti-hyperlipidemic and antioxidant effects from *Potentilla chinensis*. www.paper.edu.cn/index.php/default/releasepaper/./

Weissman, L., Garty, J. and Hochman, A. 2005. Characterization of enzymatic antioxidants in the lichen *Ramalina lacera* and their response to rehydration. Appl Environ Microbiol. **71(11)**: 6508-6514.

Weng, J.R., Tsao, L.T., Wanf, J.P., Wu, R.R. and Lin, C.N. 2004. Antiinflammatory phloroglucinols and terpenoids from *Garcinia subelliptica*. J. Nat. Prod. **67(11)**: 1796-1799.

Weng, X.C. and Gordon, M.H. 1992. Antioxidant activities of quinines extracted from tanshen (*Salvia miltiorrhiza* Bung.). J. Agric. Food Chem. **40**: 1331-1336. doi: 10.1021/jf00020a007.

Weng, X.C. and Wang, W. 2000. Antioxidant activity of compounds isolated from *Salvia plebeia*. Food Chem. **71(4)**: 489-493.

Wettasinghe, M. and Shahidi, F. 1999. Antioxidant and free-radical scavenging properties of ethanol extracts of defatted borage (*Borago officinalis* L.) seeds. Food Chem. **67**: 399-414.

Wettasinghe, M. and Shahidi, F. 2000. Scavenging of reactive oxygen species of DPPH free radicals by extracts of borage and evening primrose meals. Food Chem. **70**: 17-26.

Wetwitayaklung, P., Phaechamud, T. and Keokitichai, S. 2005. The antioxidant activity of *Caesalpinia sappan* L. heartwood in various ages. Naresuan Univ. J. **13(2)**: 43-52.

White, P.J. and Xiang, Y. 1997. Antioxidants from Cereals and Legumes. In: Shahid, F. (ed.) Natural antioxidants, Health effects and applications, AOCS Press, Champaign, IL.

Wiboonpun, N., Phuwapraisirisan, P. and Tippyang, S.2004. Identification of antioxidant compound from *Asparagus racemosus*. Phytother. Res. **18(9)**: 771-773.

Wijaya, S., Nee, T.K., Jin, K.T., Hon, L.K., San, L.H. and Wiart, C. 2011. Antibacterial and antioxidant activities of *Synedrella nodiflora* (L.) Gaertn. J.Complement. Integr.Med. **8(1)**: Article 13. DOI: 10.2202/1553-3840.1499

Wijayasiriwarddena, C., Chauhan, M.G., Sharma, P.P., Lahiri, S.K. and Shah, M.B. 2009. Anti-inflammatory activity of *Delonix elata* (L.) Gamble. J. Nat. Remedies **9(2)**: 209-215.

Wilcox, J.K. and Ash, S.L. 2004. Antioxidants and prevention of chronic disease. Crit. Rev. Food Sci. Nutr. **44**: 275-295.

Wiseman, S.A., Balentine, D.A. and Frei, B. 1997. Antioxidants in tea. Crit. Rev. Food Sci. Nutr. **37**: 705-718.

Wojcicki, J. and Samochowiec, L. 1996. *Ginkgo biloba* extract inhibits the development of experimental atherosclerosis in rabbits. Phytomedicine **3**(Suppl.): 81.

Wojcicki, J., Samochowiec, L., Gonet, B., Juzwiak, S., Dabrowska-Zamojcin, E., Katdonska, M. and Tustanowski, S. 1995. Effect of buckwheat extract on free radical generation in rabbits administered high-fat diet. Phytother. Res. **9(5)**: 323-325.

Wojcikowski, K.N.D., Stevensen, L., Leach, D., Wohlmuth, N.D.H. and Gobe, G. 2007. Antioxidant capacity of 55 medicinal herbs traditionally used to treat the urinary system: A comparison using a sequential three-solvent extraction process. J. Altenat. Complement. Med. **13**: 103-109.

Wojdylo, A., Oszmianski, J. and Czemerys. 2007. Antioxidant activity and phenolic compounds in 32 selected herbs. Food Chemistry **105**: 940-949.

Wojdylo, A., Oszmianski, J. and Sokol-Letowska, A. 2007. Comparison of antioxidative and synergistic effects of skullcap (*Scutellaria baicalensis* Georgi) in sunflower oil-in-water emulsion during storage. Electronic Journal of Polish Agric. Universities **10(2)**:

Wolfe, K., Xianzhong, W.U. and Liu, R.H. 2003. Antioxidant activity of apple peels. J. Agric. Food Chem. 51: 609-614.

Wolniak, M., Tomczykowa, M., Tomczyk, M., Gudej, J. and Wawer, I. 2007. Antioxidant activity of extracts and flavonoids from *Bidens tripartita*. Acta Poloniae Pharmaceutica **64(5)**: 441-447.

Wong, C.C., Li, H.B., Cheng, K.W. and Chen, F. 2006. A systematic study of antioxidant activity of 30 Chinese medicinal plants using the ferric reducing antioxidant power assay. Food Chem. **97(4)**: 705-711.

Wong, K.-L., Chao, H.-H., Chan, P., Chang, L.-P. and Liu, C.-F. 2005 Antioxidant activity of *Ganoderma lucidum* in acute ethanol-induced heart toxicity. Phytother. Res. **18(12)**: 1024-1026.

Wong, P.Y.Y. and Kitts, D.D. 2006. Studies on the dual antioxidant and antibacterial properties of parsley (*Petroselenium crispum*) and cilantro (*Coriandrum sativum*) extracts. Food Chem. **97**: 505-515.

Wong, S.P., Lai, P.L. and Jen, H.W.K. 2006. Antioxidant activities of aqueous extracts of selected plants. Food Chem. **99**: 775-783.

Wongsawatkul, O., Prachayasittikul, S., Isarankura-Na-Ayudhya, C., Satayavivad, J., Ruchirawat, S. and Prachayasittikul, V. 2008. Vasorelaxant and antioxidant activities of *Spilanthes acmella*. Int. J. Mol. Sci. **9(12)**: 2724-2744.

Woo, E.R. and Piao, M.S. 2004. Antioxidative constituents from *Lycopus lucidus*. Arch. Pharm. Res. **27(2)**: 173-176.

Woo, N., Kim, T., Park, C.G., Seong, H.J., Ko, S.B. and Kang. 2005. Antioxidant and antimicrobial activities of extracts from different parts of *Crotalaria sessiliflora* L. J. Korean Soc. Food Sci. Nut. 34: 948-952.

Wright, J.R., Colby H.D. and Miles, P.R. 1981. Cytosolic factors which affect microsomal lipid peroxidation in lung and liver. Arch. Biochem. Biophys. **206**: 296–304.

Wu, C., Chen, F., Wang, X., Kim, H.-J., He, G., Haley-Zitllin, V. and Huang, G. 2006. Antioxidant constituents in feverfew (*Tanacetum parthenium*) extract and their chromatographic quantification. Food Chem **96(2)**: 220-227.

Wu, J.B., Lin, W.L., Hsieh, C-C., Ho, H-Y., Tsay, H-S. and Lin, W.C. 2006. The hepatoprotective activity of kinsenoside from *Anoectochilus formosanus*. Phytother. Res. **21(1)**: 58-61.

Wu, J.-F., Chen, S.-B., Gao, J.-C., Song, H.-L., Wu, L.-J., Chen, S.-L. and Tu, P.-F. 2008. Xanthone glycosides from herbs of *Polygala hongkongensis* Hemsl. and their antioxidant activities. J. Asian Nat. Prod. Res. **10(7)**: 665-670. doi: 10.1080/1028602081967201

Wu, J.-H., Huang, C.-Y., Tung, Y.-T. and Chang, S.T. 2008. On line RP-HPLC-DPPH screening method for detection of radical-scavenging phytochemicals from flowers of *Acacia confusa*. J. Agric. Food Chem. **56**: 328-332.

Wu, K., Zhang, W., Addis, P.B., Epley, R.J., Salih, A.M. and Lehrfeld, J. 1994. Antioxidant properties of wild rice. J. Agric. Food Chem. **42(1)**: 34-37.

Wu. L., Hsu, H.W., Chen, Y.C., Chiu, C.C., Lin, Y.I. and Ho, J.A. 2006. Antioxidant and antiproliferative activities of red pitaya. Food Chem. **95**: 319-327.

Wu, L., Qiao, H., Li, Y. and Li, L. 2007. Protective roles of puerarin and Danshensu on acute ischemic myocardial injury in rats. Phytomedicine **14(10)**: 652-658.

Wu, L.C., Chang, L.-H., Chen, S.-H., Fan, N.-C. and Ho, J.A. 2009. Antioxidaant activity and melanogenesis inhibitory effect of the acetonic extract of *Osmanthus fragrans*: A potential natural and functional food flavour additive. LWT Food Sci. Tech. **42(9)**: 1513-1519. doi: 10.1016/j.lwt.2009.04.004

Wu, L.C., Ho, J.A., Shieh, M.C., Lu, I.W. 2005. Antioxidant and antiproliferative activities of *Spirulina* and *Chlorella* water extracts. J. Agric. Food Chem. **53(10)**: 4207-12.

Wu, M.J., Yen, J.H., Wang, L. and Weng, C.Y. 2004. Antioxidant activity of Porcelein berry (*Ampelopsis brevipedunculata* (Maxim.) Trautv.). Am. J. Chin. Med. **42(3)**: 681-693.

Wu, N., Fu, K., Fu, Y.J.,Zu, Y.G.,Chang, F.R.,Chen,Y.H., Liu, X.L.,Kong,Y.,Liu, W. and Gu, C.B. 2009. Antioxidant activities of extracts and main components of Pigeon pea [*Cajanus cajan* (L.) Millsp.] leaves. Molecules. **14(3)**: 1032-1043.

Wu, N., Zu, Y., Fu, Y., Kong, Y., Zhao, J., Li, X., Li, J., Wink, M. and Efferth, T. 2010. Antioxidant activities and xanthine oxidase inhibitory effects of extracts and main polyphenolic compounds obtained from *Geranium sibiricum* L. J. Agric. Food Chem. **58(8)**: 4737-4743.

Wu Q, Fu, D. X, Hou, A. J, Lei, G. Q, Liu, Z. J, Chen, J. K. and Zhou, T. S. 2005. Antioxidant phenols and phenolic glycosides from *Curculigo orchioides*. Chem. Pharm. Bull. (Tokyo). **53(8)**: 1065-1067.

Wu, Q.X., Li, Y. and Shi, Y.P. 2006. Antioxidant phenolic glucosides from *Gentiana piasezkii*.J. Asian Nat. Prod. Res. **8(5)**: 391-396.

Wu, S.J., Ng, L.T., Huang, Y.M., Lin, D.L., Wang, S.S., Huang, S.N, and Lin, C.C. 2005. Antioxidant activities of *Physalis peruviana*. Biol. Pharm. Bull. **28(6)**: 963-966.

Wu, S.J., Ng, L.T. and Lin, C.C. 2004 Antioxidant activities of some common ingredients of traditional chinese medicine, *Angelica sinensis, Lycium barbarum* and *Poria cocos*. Phytother. Res. **18(12)**: 1008-1012.

Wu, T.S., Hsu, M.Y., Kuo, P.C., Sreenivasulu, B., Damu, A.G., Su, S.R., Li, C.Y. and Chang, H.C. 2003. Constituents from the leaves of *Phellodendron amurense* var. *wilsonii* and their bioactivity. J. Nat. Prod. **66(9)**: 1207-1211.

Wu, X., Beecher, G.R., Holden, J.M., Haytowitz, D.B., Gebhardt., S.E. and Prior, R.L. 2004. Lipophilic and hydrophilic antioxidant capacities of common foods in the United States. J. Agric. Food Chem. **52**: 4026-4037.

Wu, Y. and Wang, D. 2008. Structural characterization and DPPH radical scavenging activity of an arabinoglucogalactan from *Panax notoginseng* root. J. Nat. Prod. **71(2)**: 241-245.

Wu, Y.J., Shi, J., Qu, L.B., Li, F.F., Li, X.J. and Wu, Y.M. 2006. Determination of antioxidation of the extract from Chinese medicine *Morinda officinalis* How by flow injection chemiluminescence and spectroscopy. Guang Pu Xue Yu Guang Pu Fen Xi **26(9)**: 1688-1691.

Wu, X., Beecher, G., Holden, J., Hayatowitz, D., Gebhart, S. and Prior, R. 2004. Lipophilic and hydrophilic antioxidant capacities of common food in the United States. J. Agric. Food Chem. **52**: 4026-4037.

Wu, Y., Yang, L., Wang, F., Wu, X., Zhou, C., Shi, S., Mo, J. and Y. 2007. Hepatoprotective and antioxidative effects of total phenolics from *Laggera pterodonta* on chemical-induced injury in primary cultured neonatal rat hepatocytes. Food chem. Toxicol. **45(8)**: 1349-1355.

Xanthopoulou Marianna, N., Nomikos Tzortis, Fragopoulou Elizabeth and Antonopoulou Smaragdi. 2009. Antioxidant and lipoxygenase inhibitory activities of pumpkin seed extracts. Food Res. Int. **42(5-6)**: 641-646.

Xi, M., Hai, C., Tang, H., Chen, M., Fang, K. and Liang, X. 2008. Antioxidant and antiglycation properties of total saponins extracted from Chinese medicine used to treat diabetes mellitus. Phytother. Res. **22(2)**: 228-237.

Xi, M., Hai, C., Tang, H., Wen, A., Chen, H., Liu, R., Liang, X. and Chen, M. 2010. Antioxidant and antiglycation properties of triterpenoid saponins from *Aralia taibensis* traditionally used for treating diabetes mellitus. Redox Rep. **15(1)**: 20-28.

Xiang, J., Uno, T., Ogino, F., Al, C., Duo, J. and Sankawa, U. 2005. Antioxidant constituents of *Caragana tibetica*. Chem. Pharm. Bull. **53(9)**: 1204-1206.

Xiaojun, Y., Tadahiro, N. and Xiao, F. 1996. Antioxidative activities in some common seaweeds. Plants foods for Human Nutr. **52**: 253-262.

Xie, B., Shi, H., Chen, Q. and Ho, C.T. 1993. Antioxidant properties of fractions and polyphenol constituents from green, oolong and black teas. Life Sci. **17**: 77-84.

Xie, C., Koshino, H., Esumi, Y., Takahashi, S., Yoshilkawa, K. and Abe, N. 2005. Vialinin A, a novel 2,2-diphenyl-1-picrylhydrazyl (DPPH) radical scavenger from an edible mushroom in China. Biosci. Biotechnol. Biochem. **69(12)**: 2326-2332.

Xie, G., Schepetkin, I.A, and Quinn, M.T. 2007. Immunomodulatory activity of acidic polysaccharides isolated from *Tanacetum vulgare* L. Int. Immunopharmacol. **7(13)**: 1639-50.

Xie, J. and Zhang, J. 2010. Study on antioxidant activity *in vitro* of different polysaccharides from *Momordica charantia*. Guangzhou Chemical Industry: 2010-05.

Xie, J.-H., Xie, M.-Y., Nie, S.-P., Shen, M.-Y., Wang, Y.-X. and Li, C. 2010. Isolation, chemical composition and antioxidant activities of a water soluble polysaccharide from *Cyclocarya paliurus* (Batal.) Iljinskaja. Food Chem **119(4)**: 1626-1632.

Xing, Y. and White, P.J. 1997. Identification and function of antioxidants from oat groats and hulls. J. Am. Oil Chem. Soc. 74: 303-307.

Xiong, Q., Kadota, S., Tani, T. and Namba, T. 1996. Antioxidative effects of phenylethanoids from *Cistanche deserticola*. Biol. Pharm. Bull. **19(12)**: 1580-1585.

Xu, B. and Chang, S.K. 2008. Total phenolic content and antioxidant properties of eclipse black beans (*Phaseolus vulgaris* L.) as affected by processing methods. J. Food Sci. **73(2)**: H19-27.

Xu, B. and Chang, S.K. 2009. Total phenolic, phenolic acid, anthocyanin, flavan-3-ol, and flavonol profiles and antioxidant properties of Pinto and black beans (*Phaseolus vulgaris* L.) as affected by thermal processing. J.Agric. Food Chem. **57(11)**: 4754-4764.

Xu, G., Liu, D., Chen, J., Ye, X., Ma, Y. and Shi, J. 2008. Juice components and antioxidant capacity of citrus varieties cultivated in China. Food Chem. **106**: 545-551.

Xu, G.H., Chen, J.C., Liu, D.H., Zhang, Y.H., Jiang, P. and Ye, X.Q. 2008. Minerals, phenolic compounds, and antioxidant capacity of citrus peel extract by hot water. J. Food Sci. **73(1)**: 11-18.

Xu, J., Chen, S., Hu, Q. 2005. Antioxidant activity of brown pigment and extracts from black sesame seed (*Sesamum indicum* L.), Food Chem. **91 (1)**: 79-83.

Xu, L., Zeng, Y., Long, G. and Peng, Y. 2005. The antioxidant activities and their relationship with the relative polyphenols and flavonols contents of several flowers extracts. Chinese Wild Plant resources 2005-01.

Xu, R., Zhang, C. *et al.*, 2011. Antioxidant activity from different extracts of *Syringa pubescens* Turcz. Food Sci. **32(13)**:

Xu, W., Liu, L., Hu, B., Sun, Y., Ye, H., Ma, D and Zheng, X. 2010. Antioxidant activity *in vitro* and polyphenol content of sweet potato (*Ipomoea batatas* L.) leaves of Pushu 53. Journal of Food Composition and Analysis **23**: (in press)

Xu, W., Zhang, F., Luo, Y., Ma, L., Kou, X. and Huang, K. 2009. Antioxidant activity of a water-soluble polysaccharide purified from *Pterigium aquilinum*. Carbohydr. Res. **344(2)**: 217-222.

Xu, Y., Ho, C.T., Amin, S.G., Han, C. and Chung, F.L. 1992. Inhibition of tobacco specific nitrosamine-induced lung tumorogenesis in A/J mice by green tea ar.d its major polyphenols as antioxidants. Cancer Res. **52**: 3875-3879.

Xu, Z., Tang. M., Li, Y., Liu, F., Li, X. and Dai, R. 2010. Antioxidant properties of Du-zhong (*Eucommia ulmoides* Oliv.) extracts and their effects on color stability and lipid peroxidation of raw pork patties. J. Agric. Food Chem. **58(12)**: 7289-7296.

Yadegarinia, D., Gachkar, L., Rezaei, M.B., Taghizadeh, M., Astaneh, S.A. and Rasooli, I. 2006. Biochemical activities of Iranian *Mentha piperita* L. and *Myrtus communis* L. essential oils. Phytochem. **67(12)**: 1249-1255.

Yadav, S.B. and Tripathi, V. 2004. Chemical screening of *Cuscuta reflexa* and its antioxidant activity. IUPAC International conference on Biodiversity and Natural Products Chemistry and Medical applications, New Delhi, p. P-410.

Yadav, S.B., Tripathi, V., Singh, R.K. and Pandey, H.P. 2001. Flavonoid glycosides from *Cuscuta reflexa* stems and their antioxidant activity. Indian Drugs. **38(2)**: 95-96.

Yagi, A., Kabash, A., Okamura, N., Haraguchi, H., Moustafa, S.M. and Khalifa, T.I. 2002. Antioxidant, free radical scavenging and anti-inflammatory effects of aloesin derivatives in *Aloe vera*. Planta Med. **68(11)**: 957-960.

Yaltirak, T., Aslim, B., Ozturk, S. and Alli, H. 2009. Antimicrobial and antioxidant activities of *Russula delica* Fr. Food Chem. Toxicol. 10.1016/j.f.ct.2009.05.029

Yamaguchi, F., Ariga, T., Yoshimura, Y. and Nakazawa, H. 2000a. Antioxidative and anti-glycation activity of garcinol from *Garcinia indica* fruit rind. J. Agric. Food Chem. **48**: 180-185.

Yamaguchi, F., Saito, M., Ariga, T., Yoshimura, Y. and Nakazawa, H. 2000b. Free radical scavenging activity and antiulcer activity of garcinol from *Garcinia indica* fruit rind. J. Agric. Food Chem. **48(6)**: 2320-2325.

Yamaguchi, L.F., Vassao, D.G., Kao, M.J.and Mascio, P.D. 2005. Biflavonoids from Brazilian pine *Araucaria angustifolia* as potentials protective agents against DNA damage and lipoperoxidation. Phytochem. **66**: 2238-2247.

Yamaguchi, N., Satoh-Yamaguchi, K. and Ono, M. 2009. *In vitro* evaluation of antibacterial, anticollagenase and antioxidant activities of hop components (*Humulus lupulus*) addressing acne vulgaris. Phytomedicine **16(4)**: 369-376.

Yamagushi, F., Ariga, T., Yoshimira, Y. and Nakazawa, H. 2000. Antioxidative and anti-glycation activity of garcinol from *Garcinia indica*. J. Agric. Food Chem. **48**: 80-86.

Yan, X., Suzuki, M., Ohnishi-Kameyama, M., Sada, Y., Nakanishi, T., and Nagata, T. 1999. Extraction and identification of antioxidants in the roots of yacon (*Smallanthus sonchifolius*). J. Agric. Food. Chem. **47 (11)**: 4711-4713.

Yanagimoto, K., Ochi, H., Lee, K.O. and Shibamoto, T. 2004. Antioxidative activities of fraction obtained from brewed coffee. J. Agric. Food Chem. **52(3)**: 529-596.

Yang B., Wang J., Zhao M., Liu Y., Wang W., Jiang Y. 2007. Identification of polysaccharides from pericarp tissues of litchi (*Litchi chinensis* Sonn.) fruit in relation to their antioxidant activities. Carbohydrate Research, **341(5)**: 634-638.

Yang, H., Protiva, P., Cui, B., Ma, C., Baggett, S., Hequet, V., Mori, S., Weinstein, I.B. and Kennelly, E.J. 2003. New bioactive polyphenols from *Theobroma grandiflorum* ("Cupuacu"). J. Nat. Prod. **66(11)**: 1501-1504.

Yang, H., Protiva, P., Gil, R., Jiang, B., Baggett, S., Basile, M., Reynertson, K.A., Weinstein, I.B. and Kennelly, E.J. 2005. Antioxidant and cytotoxic isoprenylated coumarins from *Mammea americana*. Planta Medica **71(9)**: 852-860.

Yang, H.L., Chen, S.C., Lin, K.Y., Wang, M.T., Chen, Y.C., Huang, H.C., Cho, H.J., Wang, L., Kumar, K.J. and Hseu, Y.C. 2011. Antioxidant activities of aqueous leaf extracts of *Toona sinensis* on free radical-induced endothelial damage. J. Ethnopharmacol. **137(1)**: 669-680.

Yang, J., Qu, Z., Xiao, Y.-L., Qu, G.-F., Zhang, T., Wu, Z.-Y., He, X.R. and Hu, X.M. 2011. Chemical composition and antioxidant activity of the essential oil of endemic *Viola tianshanica*. Nat. Prod. Res. doi: 10.1080/14786419.2010.511218

Yang, J., Wang, D., Zheng, R., Guo, L., Wang, J., Ren, Z. and Wu, C. 2008. Protective effect of an ethanolic extract of *Trollius ledebourii* on potassium dichromate-induced genotoxicity in mouse peripheral lymphocytes. J. Tradtional Med. **3(3)**:

Yang, J. *et al.*, 2009. Antioxidant capacity of some specific vegetables. J. Anhui Agric. Bull. 2009-03.

Yang, J.H., Hu, J.P., Rena, K. and Du, N.S. 2009. Structure-activity relationships of phenylethanoid glycosides in plants of *Cistanche salsa* on antioxidative activity. Zhong Yao Cai **32(7)**: 1067-1069.

Yang, J.H., Lin, H.C. and Mau, J.L. 2002. Antioxidant properties of several commercial mushrooms. Food Chem. **77**: 229-235.

Yang, S.A., Jeon, S.K., Lee, E.J., Im, N.K., Jhee, K.H., Lee, S.P. and Lee, I.S. 2009. Radical scavenging activity of the essential oil of silver fir (*Abies alba*). J. Clin. Biochem. Nutr. **44**: 253-259.

Yang, S.A., Jeon, S.K., Lee, E.J., Shim, C.H. and Lee, I.S. 2010. Comparative study of the chemical composition and antioxidant activity of six essential oils and their components. Nat. Prod. Res. **24(2)**: 140-151.

Yang, W.M., Liu, J.K., Hu, L., Dong, Z.J., Wu, W.L., Chen, Z.H. 2004. Antioxidant properties of natural *p*-terphenyl derivatives from the mushroom *Thelephora ganbajun*. Z. Naturforsch [C]. **59(5-6)**: 359-62.

Yang, W.M., Liu, J.K., Qin, X.D., Wu, W.L. and Chen, Z.H. 2004. Antioxidant activities of three dihydrochalcone glucosides from leaves of *Lithocarpus pachyphyllus*. Z. Naturforsch C. **59(7-8)**: 481-484.

Yang, X., Zhao, H.T., Wang, J., Meng, Q., Zhang, H., Yao, L., Zhang, Y.C., Dong, A.J., Ma, Y., Wang, Z.Y., Xu, D.C. and Ding, Y. 2010. Chemical composition and antioxidant activity of essential oil of pine cones of *Pinus armandii* from the Southwest region of China. J. Med. Plants Res. **4(16)**: 1668-1672.

Yang, X., Zhao, Y. and Lv, Y. 2007. Chemical composition and antioxidant activity of an acidic polysaccharide extracted from *Cucurbita moschata* Duchesne ex Poiret. J. Agric. Food Chem. **55(12)**: 4684-4690.

Yang, X., Zhao, Y., Yang, Y. and Ruan, Y. 2008. Isolation and characterization of immunostimulatory polysaccharide from an herb tea, *Gynostemma pentaphyllum* Makino. J. Agric. Food Chem. **56(16)**: 6905-9.

Yang, X.W., He, H.P., Ma, Y.L., Wang, F., Zuo, Y.Q., Lin, H., Li, S.L., Li, L. and Hao, X.J. 2010. Three new vanilloid derivatives from the stems of *Baccaurea ramiflora*. Planta Medica **76(1)**: 88-90.

Yang, X.W., Ma, Y.L., He, H.P., Wang, Y.H., Di, Y.T. Zhou, H., Li, L. and Hao, X.J. 2006. Iridoid constituents of *Tarenna attenuata*. J. Nat. Prod. **69(6)**: 971-974.

Yang, X.W., Wang, J.S., Ma, Y.L., Xiao, H.T., Zuo, Y.Q., Lin, H., He, H.P., Li, L. and Hao, X.J. 2007. Bioactive phenols from the leaves of *Baccaurea ramiflora*. Planta Med. **73(13)**: 1415-1441.

Yang, Y., Yue, Y., Runwei, Y. and Guolin, Z. 2010. Cytotoxic, apoptotic and antioxidant activity of the essential oil of *Amomum tsao-ko*. Bioresource Technol. **101**: 4205-4211.

Yang, Z., Huang, J., Wang, N. and Yao, X. 2008. New bergenin derivatives and the antioxidant activity from *Ardisia gigantifolia* Stapf. J. Shenyang Pharmaceutical Univ. **25 (1)**: 30-34.

Yangthong, M., Hutadilok-Towatana, N. and Pharmkunthong, W. 2009. Antioxidant activities of four edible seaweeds from the southern coast of Thailand. Plant Foods for Human Nutrition **64(3)**: 218-223.

Yanishlieva, N. and Marinova, E. 2003. Natural antioxidants in lipid oxidation. Bulgarian Chem. Comm. **35**: 79-91.

Yanishlieva, N., Marinova, E. and Pokorny, J. 2006. Natural antioxidant from herbs and spices. Eur. J. Lipid Sci. Tech. 108: 776-793.

Yassa, N., Sharififar, F. and Shafiee, A. 2005. *Otostegia persica* as a source of natural antioxidants. Pharmaceutical Biol. **43(1)**: 33-38.

Yasukawa, K., Kitanaka, S. and Seo, S. 2002. Inhibitory effect of stevioside on tumor promotion by 12-O-tetradecanpylphorbol-13-acetatein two-stage carcinogenesis in mouse skin. Biol. Pharm. Bull. **25**: 1488-1490.

Yazdanparast, R. and Ardestani, A. 2007. *In vitro* antioxidant and free radical scavenging activity of *Cyperus rotundus*. J. Med. Food **10(4)**: 667-674.

Yazdanparast, R., Bahramikia, S. and Ardestani, A. 2008. *Nasturtium officinale* reduces oxidative stress and enhances antioxidant capacity in hypercholesterolaemic rats. Chemico-Biological Interactions **172(3)**: 176-184.

Yeboah, E.M. and MajindAa, R.R. 2009. Radical scavenging activity and total phenolic content of extracts of the root bark of *Osyris lanceolata*. Nat. Prod. Comm. **4(1)**: 89-94.

Yele, S.U., Kulkarni, Y.A.,Gokhale, S.B. 2008. Determination of *in vitro* antioxidant activity of Kasmard (*Cassia sophera* Linn) leaves. Pharmacology online **1**: 548-554.

Yen, F.L., Wu, T.H., Lin, L.T. and Lin, C.C.2007. Hepatoprotective and antioxidant effects of *Cuscuta chinensis* against acetaminophen-induced hepatotoxicity in rats. J. Ethnopharmacol. **111(1)**: 123-128.

Yen, G.C. and Chen, H.Y. 1995. Antioxidant activity of various tea extracts in relation to their antimutagenicity. J. Agric.Food Chem. **43**: 27-32.

Yen, G.C. and Chuang, D.Y. 2000. Antioxidant properties of water extracts from *Cassia tora* L. in relation to the degree of roasting. J. Agrc. Food Chem. **48**: 2760-2765.

Yen, G.C., and Duh, P.D. 1994. Scavenging effect of methanolic extracts of peanut hulls on free-radical and active oxygen species. J. Agric. Food Chem. **42**: 629-632.

Yen, G.C., Duh, P.D. and Chuang, D.Y. 2000. Antioxidant activity of anthraquinones and anthrone. Food Chem. **70**: 437-441.

Yen, G.C., Duh, P.D. and Tsai, C.L. 1993. Relationship between antioxidant activity and maturity of peanut hulls. J. Agric. Food Chem. **41**: 67-70.

Yen, G.C. and Hsieh, C.L. 1998. Antioxidant activity of extracts from Duzhong (*Eucommia ulmoides*) toward various lipid peroxidation models *in vitro*. J. Agric. Food Chem. **46(10)**: 3952-3957.

Yen, G.-C., Lai, H.-H. and Chou, H.-Y. 2001. Nitric oxide-scavenging and antioxidant effects of *Uraria crinita* root. Food Chem. **74(4)**: 471-478.

Yen, G.C. and Wu, J.Y. 1999. Antioxidant and radical scavenging properties of extracts from *Ganoderma tsugae*. Food Chem. **65**: 375-379.

Yen, G.H. and Chen, H.Y. 1995. Antioxidant activity of various tea extract in relation to their antimutagenicity. J. Agri. Food chem. **43**: 27–32.

Yen, W.-J., Chang, L.-W. and Duh, P.-D. 2005. Antioxidant activity of peanut seed testa and its antioxidative component, ethyl protocatechuate. LWT – Food Sci. Tech. **38(3)**: 193-200.

Yeole, P.G., Wadher, S.J. and Gaikwad, N.J. 2008. Free radical scavenging activity of water extract of heartwood of *Pterocarpus marsupium*. J. Nat. Remedies **8(1)**: 89-92.

Yerra, R., Kumar S. G. P., Malya, G. and Kanti, M.U. 2005. Studies on *in vitro* antioxidant activities of methanol extracts of *Mucuna pruiens* (Fabaceae) seeds. Eur. Bull. Drug Res. **13(1)**: 31-35.

Yesilada, E., Tsuchiya, K., Takaishi, Y. and Kawazol, K. 2000. Isolation and characterization of free radical scavenging flavonoid glycosides from the flowers of *Spartium junceum* by activity-guided fractionation. J. Ethnopharmacol. **73(3)**: 471-478. doi: 10.1016/S0378-874(00)00327-5.

Yesil-Celiktas, O., Nartop, P., Gurel, A., Bedir, E., and Vardar-Sukan, F. 2007. Determination of phenolic content and antioxidant activity of extracts obtained from *Rosmarinus officinalis'* calli. J. Plant Physiol. **164(11)**: 1536-42.

Yi, O., Jovel, E.M., Towers, G.H., Wahbe, T.R. and Cho, D. 2007. Antioxidant and antimicrobial activities of native *Rosa* sp. from British Columbia, Canada. Int. J.Food Sci. Nutr. **58(3)**: 178-189.

Yiit, D., Yiit, N. and Mavi, A. 2009. Antioxidant and antimicrobial activities of bitter and sweet apricot (*Prunus armeniaca* L.) kernels. Braz. J. Med. Biol. Res. **42(4)**: 346-352.

Yildirim, A., Mavi, A. and Kara, A.A. 2001. Determination of antioxidant and antimicrobial activities of *Rumex crispus* L. extracts. J. Agric. Food Chem. **49(8)**: 4083-4089.

Yildirim, A., Mavi, A., Oktay, M., Kara, A.A., Algur, O.F. and Bilaloglu, V. 2000. Comparison of antioxidant and antimicrobial activities of Tilia (*Tilia argentea* Desf. ex DC.), Sage (*Salvia triloba* L.) and Black tea (*Camellia sinensis*) extracts. J. Agric. Food Chem. **48(10)**: 5030-5034.

Yildirin, A., Oktay, M. and Bilaloglu, V. 2001. The antioxidant activities of the leaves of *Cydonia vulgaris*. Turk. J. Med. Sci. **31**: 23-27.

Yildiz, O. and Eyduran, P. 2009. Functional components of berry fruits and their usage in food technologies. Amer. J. Agric. Res. **4(5)**: 422-426.

Yilmaz, Y. and Toledo, R.T. 2004. Major flavonoids in grape seeds and skins: Antioxidant capacity of catechin, epicatechin, and gallic acid. J. Agric. Food Chem. **52(2)**: 255-260.

Yim, T.K., Wu, W.K., Pak, W.F. and Ko, K.M. 2001. Hepatoprotective action of an oleonolic acid-enriched extract of *Ligustrum lucidum* fruits is mediated through and enhancement on hepatic glutathione regeneration capacity in mice. Phytother. Res. **15(7)**: 589-592.

Yin, J., Kwon, G.J. and Wang, M.H. 2007. The antioxidant and cytotoxic activities of *Sonchus oleraceus* L. extracts. Nutr. Res. Pract. **1(3)**: 189-194.

Yin, M.-C. and Cheng, W.-S. 1998. Antioxidant activity of several *Allium* members. J. Agric. Food Chem. **46(10)**: 4097-4101. doi: 10.1021/jf980344x.

Ying, L., Xu, P., Huang, S. and Wang, Y. 2011. Antioxidant activity of bioactive compounds extracted from *Ampelopsis grossodentata* leaves by optimized supercritical carbon dioxide. J. Med. Plants Res. **5(17)**: 4373-4381.

Yingming, P., Ping, L., Hengshan, W. and Min, L. 2004. Antioxidant activities of several Chinese medicinal herbs. Food Chem. **88**: 347-350.

Yogalatha, L., Sasidharan, S., Zuraini, Z., Saravanan, D., Suryani, S., Sangeetha, S. and Siti Aishah, M.B. 2005. Antioxidant properties of *Psophocarpus tetragonalobus*. J. Trop. Med. Plants **6(2)**: 173-178.

Yokota, J., Takuma, D., Hamada, A., Onogawa, M., Yoshioka, S., Kusunose, M., Miyamura, M., Kyotani, S. and Nishioka, Y. 2006. Scavenging of reactive oxygen species by *Eriobotrya japonica* seed extract. Biol. Pharm. Bull. **29(3)**: 467-471.

Yokozawa,., Kim, H.Y., Cho, E.J., Choi, J.S. and chung, H.Y. 2002. Antioxidant effects of isorhamnetin 3,7-di-O-beta-D-glucopyranoside isolated from mustard leaf (*Brassica juncea*) in rats with streptozotocin-induced diabetes. J. Agric. Food Chem. **50(19)**: 5490-5495.

Yoo, E.A., Kim, S.D., Lee, W.M., Park, H. J., Kim, S.K., Cho, J.Y., Min, W. and Rhee, M.H. 2008. Evaluation of antioxidant, antinociceptive and anti-inflammatory activities of ethanol extracts from *Aloe saponaria* Haw. Phytother. Res. **22 (10)**: 1389-1395.

Yoshikawa, T., Naito, Y. and Kondo, M. 1999. *Ginkgo biloba* leaf extract: review of biological actions and clinical applications. Antioxidant Redox Signal **1(4)**: 469-480.

Yoshino, K., Nishimura, M., Watanabe, A., Saito, S. and Sano, M. 2008. Preventive effects of edible mushroom (*Hypsizigus marmoreus*) on mouse type IV allergy: fluctuations of cytokine levels and antioxidant activities in mouse sera. J.Food.Sci. **73(3)**: T21-25.

Yoshiyuki, K., Michinori, K. Tadato, T, Shigeru, A. and Hiromichi, O. 1981. Studies on *Scutelariae radix* L IV, Effects on lipid peroxidation of rat liver. Chem. Pharm. Bull. **29**: 2610-2617.

Yousfi, M., Djeridane, A., Bombarda, I., Duhem, C.H.B. and Gaydou, M. 2009. Isolation and characterization of a new hispolone derivative from antioxidant extracts of *Pistacia atlantica*. Phytother. Res. **23(9)**: 1237-1242.

Yu, B.P. and Yang, Y. 1996. A critical evaluation of free radical theory of aging, A proposal for the Oxidative stress hypothesis. Ann. N.Y.Acad. Sci. **786**: 1-11.

Yu, L., Haley, S., Perret, J., Harris, M., Wilson, J. and Qian, M. 2002. Free radical scavenging properties of wheat extracts. J. Agric. Food Chem. 50: 1619-1624.

Yu, L., Zhao, M., Wang, J.S., Cui, C., Yang, B., Jiang, Y. and Zhao, Q. 2008. Antioxidant, immunomodulatory and anti-breast cancer activities of phenolic extract from pine (*Pinus massoniana* Lamb) bark. Innovative Food Sci. Emerging Technologies **9(1)**: 122-128.

Yuan, Y.V., Carrington, M.F., Walsh, N.A. 2005. Extracts from dulse (*Palmaria palmata*) are effective antioxidants and inhibitors of cell proliferation *in vitro*. Food Chem. Toxicol. **43(7)**: 1073-81.

Yuen, J.W.M. and Gohel, M.D.I. 2008. The dual roles of *Ganoderma* antioxidants and urothelial cell DNA under carcinogenic attack. J. Ethnopharmacol. **118(2)**: 324-330.

Yuan, Y.V. and Walsh, N.A. 2006. Antioxidant and antiproliferative activities of extracts from a variety of edible seaweeds. Food Chem. Toxicol. **44(7)**: 1144-50.

Yun, B.S., Lee, I.K., Ryoo, I.J., Yoo, I.D. 2001. Coumarins with monoamine oxidase inhibitory activity and antioxidative coumarino-lignans from *Hibiscus syriacus*. J. Nat. Prod. **64(9)**: 1238-1240.

Zaffir, A. and Banu, N. 2007. Remedial antioxidant action of *Withania somnifera* of restraint stress-induced oxidative damage. J. Res. Education in Indian Medicine **13(3)**: 31-38.

Zahara, R., Mehranaj, M., Farzaneh, V. and Kahzad, S. 2007. Antioxidant activity of extract from a brown alga, *Sargassum boveanum*. Afr. J. Biotch. **6**: 2740-2745.

Zahin, M., Aquil, F., Ahmad, I. 2009. The *in vitro* antioxidant activity and total phenolic content of four Indian medicinal plants. Intern. J. Pharm. Pharmaceutical Sci. **1(1)**: 88-95.

Zaibunnisa A.H., Siti Rohani K., Noor Adillah A., Shahiza M. and Noorlaila A. 2005. Effect of Ascorbic acid, Alpha-Tocopherol, Beta-Carotene and total polyphenol on antioxidant activity in selected Malaysian edible plants. Malaysian J. Sci. **24 (1)**: 219-224.

Zainol, M.K., Abdul-Hamid, A., Yusof, S. and Musa, R. 2003. Antioxidative activity and total phenolic compounds of leaf, root and petiole of four accessions of *Centella asiatica* (L.) Urban. Food Chem. **49**: 5165-5170.

Zakaria, Z., Aziz, R., Lachimanan, Y.L., Srinivasan, S. and Rathinam, X. 2008. Antioxidant activity of *Coleus blumei*, *Orthosiphon stamineus*, *Ocimum basilicum* and *Mentha arvensis* from Lamiaceae family. International J. Natural and Engineering Sciences **2(1)**: 93-95.

Zakaria, Z.A. 2007. Free radical scavenging activity of some plant available in Malaysia. Iranian J. Pharmacol. Therapeutics **6**: 87-91.

Zamble, A., Martin-Nizard, F., Sahpaz, S., Reynaert, M.L., Staels, B., Bordet, Duriez, P., Gressier, B. and Bailleul, F. 2009. Effects of *Microdesmis keayana* alkaloids on vascular parameters of erectile dysfunction. Phytother. Res. **23(6)**: 892-895.

Zamble, A., Yao, D., Martin-Nizard, F., Sahpaz, S., Offoumou, M., Bordet, R., Duriez, P., Brunet, C. and Bailleul, F. 2006. Vasoactivity and antioxidant properties of *Microdesmis keayana* roots. J. Ethnopharmacol. **104(1-2)**: 263-269.

Zanwar, A.A., Hegde, M.V. and Bodhankar, S.L. 2010. *In vitro* antioxidant activity of ethanolic extract of *Linum usitatissimum*. Pharmacology Online **1**: 683-696.

Zapesochnaya, G., Kurkin, V., Okhavov, V. and Hiroshnikov, A. 1992. Canthin-6-one and a-carboline alkaloids from *Aerva lanata*. Planta Medica **37**: 192-196.

Zaporozhets, O.A., Krushynska, O.A., Lipkovska, N.A. and Barvinchenko, V.N. 2004. A new test method for the evaluation of total antioxidant activity of herbal products. J. Agric. Food Chem. **52(1)**: 21-25.

Zaragozá, M.C., López, D. P., Sáiz, M., Poquet, M., Pérez, J., Puig-Parellada, P., Màrmol, F., Simonetti, P., Gardana, C., Lerat, Y., Burtin, P., Inisan, C., Rousseau, I., Besnard, M. and Mitjavila, M.T. 2008. Toxicity and antioxidant activity *in vitro* and *in vivo* of two *Fucus vesiculosus* extracts. J Agric Food Chem. **56(17)**: 7773-80.

Zargar, M., Azizah, A.H., Rohiyati, A.M., Fatimah, A.B., Jahansiri, F. and Pak-Dek, M.S. 2011. Bioactive compounds and antioxidant activity of different extracts from *Vitex negundo* leaf. J. Med. Plants Res. **5(12)**: 2525-2532.

Zekonis, G., Zekonis, J., Sadzeviciene, R., Simoniene, G. and Kevelaitis, E. 2008. Effect of *Perilla frutescens* aqueous extract on free radical production by human neutrophil leucocytes. Medicina (Kaunas) **44**: 699-705.

Zerbo, A., Koudou, J., Ouedraogo, N., Ouedraogo, R. and Guissou, I.P. Antioxidant and antibacterial activities of *Piliostigma reticulatum* (DC.) Hochst extracts. Afr. J. Biotech. **9(33)**: 5407-5411.

Zhang, H., Chen, F., Wang, Y. and Yao, H.Y. 2006. Evaluation of antioxidant activity of parsley (*Petroselenium crispum*) essential oil and identification of its antioxidant constituents. Food Res. Int. **39(8)**: 833-839.

Zhang, H.L, Nagatsu, A, Watanabe, T., Sakakibara, J. and Okuyama, H. 1997. Antioxidant compound isolated from safflower (*Catharanthus tinctorius* L.) oil cake. Chem. Pharm. Bull. (Tokyo). **45(12)**: 1910-1914.

Zhang, J. and Shen, X. 1997. Antioxidant activities of baicalin, green tea polyphenols and alizarin *in vitro* and *in vivo*. J. Nutr. Environ. Med. **7**: 79-89.

Zhang, L., Li, J., Hogan, S., Chung, H., Welbaum,.E. and Zhous, K. 2010. Inhibitory effect of raspberries on starch digestive enzyme and their antioxidant properties and phenolic composition. Food Chem. **119(2)**: 592-599.

Zhang, Q., Li, N., Zhou, G., Lu, X., Xu, Z. and Li, Z. 2003. *In vivo* antioxidant activity of polysaccharide fraction from *Porphyra haitanensis* (Rhodophyta) in aging mice. Pharmacol. Res. **48(2)**: 151-155.

Zhang, Q., Yu, P., Li, Z., Zhang, H., Xu, Z. and Li, P. 2003. Antioxidant activities of sulfated polysaccharide fractions from *Porphyra haitensis*. J. Appl. Phycol. **15**: 305-310.

Zhang, R., Kang, K.A., Piao, M.J., Kim, K.C., Kim, A.D., Chae, S., Park, J.S., Youn, U.J. and Hyun, J.W. 2010. Cytoprotective effect of the fruits of *Lycium chinense* Miller against oxidative stress-induced hepatotoxicity. J. Ethnopharmacol. **130(2)**: 299-306.

Zhang, S., Liu C, Bi, H., and Wang C. 2008. Extraction of flavonoids from *Rhodiola sachlinesis* A. Bor by UPE and the antioxidant activity of its extract. Nat. Prod. Res. **22(2)**: 178-87.

Zhang, F.Z. and Tan, B.W. 2000. Antihyperglycemic and antioxidant properties of *Andrographis paniculata* in normal and diabetic rats. Cli. Exp. Pharmacol. Physiol. **27**: 358-363.

Zhang, J., Yuan, K., Zhou, W., Zhou, J. and Young, P. 2011. Studies on the active components and antioxidant activities of the extract of *Mimosa pudica* Linn. from southern China. Ph. cog. Mag. **7(25)**: 35-39.

Zhang, W., Zhao, J., Wang, J., Pang, X., Zhuang, X., Zhu, X and Qu, W. 2010. Hypoglycemic effect of aqueous extract of seabuckthorn (*Hippophae rhamnoides* L.) seed residues in streptozotocin-induced diabetic rats. Phytother. Res. **24(2)**: 228-232.

Zhang, X., Thoung, P.T., Min, B.S., Ngoc, T.M., Hung, T.M., Lee, I.S., Na, N., Seong, Y.H., Song, K.S. and Bae, K.S. 2006. Phenolic glycosides with antioxidant activity from stem bark of *Populus davidiana*. J. Nat. Prod. **69(9)**: 1370-1374.

Zhang, X.F. and Tan, B.K. 2000. Antihyperglycaemic and antioxidant properties of *Andrographis paniculata* in normal and diabetic rats. Cli. Exp. Pharmacol. Physiol. **27(5-6)**: 358-363.

Zhang, Y., Li, X. and Wang, Z. 2010. Antioxidant activities of leaf extract of *Salvia miltiorrhiza* Bunge and related phenolic constituents. Food Chem. Toxicol. **48(10)**: 2656-2662.

Zhang, Y., Mills, G.L., Nair, M.G. 2002. Cyclooxygenase inhibitory and antioxidant compounds from the mycelia of the edible mushroom *Grifola frondosa*. J. Agric. Food. Chem. **50(26)**: 7581-7585.

Zhang, Y., Mills, G.L., Nair, M.G. 2003. Cyclooxygenase inhibitory and antioxidant compounds from the fruiting body of an edible mushroom, *Agrocybe aegerita*. Phytomedicine. **10(5)**: 386-90.

Zhang, Y.-P. and Yu, Y.-Z. 2007. Study on the antioxidant activity of the extracts from the peel of the *Citrus grandis*. Food Res. Dev. 2007-11.

Zhang, Z., Chang, Q., Zhu, M., Huang, Y., Ho, W.K.K. and Chen, Z. 2001. Characterization of antioxidants present in hawthorn fruits. J.Nutr. Biochem. **12**: 144-152.

Zhao, B., Li, X., He, R., Cheng, S. and Wenjuan, S. 1989. Scavenging effect of extracts of green tea and natural antioxidants on active oxygen radicals. Cell Biophys. **14**: 175-185.

Zhao, E., Zhao, X. and Fan, J. 2009. Study on the extraction and antioxidation of flavone from *Trollius chinensis* Bunge. J. Food Sci. Biotech. DOI CNKI: SUN: WXQG.0.2009-01-016.

Zhao, H., Dong, J., Lu, J., Chen, J., Li Y, Shan L, Lin Y, Fan W, and Gu G. 2006. Effects of extraction solvent mixtures on antioxidant activity evaluation and their extraction capacity and selectivity for free phenolic compounds in barley (*Hordeum vulgare* L.). J. Agric. Food Chem. **54(19)**: 7277-7286.

Zhao, H., Fan W, Dong, J., Lu, J., Chen, J., Shan L., Lin, Y. and Kong, W. 2007. Evaluation of antioxidant activities and total phenolic contents of typical malting barley varieties. Food Chem. 107(1): 296-304. doi: 10.1016/j.food.chem. 2007.08.018.

Zhao, J., Liu, T., Ma, L., Yan, M., Gu, Z., Huang, Y., Xu, F. and Zhao, Y. 2011. Antioxidant and preventive effects of extract from *Nymphaea candida* flower on *in vitro* immunological liver injury in rat primary hepatocyte cultures. Evidence Based Compl. Alt. Med. doi: 10.1093/ecam/nep003

Zhao, J., Zhang, J.-S., Yang, B., LV, G.-P. and Li, S.-P. 2010. Free radical scavenging activity and characterization of sesquiterpenoids in four species of *Curcuma* using TLC bioautography assay and GC-MS analysis. Molecules **15**: 7547-7557.

Zhao, T., Zhang, Q., Qi, H., Zhang, H., Niu, X., Xu, Z., Li, Z. 2006. Degradation of porphyran from *Porphyra haitanensis* and the antioxidant activities of the degraded porphyrans with different molecular weight. Int. J. Biol. Macromol. **38(1)**: 45-50.

Zhao, X., Xue, C.H., Li, Z.J. and Cai, Y.P. 2004. Antioxidant and hepatoproteective activities of low molecular weight sulfated polysaccharide from *Laminaria japonica*. J. Appl. Phycol. **16**: 111-115.

Zhao, X.Y., Sun, H.D., Hou, A.J., Zhao, Q.S., Wei, T.T. and Xin, W.J. 2005. Antioxidant properties of two gallotannins isolated from the leaves of *Pistacia weinmannifolia*. Biochimia et Biophysica Acta **1725**: 103-110.

Zhen, R.-Y., Zhang, H., Li, X.-Y. and Wang, H. 2009. Preparation and antioxidant activities of polysaccharides from *Limonium bicolor*. J. Tianjin Agr. Univ. 2009-4. DOI: CNKI: SUN: TJXY.0.2009-04-008.

Zhenbao, J., Fei, T., Ling, G., Guanjun, T. and Xiaolin, D. 2007. Antioxidant properties of extracts from juemingzi (*Cassia tora* L.) evaluated *in vitro*. LWT - Food Science and Technology. **40 (6)**: 1072-1077.

Zheng, W. and Wang, S. Y. **2001.** Antioxidant activity and phenolic compounds in selected herbs. *Agric. Food Chem.* **49 (11)**: 5165–5170.

Zheng, W. and Wang, S. Y. 2003. Oxygen radical absorbing capacity of phenolics in blueberries, cranberries, chokeberries, and linonberries. J. Agric. Food Chem. **51**: 502-509.

Zhi-Gang, Z., Zhi-Li, Z.L. and Zue-Xian, L. 1997. Study on the isolation, purification and antioxidation properties of polysaccharides from *Spirulina maxima*. Acta Bot. Sinica **39**: 77-81.

Zhishen, J., Mengcheng, T. and Jianming, W. 1999. The determination of flavonoids contents in mulberry and their scavenging effects on superoxide radicals. Food Chem. **64**: 555-559.

Zhongxiang, Z., Jing, J., Jinlan, R., Chenchen, Z., Chaozhan, L., Wei, F. and Yaling, C. 2007. Antioxidant flavonoid glycosides from aerial parts of the fern *Abacopteris penangiana*. J. Nat. Prod. **70(10)**: 1683-1686.

Zhou, D., Ruan J., Cai, Y., Xiong, Z., Fu, W. and Wei, A. 2010. Antioxidant and hepatoprotective activity of ethanol extract of *Arachniodes exilis* (Hance) Ching. J. Ethnopharmacol. **129(2)**: 232-237.

Zhou, G.H., Jiang, R.W., Cheng, Y., Ye, W.C., Shi, J.G., Gong, N.B. and Lu, Y. 2007. Daphnogirins A and B, two biflavones from *Daphne giraldii*. Chem. Pharm. Bull. **55(9)**: 1287-1290.

Zhou, J, Hu, N, Wu, Y.L, Pan, Y.J. and Sun, C.R. 2008. Preliminary studies on the chemical characterization and antioxidant properties of acidic polysaccharides from *Sargassum fusiforme*. J Zhejiang Univ Sci B. **9(9)**: 721-727.

Zhou, X.J., Mei, R.Q., Zhang, L., Lu, Q., Zhao, J., Adebayo, A.H. and Cheng, Y.X. 2010. Antioxidant phenolics from *Broussonetia papyrifera* fruits. J. Asian Nat. Prod. Res. **12(5)**: 399-406.

Zhou, Y. and Xu, R. 1992. Antioxidative effect of Chinese drugs. Chung Kuo Chung Yao Tsa Chih **17**: 368-369.

Zhou, Y.J., Li, C.Q. and Wang, Z.Z. 2008. Studies on the antioxidant activities of leaves from *Senecio argunensis*. Zhang Yao Cai **31(9)**: 1355-1357.

Zhu, F., Cai, Y.-Z., Sun, M., Ke, J., Lu, D., Corke, H. 2009. Comparison of major phenolic constituents and *in vitro* antioxidant activity of diverse Kudingcha genotypes from *Ilex kudingcha*, *Ilex cornuta* and *Ligustrum robustum*. J. Agric. Food Chem. **57(14)**: 6082-6089.

Zhu, N., Rabi, M.M., Dipaole, R.S., Xin, J., Chin, C.K., Badmaev, V., Ghai, G., Roseen, R.T. and Ho, C.-T. 22001. Bioactive constituents from gum guggul (*Commiphora wightii*). Phytochem. **56(7)**: 723-727. doi: 10.1016/s00311-9422(00)00485-4.

Zhu, Y.Z., Huang, S.H., Tan, B.K., Sun, J., Whiteman, M. and Zhu, Y.C. 2004. Antioxidants in Chinese herbal medicines: a biochemical perspective. Nat. Prod. Rep. **21(4)**: 478-489.

Zi, X., Mukhtar, H. and Agarwal, R. 1997. Novel cancer chemopreventive effects of a flavonoids antioxidant silymarin: Inhibition of mRNA expression of an endogenous tumor promoter TNFα. Biochem. Biophysic. Communications **239**: 334-339.

Zielinska-Przyjemska, M and Wiktorowicz, K. 2006. An *in vitro* study of the protective effect of the flavonoid silydianin against reactive oxygen species. Phytother. Res. **20(2)**: 115-119.

Zin, Z., Mohd, Abdul-Hamid, A., Osman, A. 2002. Antioxidative activity of extracts from Mengkudu (*Morinda citrifolia* L.) root, fruit and leaf. Food Chem. **78 (2)**: 227-231.

Zohera, F.T., Habib, M.R., Imam, M.Z. and Mazumder, M.E.H. and Rana, M.S. 2010. Comparative antioxidant potential of different extracts of *Celastrus paniculatus* Willd. seed. S.J.Pharm. Sci. **3(1)**: 68-71.

Zou, Y., Lu, Y. and Wei, D. 2004. Antioxidant activity of a flavonoid-rich extract of *Hypericum perforatum* L. *in vitro*. J. Agric. Food Chem. **52(16)**: 5032-5039.

Zou, Y., Lu, Y. and Wei, D.2010. Protective effects of a flavonoid-rich extract of *Hypericum perforatum* L. against hydrogen peroxide-induced apoptotis in PC12 cells. Phytother. Res. Suppl. 1: S6-S10.

Zou, Y., Qian, Z.J., Li, Y., Kim, M. M., Lee, S.H. and Kim, S. K. 2008. Antioxidant effects of phlorotannins isolated from *Ishige okamurae* in free radical mediated oxidative systems. J. Agric. Food Chem. **56(16)**: 7001-7009.

Zubia, M., Fabre, M.S., Kerjean, V., Lann, K.L., Stiger-Pouvreau, V., Fauchon, M. and Deslandes, E. 2009. Antioxidant and antitumoral activities of some Phaeophyta from Brittany coasts. Food Chem. **116(3)**: 693-701.

Zupkó, I., Hohmann, J., Redei, D., Falkay, G., Janicsak, G. and Mathe, I. 2001. Antioxidant activity of *Salvia* species in enzyme-dependent and enzyme-independent systems of lipid peroxidation and their phenolic constituents. Planta Medica **67**: 366-368.

Index to Names

English Names

Japanese Names

Hindi Names

Turkish names

Local Names

Index to Synonyms